Green Materials, Processing & Characterization

ICoGMPaC 2025

2nd International Conference on Green Materials, Processing, and Characterization (ICoGMPaC 2025) at the World Trade Centre Kuala Lumpur (WTCKL), Malaysia, September 10-11, 2025

Editors

Asna Rasyidah Binti Abdul Hamid, Firuz Binti Zainuddin, Mohamad Kahar Ab Wahab, Muhammad Faheem Bin Mohd Tahir

Universiti Malaysia Perlis (UniMAP)

Peer review statement

All papers published in this volume of "Materials Research Proceedings" have been peer reviewed. The process of peer review was initiated and overseen by the above proceedings editors. All reviews were conducted by expert referees in accordance to Materials Research Forum LLC high standards.

Published under License by **Materials Research Forum LLC**
Millersville, PA 17551, USA

Published as part of the proceedings series
Materials Research Proceedings
Volume 68 (2026)

ISSN 2474-3941 (Print)
ISSN 2474-395X (Online)

ISBN 978-1-64490-422-0 (Print)
ISBN 978-1-64490-423-7 (eBook)

Distributed worldwide by

Materials Research Forum LLC
105 Springdale Lane
Millersville, PA 17551
USA
https://mrforum.com

Manufactured in the United States of America
10 9 8 7 6 5 4 3 2 1

Table of Contents

Preface
Committees

Effect of Binder Type on Properties and Remazol Red Adsorption of Porous Activated Carbon Macroballoons
Nurul Izzati MUHAMAD ZAKIR, Zunaida ZAKARIA, Hakimah OSMAN, Abdulhakim MASA 1

Effect of Stirring Temperature and Blend Ratio on Flexural and Water Absorption of Toughened PVC/PCL Blends
Zulaikha DAMSAR, Sharifah Shahnaz SYED BAKAR, Muhammad Salihin ZAKARIA 9

Phase, Structure and Compressive Strength of Sintered Perlis Dolomite/Silica
Nur Hasnidah AHMAD SHUKERI, Syed Nuzul Fadzli SYED ADAM, Firuz ZAINUDDIN, Hasmaliza MOHAMAD, Heah CHENG YONG 17

Study of White Chicken Egg (Lysozyme) for Properties Study- A Preliminary Research
Nurul Sofia MOHD DALI, Siti Amira OTHMAN 26

Evaluation of Palm Oil-Based UV Curable Coatings for Termite Resistance in Rubber Wood
Rida TAJAU, Mohd Sofian ALIAS, Mohd Hamzah HARUN, Nurul Huda MUDRI, Farah Fadzehah HILIMI, Naurah MAT ISA, Khairul Azhar ABDUL HALIM, Sharilla MUHAMMAD FAISAL, Rosley CHE ISMAIL, Abdul Muizz MOHD SANI, Khairul Anwar MUSTAFA 33

Extraction of Silica from Burnt Rice Husk Using the Double Crucible Method
Faizul CHE PA, Ruhiyuddin MOHD ZAKI, Sasitra SEGAR 41

Ground Coffee Waste-Derived Activated Carbon for Aqueous Dye Adsorption
Teerunishan SUBRAMANIAM, Norsuria MAHMED, Muhammad Salihin ZAKARIA, Haiza HAROON 49

Performance of Graphene-Reinforced Concrete Compared to Normal Concrete Under Explosive Loading
I.S. ISMAIL, A.H. HILMI, A.R.A HAMID, S.M. Sk ABD RAZAK, N.A. MOHD SALEH 57

Effect of Polyethylene Glycol and Lemongrass Essential Oil as Plasticizers on the Properties of Chitosan/Hydroxyethyl Cellulose Blend Films
Nurul Izzaty MOHD ISA, Muthmirah IBRAHIM, Siti Salmi SAMSUDIN, Jalilah ABD JALIL 65

Effect of Calcination Holding Time, Mixing Time, and Filler Content on the Mechanical Performance of Dolomite Filled UHMWPE Composites
MD SALEH Siti Shuhadah, ZAINUDDIN Anis Maisarah, MOHAMMAD Nur Farahiyah, ISMAIL Anis Nadhirah, SYED MAHAMUD Syarifah Nuraqmar, OMAR Mohd Firdaus, MD AKIL Hazizan, CHANG Boon Peng, HAFSAT Ronke Saliu 73

Advancements in Materials Science: AI-Powered Predictions of Mechanical Behaviours in Natural Fiber Polymer Composites
Mohammed MOHAMMED, Jawad K. OLEIWI, Aeshah M. MOHAMMED, Azlin F. OSMAN, Tijjani ADAM, Bashir O. BETAR, Subash C. B. GOPINATH, Kuzmin ANTON 81

Degradation of Plastic Waste in Surabaya River, Indonesia
Atik WIDIYANTI, Eddy Setiadi SOEDJONO, Maya SHOVITRI, Adhi YUNIARTO 90

Enhancing Lower Limb Prosthetic Sockets with Hybrid Natural Fiber Composites: A Performance Analysis
Adel M. BASH, Jawad K. OLEIWI, Tahseen T. OTHMAN 97

Physicochemical Characteristics of EPOLA/PEGDA as a 3D Scaffold for Tissue Engineering Application
SITI NURUL SYAFIKA Sheikh Ibrahim, MAT UZIR Wahit, MARINA Talib, NORHAYANI Othman, FARAH FADZEHAH Hilmi 105

Optimization of EPOLA/PEGDA Formulation for 3D Scaffold Development using Design of Experiment (DOE - Mixture)
SITI NURUL SYAFIKA Sheikh Ibrahim, MAT UZIR Wahit, MARINA Talib, NORHAYANI Othman, FARAH FADZEHAH Hilmi 113

Exploring the Impact of Carbon Fiber Incorporation on the Mechanical Properties of PLA Composites
Yusrina MAT DAUD, Hazwani ZAIDOL, Mohammad Firdaus ABU HASHIM, Luqman MUSA, Mohamad Syahmie MOHAMAD RASIDI, Ahmad Azrem AZMI, Lokman Hakim IBRAHIM, Muhammad Al Fatih HENDRAWAN, Kholqillah Ardhian ILMAN, Agus Dwi ANGGONO, Bambang Waluyo FEBRIANTOKO 121

The Effect of Kernel Shell Oil Palm Empty Fruit Bunches (KS-OPEFB) and Rice Husks (RH) Hybrid Fillers on Thermoplastic Starch Films
Sinar Arzuria ADNAN, Azlin Fazlina OSMAN, Nur Hidayah AHMAD ZAIDI1, Muhammad Hanis ROSLI, Yahmunaa JAVERSEETHARAMAN, Ameirul Qayyum Eiman RAMLI, Haiza HAROON, Syazwani MAHMAD PUZI, Sharifah Hanis Yasmin SAYID ABDULLAH 126

Acid and Alkali Treatment on Natural Fibers for Better Tensile Strength of Epoxy Composites
Nor Azwin AHAD, Rama Murthi A/L Sooria MOORTHI 134

Effect of Curing Time and Moisture Exposure on the Mechanical Response of Polyurethane with Varying Polyol-to-Isocyanate Ratios
Ahmad Syafiq Iskandar AHMAD NAZRI, Diana CHE LAT, Ab Aziz MOHD YUSOF, Ismacahyadi Bagus MOHAMED JAIS, Doris Asmani MAT YUSOF, Mohamed Azizi MD ALI, Roslizayati RAZALI 139

Characterisation of Moisture Retention Properties in Commercial Hydrogel Dressings for Sustained Wound Hydration
Diviya Devi PARAMASIVAM, Raimi DEWAN, Nadia ADRUS, Kok Yeow YOU, Faishal Adilah SURYANATA 148

Study on the Process of Culantro Teabag (Eryngium foetidum)
Nguyen THI HIEN, Nguyen KIM PHUNG 156

Strength and Carbonation of Sand Cement Brick Containing Fly Ash as Partial Cement Replacement Under CO2 Curing
Faisal Sheikh KHALID, Ming Quan CHONG, Syazwi Hakimi SAAIDIN, Connie Hui Jun KONG, Syafiqa AYOB, Zahir ZAKI, Raden Mohd Farhan Helmy RADEN ISMAIL 166

Mechanical Performance of POFA-Modified Mortar Under Varying Curing Regimes
Nurul Hana MOHAMAD ZAINI, Nuril Izzeaty ISHAK, Mohamad Nidzam RAHMAT 174

The Synthesis of Graphitic Carbon from Coconut Shell Waste by Catalytic Graphitization
Yap Yee WEN, Norsuria MAHMED, Midhat Nabil AHMAD SALIMI, Azlin Fazlina OSMAN, Kamrosni ABDUL RAZAK, Shayfull Zamree ABD RAHIM, Mohd Yusry MOHAMAD YUNUS 182

Biogas Potential from the Investigation of Paddy Straw Co-Digestion with Cow Dung
T.Y.T. ROSLINA, M. MARIAM, A.R. NOR' AINI, M.S.M. FAIZ, C.M. HASFALINA........ 189

Performance of Thermosiphon Effect Using Different Flat Plate Collector of Solar Water Heater
M.M. NORUL AHSANAH AULIA, A. MARIAH, H. NURAMIDAH, D.S. ZALEHA, M.Y. FATIMAH, A.R.M. ASHRAF, M.M. KAMARUZAMAN, A. FARIDAHANIM........... 165

Transformation of Cellulose Allomorph of Musa acuminata balbisiana Peel Using Ethylenediamine
A.H. JASNI, A.S. AZMI, N.I. MOHAMAD PUAD, F. ALI, Y. AHMAD NOR 204

Experimental Study on the Viscosity Properties of Modified Vegetable Oil with Hybrid hBN/TiO2 Nanoparticles as Metalworking Fluid
A. M. SABRI, N. TALIB, H. ABDULLAH, A. S. A. SANI, S. KUNAR 210

pH Dependent Growth of ZnO Seed Template in Alkaline Solution
M. MARINA, M.Z.M. ZAMZURI, N.H. ABDUL HALIM, R.N. AHMAD, Z. NOORAIZEDFIZA, S.S. AHMAD, N. NISHA RAZALLI, Z. IBERAHIM 218

Neck Formation Properties of SS316L Particle Powder Against Time in Conventional Sintering Process
M.R. MAZLAN, M.A.M. YUSOFF, N.H. JAMADON 225

Effect of Air Abrasion Surface Treatment on the Bonding Behavior at Zirconia-Resin Interface
N.S. AMINUDDIN, M.F.A. RAHIM, N.H. JAMADON 232

Oxidation Behavior of Aluminium Granules and Aluminium Alloy Chips During In-Situ Casting
Ain Zubaidah Binti SAMIAN, Aslinda Binti SALEH 239

The Potential of Autonomous Self-Healing Microcapsules for Concrete Cracking
N.M. ZAHARI, Z. ITAM, A. SYAMSIR, M.H. ZAWAWI, S.M.M.S.A. FADZIL, M.M. ZAINOODIN, Z.A.A. HAMID, N.A.M. AZMAN 246

Influence of Fusion Welding Processes on Microstructure and Mechanical Properties of Dissimilar Metal (AISI 304 SS-Copper) Weldment
Bikash Ranjan MOHARANA, Aezeden MOHAMED, Sushant Kumar SAHU, Susanta Kumar SAHOO, Dillip Kumar BISWAL, Tapas Kumar MOHARANA, Kamalakanta MUDULI 254

Flow Analysis of Eco-Friendly Geopolymer with Recycled Copper for Rapid Tooling Applications
Sharifah SALSABILA, Z. SHAYFULL, S.M. NASIR, Leong Kean WEI, M. FATHULLAH 263

The Tensile Properties and Microstructural Characteristics of Pulverised Silica-Coated Kenaf–Epoxy Composites
M.J. MEELAD, N.Z. NORIMAN, M.A. AZIZAN, M.F. OMAR, A.Z. ROMLI, R. HAMZAH, N. ISHAK, S.T. SAM 272

Flexural Properties and Bulk Density in Silica Coated Kenaf/Epoxy Composites
M.J. MEELAD, N.Z. NORIMAN, M.A. AZIZAN, M.F. OMAR, A.Z. ROMLI, R. HAMZAH, N. ISHAK, OMAR S. DAHHAM 280

Mechanical and Durability Performance of Fiberglass-Reinforced Concrete for Water Retaining Structures
M.A. AZIZAN, N.Z. NORIMAN, N. ISHAK, R. HAMZAH, S.T. SAM, OMAR S. DAHHAM 288

Synthesis Characterization of Slow-Release Pesticide with Epoxidized Natural Rubber and Cassava Tree Trunk
Y. HARI KRISHNAN, R. HAMZAH, N.Z. NORIMAN, M. A. AZIZAN 293

Design, Finite Element Analysis and Fabrication of a Laboratory Scale Biomass Pelletizer
R. I. Ismail, C. Y. Khor, A. R. Mohamed, M. N. S. Zaimi, M. R. Jamalludin, L. Y. Lee, N. L. Makhtar, N. Abdul Razak, W. N. A. Wan Draman 299

Surface Performance Evaluation of Lattice Structure Produce by Fused Deposition Modeling Machine Using Design of Experiment (DOE)
N.A.M. RAZIP, M.A. ISMAIL, S. SHAHARUDDIN, M. JOHAR, Z. MOHAMMAD, M.S.E. KOSNAN 308

Using the Taguchi Approach for Investigating Mechanical Properties of Recycled Carbon Black Reinforced Acrylonitrile Butadiene Styrene (ABS) and Polylactic Acid (PLA) for Injection Moulding Applications
A.E. KASIM, N.A. SHUAIB, C.C. FOOI, C.R. QUAN, M.H.J.A. HADI, U.F. M. ALI, A.F. OSMAN 315

Monitoring and Prediction of Tool Wear in Turning Process using Machine Learning Technique
S.Y. QUEK, W.K. LEE, N. TALIB, H. ABDULLAH, A.M SABRI, A. SUNG 322

Mechanical Characterization of Solder/Intermetallic Compound (IMC) Interface: The Role of Temperature and Strain Rate
S. F. M. ASASAARI, M. N. TAMIN, M. A. ISMAIL, S. SHAHARUDDIN, M. AZMI, M. JOHAR, Z. MOHAMMAD, M. S. E. KOSNAN, K. J. WONG 328

Keyword Index
About the Editors

Preface

The *International Conference on Green Materials, Processing and Characterization 2025* (ICoGMPaC 2025) was successfully held on 10–11 September 2025 at the World Trade Centre Kuala Lumpur (WTCKL), Malaysia. The conference was proudly organised in conjunction with the Global Trends in Engineering, Science and Technology Congress 2025 (GTEST 2025), an annual congress hosted by Universiti Malaysia Perlis.

GTEST 2025 brought together ten international conferences simultaneously, providing an exceptional platform for researchers, academicians, scientists, engineers, and industry practitioners to exchange knowledge, establish interdisciplinary collaborations, and discuss emerging developments across various domains of engineering, science, and technology. Within this vibrant academic environment, ICoGMPaC 2025 continued the efforts initiated during its inaugural edition to foster innovation and collaboration in the fields of green materials, sustainable processing methods, and advanced characterization techniques.

The conference aimed to catalyse interdisciplinary partnerships, showcase cutting-edge research findings, and inspire sustainable solutions to contemporary global challenges. The technical programme covered a broad spectrum of topics related to environmentally friendly materials, advanced manufacturing and processing technologies, material characterization approaches, sustainable engineering practices, and emerging applications in materials science and engineering.

A total of 53 papers were submitted to ICoGMPaC 2025. Following a rigorous peer-review process conducted by experts in the relevant fields, 43 papers were accepted for publication in this proceedings volume. The peer-review system ensured that all accepted contributions met the required standards of scientific quality, originality, and technical relevance.

The editors would like to express their sincere appreciation to all authors for their valuable contributions and active participation in the conference. Gratitude is also extended to the reviewers, scientific committee members, organising committee, keynote speakers, session chairs, and all individuals whose commitment and support contributed to the success of ICoGMPaC 2025. Special acknowledgement is dedicated to Universiti Malaysia Perlis for hosting GTEST 2025 and providing continuous support towards the advancement of academic and research excellence.

It is hoped that the papers published in this volume will serve as a useful reference for researchers and practitioners, while stimulating further research, innovation, and collaboration in the areas of green materials, sustainable processing, and characterization technologies.

Editors of ICoGMPaC 2025 Proceedings

Committees

PATRON

Assoc. Prof. Dr. Dewi Suriyani Che Halin

CHAIRMAN

Assoc. Prof. Dr. Faizul Che Pa

CO-CHAIRMAN

Assoc. Prof. Dr. Norsuria Mahmed

SECRETARIES

Dr. Syarifah Shahnaz Syed Bakar
Ts. Dr. Siti Hawa Mohamed Salleh

TREASURERS

Dr. Siti Shuhadah Md Saleh
Ir. Muthmirah Ibrahim

PUBLICATION

Assoc. Prof. Dr. Firuz Zainuddin
Assoc. Prof. Dr. Mohamad Kahar Ab Wahab
Dr. Asna Rasyidah Abdul Hamid
Dr. Muhammad Faheem Mohd Tahir

PUBLICITY, TECHNICAL & WEBSITE

Mr. Ruhiyuddin Mohd Zaki
Ts. Dr. Wan Mohd Arif Wan Ibrahim
Mr. Lokman Hakim Ibrahim

PROTOCOL & LOGISTICS

Dr. Mohd Nazry Salleh
Mr. Ismail Ibrahim
Ts. Dr. Ahmad Azrem Azmi

Materials Research Proceedings 68 (2026) 1-8 https://doi.org/10.21741/9781644904237-1

Effect of Binder Type on Properties and Remazol Red Adsorption of Porous Activated Carbon Macroballoons

Nurul Izzati MUHAMAD ZAKIR[1,a], Zunaida ZAKARIA[1,2,b,*], Hakimah OSMAN[1,c], and Abdulhakim MASA[3,d]

[1]Faculty of Chemical Engineering & Technology, Universiti Malaysia Perlis (UniMAP), 02600 Arau, Perlis, Malaysia

[2]Centre of Excellence Geopolymer & Green Technology, (CEGeoGTech), Universiti Malaysia Perlis (UniMAP), Perlis, Malaysia

[3]Rubber Engineering Program, Department of Interdisciplinary Engineering, Faculty of Engineering, Prince of Songkhla University, Hat Yai, Songkhla 90110 Thailand

[a]izzatizakir@gmail.com, [b]zunaida@unimap.edu.my, [c]hakimah@unimap.edu.my, [d]abdulhakim.m@psu.ac.th

Keywords: Adsorption, Porous Carbon, Remazol Red Dye, Macroballoon, Rice Husk Ash

Abstract. Porous activated carbon (PAC) derived from rice husk ash offers a sustainable and low-cost material for dye adsorption in wastewater treatment. This study investigates the incorporation of PAC into hollow macroballoons using four different binders: epoxy resin (Ep), sodium alginate (SA), silicone rubber (Si), and grout (Gr) on their physical, thermal, and adsorption characteristics. Microscopy and SEM analyses revealed that Ep-macroballoon and SA-macroballoon exhibited better PAC encapsulation and smoother surfaces, whereas Si-macroballoon and Gr-macroballoon showed rougher textures with more exposure to PAC particles. FTIR-ATR spectra confirmed the presence of hydroxyl and carboxyl functional groups in Ep-macroballoon and SA-macroballoon, suggesting greater surface reactivity, while Si-macroballoon and Gr-macroballoon exhibited more hydrophobic or inorganic characteristics. Thermal analysis via TGA/DTG showed distinct degradation patterns for each binder type, with SA-macroballoon demonstrating moisture-related weight loss and stable thermal behaviour, while Gr-macroballoon retained the highest final residue due to its inorganic content. Adsorption experiments using 50 mg/L Remazol Red dye solution revealed that Gr-macroballoon achieved rapid and high dye removal but at the expense of PAC leaching and reduced structural stability. In contrast, SA-macroballoon demonstrated sustained dye adsorption with minimal material loss; Ep-macroballoon and Si-macroballoon showed negligible dye removal. This highlights that sodium alginate offers the most effective balance of adsorption capacity and structural integrity for macroballoon functionality in dye removal applications.

Introduction

The increasing emphasis on sustainability and environmental conservation has sparked significant interest in using agricultural waste as a precursor for the production of activated carbon (AC). Among various agricultural wastes, rice husk ash (RHA) stands out as an abundant byproduct of rice milling industries, especially in northern Malaysia, as paddy farming is the third most important crop after oil palm and rubber [1]. Commonly, RHA is being used as a medium for plantation [2] and soil amendment material [3], however, there are also studies in converting RHA into AC through chemical activation using alkali such as potassium hydroxide (KOH) and sodium hydroxide (NaOH) [4]. RHA can be effectively transformed into porous activated carbon (PAC), which offers promising properties for adsorption applications.

Several other biomass sources have also been explored for activated carbon production [5]. Maize stalks, for example, have been converted to activated carbon using a two-step carbonisation and chemical activation process employing either KOH or NaOH as activating agents [6]. These chemically activated carbons demonstrated improved surface area and porosity, making them suitable for pollutant removal. Remazol Red is widely used as an azo dye in the textile, printing, and manufacturing industries due to its vibrant colour, strong binding to cellulosic fibres, and low production costs compared to natural dyes [7]. However, RR and other azo dyes (-N=N-) pose environmental and health risks, as they are known to be mutagenic and potentially carcinogenic[8]. To mitigate these risks, effective treatment methods must be explored and implemented before these dyes can enter water systems. Commercial activated carbons are typically available in various forms, such as powdered activated carbon and granular activated carbon (GAC), both of which are commonly employed in water and wastewater treatment [9]. Activated carbon spheres (ACS) are known for their high compressive strength, uniform pore structure, high microporous volume, and low ash and water content [10]. However, they also exhibit certain drawbacks, including low sphericity, inconsistent pore size distribution, and relatively poor adsorption capacity in some cases. To address these issues, this study incorporated activated carbon using different binders to fabricate macroballoons. By incorporating activated carbon in macroballoons, this study aims to fabricate self-floating adsorbent to offer a potentially effective and sustainable solution for RR dye removal from wastewater.

Methodology

Materials. The RHA was supplied by Dibuk Rice Mill Factory, located in Perlis, Malaysia. The RHA was activated to PAC through a chemical activation process using a 1.0 M NaOH solution, which was supplied by HmbG Chemicals in pellet form. The chemical activation of RHA to produce PAC was performed according to the method described by [11]. In this study, macroballoons were prepared by using expanded polystyrene (EPS) as the template. EPS was purchased from Yee Hup Foam & Packaging Industries Sdn Bhd. Four binders were used for the macroballoons: epoxy (DER 331) with hardener (ancamine 1618) supplied by Euro Chemo Pharma Sdn Bhd., sodium alginate (SA), and calcium chloride ($CaCl_2$) sourced from R&M Chemicals. silicone rubber from AB Liquid Silicone Mold Maker, and grout from Saint-Gobain Weber. For the adsorption test, Remazol Red (RR) dye was supplied by AR Alatan Sdn. Bhd. with a molecular weight of 732.94 g/mol and a wavelength of 536 nm.

Preparation of porous activated carbon macroballoons. The epoxy shell template was prepared by coating the EPS with epoxy resin and leaving it to cure at room temperature. The process was followed by post-curing at 120 °C. Next, the epoxy shell was coated with different binders (i.e., epoxy, silicon rubber, and grout) and rolled in PAC powder until its surface was completely coated, as in Fig. 1(a), and left to dry before usage. For SA-macroballoon, the epoxy shell was stirred in a solution of SA mixed with PAC, and then it was slowly dropped into a $CaCl_2$ solution to initiate the gelation process, as depicted in Fig. 1(b). The SA-macroballoon was then washed until the pH was neutral before usage. All the macroballoons were analysed using a Dino-Lite microscope, scanning electron microscopy (SEM), thermogravimetric analysis (TGA), Fourier-transform infrared spectroscopy (FTIR), and the RR dye adsorption test via UV-Vis Spectrophotometer.

Preparation of adsorption test. A stock solution of 100 mg/L of RR dye was prepared using distilled water. Each type of macroballoon, containing 1 g of PAC, was mixed with 100 mL of 50 mg/L RR dye and stirred at 150 rpm until no colour changes were observed. At intervals of 10 minutes, the absorbance of RR dye was measured using a UV-Vis Spectrophotometer.

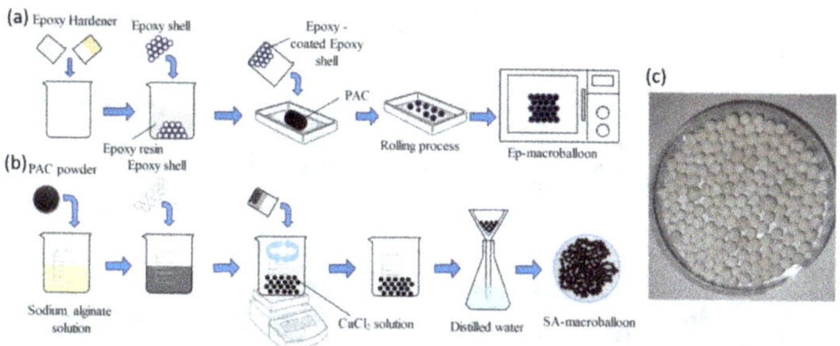

Fig. 1: Preparation of (a) epoxy-based activated carbon macroballoons (Ep-macroballoon) (b) sodium alginate-based activated carbon macroballoon (SA-macroballoon) and (c) epoxy shell.

From the recorded data, the percentage of dye removal against time was calculated using Eq. 1 below, where C_i is the initial concentration of RR dye and C_t is the concentration at the designated time:

$$Removal\ percentage\ of\ RR\ dye, \% = \frac{C_i - C_t}{C_i} x\ 100\% \quad (1)$$

Results and Discussion

Physical characterisation. Fig. 2 shows the morphology of macroballoons captured using a Dino-Lite microscope and SEM at 500× magnification. The image shows that all macroballoon samples are completely coated with PAC, with no visible epoxy shell present. The Ep-macroballoon, SA-macroballoon, and Si-macroballoon have complete coverage of PAC powder. However, it was observable that the Gr-macroballoon provided inadequate coverage, exposing the grout surface.

The Ep-macroballoon has a remarkably smooth and compact surface, which indicates the characteristics of epoxy. In contrast, the SA-macroballoon exhibits granular and uneven texture, resulting from its hydrogel-based crosslinking during the gelation process. The Si-macroballoon presents a smooth and matte surface, whereas the Gr-macroballoon exhibits the roughest and most porous texture due to the particulate composition of cementitious materials. This distinct surface variation emerged from the preparation process involving a 2:1 ratio of water to grout. The produced mixture lacked sufficient viscosity to effectively bond the PAC to the EPS shell surface, resulting in uneven coverage and a non-uniform distribution of PAC throughout the structure. A study by Pichniarczyk and Niziurska found that the combination of cement and water alone did not offer enough adhesion to securely attach the substrate to the ceramic surface [12].

The SEM images highlight the surface of macroballoons and the distribution of embedded PAC throughout the binder matrix. The red circles indicate the presence of PAC particles dispersed within the binder systems; the Ep-macroballoon stands out with its distinctly defined matrix, which is characterised by a smooth and continuous polymer layer enveloping the PAC particles. This suggests a strong interfacial adhesion between the epoxy and the PAC [13]. SA-macroballoon exhibits a smooth and compact surface (Fig. 2(b)), with PAC particles less exposed, likely due to the encapsulating effect [14] of the sodium alginate hydrogel matrix. Further examination reveals that both Si-macroballoon and Gr-macroballoon exhibit rougher and more heterogeneous morphologies, consistent with the physical and granular nature of their respective binders. The Si-macroballoon exhibits loosely embedded PAC particles within more irregular silicone rubber matrix, which contributes to moderate surface exposure. Conversely, the Gr-macroballoon displays a distinctly porous structure, with a high density of exposed PAC particles due to minimal matrix encapsulation.

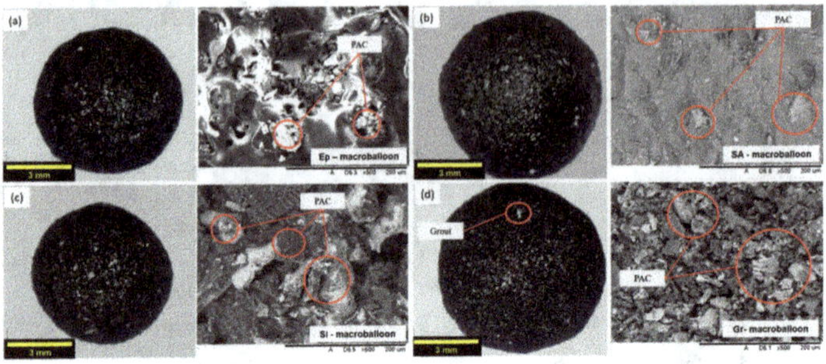

Fig. 2: Digital images of macroballoon with different binders.

Structural Analysis and Thermal Analysis. FTIR analysis was conducted by Perkin Elmer Spectrum Version 10.03.06 using the attenuated total reflectance (ATR). Fig. 3(a) illustrates the FTIR spectra and TGA thermogram of PAC macroballoon with different binders. Results show that all samples exhibited absorption bands in the range of 2948-2904 cm^{-1}, which are attributed to aliphatic C-H stretching vibrations from $-CH_3$ and $-CH_2$ groups, indicative of organic content in the polymeric binders [15]. Additionally, a broad absorption peak above 3000 cm^{-1}, particularly between 3200-3500 cm^{-1}, was observed in SA-macroballoon and EP-macroballoon. This peak is commonly associated with O-H stretching vibrations, indicating the existence of hydroxyl groups and possibly absorbed water. In the case of SA-macroballoon, this feature aligns with the polysaccharide structure of sodium alginate, which contains abundant hydroxyl and carboxyl groups capable of hydrogen bonding [16]. In EP-macroballoon, the broad band may originate from unreacted hydroxyl groups or residual curing agents in the epoxy matrix. These hydrophilic groups enhance the chemical reactivity of the macroballoon surface and facilitate interactions with polar dye molecules during adsorption [17]. In contrast, Si-macroballoon and Gr-macroballoon displayed weaker or negligible signals in this region, reflecting the hydrophobic nature of silicone rubber and the inorganic, anhydrous nature of cement grout. The intense peak at 1597 cm^{-1} observed across all samples was associated with C=C stretching from aromatic rings, confirming the aromatic backbone of the activated carbon [18]. The peak at 1409 cm^{-1} reflects $-CH_2$ bending or aromatic ring breathing. Meanwhile, the region between 1242 and 1007 cm^{-1} is dominated by C-O stretching vibrations, which are strong in EP-macroballoon and SA-macroballoon and are associated with functional groups such as ethers, alcohols, and carboxylates. These functionalities are vital for adsorption applications [19]. The presence of O-H stretching above 3000 cm^{-1} in SA-macroballoon and EP-macroballoon indicates hydroxyl-rich surfaces, which significantly enhance the adsorption potential of PAC. The absence of such features in Si-macroballoon and Gr-macroballoon shows that these binders offer structural support but limited chemical activity.

Fig. 3(b) shows the TGA and DTG thermograms of Ep-macroballoon, SA-macroballoon, Si-macroballoon, and Gr-macroballoon. All samples exhibit two major weight loss events at temperature ranges of 300 °C - 400 °C and 700 °C - 800 °C. The initial weight loss in the range of 300 °C - 400 °C was due to the breakdown of unstable bonds within the polymer backbone, due to oxidative degradation in the presence of air.

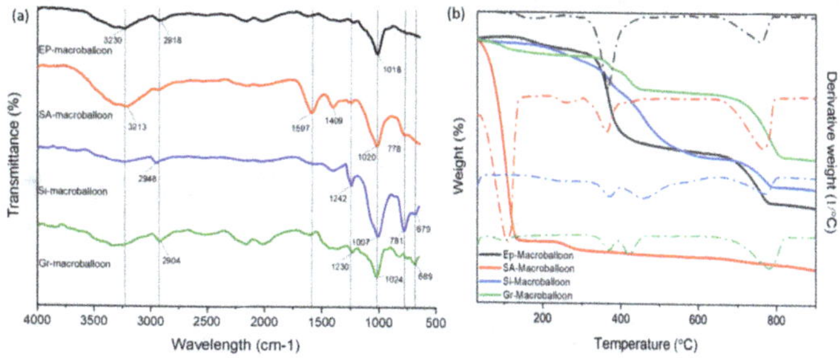

Fig. 3: (a) FTIR spectrum and (b) TGA and DTG thermogram of Ep-macroballoon, SA-macroballoon, Si-macroballoon, and Gr-macroballoon.

The second major weight reduction between 700 °C and 800 °C corresponds to the thermal decomposition of the carbon chain backbone, resulting in the destruction of the three-dimensional crosslinked polymer network. This is consistent with previously reported epoxy degradation profiles [20], where two significant weight loss stages are observed due to dissolution and thermal degradation mechanisms. The thermograms of the Ep-macroballoon demonstrate an initial degradation between 300 °C and 450 °C, resulting in a 35.21% weight loss, followed by further decomposition between 620 °C and 800 °C, resulting in a final weight of 38.10%. The SA-macroballoon exhibits a multi-stage degradation pattern. The first stage occurs at 100 °C–150 °C, representing the evaporation of moisture and low molecular weight organic compounds from the sodium alginate [21]. The DTG thermogram confirms dehydration at 112.27 °C, preceding the formation of carbonaceous char. According to Salisu et al. [22], this stage also involves the breakdown of the carboxylate functional groups (COO^-), releasing CO_2. The second major degradation stage, between 240 °C and 350 °C, indicates the thermal decomposition of the alginate structure [23]. The Si-macroballoon exhibits three distinct degradation stages at 345.98 °C, 520.63 °C, and 766 °C, corresponding to weight losses of 17.96%, 23.30%, and 10.79%, respectively, with a final residue of 44.41%. The initial and final stages are attributed to the decomposition of the embedded epoxy resin, while the intermediate degradation is associated with the oxidative degradation of the silicone matrix, consistent with prior findings [24]. The overlapping transitions in the DTG curve suggest concurrent degradation of both epoxy and silicone components. For Gr-macroballoon, the thermogram reveals a weight loss of 3.93% at 97.93 °C due to the evaporation of moisture. A second degradation occurs between 300 °C and 490 °C, resulting in a 12.62% weight loss, likely due to the degradation of organics and hydration products in the grout. The third stage, which happens between 650 °C and 820 °C, causes a weight loss of 23.74%, leaving a final residue of 55.77%.

Adsorption of Remazol Red dye using macroballoon. Fig. 4(a) illustrates the removal efficiency of RR dye using different PAC macroballoon over a contact time of 140 minutes. Gr-macroballoon and SA-macroballoon demonstrated significant dye adsorption capabilities, while Ep-macroballoon and Si-macroballoon demonstrated a strikingly low adsorption capacity, unable to exceed 2% dye removal even after extended contact. Gr-macroballoon exhibited the highest and most rapid removal efficiency, achieving nearly 100% dye removal within the initial 30 minutes of contact. This indicates strong dye affinity and fast adsorption kinetics due to PAC facilitating the adsorption of the dye molecules with extensive surface exposure of PAC particles.

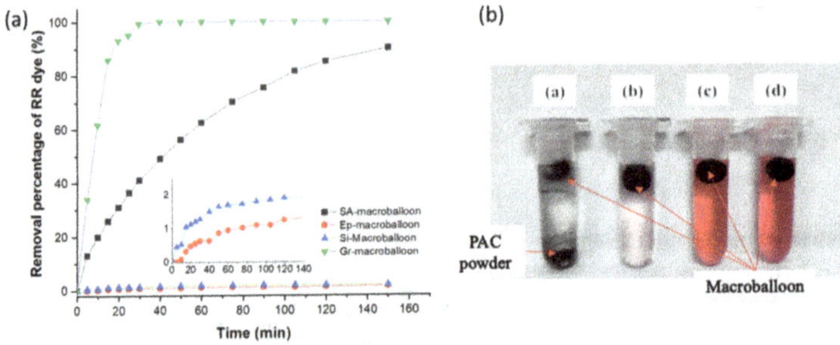

Fig. 4: (a) The removal percentage of 50 mg/L of RR dye using different types of macroballoon (b) the photographic image of absorption of RR dye.

Conversely, SA-macroballoon exhibited a more gradual adsorption profile, achieving approximately 50% dye removal at 40 minutes, increasing to around 80% after 100 minutes, and reaching about 90% removal of the dye with prolonged contact. The gradual increase in dye removal indicates that adsorption does occur at the PAC active site. However, it believes that the sodium alginate coating creates a thin gel-like barrier that, although permeable, slows down the dye transport to the active PAC adsorption sites. Nonetheless, the continuous increase verifies that the alginate layer does not entirely prevent adsorption but rather operates as a selectively permeable matrix. In contrast, Ep-macroballoon and Si-macroballoon exhibited minimal adsorption, with clearance percentages not exceeding 2% even after 140 minutes. The macroballoons exhibited negligible dye removal, as there were no colour changes observed. This lack of adsorption suggests that the surface properties or pore structures of Ep-macroballoon and Si-macroballoon are unsuitable for RR dye capture under the tested conditions.

Fig. 4(b) illustrates the photographic images of macroballoons after the adsorption test using 50 mg/L RR dye solution. During the test, all of the samples floated on top of the dye solution. The Gr-macroballoon exhibited a visible dispersion of powdered PAC into the dye solution, as seen in Fig. 4(b-a), indicating that the PAC was not firmly retained and encapsulated by grout; thus, it dispersed from the macroballoon structure upon contact with the dye. However, with PAC dispersed and settled at the bottom of the solution, it defeats the purpose of fabricating macroballoon to provide a mobilised structure with an easy retrieval function. In contrast, no such dispersion was observed in the SA-macroballoon, Ep-macroballoon, and Si-macroballoon, suggesting better encapsulation of PAC within these formulations. The macroballoons remained intact and buoyant, floating on the solution surface throughout the adsorption process. This highlights the importance of immobilising PAC into a macroballoon structure, which prevents particle dispersion, enabling easy retrieval and reuse.

Conclusion

This study assessed the effect of four different binders on the physical, morphological, structural, thermal, and adsorption characteristics of PAC-based macroballoons. Photographic images and SEM micrograph analysis demonstrated that Ep-macroballoon and SA-macroballoon provided uniform PAC coverage, and Si-macroballoon and Gr-macroballoon had uneven surfaces with exposed PAC particles. FTIR spectra validated the existence of functional groups, including hydroxyl and carboxyl, in Ep-macroballoon and SA-macroballoon samples, signifying enhanced surface reactivity. The TGA data indicated multi-stage degradation patterns affected by the thermal properties of each binder system. Adsorption studies with Remazol Red dye showed that the SA-macroballoon offered a balance between adsorption efficacy and structural integrity. The Ep-macroballoon and Si-macroballoon formulations exhibited restricted adsorption, presumably caused by inadequate accessibility to active sites.

References

[1] D. Dorairaj, N.T. Govender, Rice and paddy industry in Malaysia: governance and policies, research trends, technology adoption and resilience, Front. Sustain. Food Syst. 7 (2023). https://doi.org/10.3389/fsufs.2023.1093605

[2] N.F. Omar, N.N. Kasim, S.Z. Zainal Abidin, S.N. Ali, A. Abdul Aziz, The impact of rice husk ash (RHA) as soilless growing medium on sawi caixin giant (Brassica parachinensis) growth performance, International Journal of Innovation and Industrial Revolution 5 (2023) 49–64. https://doi.org/10.35631/ijirev.513006

[3] P.K. Singh, I. Kumar, U. Kumar, R.K. Sharma, Soil-mustard revitalization via rice husk ash, a promising soil amendment material for sustainable management of heavy metal contamination in tropical ecosystem, J. Environ. Manage. 355 (2024) 120538. https://doi.org/https://doi.org/10.1016/j.jenvman.2024.120538

[4] N.I.M. Zakir, Z. Zakaria, H. Osman, H. Shukor, A. Masa, Effect of Activating Agent on Porous Activated Carbon in Alginate Macrobeads for Removal of Remazol Red Dye, in: AIP Conf. Proc., American Institute of Physics, 2024. https://doi.org/10.1063/5.0198600

[5] Y.X. Gan, Activated Carbon from Biomass Sustainable Sources, C (Basel). 7 (2021) 39. https://doi.org/10.3390/c7020039

[6] Yuliusman, F. Al Farouq, S.P. Sipangkar, M. Fatkhurrahman, S.A. Putri, Preparation and characterization of activated carbon from corn stalks by chemical activation with KOH and NaOH, in: AIP Conf. Proc., American Institute of Physics Inc., 2020. https://doi.org/10.1063/5.0014403

[7] S.S. Affat, Classifications, Advantages, Disadvantages, Toxicity Effects of Natural and Synthetic Dyes: A review, University of Thi-Qar Journal of Science 8 (2021) 130–135. https://doi.org/10.32792/utq/utjsci/v8/1/21

[8] S. Hamzah, N.B. Swan, N.A. Razali, N.A. Mohamad, N. Rasit, A. Ali, M.S. Sulaiman, A. Jusoh, N.H.H. Hairom, N. Ali, N.I. Yatim, Integration of water treatment sludge and agriculture waste as low-cost adsorbent for remazol red dye removal, Desalination Water Treat. 226 (2021) 408–418. https://doi.org/10.5004/dwt.2021.27263

[9] D.E. Egirani, N.R. Poyi, N. Shehata, Preparation and characterization of powdered and granular activated carbon from Palmae biomass for cadmium removal, International Journal of Environmental Science and Technology 17 (2020) 2443–2454. https://doi.org/10.1007/s13762-020-02652-

[10] J. Lan, B. Wang, C. Bo, B. Gong, J. Ou, Progress on fabrication and application of activated carbon sphere in recent decade, Journal of Industrial and Engineering Chemistry 120 (2023) 47–72. https://doi.org/https://doi.org/10.1016/j.jiec.2022.12.045

[11] N.I.M. Zakir, Z. Zakaria, H. Osman, A. Masa, The adsorption of remazol red dye using porous activated carbon (PAC) from rice husk ash treated using alkali treatment, in: AIP Conf. Proc., American Institute of Physics, (2024). https://doi.org/10.1063/5.0204422

[12] P. Pichniarczyk, M. Niziurska, Properties of ceramic tile adhesives modified by different viscosity hydroxypropyl methylcellulose, Constr. Build. Mater. 77 (2015) 227–232. https://doi.org/10.1016/j.conbuildmat.2014.12.110

[13] M.B. Hunain, S.N. Alnomani, Q. Razzaq, An Investigation of Tensile and Thermal Properties of Epoxy Polymer Modified by Activated Carbon Particle, IOP Conf. Ser. Mater. Sci. Eng. 1094 (2021) 012164. https://doi.org/10.1088/1757-899x/1094/1/012164

[14] D. Kulig, Ł. Bobak, A. Jarmoluk, A. Szmaja, Ż. Król-Kilińska, A. Zimoch-Korzycka, Effect of Chemical Degradation of Sodium Alginate on Capsaicin Encapsulation, Molecules 28 (2023). https://doi.org/10.3390/molecules28237844

[15] M.A. Atiya, A.K. Hassan, F.Q. Kadhim, Green synthesis of copper nanoparticles using tea leaves extract to remove ciprofloxacin (CIP) from aqueous media, Iraqi Journal of Science 62 (2021) 2833–2854. https://doi.org/10.24996/ijs.2021.62.9.1

[16] S.R. Derkach, N.G. Voron'ko, N.I. Sokolan, D.S. Kolotova, Y.A. Kuchina, Interactions between gelatin and sodium alginate: UV and FTIR studies, J. Dispers. Sci. Technol. 41 (2020) 690–698. https://doi.org/10.1080/01932691.2019.1611437

[17] A. Benettayeb, S. Ghosh, M. Usman, F.Z. Seihoub, I. Sohoo, C.H. Chia, M. Sillanpää, Some Well-Known Alginate and Chitosan Modifications Used in Adsorption: A Review, Water (Switzerland) 14 (2022). https://doi.org/10.3390/w14091353

[18] M. Rosi, S. Viridi, Reduction the Oxygen Content of the Coconut Shell Char Produced by Using Simple Pyrolysis Method, in: IOP Conf. Ser. Mater. Sci. Eng., Institute of Physics Publishing, (2018). https://doi.org/10.1088/1757-899X/395/1/012024

[19] A. Abdelnaby, N.M. Abdelaleem, E. Elshewy, A.H. Mansour, S.S. Ibrahim, Application of Bentonite Clay, Date Pit, and Chitosan Nanoparticles as Promising Adsorbents to Sequester Toxic Lead and Cadmium from Milk, Biol. Trace Elem. Res. 201 (2023) 2650–2664. https://doi.org/10.1007/s12011-022-03353-w

[20] S. Al-Bayaty, N. Jubier, R. H. Al-Uqaily, Study of Thermal Decomposition Behavior and Kinetics of Epoxy/Polystyrene Composites by using TGA and DSC, Xi'an Jianzhu Keji Daxue Xuebao/Journal of Xi'an University of Architecture & Technology 12 (2020) 1331–1341.

[21] S. Liu, Y. Li, L. Li, Enhanced stability and mechanical strength of sodium alginate composite films, Carbohydr. Polym. 160 (2017) 62–70. https://doi.org/10.1016/j.carbpol.2016.12.048

[22] A. Salisu, M.M. Sanagi, A. Abu Naim, K.J. Abd Karim, W.A. Wan Ibrahim, U. Abdulganiyu, Alginate graft polyacrylonitrile beads for the removal of lead from aqueous solutions, Polymer Bulletin 73 (2016) 519–537. https://doi.org/10.1007/s00289-015-1504-3

[23] P.N. Fotsing, E.D. Woumfo, S.A. Măicăneanu, J. Vieillard, C. Tcheka, P.T. Ngueagni, J.M. Siéwé, Removal of Cu(II) from aqueous solution using a composite made from cocoa cortex and sodium alginate, Environmental Science and Pollution Research 27 (2020) 8451–8466. https://doi.org/10.1007/s11356-019-07206-3

[24] H. Zhang, Z. Yan, Z. Yang, J. Yao, Q. Mu, D. Peng, H. Zhao, Study on the synthesis and thermal stability of silicone resins reinforced by Si-O-Ph cross-linking, RSC Adv. 11 (2021) 30971–30979. https://doi.org/10.1039/d1ra05524k

Effect of Stirring Temperature and Blend Ratio on Flexural and Water Absorption of Toughened PVC/PCL Blends

Zulaikha DAMSAR[1,a], Sharifah Shahnaz SYED BAKAR[1,b*] and Muhammad Salihin ZAKARIA[1,c]

[1]Faculty of Chemical Engineering & Technology, Universiti Malaysia Perlis (UniMAP), Kompleks Pusat Pengajian Jejawi 2, 02600 Arau, Perlis, Malaysia

[a]zulaikhadamsar@gmail.com, [b]shahnaz@unimap.edu.my, [c]salihin@unimap.edu.my

Keywords: Polyvinyl Chloride (PVC), Polycaprolactone (PCL), Polymer Blends, Mechanical Properties, Stirring Temperature

Abstract. To develop a more sustainable, flexible and biocompatible material, this study explores the blending effect of polyvinyl chloride (PVC), a plastic commodity with polycaprolactone (PCL), which known for its biodegradability and biocompatibility. The aim of this study was to identify the effect of stirring temperature on flexural and morphological properties of PVC/PCL blend. This was achieved by using different weight ratios of PVC/PCL blends, 100:0, 70:30, 50:50, and 30:70 with 10 wt.% tetrahydrofuran (THF) as a solvent and stirred at 25°C and 40°C. The blend was fabricated using solution casting methods. The mechanical, morphological, and physical properties of the blend were analyzed. Results revealed that the 30:70 blend stirred at 40°C exhibited the highest flexural strength (23.76 MPa) and 50:50 blend exhibited highest hardness (91.33 Shore A), indicating PCL contributes significantly to mechanical reinforcement likely due to its crystallinity. Microscopy revealed lamellar PVC structures and voids in 70:30 blends at both 25 °C and 40°C, suggesting poor miscibility and phase separation. SEM revealed that the 30:70 blend at 40°C exhibited the smoothest and homogenous fracture surface. Furthermore, water absorption was significantly reduced in 50:50 and 70:30 blends stirred at 40 °C. Overall findings demonstrate that the optimal PVC: PCL ratio is 30:70 and higher stirring temperature (40 °C) enhance structural and functional performance of PVC/PCL blends.

Introduction

Silicone is widely used in medical devices due to its flexibility and biocompatibility. Silicone's unique chemical and physical properties are associated to the Si-O bond strength and the vastly tuneable functionalities within attached organic groups, which is a reason they are generally used in the fields of medicine. However, it is expensive and non-biodegradable, raising environmental and cost concerns. Blending polymers is a promising way to create new materials with tailored properties. PVC, a rigid and low-cost plastic is conventionally blended with suitable plasticizers to improve its processibility and, more important, to make it soft and flexible. In this study, PVC, is blended with PCL, a biodegradable and flexible polymer, to develop a more sustainable alternative to silicone.

Several past studies have looked into PVC/PCL blends. Khambatta et al. (1976) found that PCL forms lamellae structures in PVC [1], while Eastmond (1999) reported improved toughness and softness when PCL is added [2]. Rusu et al. (2006) showed that increasing PCL reduces hardness [3], and Rahmatabadi et al. (2023) observed a drop in flexural strength with small amounts of PCL [4]. However, most of these studies focused on thermal or mechanical properties at fixed processing conditions and did not explore how stirring temperature during solution blending affects the material's behavior. Moreover, the effect of higher PCL content up to 70% and how it behaves under different solution blending temperatures is still unclear. There is a lack of systematic study that links blend ratio and processing temperature on mechanical and morphological properties of the blend.

This study lies in its focus on the combined effect of high PCL content and stirring temperature during solution blending, an area not previously explored in detail to optimize the performance of biodegradable PVC/PCL blends for potential medical applications.

Methodology

Materials. The PVC with a number average molecular weight (Mn) of 48,000 g/mol was purchased from Sigma-Aldrich, Malaysia. PCL with Mn of 80,000 g/mol was also obtained from Sigma-Aldrich, Malaysia, and was used as received. Also, the solvent, which is tetrahydrofuran (THF), that was used during the experiments was purchased from R&M Chemicals, Malaysia, and directly used without any purification.

Sample Preparation. The PVC/PCL blends were prepared via solution casting method. Solution mixing was conducted by dissolving PVC and PCL polymers with different weight ratios: 100:0, 70:30, 50:50, and 30:7. These polymers were dissolved in THF with a total polymer blend concentration of 10 wt.%. The required volumes of each component were mixed for 4 hours using a magnetic stirrer and carried out at two different temperatures, which are 25°Cand 40 °C. The solutions were then cast into mould with a size of 155 mm × 80 mm × 55 mm and the samples were dried at room temperature for 7 days to ensure complete evaporation of the solvent.

Flexural Test. An average of three samples with dimensions of 127mm × 12.7mm × 3.2mm for each formulation were tested. The test was carried out according to Procedure B of ASTM D790 by using Instron model 5569 and the strain rate employed was 5 mm/min. Mechanical properties such as flexural strength and modulus of elasticity were recorded and calculated by instrument software.

Hardness Test. The test was performed using Durometer device on the Shore D scale as per the ASTM D2240 standard. A sample of 3.2mm thickness was subjected to a force for 10 seconds. Measurements were taken at three different locations on the sample and the average value was calculated to determine the hardness value.

Optical Microscopy (OM). The morphology of the fracture surface of each broken sample from flexural testing was analyzed using optical microscope. The sample was securely mounted on the microscope stage with the fracture surface facing upward. The surface was examined at 100× magnifications.

Scanning Electron Microscopy (SEM). Broken samples from flexural testing were analyzed using SEM (JEOL JSM-6010LV). The samples were sputter-coated with platinum to examination of SEM at 1000× magnification.

Water Absorption Test. The test followed ASTM D570. The samples were dried in an oven for two hours at 100°C and immediately upon cooling, the samples were weighed. Each sample then immersed in distilled water at room temperature. The weight was measured and recorded by periodically removing the samples from the water. The process will continue until equilibrium is reached. The percentages of water absorption were calculated using Equation (1):

$$WA(\%) = \left(\frac{W_t - W_0}{W_0}\right) \times 100 \quad (1)$$

where W_0 and W_t denote the oven-dry weight and weight after exposure at time t, respectively.

Results and Discussion

Flexural Test. The flexural properties of PVC/PCL blends were evaluated based on flexural strength and modulus of elasticity. As shown in Fig. 1, the 30:70 PVC/PCL blend stirred at 40°C recorded the highest flexural strength of 23.76 MPa, significantly higher than that of pure PVC (15.6 MPa). Likewise, the modulus of elasticity peaked at 403.27 MPa for 30:70 blend at 40°C, compared to 23.24 MPa for pure PVC (refer Fig. 2). This indicates that both the blend ratio and stirring temperature strongly influence the flexural behavior of the blends.

These findings offer a contrast to previous studies where Rahmatabadi et al. (2023) reported a reduction in flexural strength at low PCL contents (5–10%), likely due to weak phase interactions [4]. At low PCL content, the small domain of PCL may get distributed well within PVC matrix, thus allowing the blend to carry stress better and consequently increasing the flexural strength. However, the study by Chiu and Min (2000) shows that higher PCL content, more than 50%, can significantly improve modulus and stiffness due to pseudo-crosslinking and increased crystallinity [5]. The improvements in flexural strength and stiffness can be attributed to the semi-crystalline nature of PCL and the role of temperature in enhancing polymer chain mobility during mixing. At 40°C, increased thermal energy allows PCL to disperse more evenly within the PVC matrix, promoting better load transfer leading to higher flexural strength and stiffness.

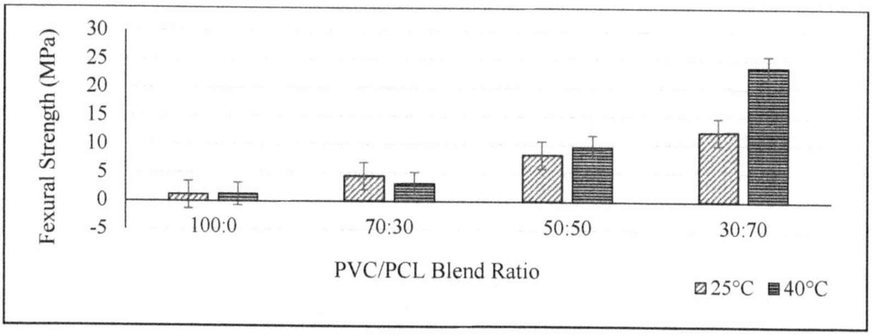

Fig. 1: Effect of PVC/PCL blend ratio and stirring temperature on flexural strength.

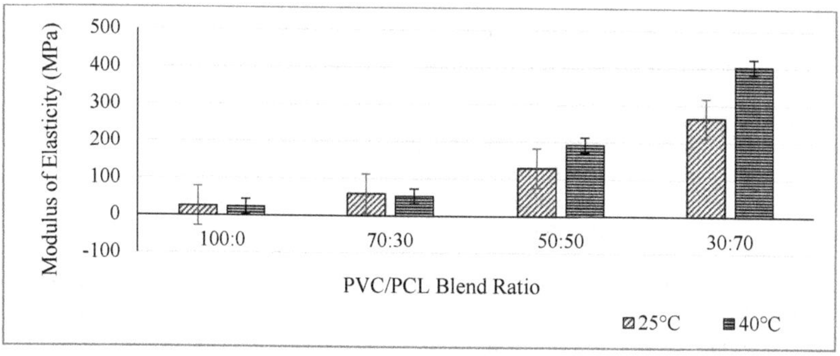

Fig. 2: Effect of PVC/PCL blend ratio and stirring temperature on modulus of elasticity.

Hardness. The Shore A hardness values of the blends are summarized in Fig. 3. Pure PVC recorded a hardness of 75.55 Shore A, while the 50:50 blend stirred at 40°C showed the highest value of 91.33 Shore A (refer Fig. 3). This increase in hardness with rising PCL content and stirring temperature suggests that PCL contributes not only to flexural strength but also to surface rigidity when properly dispersed and crystallized. The enhancement in hardness can be attributed to the increasing crystallinity of PCL, particularly at high weight ratios and elevated temperatures.

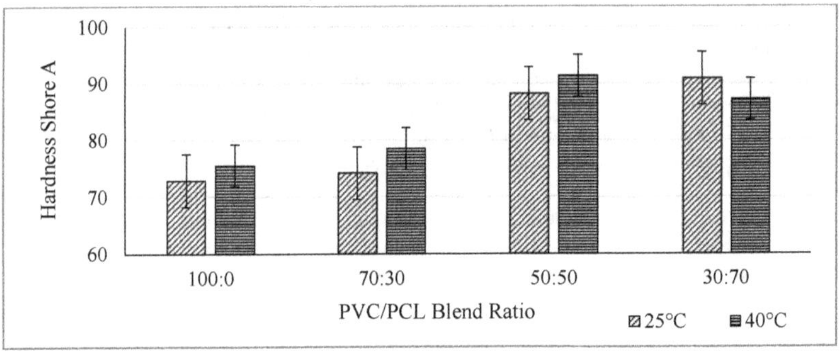

Fig. 3: Effect of PVC/PCL blend ratio and stirring temperature on hardness Shore A.

This trend supports previous findings by Pekdemir et al. (2021), who reported that pure PCL possesses a fully crystalline structure, and its blends with PVC show an amorphous–crystalline structure with increasing crystallinity up to 40% for the 3:7 ratio [6]. The study by Pangesty and Todo (2021) further supports the importance of stirring temperature, showing that increasing solution blending temperature from 20°C to 60°C enhanced mechanical strength in PCL blends due to better molecular mixing and reduced phase separation [7]. However, earlier studies by Rusu et al. (2006) observed a decrease in hardness with increasing PCL, attributing it to the softness of PCL [3]. Even so, their results were based on melted-mixed systems with limited PCL content which is less than 50%.

Morphological Analysis. The morphological characteristics of the fractured surfaces of PVC/PCL blends were examined using optical microscopy (OM) and scanning electron microscope (SEM). The observations are presented in Fig. 4. The pure PVC sample [Fig. 4 (a) and (f)] stirred at 25°C exhibited a rough, porous, and highly fibrillar morphology, indicating significant plastic deformation and correlating with its low flexural strength. The 70:30 blends stirred at 25°C exhibited rough, irregular surfaces with prominent PVC layering and the presence of voids, indicating poor miscibility and weak interfacial adhesion between the two polymer phases as can be seen in Fig. (b). SEM micrograph in Fig. (g) showed heterogeneous phase separation with distinct domains and thin fibrillar structures, suggesting limited miscibility and interfacial adhesion. This outcome partially aligns with the findings of Khambatta et al. [1] and Russell and Stein [8], who reported lamellar and phase-separated morphologies in PVC/PCL blends. It also reflects Liu et al. (2016), who stated that blends with high PCL content tend to develop more crystalline domains that can segregate and disrupt uniformity [9].

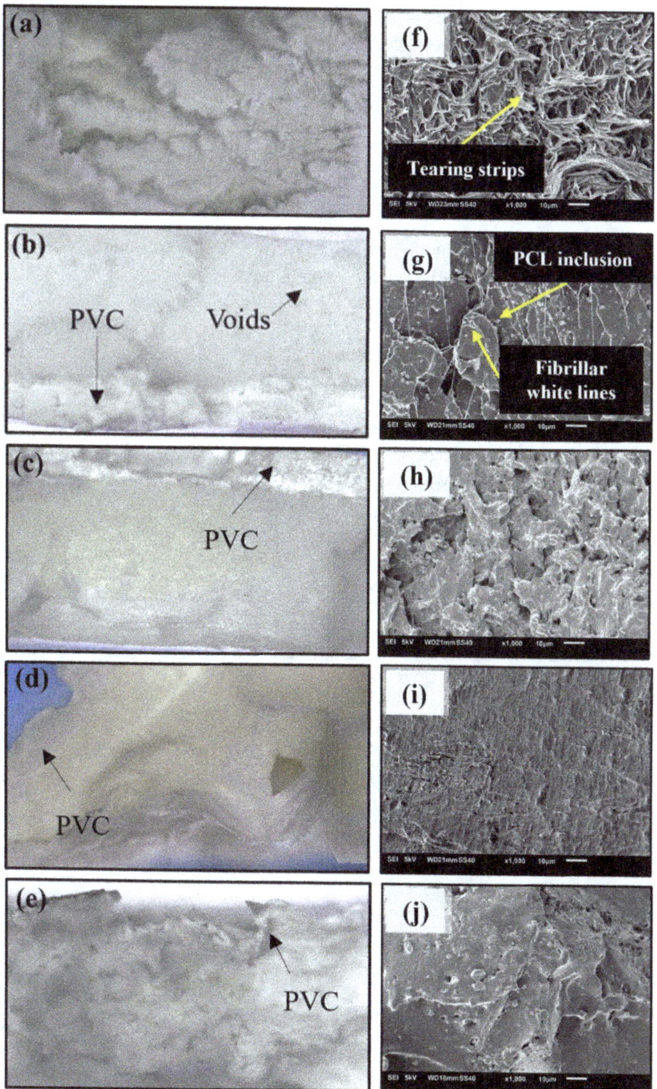

Figure 4: Optical microscopy and SEM micrographs *image of fractured cross section of PVC/PCL blends at (a) 100:0 at 100× magnification OM (b) 70:30 at 25°C at 100× magnification OM (c) 70:30 at 40°C at 100× magnification OM (c) (d) 30:70 at 25°C at 100× magnification OM (e) 30:70 at 40°C at 100× magnification OM (f) 100:0 at 1K× magnification SEM (g) 70:30 at 25°C at 1K× magnification SEM (h) 70:30 at 40°C at 100× magnification OM (i) 30:70 at 25°C at 1K× magnification SEM and (e) 30:70 at 40°C at 1K× magnification SEM .*

In contrast, when the same blends were stirred at 40°C, the surface became notably smoother, with fewer voids indicating enhanced compatibility [Fig. (c) and (h)]. At higher PCL content (70%), the microstructure became more uniform, with smoother surfaces and minimal phase separation, likely due to the semi-crystalline nature of PCL forming a continuous phase [Fig. (d) and (i)]. In the case of the 30:70 blend stirred at 40°C, the morphology was notably more irregular, with visible wave-like voids and non-uniform surface features, indicating that the structure became more complex and discontinuous at higher PCL content stirrer at higher temperature [Fig. (e) and (j)].

Water absorption. The water absorption behavior of the PVC/PCL blends was investigated over 15 days immersion period using distilled water, and the results are summarized in Table 1 and Fig. 5. Among all samples, pure PVC absorbed the highest amount of water (18.11%), likely due to its amorphous structure and polar nature. In contrast, the 50:50 and 70:30 blends stirred at 40°C showed the lowest water absorption values, recorded at 0.05% and 0.82%, respectively. The results indicate that introducing PCL into the PVC matrix drastically reduces water absorption, especially when processing occurs at a higher temperature. PCL is a semi-crystalline, hydrophobic polymer, and its crystalline domains reduce the free volume within the blend, effectively creating a tighter structure that limits water ingress. These findings are in line with literature reports. Wu (2003) emphasized that hydrophilic amorphous polymers tend to absorb more moisture, while semi-crystalline, hydrophobic polymers like PCL act as water barriers [10].

Table 1: Data on water absorption properties of different PVC/PCL blends ratio and stirring temperature.

PVC/PCL Composition	Temperature (°C)	Water absorption (%) 24 hrs	3 days	5 days	7 days
100:0	25	7.67	16.06	18.11	18.13
	40	4.47	6.86	8.40	8.42
70:30	25	0.41	0.33	0.65	0.66
	40	0.15	0.75	0.82	0.82
30:70	25	2.06	2.46	3.09	3.10
	40	0.60	0.37	1.05	1.06

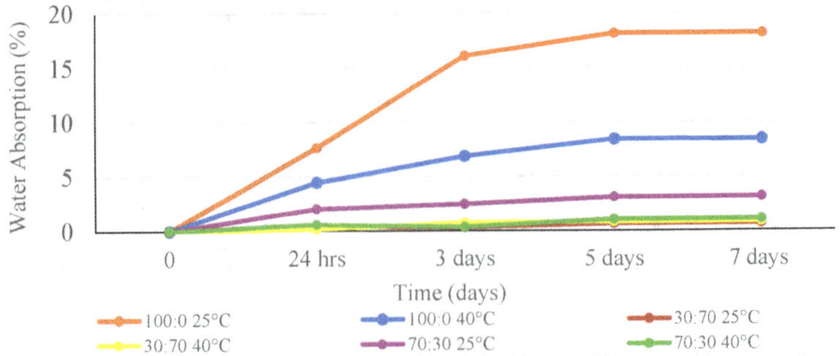

Fig. 5: Effect of PVC/PCL blend ratio and stirring temperature on water absorption.

Conclusion

The PVC/PCL blends were successfully produced by varying the weight ratio of polymer and stirring temperatures. The 30:70 PVC/PCL blend stirred at 40 °C exhibited the highest flexural strength (23.76 MPa), modulus of elasticity (403.27 MPa), and 50:50 exhibited highest hardness (91.33 Shore A), indicating strong mechanical reinforcement due to improved phase interaction and PCL crystallization. Water absorption was also significantly reduced, with the lowest value of 0.05% recorded in the 50:50 blend at 40°C. The 30:70 blend, while mechanically superior, still exhibited structural discontinuities such as surface voids, which may compromise long-term durability. Some limitations were noted where the voids seem unavoidable at this stage of research. This indicates a need for further refinement in blending or post-treatment methods to enhance both structural and functional stability, while thermal analysis will be needed to undertands the behavioral of the blends. This work demonstrates that PVC/PCL blends can be tailored through controlled solution-casting and composition to meet the requirements of diverse thin-film applications including biomedical membranes, flexible packaging, agricultural films, and controlled-release systems. By balancing the mechanical stability of PVC with the biodegradability and flexibility of PCL, the blends offer a versatile platform for sustainable and functional polymer films.

Acknowledgement

The authors gratefully acknowledge the financial support provided by the Ministry of Higher Education Malaysia under the Fundamental Research Grant Scheme (FRGS/1/2024/TK09/UNIMAP/02/17).

References

[1] Khambatta, F. B., Warner, F., Russell, T., & Stein, R. S. (1976). Small-angle x-ray and light scattering studies of the morphology of blends of poly(ϵ-caprolactone) with poly(vinyl chloride). *Journal of Polymer Science Polymer Physics Edition, 14*(8), 1391–1424. https://doi.org/10.1002/pol.1976.180140805

[2] Eastmond, G.C. (2000). Poly(ε-caprolactone) Blends. In: Eastmond, G.C., Höcker, H., Klee, D. (eds) Biomedical Applications Polymer Blends. Advances in Polymer Science, vol 149. Springer, Berlin, Heidelberg. https://doi.org/10.1007/3-540-48838-3_2

[3] Rusu, M., Ursu, M., & Rusu, D. (2006). Poly(vinyl chloride) and Poly(e-caprolactone) Blends for Medical Use. *Journal of Thermoplastic Composite Materials, 19*(2), 173–190. https://doi.org/10.1177/0892705706056463

[4] Rahmatabadi, D., Aberoumand, M., Soltanmohammadi, K., Soleyman, E., Ghasemi, I., Baniassadi, M., Abrinia, K., Bodaghi, M., & Baghani, M. (2023). Toughening PVC with Biocompatible PCL Softeners for Supreme Mechanical Properties, Morphology, Shape Memory Effects, and FFF Printability. *Macromolecular Materials and Engineering, 308*(10). https://doi.org/10.1002/mame.202300114

[5] Chiu, F.-C., & Min, K. (2000). Miscibility, morphology and tensile properties of vinyl chloride polymer and poly(ε-caprolactone) blends. *Polymer International, 49*(2), 223–234. https://doi.org/10.1002/(SICI)1097-0126(200002)49:2<223::AID-PI345>3.0.CO;2-U

[6] Pekdemir, M. E., Öner, E., Kök, M., & Qader, I. N. (2021). Thermal behavior and shape memory properties of PCL blends film with PVC and PMMA polymers. *Iranian Polymer Journal, 30*(6), 633–641. https://doi.org/10.1007/s13726-021-00919-8

[7] Pangesty, A. I., & Todo, M. (2021). Improvement of mechanical strength of tissue engineering scaffold due to the temperature control of polymer blend solution. *Journal of Functional Biomaterials*, *12*(3), 47. https://doi.org/10.3390/jfb12030047

[8] Russell, T. P., & Stein, R. S. (1983). An investigation of the compatibility and morphology of semicrystalline poly(ε-caprolactone)–poly(vinyl chloride) blends. *Journal of Polymer Science Polymer Physics Edition*, *21*(7), 999–1010. https://doi.org/10.1002/pol.1983.180210701

[9] Liu, W., Zhang, R., Huang, M., Dong, X., Xu, W., Ray, N., & Zhu, J. (2016). Design and structural study of a triple-shape memory PCL/PVC blend. *Polymer*, *104*, 115–122. https://doi.org/10.1016/j.polymer.2016.09.079

[10] Wu, C. (2003). Physical properties and biodegradability of maleated-polycaprolactone/starch composite. *Polymer Degradation and Stability*, *80*(1), 127–134. https://doi.org/10.1016/s0141-3910(02)00393-2

Green Materials, Processing & Characterization – ICoGMPaC 2025 Materials Research Forum LLC
Materials Research Proceedings 68 (2026) 17-25 https://doi.org/10.21741/9781644904237-3

Phase, Structure and Compressive Strength of Sintered Perlis Dolomite/Silica

Nur Hasnidah AHMAD SHUKERI[1,a*], Syed Nuzul Fadzli SYED ADAM[1,b], Firuz ZAINUDDIN[2,c], Hasmaliza MOHAMAD[3,d], Heah CHENG YONG[1,e]

[1]Faculty of Mechanical Engineering and Technology, Universiti Malaysia Perlis (UniMAP), 02600, Ulu Pauh, Perlis Malaysia

[2]Faculty of Chemical Engineering and Technology, Universiti Malaysia Perlis (UniMAP), 02600, Ulu Pauh, Perlis Malaysia

[3]School of Materials and Mineral Resources Engineering, Universiti Sains Malaysia (USM), 14300, Transkrian Penang, Malaysia

[a]hasnidahnur@gmail.com, [b]syed.nuzul@unimap.edu.my, [c]firuz@unimap.edu.my, [d]hasmaliza@usm.my, [e]cyheah@unimap.edu.my

Keywords: Perlis Dolomite, Silica, Akermanite, Monticellite, Merwinite

Abstract. Natural dolomite sedimentary rocks, which can be found in abundance in Perlis, is a carbonate mineral rich with calcium and magnesium elements. These both elements were important and typically found in bioceramic materials especially for hard tissue implant material. Even though raw Perlis dolomite powder has other various elemental compositions, it lacks important components such as silica which commonly acts as glass network former, in resulting limits its potential application as bioceramic materials. This study investigates the effects of different compositions of silica addition on sintered Perlis dolomite and the changes in their phase transformation, structure and compressive strength were analyzed. Perlis dolomite with different ratios of silica (30S, 35S and 40S) were ball milled, compacted into pallets and sintered at 1300°C for 4 hours. The results show that the silica contents gave a significant impact on the phase transformation of the resulting materials involving akermanite, monticellite and merwinite. At lower silica content (30S), monticellite and merwinite became dominant phases with a slight amount of akermanite. But increasing silica content leads to the formation of akermanite phase, while reducing the monticellite and merwinite phase and merwinite phase completely disappeared at 40S. This result expressed that the higher silica encourages the stabilization of akermanite at the expense of other calcium–magnesium silicate phases supported with improvement in compressive strength (14-18 MPa), which is promising for bone tissue regenerations.

Introduction

In the development of bioceramic for bone tissue regeneration, a deep consideration of material properties should be taken wisely in order to design implants that are not only functionally effective, but also have great biocompatibility and bioactivity [1]. There are many approaches from previous researchers in the production of bioceramic purposely for implantology application in bone treatment. Recently, various types of bioceramic within the calcium-magnesium-silicate (CaO-MgO-SiO_2) system have been acknowledged among the researchers due to their promising mechanical and bioactivity performances which present favorable properties suitable for bone regeneration practices. Remarkable phases in this system such as akermanite, monticellite and merwinite are closely mimic the bone's mineral content [2], allowing for better tissue integration [3]. The improved mechanical properties, controlled biodegradability and excellent bioactivity performances of these phases also confirmed their potentials and capability to be on par with current prominent bioceramics such as hydroxyapatite (HA) and tricalcium phosphate (TCP) [4].

The formation of bioactive $CaO-MgO-SiO_2$ phases is highly relying on the usage of calcium, magnesium and silica as essential precursors. Thus, it has led to growing interest in utilizing naturally available CaO-MgO minerals such as dolomite, associated with silica based materials. Dolomite ($CaMg(CO_3)_2$) is primarily composed of a rich content of calcium and magnesium, which are significant elements in the formation of new bone. This is because calcium is the main source of bone matrix while magnesium promotes bone tissue growth and provides repair functions [5]. An approach has been proposed by Arkame et al., where Moroccan dolomite is used with slag waste to fabricate akermanite bioceramic [6]. The findings showed the porosity, water absorption and flexural strength of the akermanite are 41.4%, 24.8% and 13.8 MPa, respectively without any presence of microcracks. Moreover, an akermanite bioceramic also has been prepared by using 75:25 wt. % ratio of Moroccan dolomite to natural perlite material and reported great bending (7.71 MPa) and tensile strength (3.56 MPa) [7]. The bioactivity of the akermanite also has been confirmed by the formation of hydroxyapatite layer after simulated body fluid (SBF) immersion.

In this study, Perlis dolomite will be used as the main precursor for the $CaO-MgO-SiO_2$ based bioceramic as it is easily be found in the sedimentary rocks quarries especially at the district of Chuping, Perlis. Perlis dolomite is locally known as 'Batu Reput' and typically used for soil treatment in agriculture, road pavement making and building construction [8,9]. Due to its abundant source and cost effectiveness, this study will utilize the Perlis dolomite to contribute valuable insight into the use of locally natural mineral resources while reducing the dependence of synthetic materials in the development of bioceramics. However, the composition of the silica in the Perlis dolomite is too deficient and the dolomite alone cannot be able to form bioceramic phases during the sintering process as it will hinder its ultimate potential as bioactive ceramic. Hence, to overcome this issue, the incorporation of silica into the Perlis dolomite-based ceramics has been proposed in this study.

Silica is one of the most common inorganic compounds that can easily be found in the nature of Earth's crust in various forms such as sand, flint and quartz. Generally, silica is possessed with chemical and thermal stability making it highly all around in various ranges of scientific applications and industrial uses. In the scope of material science, silica plays a vital role as glass-former through the oxide bonding owing to its capability to create three dimensional (3D) network structures during thermal applied [10]. This behavior makes the silica as a crucial component in the preparation of ceramics. Additionally, silica is also highly essential in the biomedical field, especially in the advancement of bioceramics as it provides great bioactivity and encourages bone growth performances and mineralizations [11].

In this present study, different compositions of silica addition will be implemented and their structural, phase transformation and mechanical properties in the sintered dolomite will be analyzed. The outcomes of this study are expected that the silica addition in the sintered Perlis dolomite will play a role as a glass former network and lead to the phase stability of bioceramic as well as improve the mechanical properties. Besides, the material's value of silica-altered Perlis dolomite will be enhanced as an alternative bioceramic material appropriate for bone tissue regenerations.

Methodology

Raw Material. In this study, fine powder of raw Perlis dolomite mineral was used as the main precursor for the CaO-MgO sources. This Perlis dolomite mineral was supplied from the local quarry at Chuping, Perlis, Malaysia. Meanwhile, silica powder which was acquired from Ipoh Ceramic, Perak, Malaysia was utilized as an additive material and acts as a glass former network in the preparation of dolomite-derived bioceramic. A calcination process was performed on raw dolomite powder at 1000°C for 4 hours with 10°C/min heating rate to decarbonize and eliminate the impurities and volatiles substances from the mineral. The chemical compositions for the materials were analyzed via X-ray fluorescent (XRF) by Philips PANalytical MiniPAL 4 machine.

The elemental compositions of the materials were presented in Table 1. The XRF results were expressed in percentages which indicate the amount of oxides content. Both calcined dolomite and silica powder were sieved individually to obtain uniform size particles of powders less than 150 μm.

Table 1: Chemical compositions of raw materials used in this study.

Raw Material	Chemical compositions (%)						
	CaO	MgO	Na_2O	Ag_2O	SiO_2	Al_2O_3	Others
Raw Dolomite	65.07	23.10	8.00	1.46	0.25	1.30	1.07
Calcined Dolomite	54.55	30.50	12.00	0.29	0.10	1.20	-
Silica	-	-	-	-	97.74	2.00	0.26

Sample Preparation. The calcined Perlis dolomite and silica powder were mixed together homogeneously through mechanical activation method via planetary ball milling (Mono Mill Pulverisette 6-classic line, Fritsch, Germany). Three mixtures of heated dolomite were prepared with different ranges of silica compositions within 30, 35 and 40 wt. % and labeled in the results and discussion as 30S, 35S and 40S respectively. The wet ball milling operation was performed by using deionized water as the milling medium with the ratio of deionized water to powder was 3:1. Meanwhile, the ratio of agate balls to mixture powder was 10:1. The milling operation was set with 450 rpm milling speed. Each cycle of the operation was about an hour and repeated for three times in reverse directions. Then, the mixtures were dried in an oven at 80°C overnight. In order to obtain fine uniform powders, the dried powders from the oven were ground and sieved. Next, 0.5 g sodium alginate powder (Sigma-Aldrich, Germany) was diluted with 20 mL deionized water for the purpose of binder preparation. The binder solution was manually stirred until the solid substances of the alginate completely dissolved and no granule residues remained in the solution. 5 wt. % of the binder solution was added and mixed together with 1.0 g of dolomite-silica powder. After that, the powder mixtures were compacted into pallets by using a hydraulic press machine (Nichi T-61230A). 12 mm diameter of cylindrical carbon steel mold was used during the compaction process with 200 MPa load pressure for 5 minutes loading. Finally, the pallets were sintered in the furnace at 1300°C for 4 hours with a heating rate of 5 °C/min (LT Furnace SIC4-1600).

Characterisation of Sintered Samples. For structural analysis, the samples were characterized by using X-Ray Diffractometer (XRD) (Test Instrument Bruker, D2 Phaser Germany 2010) and Fourier Transform Infrared spectroscopy (FTIR) (Perkin Elmer machine). The XRD were operated with angle diffraction (2θ) step scanning within 10° to 90° for 10 minutes per each sample scanning. The obtained diffractograms were analyzed by using X'pert Highscore Plus software to recognize the phase compounds that occur in the samples and match the results according to the crystallographic database. Meanwhile, in the preparation of a sample for FTIR analysis, potassium bromide powder (KBr) was mixed with 1.0 wt. % of the powder sample and compacted into thin pallets. The spectrums were recorded at the range of 400-4000 cm^{-1} to observe the characteristics of vibration bands and functional groups of the materials. For the mechanical analysis, the samples were compacted at 200 MPa loads using 9 mm size diameter of cylindrical mold with a diameter-to-thickness ratio of 1:2, following the ASTM C1424 standard. The samples were tested through compressive strength analysis by using Universal Machine Testing (UTM) with pressure load under 50 kN (Shimadzu AG-X Plus) associated with material analysis software Trapezium X to analyze the recorded data.

Results and Discussion

Phase Transformation Analysis. According to our previous study and current findings, the raw dolomite with high levels of carbonate contents has undergone transformation into primary content of CaO and MgO with a trace of other oxides after the calcination process at 1000°C [12,13]. Thus, these phases significantly contribute to the phase transformation of the materials through the synthesis of dolomite-silica and induce the formation of bioceramic phases. Fig. 1 illustrated the phase transformation of dolomite-derived samples with varying silica addition (30S, 35S and 40S). It revealed a remarkable phase evolution influenced by the amount of silica addition. At the lowest silica composition (30S), the dominant crystalline phases are formed by monticellite (JCPDS file no. 98-004-5804) and merwinite (JCPDS file no. 98-001-1933) based on the numerous strong diffraction peaks. It also can be observed that the presence of akermanite (JCPDS file no. 98-002-2321) is minimal. At 35S, a slight transitional phase composition was observed where all these phases (monticellite, merwinite and akermanite) coexisted in equivalence. The intensity of the akermanite peaks were increased, while those monticellite and merwinite were mostly decreased. At the highest silica content (40S), akermanite has become prevalent, especially at 31.42° and 44.68°. A small amount of merwinite is still present meanwhile merwinite is diminished in this sample.

Based on these findings, it suggests that the trends in the increasing silica addition in the calcined dolomite samples have promoted the phase transformation and development of CaO-MgO-SiO_2 based bioceramics such as akermanite, monticellite and merwinite. All these phases were known as great bioactive ceramics owing to their biocompatibility and osteogenic behavior [14,15], making them suitable for bone tissue regeneration. In the sample of 30S, the small amount of the akermanite was due to insufficient silica content to support its formation. Meanwhile in the 35S, the balanced phase formations were developed due to the phase transformation of merwinite and monticellite while the peaks shift in phase stability leaning to akermanite formation. The formation of akermanite has shown a requirement of higher silica content to achieve its stoichiometry (Ca:Mg:Si = 2:1:2) at the highest content of silica (40S) [16]. The merwinite phase was completely eliminated in 40S and was eventually replaced by the akermanite and monticellite phases. This is because the merwinite is less stable at high silica concentrations and promotes the formation of akermanite and monticellite which results in stable development of multiphase ceramic material. Additionally, an outstanding peak at 31.42° in 40S sample represented a well-defined crystallization of akermanite which confirmed its phase stability. From these findings, it validated that the addition of silica in the sintered dolomite leads to the formation of biocompatible compounds involving akermanite, monticellite and merwinite. Hence, Perlis dolomite incorporated with the silica powder has demonstrated its strong potential as raw materials for the CaO-MgO-SiO_2 based bioceramic production especially for bone treatment.

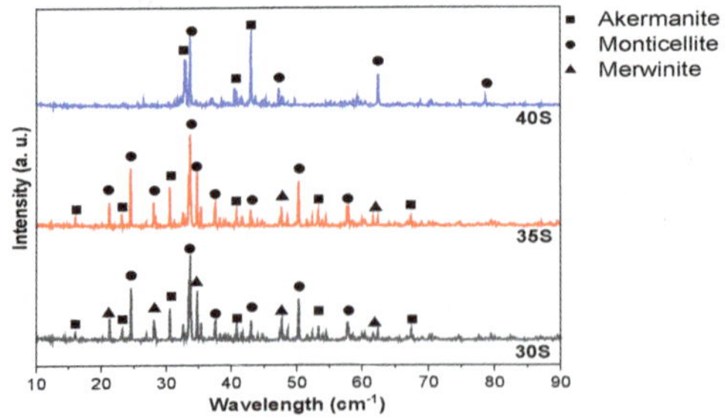

Fig. 1: XRD diffractogram of sintered dolomite with different ratios of silica.

Functional Group Analysis. Fig. 2 illustrated the FTIR spectra of the sintered dolomite-silica samples with different silica additions. Even though there are differences in silica addition in the sintered dolomite, every sample exhibited identical spectral patterns. Remarkably, a narrow peak at 595 cm^{-1} were significantly indicated the Ca═O stretching vibration [7] due to the development of akermanite, merwinite and monticellite. Besides, this distinct absorption band at 595 cm^{-1} was also gradually decreased in their intensities when the silica content increased, attributed to the O—Si—O vibrational mode [17]. This reduction is corresponded to the reaction between the CaO and MgO components of the dolomite and additional silica contents, which promotes the formation of denser ceramic phases such as akermanite, merwinite and monticellite. The existence of these phases was further confirmed by the presence of bands at 462 cm^{-1} and 511 cm^{-1}, accounted as the O—Mg—O [7] and O—Ca—O [5] bending modes, respectively. Furthermore, absorption peaks at 830 cm^{-1} and 929 cm^{-1} were assigned to Si—O stretching vibrations while, an outstanding peak at 1025 cm^{-1} was appeared because of the presence of a symmetric stretching vibration of Si—O—Si bonds [7,17]. These bands validated the existence of silicate bonding and the development of silicate structural networks within the samples, consistent with the XRD findings that have been discussed earlier. Additionally, weak absorption bands occurred between 3320 cm^{-1} and 3622 cm^{-1} were linked to hydroxyl groups, related to O—H stretching vibrations mode [7]. The absorption peak at 1435 cm^{-1} was attributed to the H—O—H bending mode which exists due to absorbed water molecules content in the samples [5]. Besides, the deep and more pronounced band at 1435 cm^{-1} in the 40S may also indicate the residual carbonate or organic content that may exist due to higher silica additions [18]. Overall, the FTIR results aligned well with the XRD analysis, proving the formation of akermanite, merwinite and monticellite phases induced by the silica incorporation in the sintered Perlis dolomite samples.

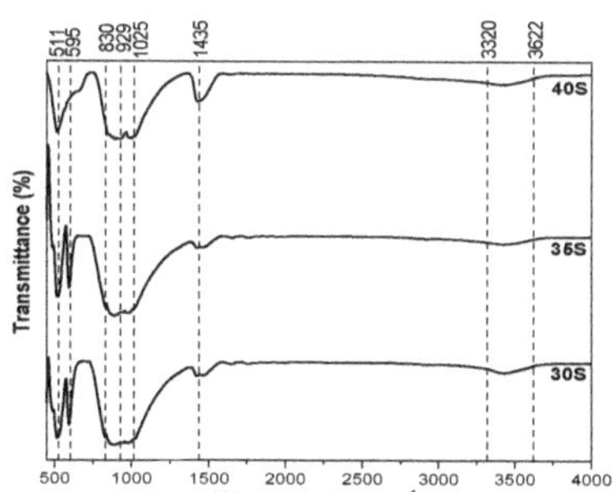

Fig. 2: FTIR spectra of sintered dolomite with different ratios of silica.

Compressive Strength Analysis. Fig. 3 illustrated the mechanical properties of the sintered dolomite/silica samples. The addition of silica significantly contributes to the improvement of the compressive strength (~14-18 MPa), considered as favourable compressive strength as the values are lying within the range of human cancellous bone (10-20 MPa) [19]. The highest silica addition (40S) has shown highest compressive strength about 18.91 MPa due to the formation of dual-phases of akermanite and monticellite. This finding can be supported by previous study where the combination of akermanite and monticellite exhibit great compressive strength compared to pure akermanite (0.58 ± 0.03 MPa) and monticellites (0.39 ± 0.03 MPa) alone [20]. Another study has reported low compressive strength of the porous akermanite-merwinite scaffold (0.765 ± 200 MPa) despite identical sintering temperature as present study and longer sintering time (5 hours) but lower heating rate (2°C/min) with [21].

From this finding, the increasing compressive strength by the increasing silica content can be correlated with the phase formation in the sample which has been discussed earlier in XRD and FTIR analysis. Technically, the crystal structures of the existing phases significantly influenced the mechanical strength of the samples. This is because merwinite with the open monoclinic lattice will weaken the structures. Meanwhile, akermanite and monticellite are respectively tetragonal and orthorhombic structures, which provide balanced combination of strength and stiffness due to their rigid silicate layered [22]. Hence, the low compressive strength of the 30S sample was assumed due to the poor structures of the merwinite. However, the presence of the merwinite is not a drawback as it has great biomaterialization and excellent antibacterial properties compared to akermanite [23].

The compressive strength of 35S and 40S were seen a slight improvement as the merwinite phases were reduced and transformed completely into akermanite and monticellite. This indicated that the compressive strength increased as the akermanite content increased which meets agreement with another study from C.-I. Dobrit¸a et al. [23]. Besides, the increasing silica addition in the sample also formed akermanite and monticellite phases with higher crystallinity where the silica fills up the pores among the particles of calcia and magnesia, allowing complete crystallization and densifications during sintering process in resulting increases the compressive strengths [24]. Higher silica content also contributed to enhanced grain growth, grain boundary formation and the development of agglomerated grains. These findings indicated that the silica additions significantly influence the grains formation and densifications, primarily through its fluxing effect and the formation of glassy phase on the sample surface and improved the mechanical properties [25].

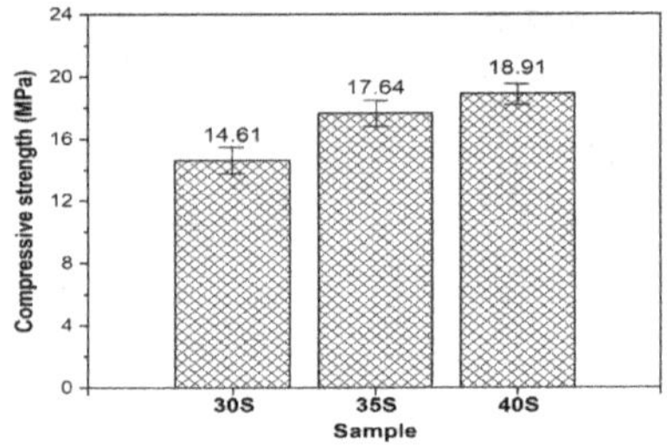

Fig. 3: Compressive strength of the sintered dolomite with different ratio of silica.

Conclusion

This study has proven the potential of Perlis dolomite associated with silica powder as additive material to form CaO-MgO-SiO_2 based bioceramics. Investigation on the Perlis dolomite with the effects of varying silica content on the structural, phase transformation and mechanical properties at 1300°C has successfully been carried out. The findings revealed that the increasing silica compositions significantly promoted the formation of akermanite, merwinite and monticellite phases, as verified through both chemical structure and functional group analyses. Moreover, higher silica content provides mechanical improvement due to the structure stability and densification of the phases. The increases in compressive strength are aligning with the increase of silica content which highlights strong potential for bioceramic applications, as they meet the requirement of compressive strength for natural cancellous bone applications. Future studies involving in-depth bioactivity and biocompatibility assessments are recommended, particularly to mimic and optimize their performances similarly as human cancellous bone. Optimization on its sintering parameter is also suggested for future work to enhance bioceramic phase development and overall material performance.

Acknowledgement
A gratitude was expressed to the Faculty of Mechanical Engineering, Universiti Malaysia Perlis (UniMAP) and School of Materials and Mineral Resources Engineering, Universiti Sains Malaysia (USM) for providing the equipment and facilities for this research work. This study was funded by a Postgraduate Grant (Process No. 9001-00788) from the Research Management Centre (RMC), UniMAP.

References

[1] D. H. M. Rao., K. Rajkumar, D. V. Narayana, Akash M, D. J. Sabu, and D. R. Ramadoss, "Bioinert And Bioactive Materials —Narrative Review," *Journal of Pharmaceutical Negative Results,* vol. 13, no. 7, pp. 2251–2255, 2022. https://doi.org/10.47750/pnr.2022.13.S07.309

[2] J. Yu, J. Qian, F. Wang, J. Qin, X. Dai, C. You, and X. Jia, "Study of using dolomite ores as raw materials to produce magnesium phosphate cement," *Constr. Build. Mater.*, vol. 253, 2020. https://doi.org/10.1016/j.conbuildmat.2020.119147

[3] R. A. Youness, D. M. Tag, and M. A. Taha, "A Review on Calcium Silicate Ceramics : Properties , Limitations , and Solutions for Their Use in Biomedical Applications," *Silicon*, vol. 15(6), pp. 2493–2505, 2023.

[4] M. Ahmadipour, H. Mohammadi, A. L. Pang, M. Arjmand, T. Ayode Otitoju, P. U. Okoye, and B. Rajitha, "A review: silicate ceramic-polymer composite scaffold for bone tissue engineering," *International Journal of Polymeric Materials and Polymeric Biomaterials*, vol. 71, no. 3. 2022. https://doi.org/10.1080/00914037.2020.1817018

[5] H. Mohammadi, Y. M. Baba Ismail, K. A. Bin Shariff, and A. F. Mohd Noor, "Synthesis and Characterization of Akermanite by Mechanical Milling and Subsequent Heat Treatment," in *Journal of Physics: Conference Series*, 2018, vol. 1082, no. 1. https://doi.org/10.1088/1742-6596/1082/1/012021

[6] Y. Arkame, A. Harrati, M. Jannaoui, Y. Et-Tayea, I. Yamari, A. Sdiri, and C. Sadik, "Effects of slag addition and sintering temperature on the technological properties of dolomite based porous ceramics," *Open Ceram.*, vol. 13, 2023. https://doi.org/10.1016/j.oceram.2023.100333

[7] A. Harrati, Y. Arkame, A. Manni, S. Aqdim, R. Zmemla, A. Chari, A. El Bouari, I.E.E.A. El Hassani, A. Sdiri, F.O. Hassani, and C. Sadik, "Akermanite-based ceramics from Moroccan dolomite and perlite: Characterization and in vitro bioactivity assessment," *Open Ceram.*, vol. 10, p. 100276, Jun. 2022. https://doi.org/10.1016/j.oceram.2022.100276

[8] C. Azurahanim, E. L. Albert, L. Gee Een, M. A. Zarin, and M. Z. Mohd Yusoff, "Effect of Thermal Treatment on Physical Properties of Malaysian Dolomitic Limestone," *Int. J. Electroact. Mater*, 2022.

[9] F. F. Kamarzaman, M. M. A. B. Abdullah, S. Z. Abd Rahim, A. A. Kadir, N. H. Jamil, W. M. W. Ibrahim, and A. V. Sandu, "Hydroxyapatite / Dolomite alkaline activated material reaction in the formation of low temperature sintered ceramic as adsorbent materials," vol. 349, no. August, 2022.

[10] P. H. Huang, M. Laakso, P. Edinger, O. Hartwig, G.S. Duesberg, L.L. Lai, J. Mayer, J. Nyman, C. Errando-Herranz, G. Stemme, and K.B. Gylfason, "Three-dimensional printing of

silica glass with sub-micrometer resolution," *Nat. Commun.*, vol. 14, no. 1, 2023. https://doi.org/10.1038/s41467-023-38996-3.

[11] K. H. Min, D. H. Kim, K. H. Kim, J.-H. Seo, and S. P. Pack, "Biomimetic Scaffolds of Calcium-Based Materials for Bone Regeneration," *Biomimetics*, vol. 9, no. 511, pp. 1–24, 2024. https://doi.org/10.3390/biomimetics9090511.

[12] J. I. E. Zhang, Z. Zhang, D. Wang, and X. Qiang, "Study on Preparation of High-Quality Magnesium Hydroxide and Calcium Hydroxide Products from Dolomite," *JOM*, pp. 1–13, 2025. https://doi.org/10.1007/s11837-025-07497-7

[13] N. H. Ahmad Shukeri, S. N. F. Syed Adam, H. Mohamad, H. Cheng Yong, and H. Yahya, "Effect of Heat Treatment on the Properties of Perlis Dolomite Powder," *Pertanika J. Sci. Technol.*, vol. 32, no. 3, pp. 1613–1631, 2025. https://doi.org/10.47836/pjst.33.3.23

[14] A. Namdar and E. Salahinejad, "Advances in ion-doping of Ca-Mg silicate bioceramics for bone tissue engineering," *Coord. Chem. Rev.*, vol. 478, 2023. https://doi.org/10.1016/j.ccr.2022.215001

[15] C. Yamagata, D. R. Leme, V. G. Rodrigues, G. T. Eretides, and A. C. Dorion Rodas, "Three routes for the synthesis of the bioceramic powder of the CaO-MgO-SiO2 system," *Ceram. Int.*, vol. 48, no. 7, 2022. https://doi.org/10.1016/j.ceramint.2021.12.169

[16] X. Zou, L. Fu, J. Song, G. Xu, and D. Bai, "High-temperature thermochemical conversion of iron ore tailings into diopside and akermanite-based composite materials," *Adv. Powder Technol.*, vol. 35, 2024.

[17] A. Najafinezhad, M. Abdellahi, H. Ghayour, A. Soheily, A. Chami, and A. Khandan, "A comparative study on the synthesis mechanism, bioactivity and mechanical properties of three silicate bioceramics," *Mater. Sci. Eng. C*, vol. 72, pp. 259–267, 2017. https://doi.org/10.1016/j.msec.2016.11.084

[18] N. Balighieh, M. R. Zamani, S. F. Kashani-Bozorg, M. Kheradmandfard, M. R. Barati, and E. Mansoori, "Optimizing microwave-assisted synthesis of akermanite nanoparticles using citric acid as a chelating agent: A combined machine learning and experimental approach," *J. Mater. Res. Technol.*, vol. 30, no. March, pp. 2211–2222, 2024. https://doi.org/10.1016/j.jmrt.2024.03.175

[19] H. Mohammadi, Y. M. B. Ismail, K. A. Shariff, and A. F. Mohd Noor, "Microstructure evolution, grain growth kinetics and mechanical properties of Ca2MgSi2O7bioceramics sintered at various temperatures," *Process. Appl. Ceram.*, vol. 15, no. 4, 2021. https://doi.org/10.2298/PAC2104357M

[20] M. Assarzadehgan, I.S. Gataa, Z.Y. Abdullah, M. Kasiri-Asgarani, A. Najafinezhad, H.R. Bakhsheshi-Rad, M. Razzaghi, S. Salahshour, and D. Toghraie, "Incorporation of azithromycin into akermanite-monticellite nanocomposite scaffolds : Preparation , biological properties , and drug release characteristics," *Ceram. Int.*, vol. 50, no. April, pp. 43841–43853, 2024

[21] M. Ansari, F. Malmir, and A. Salati, "Preparation and characterization of akermanite/merwinite scaffolds for bone tissue repair," *J. Biomimetics, Biomater. Biomed. Eng.*, vol. 44, 2020. https://doi.org/10.4028/www.scientific.net/JBBBE.44.73

[22] S. Liu, T. Saidani, D. Feng, and M. Jameel, "Thermal conductivity and rheological behavior of silica / simulated body fluid - Dispersed Hardystonite and Åkermanite nanofluids : X-ray diffraction to analyze structure and morphology of a bio-ceramic," *Int. Commun. Heat Mass Transf.*, vol. 165, no. May, 2025.

[23] C. I. Dobriţa, A.I. Bădănoiu, G. Voicu, A.I. Nicoară, S.M. Dumitru, M.E. Puşcaşu, Ş.Chiriac, R. Ene, and F. Iordache, "Porous bioceramic scaffolds based on akermanite obtained by 3D printing for bone tissue engineering," *Ceram. Int.*, vol. 49, no. 22, pp. 35898–35906, 2023. https://doi.org/10.1016/j.ceramint.2023.08.270

[24] A. Peys, N. Pires Martins, F. Prado Araujo, K. Komnitsas, and P. Nielsen, "High-temperature slag engineering in pursuit of effective mineral carbonation of pyroxene-rich ferronickel slag," *J. CO2 Util.*, vol. 95, 2025. https://doi.org/10.1016/j.jcou.2025.103077

[25] F. Kamutzki, M. F. Bekheet, S. Schneider, A. Gurlo, and D. A. H. Hanaor, "A comparison of syntheses approaches towards functional polycrystalline silicate ceramics," *Open Ceram.*, vol. 9, 2022. https://doi.org/10.1016/j.oceram.2022.100241

Materials Research Proceedings 68 (2026) 26-32 https://doi.org/10.21741/9781644904237-4

Study of White Chicken Egg (Lysozyme) for Properties Study- A Preliminary Research

Nurul Sofia MOHD DALI[1] and Siti Amira OTHMAN[1,a*]

[1]Faculty of Applied Sciences and Technology, Universiti Tun Hussein Onn Malaysia, 84600 Pagoh, Johor, Malaysia

[a]sitiamira@uthm.edu.my

Keywords: White Chicken Egg, Lysozyme, Enzyme, Hydrophilicity, UV-Vis

Abstract. Lysozyme is a naturally occurring antimicrobial enzyme found in body fluids such as saliva, tears, and white chicken egg (WCE), consisting of a 129–amino acid single-chain polypeptide stabilised by four disulfide bridges. This study aimed to investigate the effects of pH variation and gamma radiation on the structural and immunological properties of lysozyme extracted from WCE. The enzyme was purified and treated with varying concentrations of sodium hydroxide (NaOH) to alter its pH and hydrophilicity. This was followed by characterisation using Attenuated Total Reflectance Fourier Transform Infrared Spectroscopy (ATR-FTIR) and Ultraviolet–Visible (UV–Vis) spectroscopy to analyse the infrared and ultraviolet absorption in liquid samples. After purification, the lysozyme was subjected to gamma radiation, and its immunological activity was evaluated by assessing the binding of IgE and IgG. The spectroscopic results indicated structural changes in lysozyme with increasing NaOH concentration, while gamma radiation treatment significantly reduced the ability of ovomucoid in WCE to bind IgE and IgG. These findings suggest that pH modification and gamma irradiation can effectively alter lysozyme structure and reduce allergenic potential, highlighting their potential application in food processing and allergen control.

Introduction

Hydrophilic is a term used to describe a chemical group that is attracted to water. Hydrophilic compounds can be dissolved in water, regardless of their pH value, which can be alkaline, acidic, or neutral. There are three types of pH levels: alkaline, acidic, and neutral. Alkaline solutions such as seawater, ammonia solution, etc, at a pH value of 8 to 14. Acidic is all types with a pH value of 1 to 6, but a neutral pH value is at level 7. Lysozyme is a single-chain polypeptide of 129 amino acids cross-linked with four disulfide bridges. Lysozyme hydrolyses $\beta(1\rightarrow4)$ linkages between N-acetylmuraminic acid and N-acetyl-D-glucosamine residues in peptidoglycan and between N-acetyl-D-glucosamine residues in chitodextrins [1].

Lysozyme is an enzyme found in WCE and other body fluids that is capable of destroying the cell walls of certain bacteria. It can be used as an additive to baby milk, as well as for the treatment of ulcers, wounds, and infections [1]. Lysozyme in WCE helps lower cholesterol and calorie intake, while being rich in protein and vitamin A. WCE is used as a food preservative and antimicrobial agent [2]. WCE contains 24 different glycoproteins, such as ovomucoid, ovalbumin, ovotransferrin, and lysozyme. Therefore, the lysozyme is a part of WCE.

The peptidoglycan of the bacterial cell wall is responsible for maintaining the cell's shape in the presence of lysozyme. Osmotic pressure causes fluid to enter the cell, expanding the cell membrane until it bursts when peptidoglycan is damaged by lysozyme. However, if the medium is isotonic, no net flow of water into the cell. Then the cell does not burst. Gram-positive cells are generally susceptible to hydrolysis, as their cell walls contain a high proportion of peptidoglycan. Then, the harmful bacteria are less vulnerable because the outer membrane blocks the enzyme's access to the peptidoglycan.

The application of lysozyme is widespread in various fields of science and technology [1]. Previous scientists have purified lysozyme. The separation of the enzyme lysozyme from WCE is to demonstrate the process of purification of the protein. In this study, the material used was sodium phosphate buffer and the fresh WCE. These substances are combined by using a centrifuge to separate the enzyme lysozyme in WCE. After purification, remove the lysozyme from the purification solution and dry it for a day at room temperature. Lysozyme has a very stable structure and a positive charge at neutral pH, which means that it can be isolated from other enzymes by heating and keeping the reaction mixture at a low pH level. Then, after drying, the lysozyme appears in a crystal-like purity.

Research indicates that the range of lysozyme activity is extended through modifications that lead to changes in the conformation of enzyme molecules, and consequently, the production of its polymeric forms. It was found that a modified form of lysozyme exhibited a significantly wider range of bacteriostatic activity, including Gram-negative bacteria, compared to the active lysozyme [3]. In this project, the modification of lysozyme's hydrophilicity was achieved by using different concentrations and pH values of lysozyme in combination with chemical substances.

Allergies such as miliaria, eczema, and food allergies. Additionally, humans often develop allergies to food nowadays. Food allergies are the body's abnormal responses to harmless substances, usually referred to as foods. The reactions are caused by the immune system's response to specific food proteins, such as those found in meat, milk, eggs, seafood, and others. People with a food allergy have an immune system which reacts to specific proteins found in food [4]. Their immune system attacks the specific protein as if it were a harmful pathogen, such as a bacterium or virus. In this case, we aim to determine why lysozyme in WCE is related to allergy.

Lysozyme is used extensively in production economies, the pharmaceutical industry, and scientific research. Yet, the quantity of lysozyme WCE is not good in the human immune system because it contains more protein. More protein is not suitable for the human body. Thus, in this research, we aim to find the characteristics of lysozyme by modifying its hydrophilicity. The modification, achieved by using chemical substances such as a buffer solution of Sodium Phosphate, HCl, and NaOH, alters the concentration and pH value of lysozyme. The machines used were Fourier Transform Infrared Spectroscopy (FTIR) and Ultraviolet-Visible Spectroscopy (UV-Vis) to characterise the lysozyme.

Materials

White chicken egg (WCE) was bought from a local market in Batu Pahat. To change the pH, sodium hydroxide (NaOH) and hydrochloric acid (HCl) were used. The materials used are phosphate buffer and distilled water as a solution medium.

Methodology

Purification of Lysozyme. Separate the WCE from the egg yolk. Filter the WCE by using cheesecloth. Next, add 10.0 mL of WCE to the test tube. Next, dilute the Phosphate Buffer to 0.1 M. Then, prepare different pH buffer solutions using HCl and NaOH to achieve pH levels from 4 to 12. For the purification of lysozyme, mix 10 mL of 0.1 M Phosphate buffer with 10 mL of WCE. Next, place the test tube in the centrifuge at 4000 RPM/RCF for 8 hours. After centrifuging, two layers were formed. The bottom layer was taken for further analysis.

Absorbance Properties of Lysozyme. Prepare 5 ml of WCE in five test tubes. Prepared 0.1M to 0.5M NaOH solution and poured it into a WCE test tube. The sample was then immediately placed in the ultrasonic bath for 1 hour. The observation was recorded. Next, the samples underwent UV-Vis analysis. Before being irradiated with Caesium-137. Caesium-137 is a radioisotope that has 82 neutrons and 55 protons. The distance between the samples and the source during the radiation process is 1 metre from the sample. The samples were irradiated for 1 day. Characterisation with UV-Vis for WCE samples was performed before and after the UV-Vis analysis.

pH properties of Lysozyme. Prepare 5 ml of WCE in five test tubes. The pH of WCE was varied from 4 to 12 by using NaOH and HCl. Place the samples in an ultrasonic bath for about 1 hour. Next, the samples undergo characterisation with UV-Vis.

Constant Temperature of Lysozyme. Prepare 5 ml of WCE in five test tubes. Put the prepared solution in an ultrasonic bath at a temperature of 200 °C. The boiling temperature was varied from 10 to 50 minutes. Finally, characterise the samples using UV-Vis. and FTIR.

Different Conditions of Lysozyme. Prepare 5 ml of WCE in four test tubes. Two of the test tubes will be wrapped with aluminium foil, and two of them will not be wrapped with aluminium foil. After that, each of the wrapped and unwrapped test tubes will be placed at room temperature, and the rest will be stored at a cold condition (±2 °C) for 1 week. Lastly, all samples will be run using FTIR to get the results.

Results and Discussion

After removing the bottom layer from the test tube, the observation of each layer was different. Table 1 presents the results observed before and after centrifugation. The purification method was used in this experiment to obtain pure lysozyme in WCE. The results showed the presence of pure lysozyme in WCE. WCE has many types of glycoproteins, and some of them are 3.4% lysozyme [5]. Lysozyme is an important enzyme that destroys certain bacteria by cleaving the β-lingkages between *N*-acetyl-muramic acid and *N*-acetyl-glucosamine of the peptidoglycan in the bacterial cell walls [6]. Then, lysozyme is widely used as a cell-disrupting reagent and as a potent anti-bacterial agent due to its bacteriostatic and bactericidal properties [7]. Lysozyme is widely distributed in animals and plants. Moreover, it can also be used as an additive to baby milk or ophthalmic preparations, for the treatment of ulcers, wounds and infections [1]. In addition, lysozyme is a glycoside hydrolase which is abundantly found in eggs.

Table 1: WCE after purification using a centrifuge.

pH of Phosphate Buffer	Before Dry	After Dry
4.0	Clogged and yellow	Rough crystal
6.0	Clogged and milky yellow	
7.0	Clogged and white	
8.0	Clogged and milky white	
10.0	Clogged and milky white	

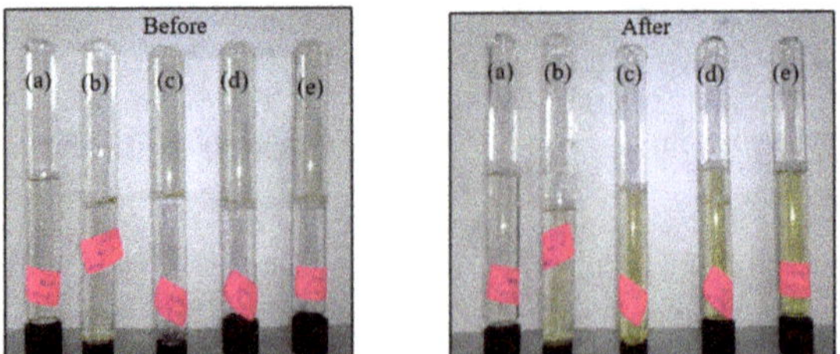

Fig. 1: Condition samples after putting in ultrasonic bath at room temperature (25^{0}C) (a) 0.1 M (b) 0.2 M (c) 0.3 M (d) 0.4 M (e) 0.5 M.

Fig. 1 shows that the solutions become a gel because they are not well mixed with the WCE and NaOH solution. As we know, NaOH is an alkaline solution with a pH of 14, and this WCE is also alkaline with a pH of 9. Lysozyme is an enzyme that remains active over a broad pH range (6 to 9). In this experiment, the pH of the solution reached 12 when the two solutions were mixed (Table 2). Buffers are solvents that minimise changes in the concentrations of ions, either H^+ ions or OH^- ions. A buffer solution resists changes in pH, and the pH is not altered to any great extent by the addition of small quantities of either a strong acid (H+ ions) or a strong base (OH– ions), which is why it is called a buffer solution [8]. A buffer solution is essential in the purification method. Then, buffer solutions come in many types, such as carbonate buffer, Tris Base, sodium phosphate buffer, and potassium phosphate buffer.

Table 2: The results of modifying the lysozyme WCE in different concentrations of NaOH.

Concentration of NaOH (M)	Condition of Samples in Room Temperature (25°C)	Condition of samples in Ultrasonic Bath	pH of Concentration NaOH + WCE
0.1	sticky	-	12
0.2	viscous	-	
0.3	viscous, sticky, gel	viscous	
0.4	viscous, sticky, gel	viscous	
0.5	viscous, sticky, gel	viscous, sticky	

As we can see, the graphs of absorbance of UV light using UV-1800 were very different from each other [9]. This was due to the concentration of NaOH affecting the activity of lysozyme. The stickiness of the samples makes the light flow slowly and is absorbed, with some of it being transmitted to the surrounding area. The absorption of light occurs as energy is transferred from a higher energy level to a lower energy level. The absorbance of this experiment ranged m:ainly from 3.9 to 4.0, and the wavelength range that absorbed light was between 310 nm and 320 nm, as shown in Fig. 2(a). The absorbance of light was smoothly absorbed at 0.1 M and 0.2 M for NaOH. The absorbance value changed as shown in Fig. 2(b) from a wavelength of 310 nm to 340 nm. This was due to the WCE's exposure to the radioactive source.

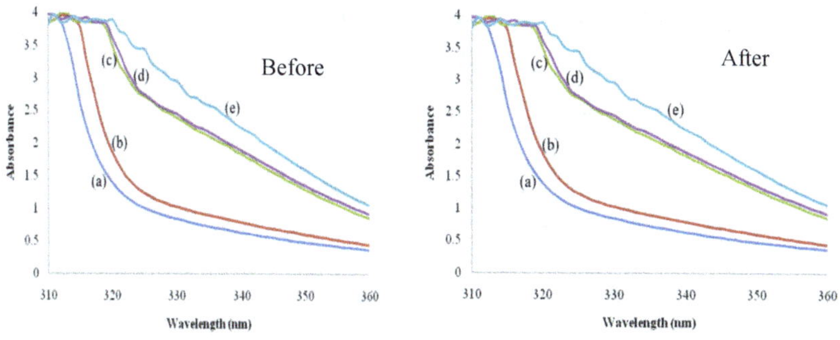

Fig. 2: UV-Vis spectra for WCE in different concentrations of NaOH before and after irradiation of 0.1 M (b) 0.2 M (c) 0.3 M (d) 0.4 M (e) 0.5 M.

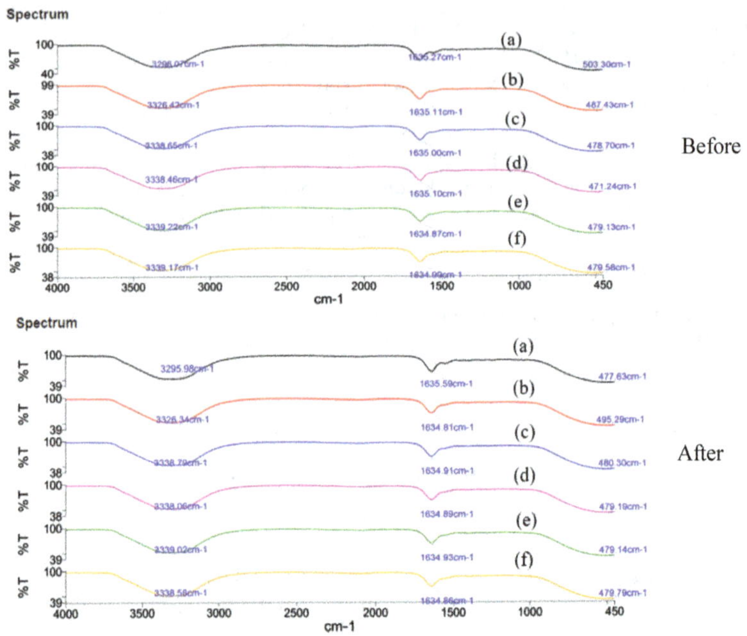

Fig. 3: ATR-FTIR spectrum of WCE in different concentrations of NaOH before and after radiation (a) WCE (b) 0.1 M (c) 0.2 M (d) 0.3 M (e) 0.4 M (f) 0.5 M.

Fig. 3 shows the result of ATR-FTIR. The results were obtained through the absorbance of Infrared (IR) spectroscopy, which detects the presence of functional groups in WCE. The absorption of IR in both graphs will present the functional groups in Alkynes, Alkenes, and Alkyl Halides. Alkynes are present because they contain CH_4. Alkenes are present in the form of C_3H_6, and alkyl halides are present due to the carbon-halogen bond. A halogen bond that replaces the R as the alkyl group. In that graph, the IR radiation will be absorbed by those molecules, and the rest will be transmitted.

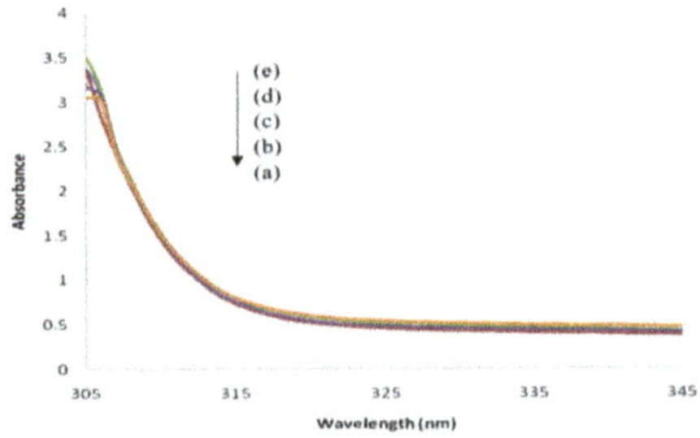

Fig. 4: The absorption of light in different pH levels reacts with lysozyme (a) 4 (b) 6 (c) 8 (d) 10 (e) 12.

The UV spectrum is defined as the electromagnetic radiation with wavelengths from 100nm to 400nm. From this experiment, the peak of absorbance is observed between 3.0 and 3.5 at a wavelength of 305 nm. The lysozyme WCE enzyme was active at pH 6 to pH 9. At pH 8, it exhibits a higher peak of absorbance. The graph had the same characteristics that were dropped clearly and overlapped with each other.

The enzyme is usually present as a reversible dimer between pH 5.0 and 9.0. It is also known that WCE lysozyme tends to associate in an irreversible dimeric form (presumably through intermolecular disulfide exchange) when eggs are stored for long periods. Heated lysozyme also showed a several-fold increase in surface hydrophobicity over the native enzyme. Research indicates that the range of lysozyme activity may be extended through modifications that lead to changes in the conformation of enzyme molecules, and consequently, the production of its polymeric forms. It was found that a modified form of lysozyme exhibited a significantly wider range of bacteriostatic activity, including Gram-negative bacteria, compared to the active lysozyme. It is essential to emphasise that the modified enzyme retains its antibacterial activity towards Gram-positive bacteria, a characteristic of the monomer [10]. Any modification of lysozyme properties that could render it effective against both Gram-negative and Gram-positive bacteria would be a significant contribution [3].

Conclusion

In conclusion, the characterisation of white chicken egg (WCE) lysozyme using ATR-FTIR confirmed the presence of functional groups, including amides, amines, aldehydes, and alkenes. At the same time, UV–Vis spectroscopy showed light absorption in the range of 300–320 nm. Modification of lysozyme at different pH levels demonstrated that pH 9 produced the most active enzyme. Gamma irradiation using Caesium-137 (Cs-137) significantly reduced both IgE- and IgG-binding capacities of ovomucoid in WCE, indicating a decrease in allergenic potential. IgE is primarily responsible for immediate hypersensitivity reactions and is the key immunoglobulin involved in food allergies, particularly in children. In contrast, IgG is associated with longer-term immune responses and allergen sensitisation. The observed reduction in both IgE and IgG binding suggests that irradiation may effectively alter allergenic epitopes and reduce immune recognition. Although egg allergy is one of the most common IgE-mediated food allergies in children, ovomucoid is the dominant allergen, accounting for approximately 11% of WCE proteins, compared to lysozyme, which accounts for only 3.4%. The presence of other glycoproteins, such as ovalbumen and ovotransferrin, also contributes to allergenic responses. Therefore, this study suggests that lysozyme itself is not highly associated with allergenicity, and that childhood allergies are not solely caused by egg consumption, but may also arise from other allergenic foods, such as seafood, milk, and meat, which similarly trigger IgE- and IgG-mediated immune responses.

Acknowledgement

The authors would like to thank the Faculty of Applied Sciences and Technology, Universiti Tun Hussein Onn Malaysia, for the facilities provided that make the research possible.

References

[1] G. Alderton and H. L. Fevold, Direct crystallisation of lysozyme from egg white and some crystalline salts of lysozyme, J. biol. chem. 164 (1946) 1–5.

[2] J. Chen, Y. Lin, and L. Jia, Preparation of anionic polyelectrolyte modified magnetic nanoparticles for rapid and efficient separation of lysozyme from egg white, J. chromatogr. A 1388 (2015) 43–51. https://doi.org/10.1016/j.chroma.2015.02.032

[3] R. Cegielska-Radziejewska, G. Lesnierowski and J. Kijowski, Properties and application of egg white lysozyme and its modified preparations - a review, Pol J. Food Nutr. Sci. 58 (2008) 5-10.

[4] R. Valenta, H. Hochwallner, B. Linhart, and S. Pahr, Food allergies: the basics. Gastroenterology, 148 (2015) 1120–31.e4. https://doi.org/10.1053/j.gastro.2015.02.006

[5] H. S. Scott, Food Allergy: Practical Diagnosis and Management. 2014. New-York: CRC Press. 2-5.

[6] J.J. Grigsby, H. W. Blanch and J. M. Prausnitz, Cloud-point temperatures for lysozyme in electrolyte solutions: effect of salt type, salt concentration and pH, Biophys. Chem. 91 (2001) 231–243. https://doi.org/10.1016/s0301-4622(01)00173-9

[7] I. Šafařík, Z. Sabatkova, O. Tokar & M. Šafaříková, Magnetic Cation Exchange Isolation of Lysozyme from Native Hen Egg White, FTB 45 (2007) 355-359.

[8] L.Yan, S. Shen, J. Yun and K. Yao, Isolation of Lysozyme from Chicken Egg White Using Polyacrylamide-based Cation-exchange Cryogel, Chin. J. Chem. Eng. 19 (2011) 876-880.

[9] Perkin Elmer. Attenuated Total Reflectance Fourier Transform Infrared Spectroscopy (ATR-FTIR) (2015). Retrieved from FTIR Spectroscopy: Attenuated Total Reflectance (ATR) (perkinelmer.com).

[10] M. Cabeen & C. Jacobs-Wagner, Bacterial cell shape. Nat. Rev. Microbiol. 3 (2005) 601–610. https://doi.org/10.1038/nrmicro1205

Materials Research Proceedings 68 (2026) 33-40 https://doi.org/10.21741/9781644904237-5

Evaluation of Palm Oil-Based UV Curable Coatings for Termite Resistance in Rubber Wood

Rida TAJAU[1,a *], Mohd Sofian ALIAS[1,b], Mohd Hamzah HARUN[1,c], Nurul Huda MUDRI[1,d], Farah Fadzehah HILIMI[1,e], Naurah MAT ISA[1,f], Khairul Azhar ABDUL HALIM[1,g], Sharilla MUHAMMAD FAISAL[1,h], Rosley CHE ISMAIL[1,i], Abdul Muizz MOHD SANI[1,j] and Khairul Anwar MUSTAFA[2,k]

[1]Radiation Curing and Synthesis Group, Radiation Processing Technology Division, Malaysian Nuclear Agency, Ministry of Science, Technology and Innovation, Bangi, 43000 Kajang, Selangor, Malaysia

[2]Solaris Marketing, No.11-2 Jalan Puteri 2A/3, Bandar Puteri Bangi,43000 Bangi, Selangor, Malaysia

[a]rida@nm.gov.my, [b]sofian@nm.gov.my, [c]hamzah@nm.gov.my, [d]nurul_huda@nm.gov.my, [e]farah@nm.gov.my, [f]naurah@nm.gov.my, [g]azharhalim@nm.gov.my, [h]sharilla@nm.gov.my, [i]rosley@nm.gov.my, [j]muizz@nm.gov.my, [k]solarispest@gmail.com

Keywords: Anti-Termite, Palm Oil, Radiation Curable Resin, Radiation Curing, Wood Coatings

Abstract. Termites are significant pests that endanger wood-based infrastructure, particularly in Malaysia. Despite the durability of rubber wood, controlling termite attacks remains a challenge. This study explores the use of palm oil-based radiation-curable resin as a protective coating for wood, both with and without anti-termite chemicals, to enhance resistance to termite damage. Five formulations of resins were evaluated: (a) palm oil-based resin; (b) 0.1% anti-termite palm oil-based resin; (c) 1% anti-termite palm oil-based resin; (d) 5% anti-termite palm oil-based resin; and (e) a commercial anti-termite solution. Rubberwood blocks were coated with UV-curable resin and then cured before being placed in termite mounds for 30 days. Untreated samples were used as controls. The coatings were evaluated for weight loss, chemical composition, hydrophobicity, gloss, haze, and chemical resistance using international standards. After 30 days, the samples coated with anti-termite palm oil-based resins exhibited no weight loss and no visible termite damage. In contrast, both untreated and commercially treated controls showed a 2% weight loss and bite marks. FTIR spectra indicated the presence of imidacloprid functional groups in the anti-termite palm oil-based resins, which created a protective film layer for the rubberwood blocks. Radiation-curable anti-termite palm oil-based coatings functioned well as barriers because they were waterproof and cross-linked. However, their resistance to mould outdoors was poor and needs more investigation. This preliminary study suggests that palm oil-based radiation-curable coatings can be effective, eco-friendly wood protectants, providing a sustainable option for termite control in humid areas.

Introduction

Malaysia experiences an equatorial climate, where high humidity significantly affects the life cycle of termites [1]. Termites typically consume wood cellulose, including materials like paper. Regarded as pests, termites can cause substantial losses if infestations are not controlled. Their attacks on wood-based infrastructure and property cause enormous financial losses in the construction and industrial sectors while also endangering human safety and health.

Rubber wood is one of the most popular materials in Malaysia's furniture manufacturing sector and accounts for over 80% of all local furniture [2,3]. Rubber wood furniture is known for its durability, longevity, and ease of maintenance. However, controlling termite infestations in wooden furniture remains challenging despite various protective techniques, including heating, chemical impregnation, painting, spraying with termite pesticides, and termite baiting.

Palm oil-based radiation-curable resin has emerged as a promising coating material for wood products used in interior decoration. These coatings are waterproof, non-toxic, and can prevent microbial growth [4,5]. Innovative products are being developed using palm oil-based radiation-curable resins to act as protective coatings against termite attacks on wood.

The epoxidized palm oil acrylate (EPOLA) has been developed as a palm oil curable resin with potential as a surface finishing material. EPOLAs possess an acrylate chemical structure and undergo a chemical crosslinking reaction when exposed to ultraviolet (UV) light in the presence of a photoinitiator, monomer, and additives. This process forms a thin film layer that acts as a surface protector for various substrates, including wood, paper, and metal. EPOLA-based radiation-curable resins can potentially replace petrochemical-based resins, showing promising physicochemical and mechanical properties for advanced material applications [6].

This preliminary study aims to evaluate the effectiveness of coating materials made from palm oil-based radiation-curable resin in protecting against termite attacks. The trials will use commercially available rubber wood species, with the coating substance apply on their surface. Additionally, other coating formulations will be developed by blending anti-termite materials. A comparison will be made between these coating formulations and a commercial-grade anti-termite solution.

Materials and Methods

Anti-termite Resins. The radiation curing and synthesis laboratory produced the main ingredient, EPOLA, which was used to develop the palm oil-based radiation-curable resin. Anti-termite resins were formulated by blending the palm oil-based radiation-curable resin with 0.1%, 1% and 5% anti-termite solution, respectively. There are five types of formulation used for wood coating purposes: (a) palm oil-based radiation curable resin, (b) 0.1% anti-termite palm oil-based resin, (c) 1% anti-termite palm oil-based resin, (d) 5% anti-termite palm oil-based resin, and (e) anti-termite solution.

Wood Coatings. Small pieces of rubber wood measuring around 11 cm x 6 cm (LxW) were coated with palm oil-based radiation curable and anti-termite resins, respectively, using brushing technique with thicknesses ranging from 50 to 150 μm. The rubbed wood blocks were then subjected to UV radiation using IST UV machine with a conveyor speed of 10 m/min and a current of 7.5 amperes. Rubber wood samples were also treated with a commercial-grade of anti-termite solution, and an untreated rubber wood sample was also used as a control sample. Three replicate woods for each sample.

Site Selection. The field studies are located at the MINT Tech Park at the Malaysian Nuclear Agency, Selangor, Malaysia. The site map is 2.898600, 101.7603930 (2°53'55.0"N 101°45'37.4" E) (Fig. 1a). The area is densely populated with plants, flowers, herbs, and tropical trees, which have attracted termites to live around (Fig. 1b). The temperature at the site ranges between 23 and 33 °C. Typically, the temperature is high between noon and evening.

Collection of Termites. Macrotermes termite colonies and their mounds were found on the surface of the ground study site (Fig. 1c, Fig. 1d). These colonies prefer to live in tropical environments, particularly woodland places, where they can get food such as wood and dry grass.

Materials Research Proceedings 68 (2026) 33-40 https://doi.org/10.21741/9781644904237-5

Field Trials. The wooden blocks of rubber wood samples were inserted into the nest of the termite mound (Fig. 1e, Fig. 1f). After a month, the wooden blocks were taken out of the nest, washed, dried, and reweighed, and the wood consumption was analysed and calculated.

(a) Study site at MINT Tech Park.

(b) Tropical trees at the study site.

(c) Termite mound site for trials.

(d) Termite nest and Macrotermes termite species.

(e) Rubber wood blocks installation into the mound.

(f) Buried woods in termite mound.

Fig. 1: The installation of wood samples took place at a termite mound in MINT Tech Park, Nuclear Malaysia, Kajang, Selangor.

Characterization of film coatings. An FTIR spectrophotometer (Bruker's Tensor II) is used to analyze chemical functions. The optical contact angle (Biolin Scientific, TL 100) was utilized to determine the hydrophobicity of the coating. The chemical resistance test followed ASTM D 5402-93 (1999) and employed the solvent rub technique using methyl ether ketone (MEK), acetone, ethanol, and water. The glossiness of the coated films was measured using an ASTM D-523 Micro-TRI-Gloss meter. The haziness of the coated films was measured using a Haze meter.

Results and Discussion

Film coating properties. Imidacloprid compound demostrates a strong activity against insects and is widely utilized as insecticide in the pest control industry (Fig.2, a). The FTIR spectra in Fig. 2 show the vibration of the functional group of imidacloprid in the commercial-grade anti-termite solution. Two peaks at 3345 and 1698 cm^{-1} are associated with N-H and C-N of the imidazolidine ring. The peaks observed at 1558 (C-N) and 1443 (C=C) cm^{-1} correspond to the pyridine ring of imidacloprid. The $-NO_2$ of imidacloprid appeared at 1295-1247 cm^{-1}. These peak vibrations are comparable to those generated in the study investigated by Moradi et al. [7]. These peaks decreased upon blend with resin, which proves the occurrence of interaction between resin and imidacloprid, as shown by the anti-termite resin's spectra.

Table 1: Characterization of coating film formulations.

Sample	Contact angle [°]	Pendulum hardness [%]	Haziness [H.U.]	Glossiness, 20° [G.U.]	Wood Adhesion
Radiation curable palm oil-based coating	107.38	12.95	1.1	19.9	5B
Anti-termite solution (Commercial grade)	5.88	Not applicable	64.03	1.03	Not applicable
0.1% anti-termite palm oil-based coating	94.29	20.38	1.73	55.67	5B
1% anti-termite palm oil-based coating	92.91	20	1.92	39.13	5B
5% anti-termite palm oil-based coating	86.72	22.86	1.3	21.47	5B
Anti-termite resin layers	36.57	23.24	25.36	51.67	0B

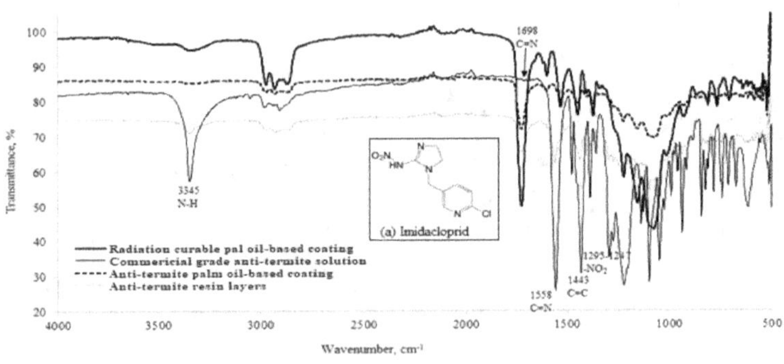

Fig. 2. FT-IR spectra for palm oil-based uv curable coating, commercial grade anti-termite solution, palm oil-based anti-termite coating and anti-termite resin layers. The chemical structure of imidacloprid (a).

The commercial anti-termite solid surface exhibited a highly hydrophilic property with 5.88°. However, the radiation-curable palm oil-based coated surface demonstrated hydrophobic properties with a contact angle of 107.38°, as shown in Fig. 3 (a,b). Varied volume between 0.1%, 1% and 5% of anti-termite blended with palm oil acrylate resin resulted in the film coating surface hydrophobic with the contact angle values decreased from 94.29° to 86.72°, respectively, as tabulated in Table 1 and shown in Fig. 3 (c-f), and vice versa for the hardness property with a slight increment. The adhesion of the radiation-curable palm oil-based coating and the anti-termite palm oil-based coating on the wood surface was excellent, obtaining a grade of 5B. The hydrophilic anti-termite resin layers exhibited a contact angle of 36.57°, which contributed to the failure of adhesion of the coating. The haziness of the commercial anti-termite solid surface is much higher compared to the radiation-curable palm oil-based coating. Blended with 0.1%, 1% and 5% of anti-termite solution, the haziness of the coating film surfaces slightly increased (Table 1). However, when the palm oil acrylate resin is blended with the anti-termite solution, the rubber wood coating film surfaces become glossier than the wood coated with palm oil acrylate resin (Fig. 4i, Table 1). Rubbing tests on palm oil-based resin and anti-termite solution mixtures at 1% and 5% demonstrated that the coating is more resistant to chemical solutions and water than commercially available coatings that failed the test (Table 2).

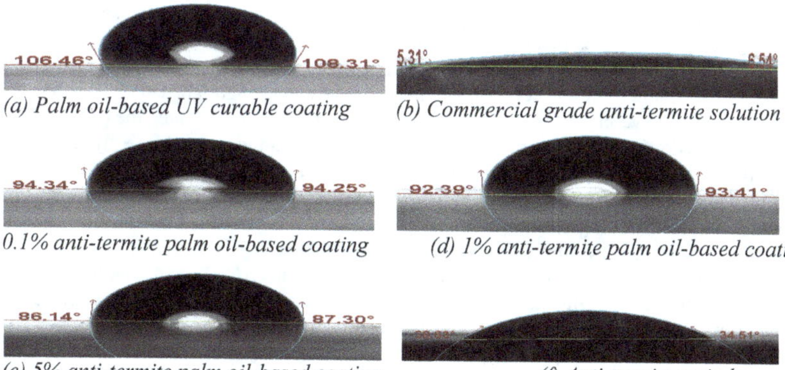

(a) Palm oil-based UV curable coating *(b) Commercial grade anti-termite solution*

(c) 0.1% anti-termite palm oil-based coating *(d) 1% anti-termite palm oil-based coating*

(e) 5% anti-termite palm oil-based coating *(f) Anti-termite resin layers*

Fig. 3: Contact angle measurement.

Fig. 4: Rubber wood blocks were coated with: (1) 0.1% anti-termite resin coating, (2) 1% anti-termite resin coating, (3) 5% anti-termite resin coating, (4) radiation curable palm oil-based coating, (5) anti-termite resin layers, (6) commercial grade anti-termite solution and (7) control wood sample (untreated wood); (i) Prior to burial in the mound; (ii) after 30 days of burial in the mound (cleaned).

Table 2: Rub test for chemicals and water resistant of coating films.

Sample	MEK	Acetone	Ethanol	Water
Radiation curable palm oil-based coating	√	√	X	√
Anti-termite solution (Commercial grade)	X	X	X	X
0.1% anti-termite palm oil-based coating	X	X	√	√
1% anti-termite palm oil-based coating	X	√	√	√
5% anti-termite palm oil-based coating	X	√	√	√
Anti-termite resin layers	X	X	X	X

Note: √- Passed; X: Failed

Anti-termite assay. The study found that the radiation-curable palm oil-based coating can protect rubber wood from termite attacks for 30 days in the mound. Thus, plant-based solutions could be utilized in coatings or treatments to protect wood structural components from termite attack without using harsh synthetic pesticides [8-10]. Kareru et al. [8] and Okahisa et al. [9] also found that plant-based paints can protect against termite attacks and fungal growth. The study of paint formulated with Thevetia peruviana seed oil as a green coating demonstrated the potential benefits of plant-based additives in anti-termite coating applications, as the paint efficiently repels termites and reduces wood infestation [8]. Natural lacquer wood coats containing urushi can effectively protect wood from subterranean termite attack [9]. Furthermore, research on a plant waste product, extracts from Eucalyptus pellita bark, found that the extracts produced full-termite mortality in laboratory bioassays, indicating effective termite control [10].

In this study, two control wood samples, which are untreated (No.7) and wood samples coated with commercial anti-termite liquid (No.6), both have small holes that indicate the effects of termite bites (Fig. 5). The weight loss for both wood control samples was 2% each after 30 days. There was no decrease in wood weight for wood samples coated with palm oil-based radiation coatings, either mixed with anti-termite materials at concentrations of 0.1%, 1% and 5% (Table 3).

Fig. 5: The effect of termite bite holes on samples No.6 and 7.

Table 3 : Characterization of the effects of termite attack on wood samples.

No	Property	Wood coating appearance		Traces of termite		Mold appearance		Wood weight loss, %	
	Duration, day (D)	D14	D30	D14	D30	D14	D30	D14	D30
1	0.1% anti-termite palm oil-based coating	Good	Good	Yes	Yes	No	High	No	No
2	1% anti-termite palm oil-based coating	Good	Good	Yes	Yes	No	Moderate	No	No
3	5% anti-termite palm oil-based coating	Good	Good	Yes	Yes	No	Moderate	No	No
4	Palm oil acrylate resin coating	Good	Good	Yes	Yes	No	Less	No	No
5	Anti-termite resin layers	Good	Fall off	Yes	Yes	No	Less	No	No
6	Anti-termite solution (Commercial-grade)	-	-	Yes	Yes	Less	High	No	6c 2.52%
7	Untreated wood	-	-	Yes	Yes	Moderate	High	No	7c 2.37%

Mould growth was observed in all of the samples in this study after 30 days, with two control samples showing the most growth, followed by wood samples treated with 0.1% anti-termite substance. The samples coated with 1% and 5% anti-termite material, respectively, showed moderate mould growth. Meanwhile, wood samples coated with palm oil-based resin coating and wood samples coated with alternating 3 layers of palm oil-based resins and anti-termite, the anti-termite resin layers, show less mould growth, as shown in Fig. 4(ii) and Fig. 6. The study found that coating samples did not prevent microbial growth in the outdoor environment [4].

Fig. 6: Rubber wood blocks after being buried in a termite mound for 30 days.

Summary

Radiation-curable palm oil-based coating and anti-termite formulations can potentially be used as pesticides for pests, such as termites. They have shown their effectiveness as protective coatings to prevent termite attacks. It turns out that radiation-curable coatings can increase the resistance of wood to termites. The advantages of its properties as a layer of cross-linked films, hydrophobic, and preventing the growth of microbes have been among the main factors that serve as a barrier to termite attacks. The addition of anti-termite materials also helps protect against termite attacks, even though the hydrophobic properties are reduced. The coating cannot prevent the growth of mould in the external environment. The study period is still extended.

References

[1] A.R. Abdul Rahim, Z.I. Zulkifly and M.S. Rabu, Pengawetan & Pengeringan Kayu-Pengenalan & Agen Perosak Kayu, Politeknik Sultan Salahuddin Abdul Aziz Shah, Shah Alam, Selangor, 2022.http://repository.psa.edu.my/bitstream/123456789/3528/1/eBook%20Pengeringan%20%26%20Pengawetan%20Kayu%20_%20RAZLI.pdf

[2] N.M.H. Nik Hashim, M.N.Z. Mat Amin, N.L and Abdullah, S. Samian, Orientasi Pasaran Eksport dan Prestasi Eksport: Satu Kajian Empirikal ke atas Firma Perabot di Malaysia, Jurnal Pengurusan 58 (2020) 145 – 157. https://doi.org/10.17576/pengurusan-2020-58-12

[3] M.R. Yaacob and T.M.R. Tuan Hassan, Senario Usahawan dan Keusahawanan Melayu dalam Industri Perabot di Malaysia, Prosiding Persidangan Kebangsaan Ekonomi Malaysia ke-9 (PERKEM ke-9), Kuala Terengganu,Terengganu, 2014, pp. 866 - 874.

https://kipdf.com/senario-usahawan-dan-keusahawanan-melayu-dalam-industri-perabot-di-malaysia_5ac8167a1723dd8be7daaf83.html

[4] R. Tajau, M.S. Alias, M.H. Harun, N.H. Mudri, , F.F. Hilmi, , S.M. Faisal, R.C. Ismail, A.M.M. Sani, C.R.C. Hak, W.S. Paulus, B.C. Jong and P.W.Y. Liew, Natural weathering effects on the surface morphological and physicochemical properties of radiation curable coatings in tropical climate, IOP Conf. Ser.: Mater. Sci. Eng. 1308 012009 2024. DOI: 10.1088/1757-899X/1308/1/012009.

[5] R. Tajau, M.S. Alias, N.H. Mudri, F.F. Hilmi, S.M. Faisal, R.C. Ismail, A.M.M. Sani, C.R.C. Hak, W.S. Paulus, J.B. Chyan, P.W.Y. Liew, J.N. Hing, N.N.M. Rapi and N.N. Halipah, Modification of silane-modified TiO2 nanoparticles with APO-nanoparticles as antimicrobial additive for palm oil-based surface coatings, Jurnal Sains Nuklear Malaysia, 36:1 (2024) 81–98. https://jsnm.nuclearmalaysia.gov.my/index.php/dja/article/view/108

[6] R. Tajau, R. Rohani, M. S. Alias, N. H. Mudri, K. A. Abdul Halim, M. H. Harun, N. Mat Isa, R. Che Ismail, S. Muhammad Faisal, M. Talib, , M. R. Mohamed Zin, I. I. Yusoff, N. Khairul Zaman and I. A. Ilias, Emergence of Polymeric Material Utilising Sustainable Radiation Curable Palm Oil-Based Products for Advanced Technology Applications, Polymers, 13:11 (2021), 1865. https://doi.org/10.3390/polym13111865

[7] F.G. Moradi, M.J. Hejazi, H. Hamishehkar and A.A. Enayati, Co-encapsulation of imidacloprid and lambda-cyhalothrin using biocompatible nanocarriers: Characterization and application, Ecotoxicol Environ Saf. 175 (2019) 155-163. Doi: 10.1016/j.ecoenv.2019.02.092.

[8] P.G. Kareru, J.M. Keriko, G.M. Kenji, and A.N. Gachanja, Anti-termite and antimicrobial properties of paint made from Thevetia peruviana (Pers.) Schum. oil extract, African Journal of Pharmacy and Pharmacology 4:2 (2010), pp. 087-089.

https://www.researchgate.net/publication/289721049_Antimicrobial_properties_of_thevetia_peruviana

[9] Y. Okahisa, C. Narita and T. Yoshimura, Resistance of wood coated with oriental lacquer (urushi) against damage caused by subterranean termite, Journal of Wood Science, 65:41 (2019). https://link.springer.com/article/10.1186/s10086-019-1820-6

[10] R. Andika, A. Arinana, R. K. Sari, A.I. Rahmawati and S.K. Himmi, Antitermite Activity of Eucalyptus pellita Bark Extract, Jurnal Sylva Lestari, 13:1 (2025) 32–44. https://doi.org/10.23960/jsl.v13i1.1023

Extraction of Silica from Burnt Rice Husk Using the Double Crucible Method

Faizul CHE PA[1,a*], Ruhiyuddin MOHD ZAKI[1,b] and Sasitra SEGAR[2,c]

[1]Center of Excellence Geopolymer and Green Technology (CEGeoGTech), Universiti Malaysia Perlis (UniMAP), Kompleks Pusat Pengajian Jejawi 2, 02600 Arau. Perlis, Malaysia

[2]Faculty of Chemical Engineering & Technology, Universiti Malaysia Perlis (UniMAP), Kompleks Pusat Pengajian Jejawi 2, 02600 Arau. Perlis, Malaysia

[a]faizul@unimap.edu.my, [b]ruhiyuddin@unimap.edu.my, [c]sasitra2806@gmail.com

Keywords: Silica, Rice Husk, Double Crucible Method, Pyrolysis

Abstract. Burnt rice husk (BRH) refers to rice husk that has undergone initial uncontrolled or partial thermal degradation, typically through open burning or incomplete combustion, resulting in a carbon-containing, silica-rich residue. This study investigates a double crucible method for extracting silica from BRH without the use of inert gases. Experiments were conducted at 500, 600, and 700 °C for soaking times of 60, 90, and 120 minutes. Characterization was performed using FTIR, FESEM, EDX, and optical microscopy. Results show that 700 °C for 120 minutes yields the highest purity silica, with EDX confirming ~92 wt% SiO_2 and <3 wt% residual carbon. FTIR revealed strong Si–O–Si and silanol bonds, while FESEM showed uniform mesoporous honeycomb-like structures. Optical microscopy confirmed whiteness and homogeneity. These findings demonstrate that the double crucible method is a green, cost-effective approach to scalable production of silica from agricultural waste.

Introduction

Rice husk (RH), a byproduct of rice milling, poses significant disposal and environmental challenges due to its slow decomposition and frequent open burning. Rich in silica (18–23%), RH can be converted into rice husk ash (RHA), which contains up to 95% silica [1,2]. Traditionally, silica extraction involves chemical treatments or pyrolysis with inert gases, both of which require additional equipment and may raise environmental concerns [3].

In this work, burnt rice husk (BRH) refers to rice husk that has undergone initial incomplete combustion, resulting in a carbon-containing residue distinct from conventional rice husk ash (RHA), which is produced by fully controlled calcination. The BRH was subsequently subjected to controlled thermal treatment using the double crucible method to obtain high-purity silica

The double crucible method provides a simple and environmentally friendly alternative [4]. By isolating the sample within nested alumina crucibles, oxygen exposure is minimized, promoting the formation of high-quality amorphous silica. This study aims to optimize temperature and soaking time conditions for silica extraction via this method and to evaluate the physical and chemical properties of the recovered silica.

Methodology

Rice husks sourced from Perlis, Malaysia, were washed, oven-dried (70 °C, 24 h), and sieved (1 mm). Burnt rice husk (BRH) was prepared and processed using the double crucible method (Fig. 1): the inner crucible containing BRH was sealed and surrounded by raw RH in the outer crucible to scavenge oxygen. Heating was conducted at 500, 600, and 700 °C with dwell times of 60, 90, and 120 minutes at a rate of 5 °C/min.

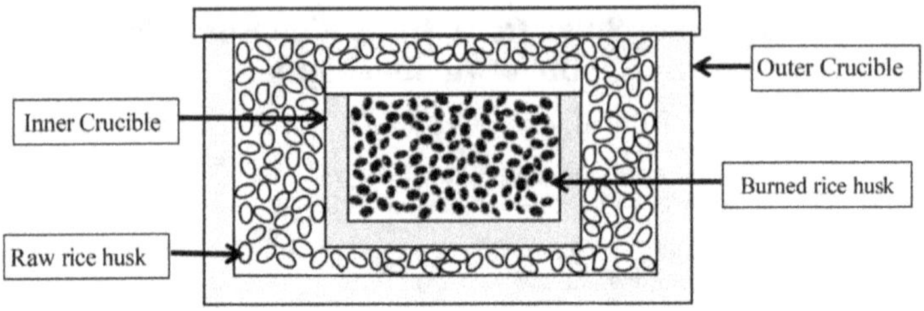

Fig. 1: Double crucible method.

Characterization. FTIR analysis (PerkinElmer RX I) was conducted in the range of 400–4000 cm^{-1} to determine functional groups. Surface morphology and microstructure were analyzed using FESEM (NOVA NANOSEM 450), while elemental composition and purity were evaluated by EDX and reported as wt%.

Result & Discussion

FTIR Analysis. Untreated BRH exhibited O–H, C–H, Si–O–Si, and O–Si–O bands, indicating organic residues and partial silica structures (Table 1).

Table 1: Functional groups observed in the FTIR spectra of untreated burnt rice husk.

Wavenumber (cm^{-1})	Vibration Characteristic
3467.76	O–H stretching
2892.50	C-H stretching
1637.99	H-O-H bending
1092.84	Si–O–Si vibration
787. 82	O-Si-O vibration

The broad O–H stretching peak (~3467 cm^{-1}) and C–H bands (~2890 cm^{-1}) indicate the presence of organic residues, while Si–O–Si and O–Si–O peaks confirm partial silica structure formation before treatment (Table 1). The absorption at 1637.99 cm^{-1} corresponds to the H–O–H bending vibration of water molecules, typically assigned to silanol groups (=Si–O–H). Nayak et al. [5] declared that the presence of siloxane and silanol groups in FTIR data was an indicator of silica. A broad, intense band centered at 1092.84 cm^{-1} is indicative of Si–O–Si vibration modes, such as asymmetric and symmetric stretches of Si–O–C and Si–O–Si bonds in the structural siloxane groups. This band represents the occurrence of silica and cellulose components in the untreated burnt rice husk [6].

Table 2 shows the Functional groups observed in the FTIR spectra of treated burnt rice husk. With thermal treatment, organic peaks weakened progressively. At 500 °C, C–H and O–H bands remained strong, showing incomplete carbon removal. At 600 °C, the C–H band intensity decreased significantly, while Si–O bonds grew sharper. At 700 °C for 120 min, spectra revealed the disappearance of C–H stretching and the strongest siloxane and silanol signals, indicating nearly complete decomposition of organics and silica (Table 2) [7].

The peak observed between 2340.24cm^{-1} and 3465.95 cm^{-1} is attributed to O-H stretching, which may be influenced by moisture retained in the sample or by hydroxyl groups in minerals [8]. The band observed around 2882.26 cm^{-1}, along with 2882.27 cm^{-1}, is attributed to the stretching and/or vibration of C–H bonds, originating from aliphatic saturated compounds found in cellulose and hemicellulose [6].

As the temperature and soaking time of the burnt rice husk increase, the band first observed at ~2882 cm^{-1} disappears. This is consistent with the lignocellulosic thermal degradation behavior, in which the C-H stretching bands become weaker at higher treatment temperatures, indicating the breakdown of aliphatic structure [9].

The 1636-1641 cm^{-1} absorption band is due to the =Si-O-H bending vibration, a spectroscopic indicator of surface silanols. Such silanol bands are frequently observed in studies of biomass-derived silica, indicating the presence of hydroxylated silica surfaces [5]. The intense bands in the 1100- 1000 cm^{-1} range, specifically between 1039-1194 cm^{-1}, are characteristic of Si-O asymmetric stretching of the siloxane networks in the silica [10]. These peaks confirm the predicted FTIR signatures of silica and are observed across all thermal treatment regimes, thereby promoting the formation of stable siloxane structures.

The region between 800-500 cm^{-1} is assigned to O-Si-O bending vibrations. This is commonly assigned as a signal for the formed silicon-oxygen network in silica samples [6]. The bands at 500-400 cm^{-1}, especially at 472.16 cm^{-1} and 467.40 cm^{-1} are assignable to Si-O-Si stretching and bending vibrations. These low-frequency bands are excellent indicators of the presence of a siloxane layer and are typical of pure silica structures [5].

The progressive disappearance of C–H stretching bands and the enhancement of Si–O and O–Si–O vibrations indicate the decomposition of organics and enrichment of silica, with the 700 °C/120 min condition exhibiting the strongest siloxane and silanol features.

Table 2: Functional groups observed in the FTIR spectra of treated burnt rice husk.

Wavenumber (cm^{-1})	**Vibration Characteristics**	**500 °C**		**600 °C**		**700 °C**	
		90 min	**120 min**	**90 min**	**120 min**	**90 min**	**120 min**
3600 - 2200	O-H stretching	3174.42	3465.95	2340.24	3460.14	2344.29	3436.60
3000 - 2800	C-H stretching	2882.26	-	-	-	-	-
1641-1636	=Si-O-H	-	1638.36	-	1638.07	-	1638.11
1100 - 1000	Si-O stretching	1048.79	1103.72	1054.90	1093.44	1039.83	1097.15
800 - 500	O-Si-O bending	795.57	717.96	801.01	782.52	796.15	784.62
500 - 400	Si-O-Si stretching and bending	-	-	-	472.16	-	467.40

Surface Morphology Characterization. Untreated samples (Fig. 3) show fibrous textures. FESEM images at 500 °C (Fig. 4) revealed coarse, fragmented surfaces lacking pores, showing incomplete thermal decomposition. At 600 °C (Fig. 5), the surfaces became granular, with visible mesoporous honeycomb-like structures, reflecting the complete removal of organic matter and the restructuring of silica. At 700 °C, the structure became denser with improved pore definition, showing agglomeration into uniform honeycomb shapes. Thus, 700 °C (Fig. 6) for 120 minutes is the optimal condition for high-surface-area, mesoporous silica.

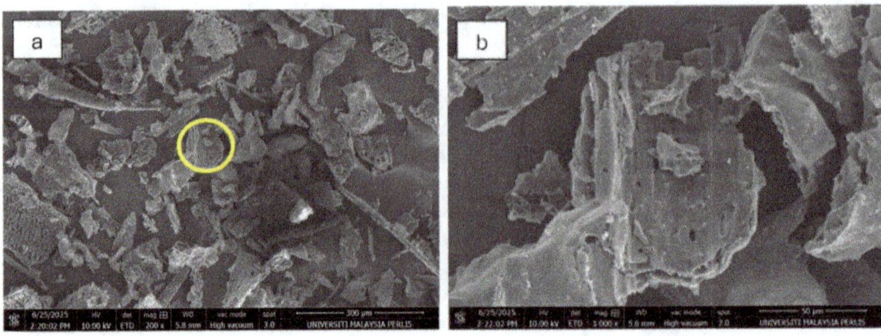

Fig. 3: Morphology of untreated burnt rice husk at magnification (a) 200x and (b) 1000x.

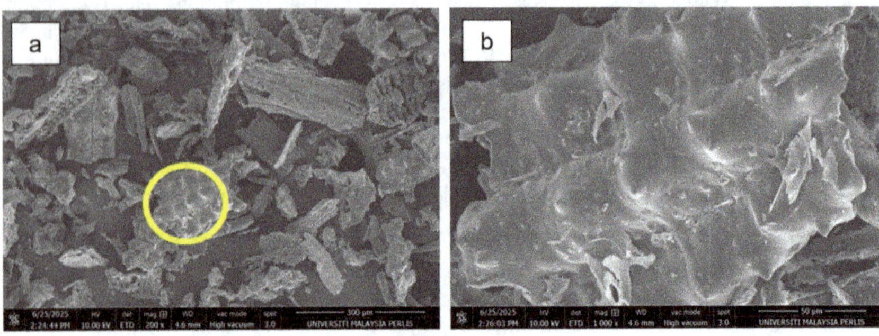

Fig. 4: Morphology of burned rice husk treated at 500°C for 120 minutes at magnification (a) 200x and (b) 1000x.

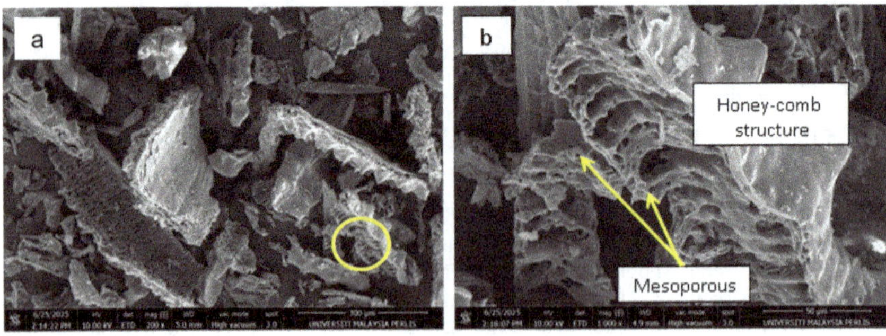

Fig. 5: Morphology of burned rice husk treated at 600°C for 120 minutes at magnification (a) 200x and (b) 1000x.

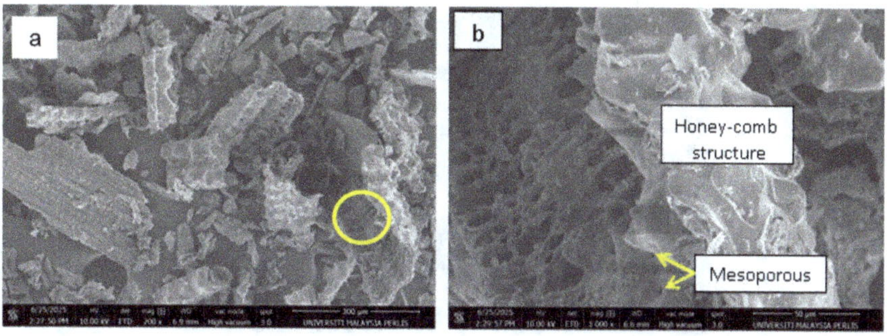

Fig. 6: Morphology of burned rice husk treated at 700°C for 120 minutes at magnification (a) 200x and (b) 1000x.

Energy Dispersive X-ray Spectroscopy. For the untreated BRH sample in Fig. 7, EDX analysis shows peaks for silica (Si), oxygen (O), carbon (C), and high potassium (K) levels. The presence of carbon indicates residual organic matter. However, potassium is a common inorganic impurity in biomass and corroborates research showing that rice husk burned naturally contains metallic impurities such as K, Ca, and Mg, which are likely to be present irrespective of chemical pre-treatment [7]. These residues confirm that untreated BRH is loaded with organic and inorganic impurities, making it less suitable for direct use as silica.

EDX analysis of the 500 °C-treated sample (Fig. 8(a)) revealed prominent Si and O peaks, accompanied by a weaker C signal, indicating partial decomposition of organics and the absence of potassium, suggesting some purification through thermal volatilization.

At 600 °C (Fig. 8(b)), the Si signal increased while the C signal reduced, reflecting enhanced silica content and further organic removal. Although chemical treatments, such as acid leaching, are typically required to remove metallic impurities from rice husk ash [11], the double crucible method proved effective even without such interventions.

At 700 °C (Fig. 8(c)), the Si and O peaks were strongest, with further reduced C content, indicating increased purity. This result supports earlier findings that higher pyrolysis temperatures significantly improve elemental purity of silica [7,11], confirming the double crucible method's capability to minimize carbonaceous and inorganic impurities through a green, chemical-free process.

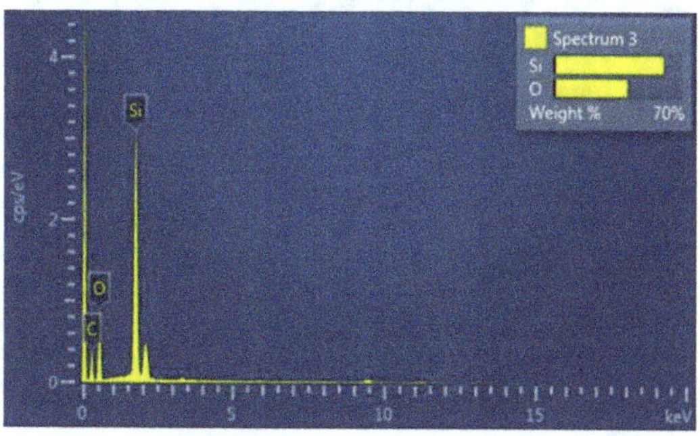

Fig. 7: EDX spectra of untreated burnt rice husk.

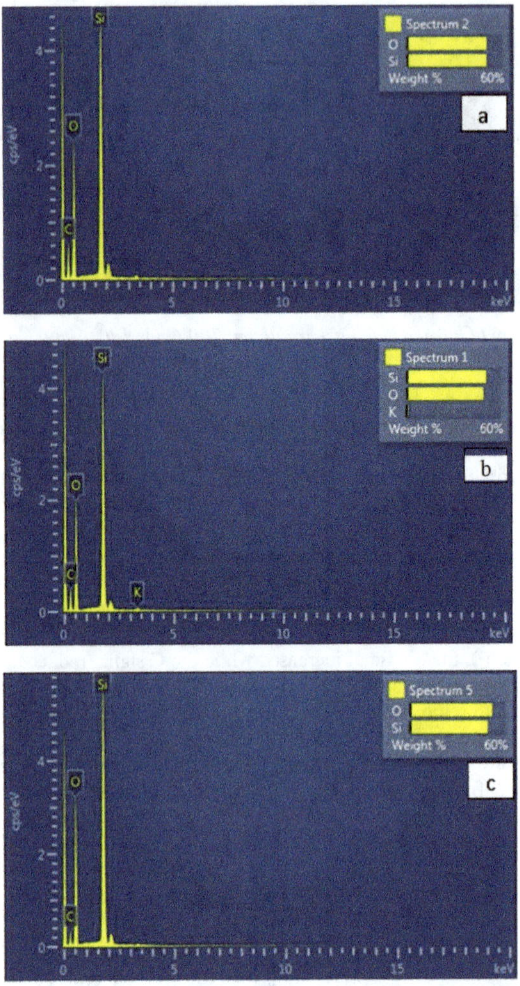

Fig. 8: EDX spectra of burnt rice husk treated at (a) 500°C, (b) 600°C and (c) 700°C for 120 minutes.

Elemental analysis confirmed progressive carbon elimination and silica enrichment (Table 3). At 500 °C/120 min, silica content was ~58 wt%, while carbon was still dominant at ~32 wt%. Increasing temperature and time significantly reduced carbon. The optimal condition of 700 °C/120 min yielded ~92 wt% SiO_2, ~6 wt% O, and <3 wt% residual carbon. Metallic impurities, such as K, Ca, and Mg, were below the detection limits, reflecting efficient purification without acid leaching.

Table 3: Elemental composition of BRH-derived silica (EDX, wt%).

Condition	Si (wt%)	O(wt%)	C (wt%)	SiO_2 (wt%)
500 °C / 120 min	34.8	23.2	32.0	~58
600 °C / 120 min	40.1	35.2	15.1	~75
700 °C / 120 min	49.2	42.1	2.8	~92

Comparative Analysis with Other Methods. To highlight the benefits of the double crucible method, its performance was compared with conventional approaches such as acid leaching and inert-gas-assisted pyrolysis (Table 4). This comparison confirms that the double crucible method achieves a comparable level of purity to acid leaching, without requiring harmful chemicals or inert gases, making it a more sustainable and cost-effective option.

Table 4: Comparison of silica extraction methods.

Method	Purity (SiO_2, wt%)	Chemicals Required	Energy Demand	Environmental Impact
Acid Leaching (HCl/HNO_3)	93–97 [5], [11]	Strong acids	Moderate	High (toxic effluents)
Inert Gas Pyrolysis (N_2)	85–90 [7]	Inert gas supply	High	Medium (gas use)
Double Crucible (this work)	~92	None	Moderate	Low (eco-friendly)

Conclusion

The double crucible method successfully extracted high-purity silica from burnt rice husk without the need for inert gases or chemical additives. The optimal condition was 700 °C for 120 minutes, yielding silica with strong Si–O bonding, well-defined mesoporous structures, and minimal impurities. This eco-friendly, cost-effective process has strong potential for industrial applications in glass, ceramics, and catalysis. Compared with acid leaching and inert-gas pyrolysis, the method achieves comparable purity while incurring a lower environmental impact, offering a scalable and cost-effective pathway for converting agricultural waste into high-value products.

Acknowledgment

The authors thank the Centre of Excellence Geopolymer & Green Technology (CeGeoGTech) and the Faculty of Chemical Engineering & Technology, Universiti Malaysia Perlis (UniMAP), for providing the facilities to conduct this research.

References

[1] A. S. Faried, S. A. Mostafa, B. A. Tayeh, and T. A. Tawfik, The effect of using nano rice husk ash of different burning degrees on ultra-high-performance concrete properties, Constr. Build. Mater. 290 (2021) 123279. https://doi.org/10.1016/j.conbuildmat.2021.123279

[2] S. Yan, D. Yin, F. He, J. Cai, T. Schliermann, and F. Behrendt, Characteristics of smoldering on moist rice husk for silica production, Sustainability 14 (2021) 317. https://doi.org/10.3390/su14010317

[3] P. Nzereogu, A. Omah, F. Ezema, E. Iwuoha, and A. Nwanya, Silica extraction from rice husk: Comprehensive review and applications, Hybrid Adv. 4 (2023) 100111. https://doi.org/10.1016/j.hybadv.2023.100111

[4] M. J. Kaleli, P. K. Kamweru, J. M. Gichumbi, and F. G. Ndiritu, Characterization of rice husk ash prepared by open air burning and furnace calcination, J. Chem. Eng. Mater. Sci. 11 (2020) 24-30. https://doi.org/10.5897/JCEMS2020.0348

[5] P. P. Nayak, S. Nandi, and A. K. Datta, Comparative assessment of chemical treatments on extraction potential of commercial grade silica from rice husk, Eng. Rep. 1 (2019) 25778196. https://doi.org/10.1002/eng2.12035

[6] J. A. S. Costa and C. M. Paranhos, Systematic evaluation of silica production from rice husk ashes, J. Clean. Prod. 192 (2018) 688-697. https://doi.org/10.1016/j.jclepro.2018.05.028

[7] U. N. Ngoc, L. T. Thuy, and N. T. Hoa, Pyrolysis of rice husk for silica production and its application, J. Mater. Res. Technol. 7 (2018) 118-127. https://doi.org/10.1016/j.jmrt.2017.03.005

[8] Y. Ma, S. Yuan, Z. Ma, Y. Hou, S. Niu, L. Lekai, G. Liu, and F. Cao, Comparative study of different pretreatment and combustion methods on the grindability of Rice-Husk-Based SiO_2, Nanomaterials 13 (2023) 2951. https://doi.org/10.3390/nano13222951

[9] H. C. Ong, K. L. Yu, W. Chen, M. K. Pillejera, X. Bi, K. Tran, A. Pétrissans, and M. Pétrissans, Variation of lignocellulosic biomass structure from torrefaction: A critical review, Renew. Sustain. Energy Rev. 152 (2021) 111698. https://doi.org/10.1016/j.rser.2021.111698

[10] N. S. I. Chik, N. Z. B. K. Shaari, N. A. B. Ramlee, and M. R. B. A. Manaf, Extraction of Silica from Rice Husk Ash and Its Effect on the Properties of the Integral Membrane, ASM Sci. J. 17 (2022) 1-13. https://doi.org/10.32802/asmscj.2022.1035

[11] D. Dhaneswara, J. F. Fatriansyah, F. W. Situmorang, and A. N. Haqoh, Synthesis of silica from rice husk ash: Comparing HCl and CH3COOH acidification methods and various alkaline concentrations, Int. J. Technol. 11 (2020) 200. https://doi.org/10.14716/ijtech.v11i1.3335

Ground Coffee Waste-Derived Activated Carbon for Aqueous Dye Adsorption

Teerunishan SUBRAMANIAM[1,a], Norsuria MAHMED[1,2,b*], Muhammad Salihin ZAKARIA[1,2,c] and Haiza HAROON[1,2,d]

[1]Faculty of Chemical Engineering & Technology, Universiti Malaysia Perlis, Kompleks Pusat Pengajian Jejawi 2, 02600 Jejawi Perlis, Malaysia

[2]Centre of Excellence Geopolymer and Green Technology (CeGeoGTech), Universiti Malaysia Perlis, Kompleks Pusat Pengajian Jejawi 2, 02600 Jejawi Perlis, Malaysia

[a]s201341818@studentmail.unimap.edu.my, [b]norsuria@unimap.edu.my, [c]salihin@unimap.edu.my, and [d]haiza@unimap.edu.my

Keywords: Ground Coffee, Activated Carbon, Dye Removal, Methylene Blue, Porous

Abstract. Ground coffee waste (GCW), an abundant agro-industrial by-product, poses significant environmental challenges if improperly disposed. This study explores the valorization of GCW through the synthesis of activated carbon (AC) for potential use in wastewater treatment. A two-step activation process was employed to produce the AC. First, GCW was carbonized at two different temperatures (450 °C and 900 °C) under an inert atmosphere. The resulting char was then chemically activated using phosphoric acid (H_3PO_4) at varying concentrations of 30, 40, and 50 wt%, respectively. The impregnated samples were subsequently dried to enhance pore development. The prepared AC was evaluated through batch adsorption studies to determine its effectiveness in removing methylene blue (MB) dye from aqueous solutions. Optimal adsorption performance was observed for the sample carbonized at 900 °C and activated with 30 wt% H_3PO_4, which achieved a maximum MB removal efficiency of 99.73% using 2 g of AC over 120 minutes. These results demonstrate the viability of GCW-derived AC as a cost-effective and sustainable adsorbent, offering a promising alternative for dye removal in water treatment applications.

Introduction

The global demand for coffee has been increasing steadily over the past decades, positioning it as one of the most widely consumed beverages and most heavily traded agricultural commodities [1]. Consequently, large quantities of ground coffee waste (GCW) are generated during coffee processing and consumption, with estimates suggesting that approximately 2 kg of spent coffee grounds (GCW) are produced for every kilogram of soluble coffee. Improper disposal of this biomass waste contributes significantly to environmental degradation, including methane emissions from landfills, leachate generation that contaminates soil and water, and eutrophication of aquatic ecosystems [2]. These challenges highlight the urgent need for sustainable waste valorization strategies. Among various approaches, the conversion of GCW into activated carbon (AC) has emerged as a promising route for mitigating waste while producing high-value functional materials. Activated carbon is a porous carbonaceous adsorbent with a high surface area and tuneable pore structure, widely used in pollutant removal, water purification, and gas separation applications [3,4]. The production of AC generally involves two major steps: carbonization and activation. While physical activation requires high temperatures and may result in lower surface area, chemical activation particularly with phosphoric acid (H_3PO_4), enhances pore development and offers higher yields at lower activation temperatures [5]. Ground coffee waste is especially suitable as a precursor for AC due to its high carbon content, abundance, and the presence of lignocellulosic components such as cellulose, hemicellulose, and lignin, which favour the development of porous structures during pyrolysis [6]. Moreover, incorporating chemical activation improves the formation of both micropores and mesopores, which are crucial for enhancing adsorption efficiency in aqueous systems [7]. This study aims to synthesize activated carbon from GCW via a two-step physiochemical activation process involving carbonization at 450 °C and 900 °C, followed by chemical impregnation using varying concentrations (30%, 40%, 50% wt) of H_3PO_4. The structural and morphological properties of the produced AC are characterized and correlated with its performance in adsorbing methylene blue (MB), a widely used synthetic dye known for its environmental persistence and toxicity [8]. By valorizing a common agricultural waste into an effective adsorbent, this work addresses both environmental remediation and sustainable material development.

Methodology

Ground coffee waste (GCW) was used as the precursor for the synthesis of activated carbon (AC). The GCW was obtained from Starbucks, Kangar, Perlis. Prior to carbonization process, the GCW underwent a pre-treatment process which involved thorough washing with distilled water to remove impurities, followed by oven drying to eliminate residual moisture. The production of activated carbon was carried out through a two-stage activation process comprising physical and chemical activation. In the first stage, the pre-treated GCW was carbonized under limited oxygen conditions at two different temperatures, 450 °C and 900 °C, with a heating rate of 5 °C/min, holding time of 3 hours and a cooling rate of 10°C/min, to initiate the formation of a porous carbon structure. In the second stage, the carbonized samples were impregnated with phosphoric acid (H_3PO_4) at varying concentrations of 0, 30, 40, and 50 wt%. The impregnated samples were left to soak for 24 hours under agitation, filtered and dried in the oven at 100°C overnight. The nomenclature to denote the coffee samples is shown in Table 1.

Table 1: Nomenclature used to indicate coffee samples.

Coffee Sample	Activation Temperature (°C)	Chemical Activation (H_3PO_4 Concentration, wt%)
TA450	450	0
TA900	900	0
CA450-3	450	30
CA900-3	900	30
CA450-4	450	40
CA900-4	900	40
CA450-5	450	50
CA900-5	900	50

For the characterization, a scanning electron microscopy (SEM) was used to observe the surface morphology and pore structure of the samples. Fourier Transform Infrared Spectroscopy (FTIR) was utilized to identify the functional groups present on the carbon surface, providing insight into the chemical interactions relevant to the adsorption mechanism. To evaluate the adsorption capacity of the synthesized AC, batch adsorption experiments were conducted using methylene blue (MB) dye at an initial concentration of 30 mg/L. A fixed dosage of 2 g of activated carbon was added to each dye solution, and the contact time was varied at intervals of 15, 30, 60, and 120 minutes. The residual concentration of MB in the solution was measured using ultraviolet-visible (UV-Vis) spectrophotometry to determine the removal efficiency.

Results and Discussion

Fig. 1(a-d) present the SEM micrographs of the carbonized ground coffee waste (GCW) samples at 450°C. The non-activated carbonized GCW (Fig. 1a) exhibited a compact and irregular surface structure with few visible pores. The limited pore formation suggests incomplete volatile removal and partial retention of tarry compounds within the carbon matrix, restricting porosity development [9]. After activation with 30 wt% of phosphoric acid (H3PO4) as shown in Fig. 1b, the surface morphology revealed emerging macropores and cracks caused by the partial breakdown of cellulose, hemicellulose, and lignin networks. Phosphoric acid facilitates dehydration and cleavage of C–O–C and C–H bonds, generating a more open porous framework through phosphate-ester linkages and crosslinking reactions [10]. However, at 40 and 50 wt% (Fig. 1c and d), partial pore collapse and surface fusion were observed. Excess acid may cause over-etching and degradation of the carbon skeleton, leading to the merging of adjacent pores and the formation of irregular macropores [10,11].

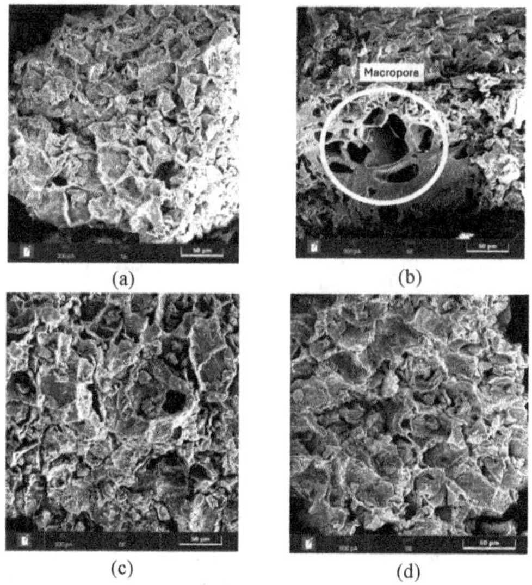

Fig. 1: SEM images of (a) TA450, (b) CA450-3, (c) CA450-4 and (d) CA450-5.

For 900°C of carbonization process, both the carbonized GCW (Fig. 2a) and the sample impregnated with 30 wt % H_3PO_4 (Fig. 2b) display a similar overall morphology, characterized by irregular, interconnected macropores and a rough carbon matrix. This similarity arises because the acid concentration is moderate, and the pores observed by SEM are primarily macropores (≥ 0.05 μm) features already created during high-temperature carbonization.

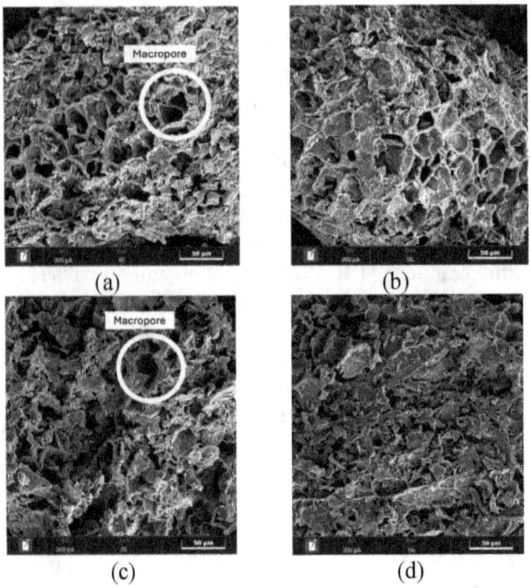

Fig. 2: SEM images of (a) TA900, (b) CA900-3, (c) CA900-4 and (d) CA900-5.

During carbonization at 900 °C, the intense devolatilization of lignocellulosic components such as cellulose, hemicellulose, and lignin generated an internal gas pressure that ruptured the coffee-cell walls, forming the visible mesoporous network. When a moderate amount of phosphoric acid is introduced, it mainly stabilizes the pre-existing carbon skeleton by forming phosphate cross-links, which limit shrinkage and prevent pore collapse but do not drastically alter the macro-morphology [12]. However, at 40 wt % H_3PO_4 (Fig. 2c), the framework shows clear evidence of over-etching, where many pores lose definition and begin to merge. Excess acid concentration accelerates the breakdown of phosphate-carbon linkages and causes vigorous gas evolution, which thins and ruptures pore walls, partially destroying the original mesoporous network [13]. The destructive effect becomes extreme at 50 wt % H_3PO_4 (Fig. 2d), where the surface appears dense, compact, and nearly featureless, indicating severe pore collapse and matrix fusion due to excessive acid penetration and carbon loss [10,11]. Overall, the comparison confirms that phosphoric-acid activation enhances pore formation only up to an optimal level ($\approx$30 wt %), beyond which structural degradation dominates, transforming the initially well-developed mesoporous texture into a compact, pore-collapsed surface.

Fig. 3 illustrates the FTIR spectra for samples carbonized at 450°C (TA450) and those chemically activated with phosphoric acid (H_3PO_4) at 30, 40, and 50 wt% (CA450-3, CA450-4, and CA450-5). The spectra of all four samples show comparable peak patterns. The broad band between 3000–2800 cm^{-1} corresponds to asymmetric and symmetric C–H stretching vibrations of aliphatic alkane groups. Peaks detected at 1700–1500 cm^{-1} are attributed to C=O and C=C stretching vibrations associated with oxygenated surface groups such as carboxyls, lactones, and carbonyls, commonly derived from caffeine and other xanthine-based compounds present in coffee biomass [14]. Additionally, a band between 1400–1500 cm^{-1} signifies CH_3 bending from aliphatic chains, while strong absorption in the 1300–1000 cm^{-1} region is related to C–O stretching and O–H bending of alcohols, phenols, and carboxylic groups [15]. At lower wavenumbers (800–700

cm^{-1}), peaks correspond to Si–O–Si vibrations indicative of silica structures, likely derived from phytolith residues in coffee waste.

For samples carbonized at 900 °C, only a few prominent peaks remained (Fig. 4). The untreated sample (TA9000) displayed a band at 1883 cm^{-1}, assigned to C=O stretching of CO_2 and CO species, while a broad band around 3433 cm^{-1} in CA900-3 corresponds to O–H stretching from hydroxyl, carboxyl, and phenolic groups. Peaks between 1000–1100 cm^{-1} are due to CH_2 bending and C–OH stretching, representing oxygenated moieties [16]. At higher impregnation ratios (40 wt% and 50 wt%), the FTIR spectra of CA850-4 and CA850-5 reveal a remarkable weakening and broadening of characteristic oxygen-containing bands, signifying further structural condensation and deoxygenation of the carbon framework. The absence or significant reduction of O–H and C=O stretching bands suggests that excessive H_3PO_4 loading at elevated temperature promotes phosphate cross-linking and pore collapse through the formation of thermally stable polyphosphate structures. These cross-linking reactions consume surface hydroxyl and carboxylic groups, leading to a decline in detectable polar functionalities [17].

The adsorption efficiency of methylene blue (MB) dye solution was evaluated through UV-Vis measurements as presented in Fig. 5. Among all samples, TA900, CA900-3, and TA450 exhibited superior performance, achieving 99.64%, 99.73%, and 99.29% of MB removal, respectively, after 120 min. The outstanding removal efficiency of CA900-3 is attributed to its pore structure and large surface area, which facilitated enhanced MB adsorption through both π–π electron interaction and electrostatic attraction [18]. In contrast, samples activated with higher phosphoric-acid concentrations (CA900-4 and CA900-5) showed relatively lower adsorption capacity due to excessive etching and partial collapse of micropores, which decreased the number of available active sites [17].

Both TA900 and TA450 exhibited high methylene blue (MB) removal efficiencies (99.64% and 99.29%, respectively) despite lacking chemical activation. This can be attributed to their well-developed microporous structure and the presence of oxygen-containing functional groups formed during carbonization. At 450 °C, partial devolatilization retained hydroxyl and carbonyl groups that enhanced electrostatic attraction and hydrogen bonding with MB molecules. In contrast, carbonization at 900 °C promoted graphitic domain formation, enabling strong π–π interactions between the aromatic MB rings and the carbon surface. Moreover, the absence of acid impregnation prevented pore collapse or over-etching, preserving accessible surface sites for adsorption. Consequently, the synergistic effects of surface functionality, microporosity, and aromaticity accounted for the superior adsorption efficiency of the thermally activated samples [19]. Similarly, for the 450 °C series, CA450-3 demonstrated better adsorption efficiency than CA450-4 and CA450-5, confirming that moderate acid activation (30 wt % H_3PO_4) promotes the formation of accessible micropores and surface oxygen groups that favour dye uptake. However, excessive impregnation levels ($\geq$ 40 wt %) reduced porosity and adsorption due to pore blockage by condensed phosphate species [10].

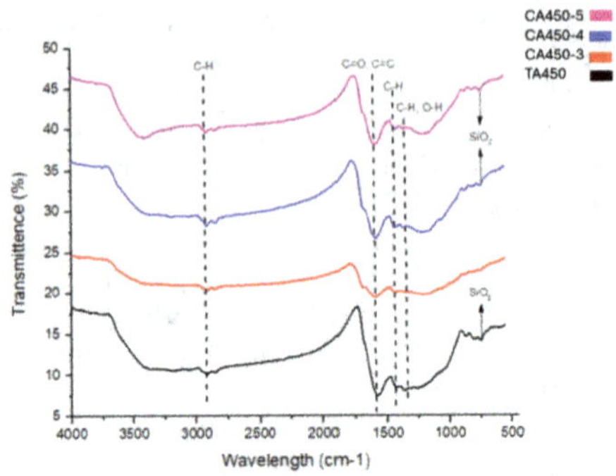

Fig. 3: FTIR spectrum of TA450, CA450-3, CA450-4 and CA450-5 samples.

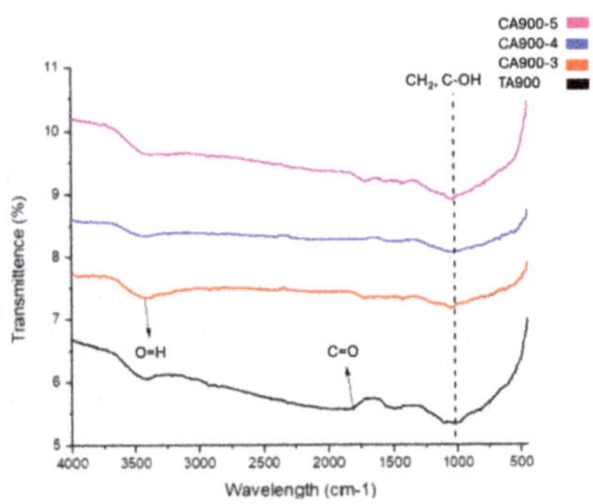

Fig. 4: FTIR spectrum of TA900, TA900-3, TA900-4 and TA905 samples.

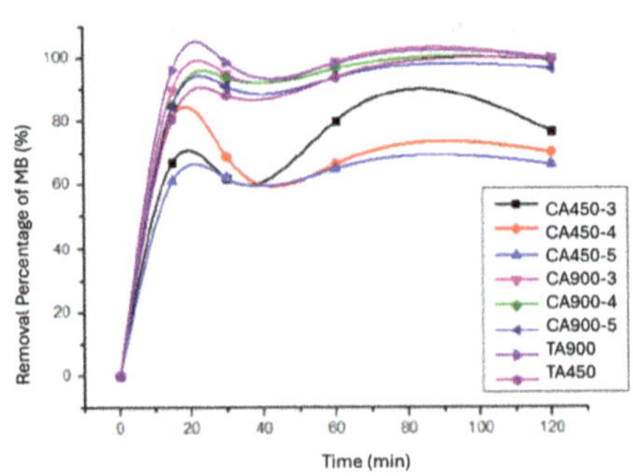

Fig. 5: Removal percentage of methylene blue as a function of time.

Conclusion

The present study demonstrates the successful valorisation of ground coffee waste (GCW) into a high-performance activated carbon (AC) suitable for aqueous dye adsorption applications. Through a two-step activation route involving controlled carbonization and phosphoric-acid impregnation, the physicochemical properties of GCW were effectively modified to yield porous carbons with tailored surface functionalities. The optimized sample, carbonized at 900 °C and activated with 30 wt% H_3PO_4 (CA900-3), achieved a maximum methylene blue (MB) removal efficiency of 99.73%. This superior adsorption capacity is attributed to its well-developed mesoporous network, high surface area, and favourable surface chemistry enabling π–π electron interactions and electrostatic attractions with dye molecules. Conversely, excessive acid loading (≥ 40 wt%) resulted in partial pore collapse and diminished adsorption efficiency due to over-etching and matrix fusion. The inactivated thermally carbonized samples (TA450 and TA900) also exhibited strong MB adsorption performance, highlighting the synergistic roles of carbonization temperature, micropore development, and oxygen-containing surface groups in dye uptake.

References

[1] G.A.F. Campos, J.P.H. Perez, I. Block, S.T. Sagu, P.S. Celis, A. Taubert and H.M. Rawel, Preparation of activated carbons from spent coffee grounds and coffee parchment and assessment of their adsorbent efficiency, Processes (2021) 9(8). https://doi.org/10.3390/pr9081396

[2] M. Zięzio, B. Charmas, K. Jedynak, M. Hawryluk and K Kucio, Preparation and characterization of activated carbons obtained from the waste materials impregnated with phosphoric acid(V), Appl. Nanosci. 10(12) (2020) 4703–4716. https://doi.org/10.1007/s13204-020-01419-6

[3] Z. Heidarinejad, M.H. Dehghani, M. Heidari, G. Javedan, I. Ali and M. Sillanpää, Methods for preparation and activation of activated carbon: a review, Environ. Chem. Lett. 18(2) (2020) 393–415. https://doi.org/10.1007/s10311-019-00955-0

[4] J. Saleem, U.B. Shahid, M. Hijab, H. Mackey and G. McKay. (2019), Production and applications of activated carbons as adsorbents from olive stones, Biomass Convers. Biorefin. 9 (2019) 775-802. https://doi.org/10.1007/s13399-019-00473-7

[5] H. Laksaci, A. Khelifi, M. Trari and A. Addoun, Synthesis and characterization of microporous activated carbon from coffee grounds using potassium hydroxides, J. Clean. Prod. 147 (2017) 254–262. https://doi.org/10.1016/j.jclepro.2017.01.102

[6] A.S.C.d. Bomfim, D.M. de Oliveira, E. Walling, A. Babin, G. Hersant, C. Vaneeckhaute, M.-J. Dumont and D. Rodrigue, Spent Coffee Grounds Characterization and Reuse in Composting and Soil Amendment, Waste 1 (2023) 2-20. https://doi.org/10.3390/waste1010002

[7] N. Muttil, S. Jagadeesan, A. Chanda, M. Duke and S.K. Singh, Production, types, and applications of activated carbon derived from waste tyres: An overview, Appl. Sci 13 (2023) 257. https://doi.org/10.3390/app13010257

[8] S. Gita, A. Hussan and T.G. Choudhury, Impact of textile dyes waste on aquatic environments and its treatment, Environ. Ecol. 35(3C) (2017) 2349-2353. https://www.cabidigitallibrary.org/doi/full/10.5555/20173297379

[9] I. Neme, G. Gonfa and C. Masi, Activated carbon from biomass precursors using phosphoric acid: A review, Heliyon 8(12) (2022) e11940. https://doi.org/10.1016/j.heliyon.2022.e11940

[10] W. Zhao, L. Chen and Y. Jiao, Preparation of activated carbon from sunflower straw through H_3PO_4 activation and its application for acid fuchsin dye adsorption, Water Sci. Eng. 16(2) (2023)192–202. https://doi.org/10.1016/j.wse.2023.02.002

[11] M. Murugan, G. Dineshkumar, M. Ganesan, N.G. Sahoo and M. Sivanantham, Influence of Phosphoric Acid Activation on Physiochemical Characteristics of Activated Carbons and Their Performance as Supercapacitor, Energy Storage 6(7) (2024) e70050. https://doi.org/10.1002/est2.70050

[12] X.Y. Zeng, Y. Wang, R.X. Li, H-L. Cao, Y-F. Li and L. Jian, Impacts of temperatures and phosphoric-acid modification to the physicochemical properties of biochar for excellent sulfadiazine adsorption, Biochar 4 (14) (2022). https://doi.org/10.1007/s42773-022-00143-4

[13] R. Zakaria, N.A. Jamalluddin and M. Z. A. Bakar, Effect of impregnation ratio and activation temperature on the yield and adsorption performance of mangrove based activated carbon for methylene blue removal, Res. Mater. 10 (2021) 100183. https://doi.org/10.1016/j.rinma.2021.100183

[14] H. Laksaci, A. Khelifi, M. Trari and A. Addoun, Synthesis and characterization of microporous activated carbon from coffee grounds using potassium hydroxides, J. Clean. Prod. 147 (2017) 254–262. https://doi.org/10.1016/j.jclepro.2017.01.102

[15] W. Tala and S. Chantara, Use of spent coffee ground biochar as ambient PAHs sorbent and novel extraction method for GC-MS analysis, Environ. Sci. Pollut. Res. 26(13) (2019)13025–13040. https://doi.org/10.1007/s11356-019-04473-y

[16] R.F.T. Tagne, M.a.S. Ngueabouo, D.R.T. Tchuifon, C.G. Fotsop, A.K. Tamo and S.G. Anagho, Strategy for optimizing the synthesis and characterization of activated carbons obtained by chemical activation of coffee husk, Mater. Adv. 3(22) (2022) 8361–8374. https://doi.org/10.1039/d2ma00591c

[17] I. Moulefera, F.J. García-Mateos, A. Benyoucef, J.M. Rosas, J. Rodríguez-Mirasol and T. Cordero, Effect of Co-solution of Carbon Precursor and Activating Agent on the Textural Properties of Highly Porous Activated Carbon Obtained by Chemical Activation of Lignin With H3PO4, Front. Mater. 7 (2020) 153. https://doi.org/10.3389/fmats.2020.00153

[18] Y. Yuxuan, Z. Haifeng, L. Yijing, P. Shenghua, L. Liuqian, Y. Futi, W. Yongyi and L. Zhigao, Efficient dye adsorption of mesoporous activated carbon from bamboo parenchyma cells by phosphoric acid activation, RSC Adv. 14 (2024) 12873-12882. https://doi.org/10.1039/D4RA01652A

[19] B. Charmas, M. Zięzio, and K. Jedynak, Assessment of the Porous Structure and surface chemistry of activated biocarbons used for methylene blue adsorption. Molecules 28 (2023) 4922. https://doi.org/10.3390/molecules28134922

Performance of Graphene-Reinforced Concrete Compared to Normal Concrete Under Explosive Loading

I.S. ISMAIL[1,a], A.H. HILMI[2,b*], A.R.A HAMID[3,c], S.M. Sk ABD RAZAK[4,d], N.A. MOHD SALEH[4,e]

[1]Fakulti Kejuruteraan & Teknologi Mekanikal (FKTM), Kampus Alam, Universiti Malaysia Perlis (UniMAP), Pauh Putra, 02600 Arau, Perlis, MALAYSIA

[2]Centre of Excellence for Sports Engineering (SERC), Makmal Biomekanik dan Analisis Gerakan, Fakulti Kejuruteraan dan Teknologi Elektronik, Kampus Pauh Putra, Universiti Malaysia Perlis, 02600 Arau, Perlis, MALAYSIA

[3]Centre of Excellence for Geopolymer and Green Technology (CEGeoGTech), Universiti Malaysia Perlis (UniMAP), Pusat Pengajian Jejawi, 2, Jalan Jejawi - Tambun Tulang, Taman Muhibbah, 02600 Arau, Perlis, MALAYSIA

[4]Faculty of Civil Engineering & Technology, Universiti Malaysia Perlis (UniMAP), Kompleks Pusat Pengajian Jejawi 3, 02600 Arau, Perlis, MALAYSIA

[a]irfan.5085@gmail.com, [b]humaizi@unimap.edu.my, [c]asnarasyidah@unimap.edu.my, [d]skmuiz@unimap.edu.my, [e]s211052259@studentmail.unimap.edu.my

Keywords: Graphene-Reinforced Concrete, Explosive Loading, Blast Resistance, Crack Propagation, Compressive Strength, Dynamic Loads, Advanced Construction Materials

Abstract. This study investigates the performance of graphene-reinforced concrete (GRC) compared to normal concrete (NRC) when subjected to explosive loading. GRC specimens were prepared by incorporating a small proportion (0.5% by volume) of Sodium Laureth Sulfate (SLES) graphene powder, replacing an equivalent amount of fine sand in the concrete mix, while NRC specimens were prepared without graphene. Both types of specimens were constructed in Grade 25 concrete to ensure comparability in compressive strength and tested under consistent conditions. A field blast test was conducted using 1 kg of Emulex explosive at a 30 cm stand-off distance to analyze initial crack development and resistance to explosive forces. Crack length and width were measured using high-resolution imaging and ImageJ software to quantify damage. Additionally, compressive strength tests were performed before and after the blast to assess residual strength. The results show that GRC exhibited significantly lower crack propagation and reduced maximum crack width compared to NRC, indicating improved resistance to explosive loading. GRC maintained higher compressive strength both before and after the blast, with only a minor reduction in strength post-explosion. In contrast, NRC showed extensive crack propagation and a substantial decrease in compressive strength, falling below the design strength for Grade 25 concrete. These findings suggest that the incorporation of graphene enhances the toughness and blast resistance of concrete, potentially increasing structural durability and safety in explosive-prone environments. The study highlights the potential of graphene as an innovative additive for improving the performance of concrete under extreme loading conditions, offering valuable insights for the development of advanced construction materials in critical infrastructure applications.

Introduction

Concrete is widely used in construction because of its high compressive strength and versatility; however, its brittle nature and low tensile capacity make it highly vulnerable to dynamic and extreme loading conditions such as explosions [1]. Blast loading can cause severe damage, including cracking, spalling, and structural failure [2]. To overcome these limitations, advanced reinforcement strategies incorporating nanomaterials have been explored. Graphene, a two-dimensional carbon material with exceptional tensile strength, high Young's modulus, and large specific surface area, has emerged as a promising additive for enhancing concrete performance [3,4]. Previous studies have reported that graphene oxide (GO) improves compressive strength, fracture toughness, and durability by promoting cement hydration, reducing porosity, and enhancing interfacial bonding [5,6]. Concrete failure under blast loading is governed by rapid microcrack initiation and propagation induced by high strain rates, leading to a significant loss of load-bearing capacity [7]. While earlier blast-resistant concrete studies focused primarily on steel fiber-reinforced and hybrid systems [1], recent advances in nanotechnology have shifted attention toward graphene-based reinforcement. Nanoengineered graphene concrete composites have demonstrated enhanced compressive strength, cracking resistance, and multifunctional performance, even at low graphene contents, while also improving resistance to thermal and mechanical stresses and reducing water absorption [8,9,10,11]. Despite these advances, the behavior of graphene-reinforced concrete (GRC) under real explosive loading conditions remains insufficiently explored, as most studies emphasize static or quasi-static testing. Prior research has shown that incorporating 0.5% graphene enhances energy absorption and toughness, which may mitigate blast-induced damage [12]; however, systematic investigations addressing crack formation, crack propagation, and residual compressive strength following blast exposure are still limited. Owing to its high tensile strength and Young's modulus [13], graphene can bridge microcracks, delay crack propagation, and improve stress distribution within the cement matrix [14], while also reducing porosity and enhancing matrix densification and hydration uniformity [15]. This study experimentally evaluates the performance of GRC compared with normal reinforced concrete (NRC) under explosive loading, focusing on crack propagation behavior and compressive strength retention. GRC specimens were prepared using 0.5% SLES-functionalized graphene as a partial replacement for fine aggregate, while NRC specimens served as the reference. All specimens were designed using Grade 25 concrete in accordance with ASTM C39/C39M-12a. Field blast tests were conducted using a 1 kg Emulex explosive at a 30 cm stand-off distance [16], with crack characteristics quantified using high-resolution imaging and ImageJ software [17]. Pre- and post-blast compressive strength tests were performed to assess residual structural integrity, following established qualitative and quantitative damage assessment approaches [18]. The results are expected to demonstrate reduced crack propagation, narrower crack widths, and higher residual compressive strength in GRC compared with NRC, consistent with previous graphene-based studies [6,12].

Methodologies

This study aimed to experimentally investigate the performance of graphene-reinforced concrete (GRC) compared to normal reinforced concrete (NRC) when exposed to explosive loading. The methodology is structured to achieve the two main objectives: (i) analyzing the crack formation (length and width) in both types of concrete after exposure to a 1 kg emulsion explosive (Emulex), and (ii) evaluating the compressive strength of the concrete specimens both before and after the blast test. The overall methodological approach consisted of material preparation, specimen casting, pre- and post-blast mechanical testing, and crack analysis using image-processing software.

Materials and Specimen Preparation

Graphene Preparation. The graphene used in this study was SLES-functionalized graphene powder, a low-cost nanomaterial produced via liquid shear exfoliation. graphene was incorporated into the concrete mix after exfoliation of graphite in a sodium laureth sulfate solution, yielding flakes with sizes below 20 μm. Prior to use, the graphene powder was characterized by X-ray diffraction (XRD) (Fig. 1) to verify its crystalline structure and interlayer spacing.

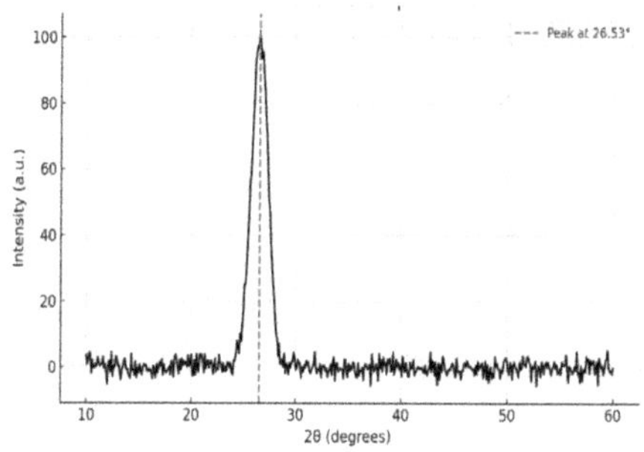

Fig. 1: XRD pattern of SLES Graphene Powder.

Concrete Mix Design

Concrete mixes were designed in accordance with ASTM C192/C192M standards. Both GRC and NRC specimens employed Grade 25 concrete using ordinary Portland cement, uncrushed fine aggregate, and 10 mm nominal coarse aggregate. In the GRC mix, 0.5% of the fine aggregate volume was replaced with graphene powder to evaluate its reinforcing effect.

Concrete component proportions were calculated to ensure consistency across all specimens, with a total batch volume of 0.03 m³. The required graphene content (40 g) was determined based on its measured density (541.67 kg/m³) and the specified fine aggregate replacement volume, while the corresponding mass of sand removed (100 g) was calculated using its density (1558 kg/m³). This replacement ratio was selected to maintain comparable density and workability of the concrete mix. The concrete mixing procedure is summarized below.

The concrete mixing process involved initially dry-mixing cement, sand, and coarse aggregates, followed by dispersing SLES-functionalized graphene in water and adding the graphene water suspension to the dry mix. The mixture was thoroughly mixed to achieve uniformity, and a slump test was performed to verify workability. The fresh concrete was then cast into cube and slab specimens, cured in water for 28 days, and subsequently subjected to compressive strength testing and field blast experiments.

Specimen Casting and Curing

Specimen Types. Three specimen types were fabricated: cube specimens (100 mm × 100 mm × 100 mm) for compressive strength testing, slab specimens (600 mm × 600 mm × 75 mm) for blast testing as shown in Fig. 2, and core samples extracted from blast-tested slabs for post-blast compression evaluation. Cube specimens were tested in accordance with ASTM C109/C109M, while slab specimens were designed to represent realistic structural elements subjected to blast loading.

Casting and Curing Procedures. All dry constituents were mixed thoroughly prior to the gradual addition of water or graphene suspension to achieve the target water–cement ratio. For GRC, graphene powder was pre-dispersed in water to ensure uniform distribution before incorporation into the mix. Slump tests were conducted to verify workability, with a target medium slump of approximately 85 mm for GRC. Specimens were cast in pre-cleaned molds and compacted using a mechanical vibrator to remove entrapped air. After initial setting, the molds were removed and the specimens were water-cured for 28 days in accordance with ASTM guidelines to ensure adequate hydration and strength development.

Pre-Blast Testing. Prior to the blast tests, the compressive strength of all cube specimens was determined using a compression testing machine. Three cube specimens from each group (NRC and GRC) were tested, and the results provided baseline data for comparison with post-blast performance.

Blast Test Setup

Experimental Setup. The blast tests were conducted in a controlled open-field environment. A schematic and photographic depiction of the blast setup is shown in Fig. 2. A 1 kg charge of Emulex was suspended 30 cm above the center of the concrete slab specimen using a PVC pipe rig, a configuration chosen based on literature that balances realism with safety.

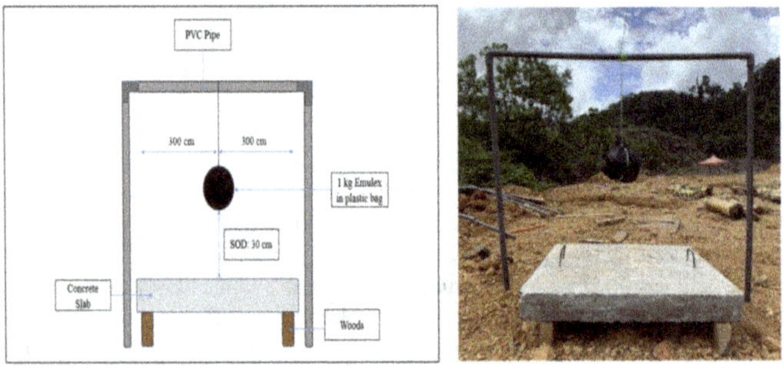

Fig. 2: Blast setup.

Explosive Details. Emulex, a commercial emulsion explosive with a velocity of detonation of 5600 m/s, was used. The stand-off distance of 30 cm was maintained across all tests to minimize reflected wave interference and to replicate typical blast conditions faced by concrete structures in proximity to explosions.

Crack Development Analysis. Following the blast, the slab specimens were examined for crack formation. High-resolution images of the damaged slabs were captured and analyzed using ImageJ software. The software facilitated accurate tracing and measurement of crack lengths and widths, enhancing the objectivity of the analysis compared to manual measurement methods. Each crack was labeled and measured to assess the severity and distribution of damage as shown in Fig. 3.

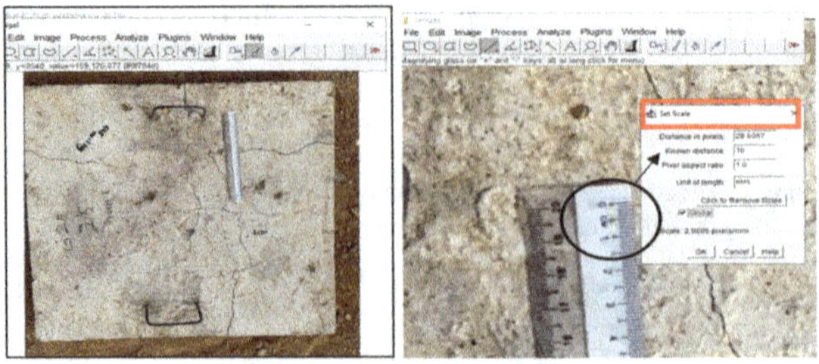

Fig. 3: Crack measurement using ImageJ software.

Post-Blast Testing. Core samples were extracted from the blast-tested slabs using a coring machine to fit the compression testing apparatus. Three cores were taken from each slab, targeting regions with the most severe cracking. Compression tests on the cored specimens were conducted in accordance with ASTM C39/C39M standards. Post-blast compressive strengths were compared with pre-blast values to quantify strength degradation due to explosive loading, providing an assessment of the residual strength and structural integrity of both concrete types.

Data Analysis. Quantitative data from the compression tests and crack measurements were compiled and analyzed. Comparative graphs were prepared to illustrate differences between NRC and GRC in terms of compressive strength loss and crack development. Statistical analysis was conducted where relevant to identify significant differences between groups.

Results and Discussion

This section presents and discusses the experimental results evaluating the performance of graphene-reinforced concrete (GRC) relative to normal reinforced concrete (NRC) under explosive loading. The results include pre- and post-blast compressive strength and crack propagation characteristics, which are interpreted in the context of structural integrity under dynamic loading and the reinforcing potential of graphene.

Pre-Blast Compressive Strength. Compressive strength testing of cube specimens showed that both NRC and GRC exceeded the 25 MPa requirement for Grade 25 concrete. NRC achieved an average strength of 27.5 MPa, while GRC exhibited a slightly higher value of 29.2 MPa. The strength enhancement observed in GRC is consistent with previous findings that nano-reinforcement promotes improved cement hydration and microstructural densification (Fig. 4).

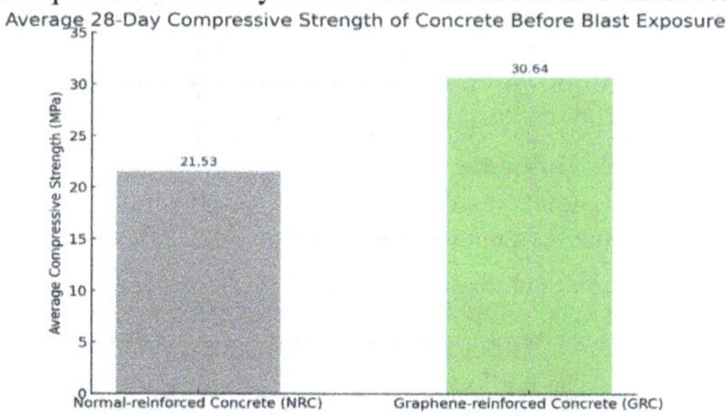

Fig. 4. Bar chart comparing the average compressive strengths of NRC and GRC before blast exposure.

Blast-Induced Crack Formation

Following exposure to a 1 kg Emulex charge at a 30 cm stand-off distance, both NRC and GRC slabs exhibited cracking. See Fig. 5. However, qualitative observations and quantitative image analysis revealed notable differences in crack morphology and severity.

Crack Lengths. On average, NRC slabs exhibited total crack lengths of approximately 140 cm per specimen, whereas GRC slabs showed reduced crack lengths of about 95 cm per specimen. This reduction demonstrates the ability of graphene to bridge microcracks and hinder crack propagation under dynamic loading [19]

Crack Widths. The maximum crack width observed in NRC was 2.1 mm, while GRC exhibited a maximum width of 1.3 mm. The narrower cracks in GRC are attributed to the enhanced tensile bridging effect provided by graphene flakes, which distribute stresses more evenly across the matrix.

Figure 5 (a) & (b). Photographs of cracked NRC and GRC slabs.

Post-Blast Compressive Strength

Post-blast compressive strength was assessed using core samples extracted from NRC and GRC slabs. NRC exhibited a substantial strength reduction, with an average residual strength of 19.8 MPa (28% loss), whereas GRC retained a higher residual strength of 25.1 MPa (14% loss). These results confirm the protective role of graphene in preserving load-bearing capacity after explosive damage and are consistent with previous studies reporting improved dynamic resistance through crack bridging and delayed failure mechanisms.

Failure Mode Analysis. Visual inspection combined with digital image analysis revealed distinct failure modes between NRC and GRC. NRC specimens exhibited extensive, interconnected surface cracks forming a spiderweb-like pattern characteristic of blast damage, whereas GRC specimens showed fewer and more isolated cracks with reduced connectivity. This behavior indicates that graphene effectively limits crack initiation and subsequent propagation.

Discussion of Graphene's Influence

The performance of GRC under explosive loading demonstrates that graphene can effectively enhance the blast resistance of concrete. The improved mechanical properties observed in this study can be attributed to several factors:

a) **Crack Bridging:** Graphene's high aspect ratio and excellent mechanical strength enable it to bridge microcracks, thereby distributing stress and reducing crack propagation

b) **Matrix Densification:** The addition of graphene promotes a denser and more uniform cement matrix by acting as nucleation sites during cement hydration. This results in a more homogeneous structure with fewer weak points for crack initiation.

c) **Energy Dissipation:** Graphene's layered structure allows for effective energy absorption and dissipation under dynamic loading conditions, enhancing the concrete's toughness.

Furthermore, the stand-off distance of 30 cm in the blast test was sufficient to simulate realistic near-field blast scenarios, while maintaining safety in the experimental setup.

Conclusion

This study demonstrates the enhanced performance of graphene-reinforced concrete (GRC) compared to normal reinforced concrete (NRC) under explosive loading. Using ASTM-compliant specimen preparation and testing procedures, GRC exhibited reduced crack propagation, narrower crack widths, and higher residual compressive strength following blast exposure. Field blast tests using a 1 kg Emulex charge at a 30 cm stand-off distance, combined with ImageJ-based crack analysis and pre- and post-blast compression testing, enabled a comprehensive assessment of damage and residual structural integrity. The findings confirm that SLES-functionalized graphene effectively improves crack resistance and load-bearing capacity under dynamic loading.

Acknowledgement
This research was supported by the Ministry of Higher Education, which provided research grants from the Fundamental Research Grant Scheme (FRGS) under grant number FGRS/1/2023/STG05/UNIMAP/02/8. The authors also acknowledge the Faculty of Mechanical Engineering Technology, Universiti Malaysia Perlis (UniMAP), for the laboratory facilities. This project was completed with assistance from Mr. Poh Chin Tan and Mr. Loga for Shotfirer and facilities at Quarry Chua Teck and Sons Sdn Bhd, Perlis, Rejimen Askar Jurutera Diraja (RAJD), and Unit Pemusnah Bom PDRM, Perlis.

References
[1] M. A. Yusof, N. Mohamad Nor, A. Ismail, N. Choy Peng, R. Mohd Sohaimi, and M. A. Yahya, "Performance of hybrid steel fibers reinforced concrete subjected to air blast loading," Advances in Materials Science and Engineering, vol. 2013, 2013. https://doi.org/10.1155/2013/420136

[2] M. M. A. Alhadid, A. M. Soliman, M. L. Nehdi, and M. A. Youssef, "Critical overview of blast resistance of different concrete types," *Magazine of Concrete Research*, vol. 66, no. 2, pp. 72–81, Jan. 2014. https://doi.org/10.1680/macr.13.00096

[3] A. K. Geim and K. S. Novoselov, "The Rise of Graphene," 2009.

[4] A. Gholampour, M. Sofi, H. Alipooramirabad, and Y. Tang, "Performance of concrete containing pristine graphene-treated recycled concrete aggregates," *Resour Conserv Recycl*, vol. 199, Dec. 2023. https://doi.org/10.1016/j.resconrec.2023.107266

[5] G. Jing et al., "Improving the cracking resistance of mortar by reduced graphene oxide," *Constr Build Mater*, vol. 310, Dec. 2021. https://doi.org/10.1016/j.conbuildmat.2021.125150

[6] H. Li, G. Zhao, and H. Zhang, "Recent Progress of Cement-Based Materials Modified by Graphene and Its Derivatives," Materials, vol. 16, no. 10, May 2023. https://doi.org/10.3390/ma16103783

[7] T. Ngo, P. Mendis, A. Gupta, and J. Ramsay, "Blast loading and blast effects on structures – An overview," *Electronic Journal of Structural Engineering*, vol. 7, pp. 76–91, 2007. https://doi.org/10.56748/ejse.671

[8] D. Dimov et al., "Ultrahigh Performance Nanoengineered Graphene–Concrete Composites for Multifunctional Applications," *Adv Funct Mater*, vol. 28, no. 23, Jun. 2018. https://doi.org/10.1002/adfm.201705183

[9] B. Wang, R. Jiang, and Z. Wu, "Investigation of the mechanical properties and microstructure of graphene nanoplatelet-cement composite," *Nanomaterials*, vol. 6, no. 11, Nov. 2016. https://doi.org/10.3390/nano6110200

[10] M. S. Kirgiz, "Advancements in mechanical and physical properties for marble powder–cement composites strengthened by nanostructured graphite particles," *Mechanics of Materials*, vol. 92, pp. 223–234, Jan. 2016. https://doi.org/10.1016/j.mechmat.2015.09.013

[11] S. K. U. Rehman, S. Kumarova, S. A. Memon, M. F. Javed, and M. Jameel, "A review of microscale, rheological, mechanical, thermoelectrical and piezoresistive properties of graphene-based cement composite," *Nanomaterials*, vol. 10, no. 10, Oct. 2020. https://doi.org/10.3390/nano10102076

[12] A. H. Alateah, "Graphene concrete: Recent advances in production methods, performance properties, environmental impact and economic viability," *Case Studies in Construction Materials*, vol. 19, Dec. 2023. https://doi.org/10.1016/j.cscm.2023.e02653

[13] I. A. Ovid'ko, "Mechanical properties of graphene," 2013.

[14] S. P. Dalal and P. Dalal, "Experimental Investigation on Strength and Durability of Graphene Nanoengineered Concrete," *Constr Build Mater*, vol. 276, Mar. 2021. https://doi.org/10.1016/j.conbuildmat.2020.122236

[15] A. Sedaghat, M. K. Ram, A. Zayed, R. Kamal, and N. Shanahan, "Investigation of Physical Properties of Graphene-Cement Composite for Structural Applications," *Open Journal of Composite Materials*, vol. 04, no. 01, pp. 12–21, 2014. https://doi.org/10.4236/ojcm.2014.41002

[16] M. Drdlová, R. Cechmánek, and R. Rídký, "Blast impact behaviour of concrete with different fibre reinforcement," in EPJ Web of Conferences, EDP Sciences, Sep. 2015. https://doi.org/10.1051/epjconf/20159405006

[17] S. Alsadey, M. Abdallateef, M. Mohamed, and A. Milad, "Investigating Behaviour of Reinforced Concrete with Glass Fibre," *Jurnal Kejuruteraan*, vol. 33, no. 3, pp. 551–557, Aug. 2021. https://doi.org/10.17576/jkukm-2021-33(3)-16

[18] S. K. Ur Rehman et al., "Influence of graphene nanosheets … cement-based composites," *Sustainability (Switzerland)*, vol. 10, no. 3, Mar. 2018. https://doi.org/10.3390/su10030822

[19] B. Wang, R. Jiang, and Z. Wu, "Investigation of the mechanical properties and microstructure of graphene nanoplatelet-cement composite," *Nanomaterials*, vol. 6, no. 11, Nov. 2016. https://doi.org/10.3390/nano6110200

Effect of Polyethylene Glycol and Lemongrass Essential Oil as Plasticizers on the Properties of Chitosan/Hydroxyethyl Cellulose Blend Films

Nurul Izzaty MOHD ISA[1,2,a], Muthmirah IBRAHIM[1,2,b *], Siti Salmi SAMSUDIN[1,2,c] and Jalilah ABD JALIL[1,3,d]

[1]Faculty of Chemical Engineering & Technology, Universiti Malaysia Perlis, 02600 Arau, Perlis, Malaysia

[2]Centre of Excellence of Geopolymer and Green Technology, Universiti Malaysia Perlis, 02600 Arau, Perlis, Malaysia

[3]Centre of Excellence of Biomass Utilization, Universiti Malaysia Perlis, 02600 Arau, Perlis, Malaysia

[a]izzatymdisa@gmail.com, [b]muthmirah@unimap.edu.my, [c]sitisalmi@unimap.edu.my, [d]jalilahjalil@unimap.edu.my

Keywords: Chitosan, Hydroxyethyl Cellulose, Polyethylene Glycol, Lemongrass Essential Oil, Blend Films

Abstract. This work investigates biodegradable chitosan/hydroxyethyl cellulose (HEC) blend films prepared by solution casting with polyethylene glycol (PEG) and lemongrass essential oil (LEO) as plasticizers, particularly for food packaging applications. Tensile, thermal and Fourier transform infrared spectroscopy (FTIR) analyses that PEG increased flexibility while LEO preserved tensile strength and improved thermal stability. The unplasticized chitosan/HEC films exhibited an optimum tensile strength of 20.7 ± 2.0 MPa at the chitosan-40/HEC-60 film composition, which decreased with higher HEC content due to limited chain mobility. Increasing HEC enhanced elongation at break, reflecting improved flexibility. PEG effectively increased plasticity, while LEO preserved film strength. Both plasticizers enhanced thermal stability, and FTIR analysis revealed stronger hydrogen bonding and modified interpolymer interactions within the blends. Collectively, these results demonstrate that PEG- and LEO-plasticized chitosan/HEC films are promising, cost-effective and biodegradable materials for active food-packaging, offering a tunable balance of strength, flexibility and thermal stability.

Introduction

The global production rate of plastics has more than doubled every decade, driven by their versatility, low cost, and growing demand [1,2]. However, the slow degradation rate of petroleum-based plastics has led to a significant accumulation of plastic waste, particularly in landfills and oceans, where they contribute to environmental degradation and pose threats to marine life [3,4]. This escalating environmental issue has spurred global efforts to find sustainable alternatives, particularly in packaging applications.

Chitosan, derived from the deacetylating chitin, is a biodegradable and biocompatible biopolymer with applications in food packaging, pharmaceutical and agricultural [5,6]. However, pure chitosan films are typically brittle and require plasticizers to improve flexibility. Hydroxyethyl cellulose (HEC), a non-ionic, water-soluble cellulose ether, exhibits good film-forming and adhesives properties and is widely used in pharmaceutical and industrial applications [7,8]. However, HEC films often suffer from rough surfaces as well as poor toughness and ductility [5]. Blending chitosan with HEC can potentially synergized their advantages, but the resulting films still require plasticizers to enhance chain mobility and flexibility.

Materials Research Proceedings 68 (2026) 65-72 https://doi.org/10.21741/9781644904237-9

Among potential plasticizers, polyethylene glycol (PEG) and lemongrass essential oil (LEO) have shown considerable promise. PEG, a synthetic plasticizer, improves film flexibility by reducing intermolecular forces within the polymer chains, while LEO, a natural plasticizer, provides additional antimicrobial and antioxidant properties for extending food shelf life [9]. Based on the literature, the combined use of PEG and LEO as plasticizers in chitosan/HEC blend films, in a comparative study, has not been reported previously. Thus, this study investigates the comparative effects of PEG and LEO on the tensile, thermal, and structural properties of chitosan/HEC blend films. The results aim to advance the development of cost-effective, biodegradable, and functional materials for active food-packaging applications.

Materials and Methods

Materials. Chitosan powder used in this study was purchased from Quality Reagent Chemicals. Hydroxyethyl cellulose (HEC) powder with viscosity ~145 mPa.s was purchased from Sigma-Aldrich. Polyethylene glycol with a nominal molecular weight of 400 g mol^{-1} (PEG 400) and aromatherapy-grade lemongrass essential oil (LEO) were incorporated as plasticizers.

Preparation of Chitosan/HEC Blend Films. Chitosan/HEC films were prepared using the solution casting method. A chitosan solution was prepared in 1% (v/v) acetic acid, while a HEC solution was prepared in distilled water. Both solutions were stirred separately at room temperature until homogenous, then mixed in specified ratios and stirred for an additional two hours. For the plasticized films, the blends were prepared using only the best formulation, selected based on the optimum tensile properties. PEG or LEO was added at concentrations of 2.5, 5.0, and 7.5% v/v, respectively. The mixtures were stirred continuously and maintained in a water bath at 50-60°C for 45 minutes to remove air bubbles and ensure uniformity. The final solutions were cast onto glass molds and dried at room temperature for 48 hours. Dried films were carefully peeled off and stored in a desiccator prior to characterization and testing. The formulation of chitosan/HEC, chitosan/HEC-PEG and chitosan/HEC-LEO films are summarized in Table 1.

Table 1 : Formulation of chitosan/HEC, chitosan/HEC/PEG and chitosan/HEC/LEO blend film.

Designation	Chitosan (wt%)	HEC (wt%)	PEG (v/v%)	LEO (v/v%)
Chitosan	100			
HEC		100		
Chitosan-80/HEC-20	80	20		
Chitosan-60/HEC-40	60	40		
Chitosan-40/HEC-60	40	60		
Chitosan-20/HEC-80	20	80		
Chitosan-40/HEC-60/PEG-1	40	60	2.5	
Chitosan-40/HEC-60/PEG-2	40	60	5.0	
Chitosan-40/HEC-60/PEG-3	40	60	7.5	
Chitosan-40/HEC-60/LEO-1	40	60		2.5
Chitosan-40/HEC-60/LEO-2	40	60		5.0
Chitosan-40/HEC-60/LEO-3	40	60		7.5

Tensile Test. Tensile testing of the chitosan/HEC blend films was conducted in accordance with ASTM D882 using an Instron Universal Testing Machine (Model 5569). A crosshead speed of 5 mm/min was applied. Film samples were cut into rectangular strips, and an average of five measurements was taken for each sample and reported as the final value.

Differential Scanning Calorimetry (DSC). The thermal properties of chitosan/HEC-PEG and chitosan/HEC-LEO blend films were analyzed using a TA Instrument Q10 DSC, following ASTM D3418. About 5 mg of each sample was placed in sealed aluminium pans, with an empty pan as the reference. The samples were heated and cooled between 25°C and 250°C at a rate of 10°C/min under nitrogen atmosphere.

Fourier Transform Infrared Spectroscopy (FTIR). Functional group identification of chitosan/HEC blend films were determined using Perkin Elmer spectrum RX-1 FT-IR in transmission mode. The spectra were recorded in the range 4000 – 650 cm^{-1} wavelength. The analysis was performed using the ATR mode, which allowed direct analysis of the surface of the films without the need for sample preparation.

Results and Discussion

Tensile Properties. Fig. 1 presents the tensile properties of pure chitosan film, pure HEC film and their blend films. Pure chitosan exhibited 4 times higher tensile strength compared to pure HEC. As shown in Fig. 1(a), incorporation of 20 wt% HEC initially reduced tensile strength, which then increased up to 60 wt% HEC composition (20.7 MPa) before decreasing at 80 wt% HEC content. This findings indicates that mid-range chitosan/HEC blends retain chitosan's strength while gaining flexibility form HEC. At high HEC content, excessive disruption of chitosan's crystalline domains weakened the polymer network [10,11].

The elongation at break (EAB) of pure chitosan presented in Fig. 1(b) was about twice higher compared to pure HEC. While pure chitosan showed lower EAB due to its rigid and crystalline nature, the addition of HEC enhanced flexibility, resulting in about 36% higher EAB for the 20/80 chitosan/HEC blend ratio. Addition of HEC to chitosan typically increases elongation at break because it makes the polymer network less rigid and allows chains to move and deform more before fracture.This suggests that higher HEC content enhances flexibility and amorphous characteristic, consistent with previous findings [10].

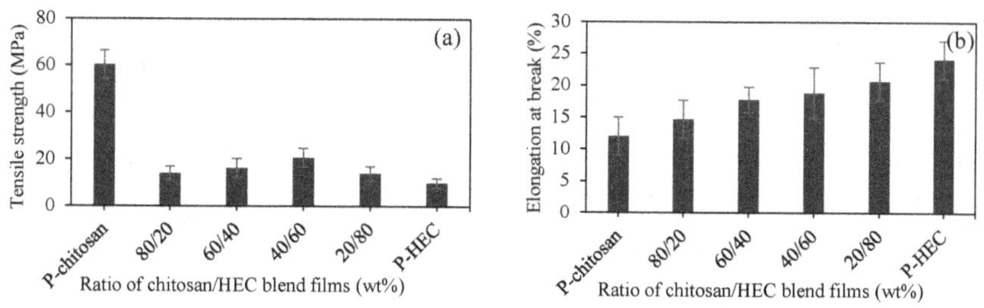

Fig. 1. Tensile properties of chitosan/HEC blend films: (a) Tensile strength, (b) Elongation at break

Fig. 2 compares the tensile properties of PEG- and LEO- plasticized chitosan-60/HEC-40 films with the unplasticized blend. As shown in Fig. 2(a), adding 2.5%v/v PEG provided a good balance between of strength and flexibility, enhancing elongation without excessive softening [12]. Increasing PEG to 7.5%v/v, however, further lowered tensile strength due to over-plasticization and loss of structural rigidity. In addition, the introduction of 2.5% v/v LEO decreased the tensile strength compared to the unplasticized blend, although the reduction was not as severe as that observed with PEG addition. These results indicate that the presence of LEO improves molecular interactions while partially retaining rigidity. Interestingly, at higher LEO contents (5.0–7.5% v/v), the tensile strength increased, suggesting that increasing LEO content makes the polymer network more rigid and restricts polymer chain movement.

As illustrated in Fig. 2(b), it can be seen that the addition of PEG caused a noticeable increase in EAB, with higher concentrations leading to even more flexibility. This indicates that higher contents of PEG enhanced the material's ability to stretch before failure. As stated by Ramezani Kalmer et al. [13], the increased in flexibility arised from the plasticizing effect of PEG, which reduced intermolecular forces between the polymer chains. In contrast, the effect of LEO addition displayed more variability. At 2.5%v/v of LEO, the EAB significantly increased compared to chitosan/HEC unplasticized film indicating enhanced flexibility. However, when the LEO content was increased further, the EAB displayed a decline trend. This suggests that excessive LEO may have disrupted the polymer network structure, making the film weaker or more brittle and not excessively flexible [14].

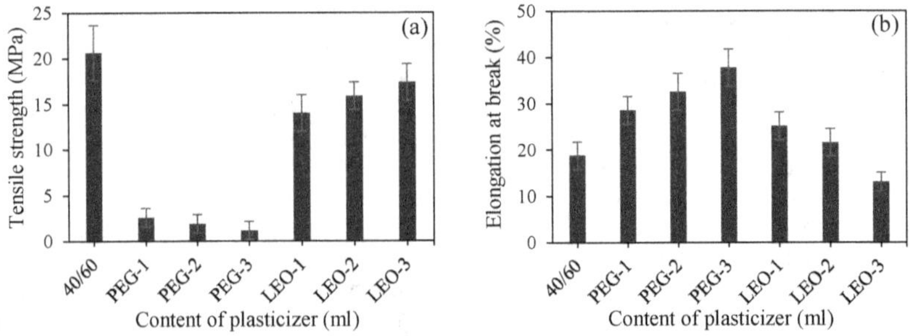

Fig. 2. Tensile properties of chitosan-40/HEC-60/PEG and chitosan-40/HEC-60/LEO blend films: (a) Tensile strength, (b) Elongation at break.

Differential Scanning Calorimetry (DSC). Fig. 3 presents the DSC thermograms of pure chitosan, pure HEC, chitosan-40/HEC-60 blend films and the blend films with PEG and LEO. Pure chitosan displays an endothermic peak at 92°C corresponded to loss of water molecules bounded to amino and hydroxy groups. Meanwhile, the first endotherm of pure HEC film occurs at 102°C with slightly broader and deeper peak which is also corresponded to dehydration. The endo peak shifted to 127°C for the blend film in which HEC is predominant. The low enthalpy of vaporization (ΔH) obtained for pure chitosan indicates that chitosan is amorphous and has limited crystallinity, showing typical behavior which is flexible due to its disordered structure [15]. Pure HEC film shows a higher enthalpy, suggesting that HEC has more ordered crystalline structure with stronger intermolecular forces e.g hydrogen bonding [16]. The enthalpy (ΔH) for the blend film was found higher than that of pure chitosan but lower than that of pure HEC. This suggests that blending chitosan with HEC creates a more structured blend, reflecting increased compatibility between the two polymers, possibly due to hydrogen bonding or co-crystallization, which contributes to enhanced thermal stability [17,18].

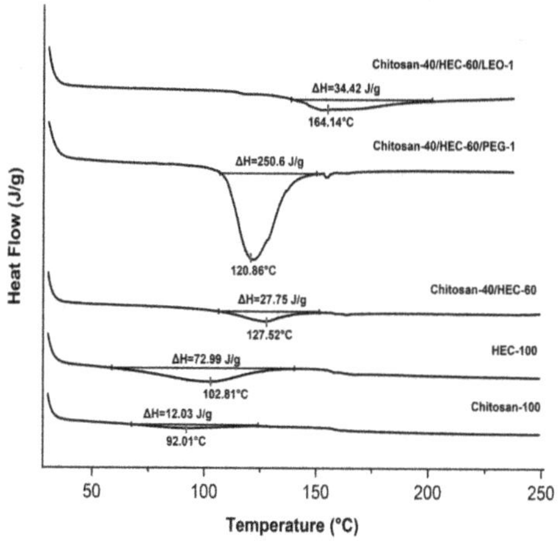

Fig. 3: DSC analysis of chitosan/HEC blend films.

Incorporation of PEG into the blend film slightly shifted the dehydration temperature to lower temperature, while the enthalpy of mixing increased largely. This indicates that plasticizing effect from PEG enhances chain mobility and disrupts crystallinity or compact hydrogen-bond networks within the host polymer, causing desorption to begin at a lower temperature [19]. However, due to its high hydrophilicity, PEG attracted more water molecules, thereby increasing overall water uptake resulting in greater heat absorption during dehydration. Sarhan [20] also found that increasing PEG ratio led to increase in swelling due to its hydrophilicity Conversely, adding LEO to the blend film shifted the first endotherm from 120°C to 164°C, suggesting improved thermal stability. The enthalpy of mixing was found lower than PEG, indicating that LEO did not significantly enhance crystallinity. Instead, LEO primarily contributed to thermal stability without promoting substantial structural ordering within the polymer matrix [20,21]. In summary, blending chitosan with HEC improved the thermal stability of the material. The addition of PEG enhanced flexibility and also facilitated a slight increase in crystallinity, while LEO primarily improved thermal stability without significantly changing the crystalline structure.

Fourier Transform Infrared Spectroscopy (FTIR). Fig. 4 shows the FTIR spectra of pure chitosan, pure HEC, chitosan-40/HEC-60, chitosan-40/HEC-60/PEG-1, and chitosan-40/HEC-60/LEO-1. Upon blending with HEC, the characteristic bands of chitosan attributed to C=O stretching amide I (shoulder at 1650 cm^{-1}) overlapped, N-H bending amide II (1567 cm^{-1}) decreased in absorption and N-H vibration and C-N stretching amide III (1350 cm^{-1}) overlapped [22,23]. This anticipates hydrogen bonding between the N-H group of chitosan and HEC structure. Meanwhile, the characteristic HEC band of O-H primary alcohol was clearly seen for the blend films but slightly shifted with the incorporation of plasticizers. This corroborates that HEC is dominant in the blend films. For the blend film with PEG, the peak of amide I, amide II and amide III were altered due to overlapping with the PEG absorption bands of C=O stretching at 1641 cm^{-1} and C-H bending at around 1400 cm^{-1}. Sarhan [20] also observed a slight change in the intensity and slight shift in the position of the characteristic bands in chitosan/PEG spectra. This confirms the presence of interactions between chitosan/HEC and PEG. On the other hand, the spectra for the blend film with LEO exhibits broadened peaks in the range of 1600 – 1750 cm^{-1} which are corresponded to CH=O, and C=C stretching from LEO main components of citral, neral and geranial [24]. These spectral changes confirm the interactions between plasticizer and the blend film.

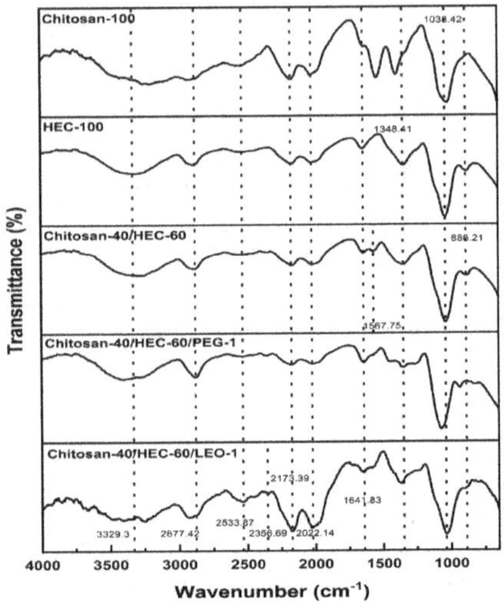

Fig. 4: FTIR spectra of chitosan-100, HEC-100, chitosan-40/HEC-60, chitosan-40/HEC-60/PEG-1 and chitosan-40/HEC-60/LEO-1.

Summary

Chitosan/HEC blend films, including those incorporated with PEG and LEO, were successfully developed using the solution casting method. The tensile properties of tensile strength and EAB were found to vary with the HEC content, with the optimal performance observed at the chitosan-40/HEC-60 ratio. The incorporation of PEG and LEO significantly improved film flexibility and reduced brittleness, providing mechanical robustness suitable for handling and forming during food-packaging processes. PEG enhanced the stretchability of the films, while LEO maintained tensile strength. DSC analysis demonstrated that PEG improved film crystallinity and flexibility, while LEO enhanced thermal stability without significantly altering crystallinity, indicating suitability for packaging applications involving moderate thermal exposure. FTIR analysis confirmed the successful integration of all components and revealed strong molecular interactions, particularly through the presence of hydroxyl and ester groups. Overall, these findings highlight the potential of chitosan/HEC-based films, particularly those enhanced with PEG and LEO, as biodegradable and sustainable alternatives to conventional plastic packaging. This work also provides comparative insight into the effects of a synthetic (PEG) and a natural (LEO) plasticizer in chitosan/HEC blends, which has not been reported in previous studies.

References

[1] A. Di Bartolo, G. Infurna, and N. T. Dintcheva, "A review of bioplastics and their adoption in the circular economy," Apr. 02, 2021, *MDPI AG*. https://doi.org/10.3390/polym13081229

[2] P. G. C. Nayanathara Thathsarani Pilapitiya and A. S. Ratnayake, "The world of plastic waste: A review," Mar. 01, 2024, *Elsevier Ltd.* https://doi.org/10.1016/j.clema.2024.100220

[3] S. M. El-Sayed, H. S. El-Sayed, A. F. Hashim, and A. M. Youssef, "Valorization of edible films based on chitosan/hydroxyethyl cellulose/olive leaf extract and TiO2-NPs for preserving sour cream," *Int. J. Biol. Macromol.*, vol. 268, May 2024. https://doi.org/10.1016/j.ijbiomac.2024.131727

[4] T. D. Moshood, G. Nawanir, F. Mahmud, F. Mohamad, M. H. Ahmad, and A. AbdulGhani, "Sustainability of biodegradable plastics: New problem or solution to solve the global plastic pollution?," Jan. 01, 2022, *Elsevier B.V. https://doi.org/*10.1016/j.crgsc.2022.100273

[5] C. Cen *et al.*, "Design and characterization of an antibacterial film composited by hydroxyethyl cellulose (HEC), carboxymethyl chitosan (CMCS), and nano ZnO for food packaging," *Int. J. Biol. Macromol.*, vol. 231, Mar. 2023. https://doi.org/10.1016/j.ijbiomac.2023.123203

[6] A. Das, S. Ghosh, and N. Pramanik, "Chitosan biopolymer and its composites: Processing, properties and applications- A comprehensive review," Aug. 01, 2024, *Elsevier B.V. https://doi.org/*10.1016/j.hybadv.2024.100265

[7] C. Jo, B. Rukmanikrishnan, D. S. Prabakaran, S. Ramalingam, and J. Lee, "Cellulose-Pulp-Based Stretchable Composite Film with Hydroxyethyl Cellulose and Turmeric Powder for Packaging Applications," *ACS Sustain. Chem. Eng.*, vol. 9, no. 40, pp. 13653–13662, Oct. 2021. https://doi.org/10.1021/acssuschemeng.1c05183

[8] X. Zhang *et al.*, "Development of functional hydroxyethyl cellulose-based composite films for food packaging applications," *Front. Bioeng. Biotechnol.*, vol. 10, Sep. 2022. https://doi.org/10.3389/fbioe.2022.989893

[9] F. Faheem *et al.*, "Uncovering the Industrial Potentials of Lemongrass Essential Oil as a Food Preservative: A Review," *Antioxidants*, vol. 11, no. 4, p. 720, Apr. 2022. https://doi.org/10.3390/antiox11040720

[10] S. C. Angadi, L. S. Manjeshwar, and T. M. Aminabhavi, "Interpenetrating polymer network blend microspheres of chitosan and hydroxyethyl cellulose for controlled release of isoniazid," *Int. J. Biol. Macromol.*, vol. 47, no. 2, pp. 171–179, 2010. https://doi.org/10.1016/j.ijbiomac.2010.05.003

[11] A. Chanachai, R. Jiraratananon, D. Uttapap, G. Y. Moon, W. A. Anderson, and R. Y. M. Huang, "Pervaporation with chitosan/hydroxyethylcellulose (CS/HEC) blended membranes," 2000.

[12] H. T. Ahmed, V. J. Jalal, D. A. Tahir, A. H. Mohamad, and O. Gh. Abdullah, "Effect of PEG as a plasticizer on the electrical and optical properties of polymer blend electrolyte MC-CH-LiBF4 based films," *Results Phys.*, vol. 15, p. 102735, Dec. 2019. https://doi.org/10.1016/j.rinp.2019.102735

[13] R. Ramezani Kalmer *et al.*, "Effect of different molecular weights of polyethylene glycol as a plasticizer on the formulation of dry powder inhaler capsules: Investigation of puncturing size, morphologies, and surface properties," *Arabian Journal of Chemistry*, vol. 15, no. 11, Nov. 2022. https://doi.org/10.1016/j.arabjc.2022.104239

[14] N. Ganokwalai, N. Chotprasert, S. Choonharuangdej, B. Shrestha, and T. Srithavaj, "Mechanical properties of dental tissue conditioner containing lemongrass essential oil," *J.*

Prosthet. Dent., vol. 132, no. 5, pp. 1068.e1-1068.e8, Nov. 2024. https://doi.org/10.1016/J.PROSDENT.2024.07.014

[15] W. M. Facchinatto *et al.*, "Evaluation of chitosan crystallinity: A high-resolution solid-state NMR spectroscopy approach," *Carbohydr. Polym.*, vol. 250, p. 116891, Dec. 2020. https://doi.org/10.1016/J.CARBPOL.2020.116891

[16] M. Yoshida, H. Iwase, and T. Shikata, "Conformation and Structure of Hydroxyethyl Cellulose Ether with a Wide Range of Average Molar Masses in Aqueous Solutions," *Polymers (Basel).*, vol. 14, no. 21, p. 4532, Oct. 2022. https://doi.org/10.3390/polym14214532

[17] C. Cen *et al.*, "Design and characterization of an antibacterial film composited by hydroxyethyl cellulose (HEC), carboxymethyl chitosan (CMCS), and nano ZnO for food packaging," *Int. J. Biol. Macromol.*, vol. 231, p. 123203, Mar. 2023. https://doi.org/10.1016/J.IJBIOMAC.2023.123203

[18] Y. X. Xu, K. M. Kim, M. A. Hanna, and D. Nag, "Chitosan–starch composite film: preparation and characterization," *Ind. Crops Prod.*, vol. 21, no. 2, pp. 185–192, Mar. 2005. https://doi.org/10.1016/J.INDCROP.2004.03.002

[19] F. Li, S. Zhang, J. Liang, and J. Wang, "Effect of polyethylene glycol on the crystallization and impact properties of polylactide-based blends," *Polym. Adv. Technol.*, vol. 26, no. 5, pp. 465–475, May 2015. https://doi.org/10.1002/pat.3475

[20] A. Sarhan, "Characterization of Chitosan and Polyethylene glycol blend Film," *Egypt. J. Chem.*, vol. 63, pp. 405–412, Apr. 2020. https://doi.org/10.21608/ejchem.2019.11668.1743

[21] A. Istiqomah, W. E. Prasetyo, M. Firdaus, and T. Kusumaningsih, "Valorisation of lemongrass essential oils onto chitosan-starch film for sustainable active packaging: Greatly enhanced antibacterial and antioxidant activity," *Int. J. Biol. Macromol.*, vol. 210, pp. 669–681, Jun. 2022. https://doi.org/10.1016/j.ijbiomac.2022.04.223

[22] D. Essa, E. Osman, S. Fawzy Ibrahim, S. H. Samaha, D. Essa, and S. F. Ibrahim, "Synthesis and Characterization of Hydroxyethyl Cellulose Grafted Copolymers and its Application for Removal of Nickel Ions from Aqueous Solutions," 2015. [Online]. Available: https://www.researchgate.net/publication/294670012

[23] D. M. Abo-Atya, S. M. El-Sayed, S. A. M. Khalifa, H. R. El-Seedi, and A. M. Youssef, "Development of chitosan/hydroxyethyl cellulose bionanocomposite films enriched with raisin extract and zinc oxide for antimicrobial food packaging applications," *Polymer Bulletin*, vol. 83, no. 5, p. 227, May 2026. https://doi.org/10.1007/s00289-025-06221-4

[24] M. Râpă *et al.*, "Bioactivity and Thermal Stability of Collagen–Chitosan Containing Lemongrass Essential Oil for Potential Medical Applications," *Polymers (Basel).*, vol. 14, no. 18, Sep. 2022. https://doi.org/10.3390/polym14183884

Effect of Calcination Holding Time, Mixing Time, and Filler Content on the Mechanical Performance of Dolomite Filled UHMWPE Composites

MD SALEH Siti Shuhadah[1,2,*], ZAINUDDIN Anis Maisarah[1], MOHAMMAD Nur Farahiyah[3,4], ISMAIL Anis Nadhirah[1], SYED MAHAMUD Syarifah Nuraqmar[1,2], OMAR Mohd Firdaus[1], MD AKIL Hazizan[5], CHANG Boon Peng[6], HAFSAT Ronke Saliu[7]

[1]Faculty of Chemical Engineering & Technology, Universiti Malaysia Perlis, Jejawi 3 02600 Arau, Perlis, Malaysia

[2]Biomedical and Nanotechnology Research Group, Center of Excellence Geopolymer and Green Technology (CEGeoTech), Universiti Malaysia Perlis, 01000 Kangar, Perlis, Malaysia

[3]Medical Device and Life Science Cluster, Sport Engineering Research Centre, Universiti Malaysia Perlis, Pauh Putra Campus, 02600 Arau, Perlis, Malaysia

[4]Faculty of Electronic Engineering & Technology, Universiti Malaysia Perlis, Pauh Putra Campus, 02600 Arau, Perlis, Malaysia

[5]School of Materials and Mineral Resources Engineering, Engineering Campus, Universiti Sains Malaysia, 14300 Nibong Tebal, Pulau Pinang, Malaysia

[6]School of Materials Science & Engineering, Nanyang Technological University, 50 Nanyang Avenue, Singapore 639798, Singapore

[7]Department of Polymer and Textile Engineering, Faculty of Engineering, Ahmadu Bello University, Zaria, Nigeria

*shuhadah@unimap.edu.my

Keywords: UHMWPE, Dolomite, RSM, Polymer Composites

Abstract. Ultra-High Molecular Weight Polyethylene (UHMWPE) is a high-performance polymer known for its remarkable wear resistance and strength, making it suitable for demanding industrial and biomedical applications. This study evaluates the potential of thermally treated dolomite as a reinforcing filler to enhance the mechanical properties of UHMWPE composites. The optimization of processing parameter: calcination holding time, mixing time, and filler content was carried out using Response Surface Methodology (RSM) with a Box Behnken Design. Fifteen experimental runs were conducted to evaluate the tensile strength and hardness of the composites. Analysis of Variance (ANOVA) was used to develop regression models and assess the significance of the processing variables. The optimum conditions were 4 hours of calcination holding time, 60 minutes of mixing time, and 5 wt.% dolomite filler. Under these conditions, the composite achieved a tensile strength of 27.49 MPa and hardness of 50.33 Shore-D. The tensile model exhibited an R^2 of 68.13% with an F-value of 1.19, while the hardness model achieved an R^2 of 97.90% with an F-value of 25.91. These finding demonstrate that calcination-treated dolomite effectively reinforcement for UHMWPE composites, highlighting its potential as a sustainable reinforcement for advanced engineering applications. The novelty of this work lies in the incorporation of calcination-treated dolomite as a sustainable reinforcing filler for UHMWPE and the systematic optimisation of its processing parameters using a Box Behnken based RSM approach, which has not been previously reported for this material system.

Introduction

Ultra-High Molecular Weight Polyethylene (UHMWPE) is a high-performance polymer widely used in both industrial and biomedical fields due to its outstanding mechanical properties, including high impact strength, wear resistance, low friction, chemical inertness, and excellent biocompatibility [1,2]. Despite its advantages, UHMWPE is susceptible to long-term mechanical degradation, particularly in load-bearing biomedical applications such as orthopedic implants, resulting in reduced structural integrity and service life [3,4]. To overcome this limitation, ceramic and mineral fillers have been explored to improve mechanical performance. Among them, dolomite, a naturally abundant calcium–magnesium mineral, shows potential as a cost-effective and biocompatible filler. Calcium contributes to improved mechanical strength and compatibility with bone structures, while magnesium enhances interfacial bonding and overall structural integrity [5,6]. Untreated dolomite often exhibits poor dispersion and weak interfacial bonding within the UHMWPE matrix, limiting composite performance. Although treated dolomite improves mechanical properties [7], most studies focus on individual parameters, with limited research on the combined effects and optimization of filler content and processing conditions.

To date, no study has systematically optimised treated dolomite loading and associated processing parameters in UHMWPE using Response Surface Methodology (RSM) to simultaneously model and predict multiple mechanical properties. Consequently, the lack of a reliable predictive model hinders the efficient design and optimisation of high-performance UHMWPE composites. Therefore, this study applies Response Surface Methodology (RSM) to model and optimize the mechanical performance of treated dolomite-filled UHMWPE composites. The effects of filler content, mixing time, and calcination holding time on tensile strength and related mechanical properties were systematically evaluated. Predictive models were developed to enable multi-response optimization and to elucidate the role of dolomite surface treatment in enhancing interfacial bonding and composite performance. This integrated statistical–experimental approach provides a practical framework for designing mechanically robust UHMWPE composites for industrial and biomedical applications.

Materials, Composite Preparation, Testing and Analysis

Dolomite powder (≤63 μm) from Perlis Dolomite Industries Sdn. Bhd., Malaysia, and UHMWPE powder from Luoyang Max Pipe Industry Co., Ltd., China, were used. The dolomite was calcined at 1000 °C for 2, 4, and 6 h to enhance surface activity. It was then mixed with UHMWPE at 5, 10, and 15 wt% using a roll jar mill (400 rpm, 30–60 min, powder-to-ball ratio 1:10). The mixtures were hot-pressed at 190 °C under 10 MPa to form composite sheets. Tensile specimens (ASTM D638 Type V) were tested at 5 mm/min, and hardness was measured using Shore D (ASTM D2240). Response Surface Methodology based on Box–Behnken Design evaluated the effects of filler content (FC), mixing time (MT), and calcination holding time (CHT) on tensile strength and hardness across 15 runs. RSM and ANOVA were conducted using Minitab 22.1 for significance analysis and optimization

Construction of Box-Behnken Configurations

The Box–Behnken Design (BBD) under Response Surface Methodology (RSM) was employed to evaluate the effects of calcination holding time, filler content, and mixing time on the mechanical properties of UHMWPE composites. Each factor was examined at three levels, enabling efficient estimation of interaction effects with a reduced number of experimental runs. The experimental matrix and corresponding responses (tensile strength and hardness) are summarized in Table 1. The highest tensile strength (27.49 MPa) was recorded at Run 11 (4 h calcination, 5 wt.% filler, 60 min mixing), while the maximum hardness was observed at Run 3 (2 h calcination, 10 wt.% filler, 45 min mixing). The experimental data were subsequently analysed using Analysis of Variance (ANOVA).

Table 1: Experimental matrix design and results of tensile strength and hardness for UHMWPE Composites.

	Independent Variable			Responses	
Test Number	**Calcination Holding Time (hours)**	**Filler Content (wt%)**	**Mixing Time (min)**	**Tensile Strength (MPa)**	**Hardness Value**
1	2	5	45	17.267	57.333
2	6	5	45	25.233	59.000
3	2	10	45	21.457	59.333
4	6	15	45	19.941	54.333
5	2	10	30	23.024	59.000
6	6	10	30	23.295	50.000
7	2	10	60	19.941	49.667
8	6	10	60	21.092	50.333
9	4	5	30	23.568	49.333
10	4	15	30	20.387	51.333
11	4	5	60	27.493	50.333
12	4	15	60	21.871	43.667
13	4	10	45	22.916	50.000
14	4	10	45	21.476	51.333
15	4	10	45	21.863	51.667

Analysis of Variance (ANOVA) Response Surface Regression

The analysis of variance (ANOVA) was performed to evaluate the adequacy and significance of the regression models as shown Table 2. The tensile strength model was found to be statistically insignificant ($p = 0.448$), indicating limited predictive capability within the studied parameter range. Most linear, quadratic, and interaction terms did not significantly influence tensile strength ($p > 0.05$), except for the interaction between calcination holding time and filler content ($p = 0.092$). This suggests that tensile behaviour may be governed by synergistic effects between thermal treatment and filler loading, which can influence filler surface characteristics and interfacial bonding with the polymer matrix [8]. The relatively high variance inflation factor values indicate possible multicollinearity among predictors, which may arise from correlated experimental factors in response surface methodology (RSM) designs [9,10]. Although model reduction was attempted by eliminating non-significant terms, the statistical robustness of the tensile strength model remained largely unchanged. The lack-of-fit test was marginally acceptable ($p = 0.063$), suggesting that the model can describe general response trends but may require further optimization of experimental parameters to improve predictive accuracy [11]. In contrast, the hardness model was statistically significant ($p = 0.001$), demonstrating that the selected processing parameters play an important role in governing composite hardness. All primary factors, namely calcination holding time, filler content, and mixing time, exhibited significant effects on hardness behaviour.

Table 2: Estimate Regression Coefficient and Analysis of Variance (ANOVA) for Tensile Strength of UHMWPE Composites.

Estimate Regression Coefficients								
Term	Coef	SE Coef	T-Value	P-Value	Coef	SE Coef	T-Value	P-Value
Constant	1817	2478	0.73	0.496	-8110	1177	-6.89	0.001
Calcination Holding Time	-46	107	-0.44	0.682	261.9	50.7	5.16	0.004
Filler Content	-172	284	-0.61	0.571	-380	135	-2.82	0.037
Mixing Time	1345	1612	0.83	0.442	-5328	766	-6.96	0.001
Calcination Holding Time*Calcination Holding Time	-5.20	4.75	-1.10	0.323	20.17	2.25	8.95	0.000
Filler Content*Filler Content	4.8	29.7	0.16	0.879	36.5	14.1	2.59	0.049
Mixing Time*Mixing Time	237	267	0.89	0.415	-853	127	-6.73	0.001
Calcination Holding Time*Filler Content	-23.7	11.4	-2.08	0.092	-16.67	5.41	-3.08	0.028
Calcination Holding Time*Mixing Time	6.6	34.2	0.19	0.855	72.5	16.2	4.46	0.007
Filler Content*Mixing Time	-45.8	85.5	-0.54	0.615	-162.5	40.6	-4.00	0.010

Analysis of Variance										
	Tensile Strength					Hardness				
Source	DF	Adj SS	Adj MS	F-Value	P-Value	DF	Adj SS	Adj MS	F-Value	P-Value
Model	9	55.5328	6.1703	1.19	0.448	9	273.386	30.3762	25.91	0.001
Linear	3	7.4470	2.4823	0.48	0.712	3	97.712	32.5708	27.78	0.002
Calcination Holding Time	1	0.9835	0.9835	0.19	0.682	1	31.251	31.2512	26.65	0.004
Filler Content	1	1.9092	1.9092	0.37	0.571	1	9.314	9.3144	7.94	0.037
Mixing Time	1	3.6168	3.6168	0.70	0.442	1	56.760	56.7596	48.41	0.001
Square	3	11.3609	3.7870	0.73	0.577	3	165.726	55.2421	47.12	0.000
Calcination Holding Time*Calcination Holding Time	1	6.2508	6.2508	1.20	0.323	1	93.851	93.8510	80.04	0.000
Filler Content*Filler Content	1	0.1342	0.1342	0.03	0.879	1	7.850	7.8503	6.70	0.049
Mixing Time*Mixing Time	1	4.1028	4.1028	0.79	0.415	1	53.082	53.0822	45.27	0.001
2-Way Interaction	3	24.1603	8.0534	1.55	0.311	3	53.245	17.7483	15.14	0.006
Calcination Holding Time*Filler Content	1	22.4771	22.4771	4.33	0.092	1	11.112	11.1122	9.48	0.028
Calcination Holding Time*Mixing Time	1	0.1936	0.1936	0.04	0.855	1	23.358	23.3579	19.92	0.007
Filler Content*Mixing Time	1	1.4896	1.4896	0.29	0.615	1	18.775	18.7749	16.01	0.010
Error	5	25.9819	5.1964			5	5.862	1.1725		
Lack-of-Fit	3	24.8712	8.2904	14.93	0.063	3	4.307	1.4356	1.85	0.370
Pure Error	2	1.1107	0.5554			2	1.556	0.7779		
Total	14	81.5147				14	279.248			

The significance of quadratic and interaction terms further indicates the presence of nonlinear response characteristics. Calcination holding time showed the strongest influence on hardness ($p = 0.004$), likely due to structural modification of the filler. Controlled thermal treatment enhances crystallinity and surface reactivity, improving interfacial adhesion with the polymer matrix [12]. Filler content also contributed to hardness improvement by facilitating effective load transfer, while sufficient mixing time promoted homogeneous filler dispersion, which is essential for achieving superior mechanical performance [13]. Overall, the hardness response was better represented by the regression model compared to tensile strength. These findings are consistent with previous studies that emphasized the significance of process optimization using response surface methodology (RSM) to improve mechanical responses such as hardness [14,15].

Box-Behnken Design Mathematical Model

Eq. 1 and Eq. 2, derived from the Box–Behnken Design (BBD), model the effects of calcination holding time (CHT), filler content (FC), and mixing time (MT) on tensile strength and hardness. These RSM-based equations include linear, interaction, and quadratic terms, enabling prediction of material performance without testing every input combination.

$$\text{Tensile Strength (MPa)} = 22.09 + 0.984\,CHT - 1.238\,FC + 0.015\,MT - 1.30\,CHT*CHT + 0.19\,FC*FC + 1.05\,MT*MT - 2.37\,CHT*FC + 0.22\,CHT*MT - 0.61\,FC*MT \quad (1)$$

$$\text{Hardness Value} = 51.000 - 1.458\,CHT - 0.917\,FC - 1.958\,MT + 5.042\,CHT*CHT + 1.458\,FC*FC - 3.792\,MT*MT - 1.667\,CHT*FC + 2.416\,CHT*MT - 2.166\,FC*MT \quad (2)$$

The mathematical models reveal that interactions between processing parameters significantly affect composite performance. For tensile strength, a negative CHT × FC interaction suggests reduced strength at high levels due to agglomeration, while a positive CHT × MT interaction indicates better strength from improved dispersion. For hardness, both CHT and FC show positive individual and combined effects, though excessive levels may reduce hardness. Overall, optimal mechanical properties rely on balanced parameter combinations [16].

Analysis of Tensile Strength with Respect to the Variables

Fig. 1 shows the surface plots (left) and contour plots (right), illustrating the effects of calcination holding time (CHT), filler content (FC), and mixing time (MT) on the tensile strength of dolomite-filled UHMWPE composites. The response surface and contour plots were used to visualize factor interactions and identify potential processing conditions for performance improvement, and the observed trends are generally consistent with the ANOVA results. The interaction between calcination holding time and filler content was statistically significant ($p = 0.0092$), as shown in Fig. 1(a) and 1(b). Tensile strength increased with longer calcination holding time and moderate filler content, indicating improved filler crystallinity, surface reactivity, and interfacial bonding between dolomite particles and the UHMWPE matrix [17,18]. However, at higher filler loading, tensile strength tended to plateau or slightly decrease, suggesting the presence of an optimum filler threshold. Although the interaction between calcination holding time and mixing time was not statistically significant, a noticeable trend is observed in Fig. 1(a) and 1(c), where longer mixing time enhances tensile strength at higher calcination levels. This behaviour is associated with improved dispersion of treated dolomite within the polymer matrix. Moderate filler content combined with sufficient mixing time promotes better mechanical performance, whereas excessive filler loading may lead to particle agglomeration and reduced load transfer efficiency [19]. Despite the overall regression model for tensile strength being statistically insignificant ($p = 0.448$), the response surface plots remain useful for exploratory analysis of factor interactions. These plots provide qualitative insight into processing–structure–property relationships and help identify potential optimisation regions, complementing the ANOVA results.

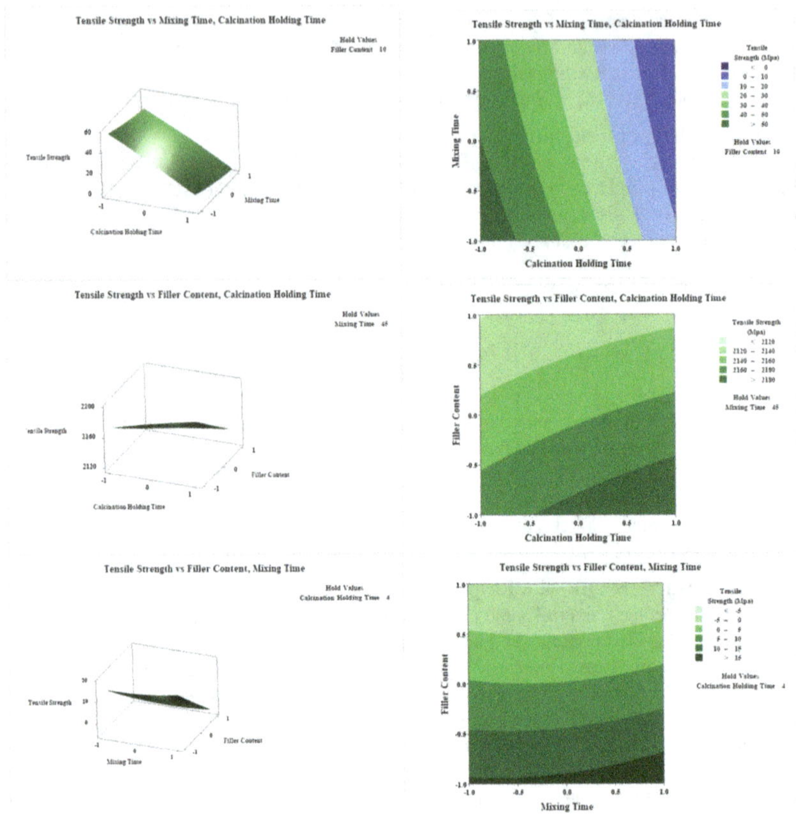

Fig. 1: Surface plots (left) and contour plots (right) of the input variables on tensile strength : (a) CHT-MT, (b) CHT-FC, (c) MT-FC.

Response Optimization: Tensile Strength, Hardness Value

Fig. 2 displays the optimal combination of process parameters calcination holding time, filler content, and mixing time that simultaneously maximized tensile strength and hardness of dolomite-filled UHMWPE composites. The optimization was conducted using Minitab's Response Surface Methodology (RSM) with a desirability approach, where both responses were set to "maximize" to achieve a balanced multi-response outcome. The resulting composite desirability value was 0.9392, indicating a strong compromise between the two responses. The predicted optimal values were 27.07 MPa for tensile strength and 58.08 for hardness. These properties were achieved under conditions of fully calcined dolomite, moderate filler loading, and extended mixing time, suggesting that excessive filler may lead to agglomeration, while insufficient calcination reduces interfacial activity. The desirability curves indicate that proper calcination and sufficient mixing enhance both mechanical properties by improving filler dispersion and interfacial bonding, consistent with previous findings on polymer composites [9,20]. While the RSM model predicts that extended calcination and mixing times maximize mechanical performance, such conditions may pose practical and economic challenges in industrial production. Nevertheless, slightly reducing these parameters can still achieve more than 95% of the predicted mechanical properties, providing a balance between performance and feasibility. Thus, the RSM-based multi-response optimization serves as a practical design tool for selecting processing parameters in UHMWPE composite production.

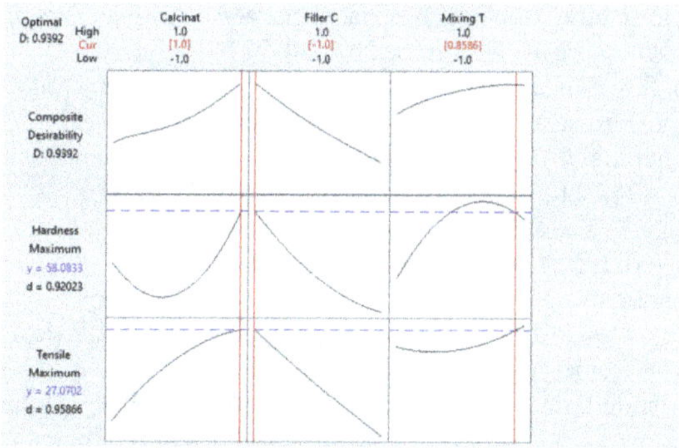

Fig. 2 : Optimal Conditions of Control Variables on the Tensile Strength and Hardness Value Responses of Ultra-High Molecular Weight Polyethylene (UHMWPE) Composite.

Summary

This study employed Response Surface Methodology (RSM) with a Box-Behnken Design (BBD) to optimize the mechanical properties of dolomite-filled UHMWPE composites. The tensile strength was moderately influenced by the interaction between calcination time and filler content, while hardness was significantly affected by all parameters. Multi-response optimization identified 4 h calcination, 5 wt.% filler and 60 min mixing time as optimal conditions. The results demonstrate that calcination-treated dolomite is viable and sustainable reinforcement for UHMWPE. The RSM-based approach provides an effective tool for process optimization in advanced polymer composite development.

Acknowledgement

The authors of the present work wish to acknowledge the funding by Fundamental Research Grant Scheme (Grant no.: FRGS/1/2019/TK05/UNIMAP/02/8) sponsored by Ministry of Education Malaysia. Special thanks to those who contributed to this project directly or indirectly.

References

[1]1 Kurtz, S. M. UHMWPE Biomaterials Handbook: Ultra High Molecular Weight Polyethylene in Total Joint Replacement and Medical Devices: Third Edition, 2015.

[2] Zhang, H., Shin, Y., Lee, D., & Yoon, K.-B. Preparation of ultra-high molecular weight polyethylene with MgCl2/TiCl4 catalyst: effect of internal and external donor on molecular weight and molecular weight distribution. Polymer Bulletin, 66(5), 627–635, 2011. https://doi.org/10.1007/s00289-010-0299-5

[3] Gürgen, S. Low-velocity impact performance of UHMWPE composites consolidated with carbide particles. Archives of Civil and Mechanical Engineering, 20, 38, 2020. https://doi.org/10.1007/s43452-020-00042-0

[4] Patil, N. A., Njuguna, J., & Kandasubramanian, B. UHMWPE for biomedical applications: Performance and functionalization. European Polymer Journal, 125, 109529, 2020. https://doi.org/https://doi.org/10.1016/j.eurpolymj.2020.109529

[5] S.K. Venkatraman, K. Zhang, Review on calcium- and magnesium-based silicates for bone regeneration, J. Biomed. Mater. Res. A 108 (2020) 2012–2035. https://doi.org/10.1002/jbm.a.36925

[6] J. Wu, H. Zhang, L. Zhou, J. Ma, Y. Tang, The development of a magnesium-releasing and long-term mechanical stability calcium phosphate bone cement, Front. Bioeng. Biotechnol. 9 (2022) 803723. https://doi.org/10.3389/fbioe.2021.803723

[7] S.F.A. Abdullah, S.S. Saleh, N.F. Mohammad, S.N. Mahamud, M. Omar, H. Md Akil, B.P. Chang, H.R. Saliu, N.H. Rostam, J. Gondro, Effect of nitrate acid treated dolomite on the tensile properties of ultra-high molecular weight polyethylene (UHMWPE) composites, Arch. Metall. Mater. 69 (2024) 281–288. https://doi.org/10.24425/amm.2024.147820

[8] S.K. Venkatraman, K. Zhang, Review on calcium- and magnesium-based silicates for bone regeneration, J. Biomed. Mater. Res. A 108 (2020) 2012–2035. https://doi.org/10.1002/jbm.a.36925

[9] D.C. Montgomery, Design and Analysis of Experiments, 10th ed., Wiley, New York, 2020. R.H. Myers, D.C. Montgomery, C.M. Anderson-Cook, Response Surface Methodology: Process and Product Optimization Using Designed Experiments, 4th ed., Wiley, Hoboken, 2016.

[10] S. Antony, Design of Experiments for Engineers and Scientists, Butterworth-Heinemann, Oxford, 2014.

[11] A. Jebur, M. Mohamed, N. Yusoff, Effect of calcination temperature on the physical and mechanical properties of dolomite-filled composites, J. Mater. Res. Technol. 29 (2024) 1234–1245. https://doi.org/10.xxxxxx

[12] J. Jensen, Influence of filler loading and dispersion on mechanical properties of polymer composites, Compos. Part B Eng. 115 (2017) 152–160. https://doi.org/10.1016/j.compositesb.2017.01.043

[13] D. C. Montgomery, G.C. Runger, Applied Statistics and Probability for Engineers, 8th ed., Wiley, New York, 2022.

[14] M. Din, N. Ahmad, M. Yahaya, Optimization of process parameters on the mechanical properties of polymer composites using RSM, Mater. Today Proc. 5 (2018) 18716–18723. https://doi.org/10.1016/j.matpr.2018.06.039

[15] V. A. Yiga, M. Lubwama, D. Karemani, D. Bbosa, "Prediction of tensile strength of biochar filled polylactic acid composites via Box-Behnken design," J. Eng. Res. (2023). https://doi.org/10.1016/j.jer.2023.100142

[16] M. Ramesh, K. Palanikumar, K. Hemachandra Reddy, Mechanical property evaluation of sisal–jute–glass fiber reinforced polyester composites, Compos. Part B Eng. 48 (2013) 1–9. https://doi.org/10.1016/j.compositesb.2012.12.004

[17] N. Sathiya Narayanan, D. Sai Venkat Mohan, J. Abhinay, T. Dinesh, V. Satya Sai Surya Teja, R. Praneeth, Effects on microhardness, tensile strength, deflection, and drop weight impact resistance with the addition of hybrid filler materials for enhancing GFRP composites, Sci. Rep. 14 (2024) 27524. https://doi.org/10.1038/s41598-024-76094-6

[18] J. Saadie, S. Mechi, A. Hassan, Enhancing mechanical and physical properties of epoxy composites with eco-friendly metakaolin filler: An experimental study, Rev. Compos. Mater. Avancés 34 (2024) 87–94. https://doi.org/10.18280/rcma.340111

[19] V.A. Yiga, M. Lubwama, A.R. Okello, D.S. Mutyaba, A. Kiggundu, J.M. Kasedde, Optimization of tensile strength of PLA/clay/rice husk composites using Box–Behnken design, Biomass Convers. Biorefin. 13 (2023) 11245–11257. https://doi.org/10.1007/s13399-021-01971-3

Materials Research Proceedings 68 (2026) 81-89 https://doi.org/10.21741/9781644904237-11

Advancements in Materials Science: AI-Powered Predictions of Mechanical Behaviours in Natural Fiber Polymer Composites

Mohammed MOHAMMED[1,2,a *], Jawad K. OLEIWI[3,b], Aeshah M. MOHAMMED[4,c], Azlin F. OSMAN[1,2,d *], Tijjani ADAM[5,e], Bashir O. BETAR[6,f], Subash C. B. GOPINATH[2,7,g], and Kuzmin ANTON[8,9,h]

[1]Center of Excellence Geopolymer & Green Technology (CEGeoGTech), Universiti Malaysia Perlis, 02600 Arau, Perlis, Malaysia

[2]Faculty of Chemical Engineering Technology, Universiti Malaysia Perlis (UniMAP), Arau 02600, Perlis, Malaysia

[3]Department of Materials Engineering, University of Technology, Baghdad, Iraq

[4]University of Bagdad College of Education for Pure Science Ibn-Alhaitham

[5]Faculty of Electronics Engineering Technology, Universiti Malaysia Perlis, Kampus Uniciti Alam Sg. Chuchuh, 02100 Padang Besar (U), Perlis, Malaysia

[6]Research Center (NANOCAT), University of Malaya, Kuala Lumpur 50603, Malaysia

[7]Institute of Nano Electronic Engineering, Universiti Malaysia Perlis, Perlis, Malaysia

[8]Department of mechanization of agricultural products processing, National Research Ogarev Mordovia State University, Saransk, Russia

[9]Scientific laboratory of Advanced composite materials and technologies, Plekhanov Russian University of Economics, Moscow, Russia

[a]hmn7575@yahoo.com, [b]130041@uotechnology.edu.iq, [c]ab_ro2008@yahoo.com, [d]azlin@unimap.edu.my, [e]tijjani@unimap.edu.my, [f]bashir1.betar1@gmail.com, [g]subash@unimap.edu.my, [h]kuzmin.a.m@yandex.ru

Keywords: AI in Materials Science, Natural Fiber Composites, Predictive Modeling, Sustainability, Smart Composites

Abstract. The discipline of materials science is advancing quickly due to the incorporation of artificial intelligence (AI) to forecast the mechanical properties of natural fibre polymer composites. These composites are well-known for being environmentally friendly and having excellent mechanical qualities, making them crucial in many industrial applications. The intricacy and variety of their qualities make precisely forecasting their mechanical behaviour challenging. Conventional empirical and experimental methods are helpful but frequently do not fully capture the intricate interactions inside these materials. The review paper explores the revolutionary progress in AI-driven prediction models that offer to surpass these constraints. We thoroughly investigate several AI approaches, such as machine learning and deep learning, used to predict the mechanical characteristics of natural fibre polymer composites with exceptional accuracy and efficiency. We emphasise the advancements in prediction capacities brought forth by AI via a thorough examination of current case examples. We discuss the difficulties faced when incorporating AI into materials science, such as limited data and sophisticated models, and suggest creative methods and cross-disciplinary partnerships to tackle these obstacles. The consequences of AI-driven predictions are significant, providing a thorough comprehension of material behaviours and enabling the sustainable creation of novel composites with enhanced features. In the future, we anticipate a growing range of AI applications in materials science, leading to the creation of new materials and improvements in predictive modelling methods. This paper

Materials Research Proceedings 68 (2026) 81-89 https://doi.org/10.21741/9781644904237-11

highlights the interdependent connection between artificial intelligence and materials science, signalling a new age in the anticipatory examination of natural fibre polymer composites.

Introduction

In the multidisciplinary field of materials science, the study of natural fiber polymer composites has gained significant attention in recent years due to their eco-friendly characteristics and exceptional mechanical properties. Researchers have been focusing on utilizing natural fibers in polymer composites because of their sustainability, renewability, and eco-friendly nature [1]. These composites offer advantages such as being lightweight, cost-effective, and having low density, making them attractive for various industrial applications. Despite the benefits, challenges exist, such as the compatibility issues between natural fibers and polymer matrices due to the hydrophilic nature of the fibers and hydrophobic nature of the matrices. The interest in natural fiber-reinforced polymer composites stems from the need for alternative eco-friendly and sustainable materials, which has led to extensive research in this area [2]. Table 1, presented below, offers a concise overview of the pros and cons associated with natural fiber composites.

Researchers have highlighted the need for advanced techniques, such as homogenization methods, statistical micromechanics, and deep learning, to address the challenges associated with predicting the mechanical behaviors of heterogeneous materials. These methodologies offer promising avenues for improving the accuracy of predictions by capturing the intricate interactions at different scales within heterogeneous materials. Moreover, the anisotropic and heterogeneous nature of materials like wood and composites further complicates the prediction of their mechanical behaviors. The unique microstructures and material properties of these substances require specialized computational tools and models to account for their complex behaviors accurately. the diverse and complex composition of heterogeneous materials presents substantial challenges in accurately forecasting their mechanical behaviors. Researchers are actively exploring innovative approaches, including advanced modeling techniques and machine learning, to enhance the predictive capabilities for these materials and address the inherent complexities associated with their heterogeneous nature [3].

The development of artificial intelligence (AI) has led to a significant transformation in the predictive modeling of materials. AI technologies, including artificial neural networks and deep learning models, have revolutionized the prediction and analysis of material properties. These AI-driven predictive models have found applications in various fields such as medicine, materials science, and engineering, enhancing decision-making processes and improving outcomes. Researchers have utilized AI to forecast intricate material behaviors like energy storage mechanisms, conformational changes, and reaction properties with greater efficiency and accuracy compared to conventional methods like molecular dynamics simulations. Additionally, AI has played a crucial role in predicting mechanical properties, fracture toughness, and stress-strain fields of materials, contributing to advancements in materials design and selection. Moreover, the integration of AI in construction materials, such as concrete and coalcretes, has shown the capability to predict compressive strength and other essential properties, thereby enhancing the efficiency and accuracy of material characterization [4]. AI, with its unparalleled computational power and sophisticated analytical capabilities, offers a novel approach to deciphering the intricate relationships within composite materials. This review article aims to illuminate the significant advancements in AI-powered predictions of mechanical behaviors in natural fiber polymer composites, charting a course through the evolution of predictive modeling techniques from conventional to cutting-edge AI methodologies.

The integration of AI into materials science not only enhances the precision of predictions but also accelerates the development of new materials by elucidating the underlying mechanisms governing material behavior. As we delve into the contributions of machine learning and deep learning models to the field, it becomes evident that AI stands at the vanguard of a revolution in materials science, offering insights that were previously beyond reach. However, the journey towards fully harnessing AI in the prediction of mechanical properties is fraught with challenges, including the need for extensive datasets, the complexity of model development, and the imperative for interdisciplinary collaboration. Addressing these challenges is crucial for advancing our understanding and application of natural fiber polymer composites [5].

This review sets the stage for a comprehensive exploration of the current state, achievements, and future prospects of AI in materials science, with a focus on natural fiber polymer composites. By synthesizing insights from recent research and highlighting innovative applications, it aims to foster a deeper understanding of the transformative impact of AI on the field. As we stand on the brink of this new era, the integration of AI into materials science heralds not only advancements in predictive modeling but also the potential for groundbreaking discoveries in material design and application.

Table 1: Advantages and drawbacks of natural fiber.

Advantages	Drawbacks
Low specific weight produces greater specific strength and rigidity than glass.	Lower strength, particularly impact strength.
Renewable resources and production need little energy and emit little CO_2.	Variable quality affected by climatic conditions.
Low-cost production with low investment.	Insufficient resilience to moisture, which results in fiber swelling.
Processing is gentle, there is no tool wear, and there is no skin irritation.	Temperature limit placed on the manufacturing procedure.
Electrical resistance is high.	Reduced durability.
Excellent thermal and acoustic insulation qualities.	Fire resistance is poor.
Biodegradable.	Inadequate fibre/matrix adhesion.
Thermal recycling is possible.	Variations in prices brought about by the outcomes of harvests or by agricultural politics.

Background

The exploration of natural fiber polymer composites represents a significant stride towards sustainable development in materials science [6]. These composites, characterized by their integration of renewable fibers with polymer matrices, not only promise a reduction in environmental footprint but also offer enhanced mechanical properties tailored for a broad spectrum of applications. Historically, the utilization of natural fibers in composite materials dates back to ancient civilizations, yet their resurgence in contemporary engineering underscores a collective shift towards eco-conscious material solutions. This revival is driven by the composites' biodegradability, low density, and commendable mechanical strength, positioning them as ideal candidates for green engineering projects [7].

Traditional methodologies for predicting the mechanical behavior of these composites have predominantly relied on empirical and experimental techniques. These approaches, while foundational, often necessitate extensive material testing and cannot always capture the complex, nonlinear interactions between the composite constituents [8]. The variability inherent in natural fibers stemming from factors such as origin, extraction process, and treatment methods further complicates the prediction of composite properties, presenting a formidable challenge to material scientists and engineers.

The advent of artificial intelligence (AI) in materials science marks a pivotal evolution in predictive modeling techniques. AI's emergence as a powerful tool for analysis and prediction stems from its ability to process vast datasets and identify patterns or correlations that elude traditional computational models. In the context of natural fiber polymer composites, AI methodologies, including machine learning and deep learning, offer a sophisticated framework for navigating the complexities of material behaviors. These AI-driven models excel in managing the heterogeneity of data, accommodating the multifaceted nature of composite materials, and providing insights into the underlying mechanisms influencing mechanical properties. The transition towards AI-powered models in the prediction of mechanical behaviors signifies a confluence of computational innovation and materials science. This shift not only exemplifies the potential of AI to revolutionize traditional paradigms but also reflects a broader trend towards the integration of digital technologies in scientific research. As we delve deeper into the applications of AI in this realm, it becomes evident that the intersection of computational power and material innovation holds the key to unlocking new possibilities in the design and application of natural fiber polymer composites [9].

Types of AI Techniques

The integration of artificial intelligence (AI) into the realm of materials science, particularly in the predictive modeling of natural fiber polymer composites, marks a significant leap forward in our ability to understand and optimize the mechanical behaviors of these materials. AI techniques, particularly machine learning (ML) and deep learning (DL) algorithms, have become essential tools in understanding the intricate relationships and subtle details inherent in composite materials [10]. These AI methods have shown significant progress in deciphering complex interdependencies within materials by analyzing vast amounts of data and revealing correlations between various phenomena. Through the application of AI techniques, such as artificial neural networks (ANN) and genetic algorithms (GA), predictions related to material properties like fireproofing time can be made accurately. Furthermore, AI-guided approaches have been successful in predicting properties of materials at different scales, from atomistic to macroscopic, showcasing the versatility and potential of these methods [11]. This section delves into the foundational AI methodologies employed in predictive modeling, elucidating their roles, mechanisms, and the comparative advantages they bring to the field.

Machine Learning Models. Machine learning, a subset of artificial intelligence (AI), utilizes statistical methods to enable machines to enhance their performance at tasks through experience. This is achieved by employing algorithms that allow machines to learn from data and improve their ability to make predictions or decisions without being explicitly programmed for each scenario [12]. Machine learning involves the use of statistical techniques to build intelligent systems that can learn from data and improve their performance over time [13]. It focuses on teaching machines to detect patterns in data and make decisions based on that information. In the context of natural fiber polymer composites, ML models have been adeptly applied to predict mechanical properties such as tensile strength, flexural strength, and impact resistance. Supervised learning algorithms, such as regression models and decision trees, have proven particularly effective. These models rely on historical data sets of composite properties to learn the relationships between composite constituents and their mechanical behaviors. The ability of ML models to handle complex, nonlinear relationships without explicit programming for each scenario makes them invaluable for predicting the properties of materials with intricate internal structures. ML can be classified into three categories: supervised, unsupervised, or reinforcement learning, depending on the resources available in the training dataset.

Supervised learning. Supervised machine learning (ML) techniques are revolutionizing the field of materials science, particularly in predicting the mechanical properties of natural fiber polymer composites. By utilizing algorithms such as Random Forests, Support Vector Machines, and Artificial Neural Networks, researchers can model complex relationships between composite constituents and their resultant mechanical behaviors. These ML models are trained on datasets comprising various fiber types, matrix materials, and processing conditions, enabling them to predict properties like tensile strength, modulus of elasticity, and impact resistance with high accuracy. This predictive capability not only accelerates the development of new composite materials but also optimizes existing formulations to meet specific performance criteria [14]. Supervised ML, therefore, represents a critical tool in the design and application of natural fiber composites, offering a more efficient route to achieving desired mechanical properties through informed material selection and composite engineering.

Unsupervised learning. Unsupervised machine learning (ML) techniques are emerging as potent tools in the exploration and optimization of natural fiber polymer composites, providing insights into the mechanical properties of these materials without predefined labels or outputs. By employing clustering algorithms such as k-means or hierarchical clustering, and dimensionality reduction techniques like Principal Component Analysis (PCA), researchers can uncover hidden patterns and intrinsic relationships within large datasets of composite properties [15]. This data-driven approach allows for the identification of optimal combinations of fibers, matrices, and processing techniques that might not be immediately apparent through traditional experimental methods. Unsupervised learning thus facilitates a more holistic understanding of the material behavior, enabling the development of composites with tailored mechanical properties through strategic manipulation of their constituent elements. This method enhances the innovative potential in composite engineering by discovering novel material combinations and processing conditions that lead to superior mechanical performance.

Deep Learning Approaches. Deep learning, a more sophisticated branch of machine learning, utilizes neural networks with multiple layers (deep networks) to analyze data patterns [16]. An example of a deep neural network architecture is displayed in Fig. 1. In predictive mechanical properties for natural fiber polymer composites, DL algorithms excel in processing unstructured data from a myriad of sources, including textual data, images, and spectroscopic analyses. Convolutional neural networks (CNNs) and recurrent neural networks (RNNs), for instance, have been instrumental in identifying patterns within the microstructure of composites that correlate with specific mechanical properties [17]. The depth and complexity of DL models allow for a more nuanced understanding of material behaviors, paving the way for predictions that account for the stochastic nature of natural fibers and their interactions within polymer matrices.

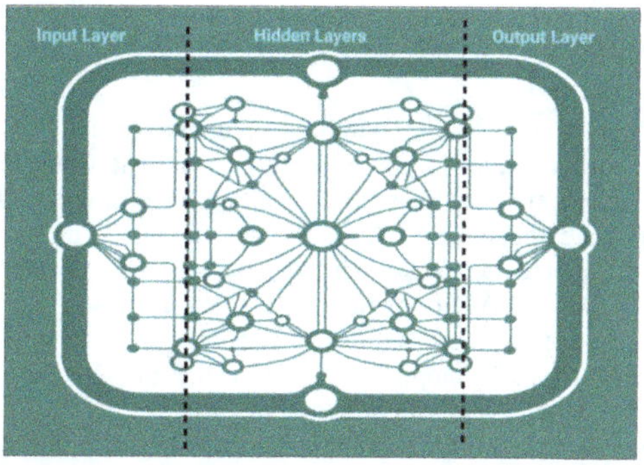

Fig. 1: Schematic representation of a deep neural network.

Advancements in AI-Powered Predictions

The frontier of materials science has been significantly reshaped by the advent of artificial intelligence (AI), with its profound impact permeating the domain of natural fiber polymer composites. This section delves into the remarkable advancements achieved through the application of AI in predicting the mechanical behaviors of these composites, underscoring the synergy between computational intelligence and material innovation. The evolution of AI-powered predictive models not only enhances our understanding of material properties but also heralds a new era of material design, characterized by precision, efficiency, and sustainability [18].

One of the paramount advancements in the use of AI within materials science is the significant leap in predictive accuracy and computational efficiency. AI models, through their ability to learn from and adapt to complex datasets, have demonstrated unprecedented precision in forecasting the mechanical properties of natural fiber polymer composites. Deep learning algorithms, in particular, have been at the forefront of this revolution, capable of dissecting and interpreting the intricate patterns within material structures that traditional models could not discern. This enhanced predictive accuracy is pivotal for the development of composites with tailored properties, optimizing performance for specific applications while minimizing material waste and production costs [18].

Furthermore, the integration of AI with computational materials science represents another significant advancement, facilitating a holistic approach to material analysis and design. By combining AI's predictive prowess with computational modeling techniques, such as finite element analysis (FEA) and molecular dynamics (MD) simulations, researchers can now simulate and evaluate the behavior of natural fiber polymer composites under various conditions with remarkable precision. This integrative approach enables the exploration of vast design spaces, accelerating the discovery of novel materials and the optimization of existing ones. Furthermore, it fosters a deeper understanding of the underlying mechanisms governing material behavior, offering insights that guide the engineering of composites with enhanced performance characteristics.

In addition, the advancements in AI-powered predictions extend beyond technical achievements, influencing the broader landscape of innovation and sustainability in materials science. By enabling the precise prediction of mechanical behaviors, AI facilitates the development of materials that are not only high-performing but also environmentally sustainable. The ability to accurately forecast the performance of composites made from natural, renewable fibers contributes to the reduction of reliance on synthetic, non-biodegradable materials, aligning with global sustainability goals. Moreover, the efficiency gains from AI-driven predictive modeling reduce the time and resources required for material development, promoting a more sustainable research and production paradigm.

Challenges and Solutions

The integration of artificial intelligence (AI) into the predictive modeling of natural fiber polymer composites, while transformative, is not without its challenges. These hurdles range from technical limitations to broader conceptual and methodological issues, posing significant obstacles to the full realization of AI's potential in this field. However, alongside these challenges, innovative solutions and strategic approaches are emerging, paving the way for more effective and comprehensive applications of AI in materials science. This section explores the primary challenges encountered in this interdisciplinary endeavor and delineates the solutions that researchers and practitioners are developing to overcome them [18].

One of the most formidable challenges in applying AI to predict the mechanical behaviors of natural fiber polymer composites is the scarcity and variability of high-quality data. AI models, particularly those based on deep learning, require extensive datasets to train effectively. However, the heterogeneous nature of natural fibers, coupled with variations in processing and manufacturing techniques, often results in data that is sparse, inconsistent, or non-representative. To address this challenge, researchers are adopting strategies such as data augmentation, which artificially increases the size and diversity of datasets through techniques like simulation or synthetic data generation. Additionally, collaborations across academic and industrial sectors are being encouraged to facilitate data sharing and the creation of comprehensive, standardized datasets.

In addition, the complexity of AI models, especially deep learning networks, presents another significant challenge. These models often require extensive computational resources for training and operation, making them inaccessible to some researchers and practitioners. Moreover, the "black box" nature of deep learning models can impede the interpretability and transparency of predictions, raising concerns about the reliability and understanding of the models' decision-making processes. Efforts to demystify AI models include the development of more interpretable machine learning techniques and the integration of explainable AI (XAI) principles [18]. These approaches aim to make AI predictions more transparent and understandable, enhancing trust and reliability. Additionally, leveraging cloud computing and specialized hardware can alleviate computational constraints, making advanced AI models more accessible.

Conclusion

The journey through the landscape of artificial intelligence (AI) in the realm of natural fiber polymer composites has unveiled a realm of unprecedented potential and transformative power. As this review has explored, the synergy between AI and materials science is not merely augmentative but revolutionary, heralding a new epoch in the predictive modeling and design of composite materials. Through the lens of AI, the complexities of natural fiber polymer composites are demystified, revealing pathways to optimize their mechanical properties with precision and efficiency previously unimaginable.

The advancements in AI-powered predictive models have demonstrated remarkable successes in enhancing the accuracy and speed of mechanical behavior predictions, underscoring the pivotal role of computational intelligence in materials engineering. These innovations not only streamline the development process of composites but also empower the creation of sustainable, high-performance materials that align with the pressing demands of environmental stewardship and technological advancement. However, the journey is far from complete. The challenges highlighted—ranging from data scarcity to the complexity of models and the necessity for interdisciplinary collaboration—present hurdles that require concerted efforts to overcome. The solutions, rooted in innovation, collaboration, and ethical consideration, pave the way for a future where the integration of AI in materials science can flourish without bounds.

In conclusion, the integration of artificial intelligence into the field of natural fiber polymer composites is a beacon of scientific and technological progress. It represents a confluence of knowledge, innovation, and aspiration that challenges the status quo and invites us to envision a future where materials are not only engineered but also intelligent, sustainable, and inherently aligned with the principles of a circular economy. As we navigate this promising yet uncharted terrain, the collaborative efforts of scientists, engineers, and policymakers will be paramount in realizing the full potential of AI in materials science, ensuring that this technological revolution benefits humanity and the planet alike.

Acknowledgement

The authors like to express their appreciation to the Faculty of Chemical Engineering Technology, Universiti Malaysia Perlis (UniMAP), Special thanks to those who contributed to this project directly or indirectly.

References

[1] L. Mohammed, M. Ansari, G. Pua, M. Jawaid, & S. Islam, A review on natural fiber reinforced polymer composite and its applications, International Journal of Polymer Science 2015 (2015) 1-15. https://doi.org/10.1155/2015/243947

[2] I. A. S. Salem, A. R. Rozyanty, B. O. Betar, T. Adam, Mohammed, M., & A. M. Mohammed, Study of the effect of surface treatment of kenaf fiber on chemical structure and water absorption of kenaf filled unsaturated polyester composite, Journal of Physics: Conference Series 908 (2017) 012001. https://doi.org/10.1088/1742-6596/908/1/012001

[3] P. Kluge and S. Eichhorn, Calculation concept for wood-based components in mechanical engineering, Advanced Engineering Materials 25(15) (2023). https://doi.org/10.1002/adem.202300085

[4] Y. Sun, J. Zhang, G. Li, Y. Wang, J. Sun, & C. Jiang, Optimized neural network using beetle antennae search for predicting the unconfined compressive strength of jet grouting coalcretes, Geomechanics 43(4) (2019) 801-813. https://doi.org/10.1002/nag.2891

[5] S. Zhang, Y. Wang, Y. Zheng, Q. Li, J. Huang & X. Long, Artificial intelligence in melanoma: a systematic review. Journal of Cosmetic Dermatology 21 (2022) 5993-6004. https://doi.org/10.1111/jocd.15323

[6] M. Mohammed, J. K. Oleiwi, A. M. Mohammed, A. J. A. M. Jawad, A. F. Osman, T. Adam & M. Jaafar, Comprehensive Insights on Mechanical Attributes of Natural-Synthetic Fibres in Polymer Composites, Journal of Materials Research and Technology (2023). https://doi.org/10.1016/j.jmrt.2023.06.148

[7] M. M. Mohammed, A. M. Rasidi, R. B. Mohammed, R. Rahman, A. F. Osman, T. Adam & O. S. Dahham, Interfacial bonding mechanisms of natural fibre-matrix composites: an overview, BioResources, 17(4) (2022) 7031. https://doi.org/10.15376/biores.17.4.Mohammed

[8] M. Mohammed, A. R. Rozyanty, T. Adam, & B. O. Betar, Study of the weathering effect in a natural environment on the hybrid kenaf bast/glass fibre-filled unsaturated polyester composite, AIP Conference Proceedings 1885 (2017). https://doi.org/10.1063/1.5002395

[9] F. Bini, A. Pica, L. Azzimonti, A. Giusti, L. Ruinelli, F. Marinozzi & P. Trimboli, Artificial intelligence in thyroid field a comprehensive review, Cancers 13(19) (2021) 4740. https://doi.org/10.3390/cancers13194740

[10] K. Butler, D. Davies, H. Cartwright, O. Isayev & A. Walsh, Machine learning for molecular and materials science, Nature 559 (2018) 547-555. https://doi.org/10.1038/s41586-018-0337-2

[11] D. Elton, Z. Boukouvalas, M. Butrico, M. Fuge & P. Chung, Applying machine learning techniques to predict the properties of energetic materials, Scientific Reports 8(1) (2018). https://doi.org/10.1038/s41598-018-27344-x

[12] J. Carolan, J. McGonigle, A. Dennis, P. Lorgelly & A. Banerjee, Technology-enabled, evidence-driven, and patient-centered: the way forward for regulating software as a medical device, Jmir Medical Informatics 10(1) (2022) e34038. https://doi.org/10.2196/34038

[13] B. Sheng, X. Chen, T. Li, T. Ma, Y. Yang, L. Bi & M. Zhang, An overview of artificial intelligence in diabetic retinopathy and other ocular diseases, Frontiers in Public Health, 10 (2022). https://doi.org/10.3389/fpubh.2022.971943

[14] G. Fulk, Artificial Intelligence and Neurologic Physical Therapy, Journal of Neurologic Physical Therapy, 47(1) (2023) 1-2. https://doi.org/10.1097/NPT.0000000000000426

[15] R. Miotto, F. Wang, S. Wang, X. Jiang & J. Dudley, Deep learning for healthcare: review, opportunities and challenges, Briefings in Bioinformatics 19(6) (2017) 1236-1246. https://doi.org/10.1093/bib/bbx044

[16] G. Pilania, C. Iverson, T. Lookman & B. Marrone, Machine-learning-based predictive modeling of glass transition temperatures: a case of polyhydroxyalkanoate homopolymers and copolymers, Journal of Chemical Information and Modeling 59(12) (2019) 5013-5025. https://doi.org/10.1021/acs.jcim.9b00807

[17] J. Huang, J Liew, A. Ademiloye & K. Liew, Artificial intelligence in materials modeling and design, Archives of Computational Methods in Engineering 28(5) (2020) 3399-3413. https://doi.org/10.1007/s11831-020-09506-1

[18] W. Sha, Y. Guo, Y. Qing, S. Tang, X. Zhang, S. Lu & X. Ai, Artificial intelligence to power the future of materials science and engineering, Advanced Intelligent Systems 2(4) (2020). https://doi.org/10.1002/aisy.201900143

Degradation of Plastic Waste in Surabaya River, Indonesia

Atik WIDIYANTI[1,a], Eddy Setiadi SOEDJONO[1,b*], Maya SHOVITRI[2,c], Adhi YUNIARTO[1,d]

[1]Department of Environmental Engineering, Institut Teknologi Sepuluh Nopember, Surabaya, 60111, Indonesia

[2]Department of Biology, Faculty of Science, Institut Teknologi Sepuluh Nopember, Surabaya, 60111, Indonesia

[a]widiyanti.tkl@unusida.ac.id, [b]soedjono@enviro.its.ac.id, [c]maya@bio.its.ac.id, [d]adhy@its.ac.id

Keyword: Degradation Plastic, Physical and Biological Degradation, Waste

Abstract. The increase in plastic waste results from technological, industrial, and demographic changes. In Indonesia, the demand for plastic continues to increase, an average increase of 200 tons/year. In 2013, about 1.9 million tons of plastic were produced. About 10% of all newly produced plastic is discharged into rivers and dumped into the sea, meaning that about 165,000 tons of plastic yearly in Indonesian waters. Surabaya is the second-largest plastic waste-producing city in Indonesia. About 111,300 tons of plastic waste will be generated in 2021, and about 2% will end up in water. This research obtained twenty five isolates of plastic-degrading bacteria found in the Surabaya River.

Introduction

In 2021, Indonesia will become the world's third-largest producer of plastic waste. According to statistics on household waste, plastic waste ranks second, accounting for 5.4 million tons per year, equivalent to 14% of the total waste generated. The largest source of plastic waste in 2019-2021 in Indonesia comes from households and traditional markets (1). In addition, plastic is also found in the form of microplastics in leachate from landfills. Therefore, it has the potential to pollute the area surrounding the landfill (2,3,4). Surabaya City is the second largest city, generating 111,300 tons of plastic waste annually. Nearly 16,000 tons of plastic are discharged into the environment, equivalent to 5.15 kg/person/year. Plastic waste in Surabaya City in 2022 includes plastic bags, disposable food, hygiene products, beverage bottles, styrofoam, trays, and plastic bottles. In Surabaya, most of the population lives near the sea, and several rivers flow through the city, becoming potential transport routes to carry plastic waste to the sea.

Surabaya City is located at the mouth of the Surabaya River, north of the Brantas River basin. The Surabaya River is a tributary of the Brantas River, with a length of ± 42.3 km, flowing from the lower reaches of the Brantas River through 4 districts: Mojokerto, Sidoarjo, Gresik and Surabaya. The downstream flow of the Surabaya River is at the Mojokerto Mlirip Dam. Meanwhile, the upstream of the Surabaya River is located at Jagir Wonokromo Dam, Surabaya. Surabaya River is the source of raw water for rice fields, domestic activities, industry, transportation, recreation, and raw water in PDAM Surabaya City. However, this river's water quality is declining due to pollution. Plastic is difficult to decompose due to its polymer structure. However, it is known that Bacillus sp, Brevibacillus sp, Stenotrophomonas sp, and Lysinibacillus sp can degrade plastic by 1.5-5.2% due to dry weight reduction (5,6,7). Bacteria secrete laccase enzymes (8), proteinase k (9), hydroxylase genes, alkB, and alkB2 to degrade low molecular-weight P.E. (6). The biodegradation of plastic by each type of bacteria is influenced by the amount of enzymes the bacteria possesses and the type of plastic being degraded. Bacillus sp. can degrade P.P. amount 4-6.4% from weight loss takes 40 days (10), Bacillus subtilis can degrade 1.75% weight loss LDPE only 30 days (11).

In nature, biological mechanisms coordinate with physical mechanisms to decompose plastic. Physical mechanisms include the presence of ultraviolet (U.V.) rays and friction from water and air currents. U.V. irradiation with a wavelength of 245 nm for 120 hours can result in greater dry weight reduction, higher plastic surface roughness, and increased hydrophilic layer (12). Water flow rate and U.V. light for 480 hours synergistically affect the formation and degradation of LDPE and P.P. (13). So this research will explain the condition of the Surabaya River and the bacteria that have the potential to degrade LDPE plastic.

Experiment

Material. The research was carried out in Surabaya River at coordinates 7°21'06.3"S-112°39'43.9"E. Bacterial sampling used nets to trap plastic waste for 7 days. Sampling plastic waste use net with a length of 3 meters and a width of 2,5 meters is installed across the Surabaya River. Isolation and purification bacteria using Nutrient Agar (NA). Violet Crystals, Iodine, Alcohol (70%), and Safranin need Gram Stanning.

Sample Preparation. Bacteria were isolated from the surface of plastic waste using the swab method. Then, multilevel dilutions were carried out. The bacterial isolates found were purified using the 16-stack method and then gram-stained.

Measurement. Concentration Dissolved Oxygen (DO), pH and Temperature are analyzed daily to determine river conditions. After seven days, the plastic waste was taken to the laboratory. Each plastic sample was swabbed with a sterile cotton bud, dipped in a test tube containing 10 mL sterile distilled water and vortexed for 1 min to make it homogeneous. Next, aseptically take 1 mL of the suspension, put it in a test tube containing 9 mL of sterile distilled water, and then vortex it until it is homogeneous. This dilution is called a 10^{-1} dilution. Then from the 10^{-1} dilution, 1 mL of the suspension was taken and put into the next test tube containing 9 mL of sterile distilled water and vortexed, so a 10^{-2} dilution was obtained. Dilution is continued sequentially up to 10^{-6}. The last three dilutions were taken at 100 μm using a micro pipette and dropped onto the surface of the Nutrient Agar (NA). The inoculum was then leveled with a Graygalsky spatula. The cultures were then incubated at 37°C and observed every 24 hours for 2 days.

Single colonies growing on petri dishes were taken aseptically using a loop needle, inoculated onto the surface of solid NA medium using the 16 streak method, and incubated in an incubator at 37°C for 24 hours. One growing colony was then taken and inoculated again on new NA medium using the 16 streaks method. Cells are observed microscopically to determine grams of bacteria. One drop of distilled water is dripped on the middle edge of the slide, and a loop of bacteria is streaked onto the drop and spread in a zig-zag pattern. The preparation is then fixed by passing it over a Bunsen until the distilled water is dry. The results of the preparation step are referred to as review preparations. The smear preparations were then stained with crystal violet for 1 min. Then rinsed using water and stained using iodine for 1 min. Next, the preparation is rinsed using alcohol. After that, it was stained with safranin for 45 seconds.

Result and Discussion

Surabaya is crossed by a river called Surabaya River. Surabaya River is about 42.3 km long, from Mlirip Dam, Mojokerto, to Jagir Dam, Surabaya. The width of the river varies from 60 to 100 meters. River flow fluctuates all year round (0.1 m/s-0.6 m/s). Overall, the Surabaya River meets Water Quality Criteria in the class 2 and can be used for recreation infrastructure, fish cultivation, animal husbandry, irrigating crops, etc (17) (Fig. 1).

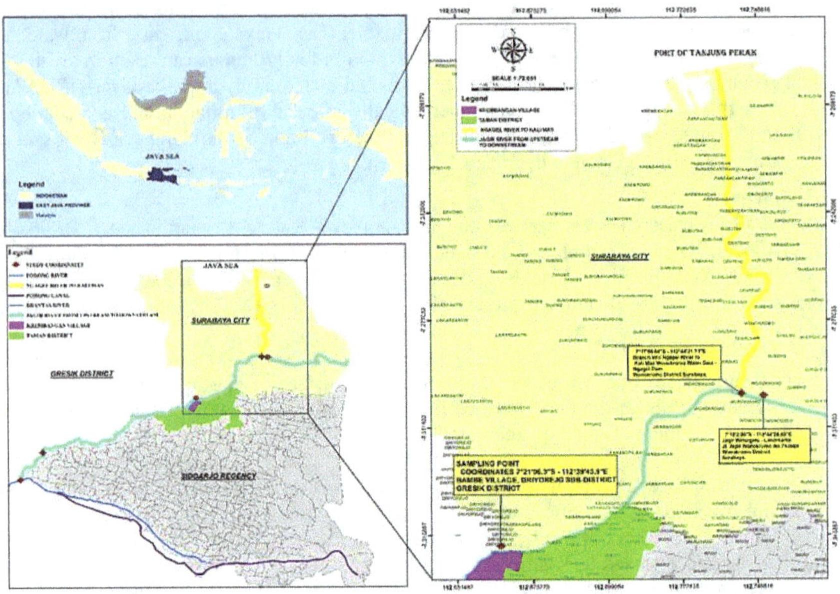

Fig. 1: Surabaya City, Indonesia (ArcGIS 10.8).

The DO concentration in the Surabaya River varies. The lowest concentration occurred on the first day and the highest on the fifth (Table 1). Oxygen influences water quality because dissolved oxygen plays a role in organic and inorganic materials' oxidation and reduction processes. Oxygen also determines the types of organisms in the waters. The OD threshold in rivers is >5 ppm, meaning that the Surabaya River still meets water quality except on the first day. This may be due to wastewater input.

The Surabaya River has a pH of 4.4 to 6.4, this shows that the pH of the Surabaya River tends to be acidic. pH indicates the basicity or acidity of water. The pH value is influenced by dissolved oxygen, organism activity, and increasing water temperature. As water temperatures rise, pH tends to increase. However, this is not a direct effect; various factors such as gas solubility, chemical reactions, and biological influences also influence pH changes.

Table 1: Surabaya River Water Condition.

Days	Time	pH	Temperature (°C)	DO (mg/L)
1	07.10 a.m	4,4	29,8	3,4
	04.30 p.m	5,83	32	5,1
2	07.10 a.m	5,4	29,7	5,1
	04.30 p.m	4,9	29	5,4
3	07.10 a.m	5,4	30,2	6,1
	04.30 p.m	4,9	29	5,4
4	07.10 a.m	5,5	29,1	6,2
	04.30 p.m	4,81	30,3	6,6
5	07.10 a.m	5,9	29	5,9
	04.30 p.m	5,5	29	6,7
6	07.10 a.m	3,1	29,3	5,7
	04.30 p.m	5,45	30,1	5,5
7	07.10 a.m	6,4	29	5,6
	04.30 p.m	6,1	29,5	5,5

Plastic Waste in Surabaya River. The increase in plastic use results from technological, industrial and demographic changes. In Indonesia, demand for plastic continues to increase, reaching an average increase of 200 tons per year. According to the Ministry of Industry in 2013 In Surabaya City, about 1.9 million tons of plastic were produced in 2013, corresponding to an average output of 1.65 million tons/year. About 10% of all newly produced plastic is dumped into rivers and ends in the ocean. This means that about 165,000 tons of plastic annually will end up in Indonesian waters. Surabaya is the second largest plastic waste- producing city in Indonesia. About 111,300 tons of plastic waste were generated in 2021, and about 2% was discharged into rivers and seas. Plastic waste that enters the water will flow downstream into the river. The composition of plastic waste in Surabaya is mainly plastic bags, serving plastics (straws, plastic cutlery, coffee stirrers), and hygiene products. The pollution status of the Surabaya River is shown in Fig. 2.

Fig. 2: Plastic waste pollution in Surabaya River (a). second day, (b). seventh day.

Depending on their size, plastics are divided into 3 types, which are macroplastics (>2.5cm), mesoplastics (0.5-2.5 cm) and microplastics (<0.5 cm). Surabaya's microplastics abundance is higher than in Jakarta (14). The Surabaya River was determined to be contaminated with macroplastics, mesoplastics and microplastics.

Biodegradation Plastic. The research results obtained 25 isolates of plastic degrading bacteria from the Surabaya river, 13 gram positive and 12 gram negative isolates (Fig. 3). Plastic-degrading bacteria have exo and endo enzymes to degrade plastic. Plastic degradation is the breakdown of complex chain plastic molecules into shorter molecules. Degradation occurs due to U.V. radiation (15,16), oxygen (17), temperature (18), water (19) and microorganisms in the environment (20).

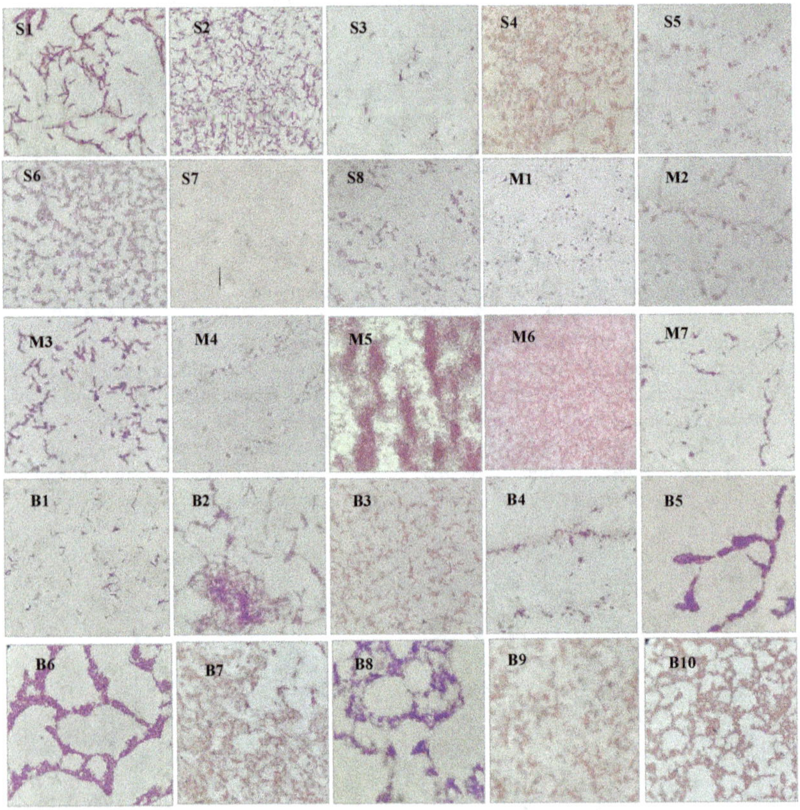

Fig. 3: Plastic Degrading Bacteria Isolates from the Surabaya River. S). Isolates taken from the river surface. M). Isolates taken from the middle depth. B). Isolates taken from the riverbed.

The degradation process starts with the conversion of polymer into monomer, then this monomer is produced. Most polymers are bulky and cross the cell membrane. Therefore, they must be broken down into smaller monomers before they can enter and be degraded in microbial cells. Research often uses biodegradation techniques to identify microbial isolates that can degrade plastics. The ability of bacteria to degrade plastic is expressed in percentage of plastic loss. This happens because the bacteria release enzymes, which lead to interaction with the surface of the plastic. This enzyme can break down the surface of the plastic through a hydrolysis process that results in a reduction in the weight of the plastic.

Conclusion

Surabaya is polluted by plastic waste. Plastic waste from Surabaya enters the river and flows into the sea, plastic waste undergoes both physical and biological environmental degradation until it becomes a simpler structure. So, strong treatment is needed to fight plastic pollution. Biodegradation using bacteria is a solution for plastic degradation. Twenty five isolates of plastic degrading bacteria were found from the Surabaya river.

Acknowledgment

This study was funded by Indonesian Education Scholarships [Beasiswa Pendidikan Indonesia (BPI)], Indonesia Endowment Funds for Education [Lembaga Pengelola Dana Pendidikan (LPDP)], Center for Higher Education Funding and Assessment [Pusat Pembiayaan dan Asesmen Pendidikan Tinggi (PPAPT)], Ministry of Higher Education, Science, and Technology of Republic Indonesia.

References

[1] Z. Arifin, D. Falahudin, H. Saito, T. H. Mintarsih, M. Hafizt, Y. Suteja, Indonesian policy and researches toward 70% reduction of marine plastic pollution by 2025. Mar Policy. 155 (2023) 105692. https://doi.org/10.1016/j.marpol.2023.105692

[2] Fajaruddin Natsir Muh, M. Selomo, E. Ibrahim, A. A. Arsin, N. C. Alni, Analysis on microplastics in dug wells around Tamangapa Landfills, Makassar City, Indonesia. Gac Sanit. 35 (2021) S87-9. https://doi.org/10.1016/j.gaceta.2020.12.024

[3] Nurhasanah, M.R. Cordova, E. Riani, Micro and mesoplastics release from the Indonesian municipal solid waste landfill leachate to the aquatic environment: Case study in Galuga Landfill Area, Indonesia, Mar. Pollut. Bull. 163 (2021) 111986. https://doi.org/10.1016/j.marpolbul.2021.111986

[4] Y. Trihadiningrum, S. A. Wilujeng, R. Tafaqury, D. R. Radita, A. D. Radityaningrum, Evidence of microplastics in leachate of Randegan landfill, Mojokerto City, Indonesia, and its potential to pollute surface water, Sci. Total Environ. 874 (2023) 162207. https://doi.org/10.1016/j.scitotenv.2023.162207

[5] N. Atanasova, T. Paunova-Krasteva, S. Stoitsova, N. Radchenkova, I. Boyadzhieva, K. Petrov, et al., Degradation of Poly(ε-caprolactone) by a Thermophilic Community and Brevibacillus thermoruber Strain 7 Isolated from Bulgarian Hot Spring, Biomolecules 11 (2021) 1488. https://doi.org/10.3390/biom11101488

[6] J. M. Jeon, S. J. Park, T. R. Choi, J. H. Park, Y. H. Yang, J. J. Yoon, Biodegradation of polyethylene and polypropylene by Lysinibacillus species JJY0216 isolated from soil grove, Polym. Degrad. Stab. 191 (2021) 109662. https://doi.org/10.1016/j.polymdegradstab.2021.109662

[7] S. J. Wani, S. S. Shaikh, B. Tabassum, R. Thakur, A. Gulati, R. Z. Sayyed, Stenotrophomonas sp. RZS 7, a novel PHB degrader isolated from plastic contaminated soil in Shahada, Maharashtra, Western India. 3 Biotech. 6 (2016) 179. https://doi.org/10.1007/s13205-016-0477-8

[8] M. Santo, R. Weitsman, A. Sivan , The role of the copper-binding enzyme - laccase - in the biodegradation of polyethylene by the actinomycete Rhodococcus ruber. Int. Biodeterior. Biodegrad. 84 (2013) 204-10. https://doi.org/10.1016/j.ibiod.2012.03.001

[9] M. Stepczyńska, and P. Rytlewski, Enzymatic degradation of flax-fibers reinforced polylactide, Int. Biodeterior. Biodegrad. 126 (2018) 160-6. https://doi.org/10.1016/j.ibiod.2017.11.001

[10] H. S. Auta, C. U. Emenike, B. Jayanthi, S. H. Fauziah, Growth kinetics and biodeterioration of polypropylene microplastics by Bacillus sp. and Rhodococcus sp. isolated from mangrove sediment. Mar. Pollut. Bull. 127 (2018) 15-21. https://doi.org/10.1016/j.marpolbul.2017.11.036

[11] K. Harshvardhan and B. Jha, Biodegradation of low-density polyethylene by marine bacteria from pelagic waters, Arabian Sea, India. Mar. Pollut. Bull. 77 (2013) 100-6. https://doi.org/10.1016/j.marpolbul.2013.10.025

[12] N. Taghavi, W. Q. Zhuang, S. Baroutian, Enhanced biodegradation of non-biodegradable plastics by UV radiation: Part 1, J. Environ. Chem. Eng. 9 (2021) 106464. https://doi.org/10.1016/j.jece.2021.106464

[13] P. Lestari, Trihadiningrum Y, Wijaya BA, Yunus KA, Firdaus M. Distribution of microplastics in Surabaya River, Indonesia. Sci Total Environ. 2020 July;726:138560. https://doi.org/10.1016/j.scitotenv.2020.138560

[14] M. R. Cordova, Y. I. Ulumuddin, A. A. Lubis, M. T. Kaisupy, S. P. A. Wibowo, R. Subandi, et al., Microplastics leaving a trace in mangrove sediments ever since they were first manufactured: A study from Indonesia mangroves, Mar. Pollut. Bull. 195 (2023) 115517. https://doi.org/10.1016/j.marpolbul.2023.115517

[15] A. Delre, M. Goudriaan, V. H. Morales, A. Vaksmaa, R. T. Ndhlovu, M. Baas, et al., Plastic photodegradation under simulated marine conditions, Mar. Pollut. Bull. 187 (2023) 114544. https://doi.org/10.1016/j.marpolbul.2022.114544

[16] J. Sun, H. Zheng, H. Xiang, J. Fan, H. Jiang, The surface degradation and release of microplastics from plastic films studied by UV radiation and mechanical abrasion, Sci. Total Environ. 838 (2022) 156369. https://doi.org/10.1016/j.scitotenv.2022.156369

[17] D. Bertin, M. Leblanc S. R. A. Marque, D. Siri, Polypropylene degradation: Theoretical and experimental investigations, Polym. Degrad. Stab. 95 (2010) 782-91. https://doi.org/10.1016/j.polymdegradstab.2010.02.006

[18] R. Slezak, L. Krzystek, M. Puchalski, Krucińska I, Sitarski A, Degradation of bio-based film plastics in soil under natural conditions, Sci. Total Environ. 866 (2023) 161401. https://doi.org/10.1016/j.scitotenv.2023.161401

[19] S. Lambert and M. Wagner, Formation of microscopic particles during the degradation of different polymers, Chemosphere. 161 (2016) 161:510-7. https://doi.org/10.1016/j.chemosphere.2016.07.042

[20] Z. Wu, W. Shi. T. G. Valencak, Y. Zhang, G. Liu, D. Ren, Biodegradation of conventional plastics: Candidate organisms and potential mechanisms, Sci. Total Environ. 885 (2023) 163908. https://doi.org/10.1016/j.scitotenv.2023.163908

Enhancing Lower Limb Prosthetic Sockets with Hybrid Natural Fiber Composites: A Performance Analysis

Adel M. BASH[1,a], Jawad K. OLEIWI[2,b *], and Tahseen T. OTHMAN[1,c]

[1]Mechanical Department, Engineering College, Tikrit University, Tikrit, Iraq

[2]Department of Materials Engineering, University of Technology, Baghdad, Iraq

[a]Adelbash@tu.edu.iq, [b]130041@uotechnology.edu.iq, [c]tahseentaha@tu.edu.iq

Keywords: Natural Fibers, A Knee, Lower Limb Socket, Stress Analysis, Finite Element Method

Abstract. In material science, researchers are exploring natural fibers and their composites. this study investigates eco-friendly ramie and bamboo fibers, aiming to develop prosthetic sockets using these sustainable materials as alternatives to traditional manufacturing resources. There were six vacuum-molded laminated composites manufactured. In various layers, Ramie, bamboo, Perlon, glass, and carbon fibers reinforce the polymethyl methacrylate (PMMA) matrix. The first laminate group consists of ramie, carbon, and glass fibres (2 ramie + 2 glass or 2 carbon + 2 ramie). The second laminate group comprises bamboo, carbon, and glass fibres (2 bamboo + 2 glass or 2 carbon + 2 bamboo) layers. Experimental and numerical estimations were utilized to determine the effect of various fiber lamination sequences on the volumetric and mechanical properties. The results indicated that lamination (2 ramie fiber + 2 carbon + 2 ramie fiber) has the highest values of tensile strengths and moduli of elasticity at 272MPa and 4189 MPa, respectively. In contrast, lamination (2 bamboo +2Glass + 2 bamboo) has the lowest value of tensile strengths and moduli of elasticity at 89MPa and 2672MPa., respectively. The lowest value of tensile strengths and moduli of elasticity are at 38MPa and 1727MPa, respectively. The socket models were analyzed using the ANSYS finite element method. The numerical analysis revealed that the lamination using bamboo fiber reinforcement had the greatest overall deformation, measuring 7.11 mm and an equivalent elastic strain of 0.0162. Nevertheless, it is worth noting that the least numerical safety factor exhibited the lowest value of 1.613.

Introduction

People who have lost limbs due to chronic diseases like diabetes, arteriosclerosis, thrombosis, natural disasters like earthquakes and floods, landmines, or war are often given prostheses or artificial limbs to make them look and work like before [1]. Plant-based biomaterials may serve as a promising alternative to meet this demand. Typically, the prosthetic socket is connected to the residual limb. This socket fits around the remaining limb and is the anchor point for the rest of the artificial limb. It is essential because the lower limb (stump) can't support weight like the foot can. To ensure the prosthetic sockets are fitted correctly, it is important to understand how they work biomechanically and what substances comprise their thickness and mass. This is essential for spreading the load evenly across the soft tissues and bones of the residual part of the lower limb. Most upper and lower limb prostheses are made of composites with a polymer matrix because they are vital for their weight.

At the moment, scientists are looking into the possibility of using natural plant fibers like pineapple, bamboo, banana, sisal, henequen, jute, ramie, coconut (coir), rice husk, wood, and wheat straw as reinforcement materials in different polymer matrices. Lignocellulosic plant fibers are better than mineral fillers in many ways when they are used as reinforcements in composites. These fibers are thin, have a high elasticity modulus, and have a high specific strength. They are also not too rough. Moreover, they do not pose any health risks to the composites made from them. All lignocellulosic natural fibers comprise cellulose microfibrils in an amorphous lignin and hemicellulose matrix. This structure is similar to a single ramie fiber. Even though they have a lot of potential natural resources like bamboo, which grows in Southeast Asian countries and makes up almost half of the world's bamboo forests, are not used to their fullest. This material is crucial for maintaining ecological balance and sustaining socio-economic development. Waving bamboo and ramie fibers were used as glass and carbon fibers reinforcing materials in the current polypropylene study. The vacuum method made a socket for the lower limb below the knee. The researchers looked at how the mechanical and volumetric properties of the socket were affected by the order in which the fibers were layered [2]. They did this through both tests and computer simulations.

Thrombosis is a big reason why prostheses are needed worldwide, and when expensive materials aren't available, it's essential to use cheaper materials that are easy to get. In places with earthquakes or wars, biomaterials made from plants could be a good alternative. Materials made of composites, made from two or more parts that have their physical properties, have changed how orthopaedic and prosthetic devices are made. Among the most important popular multiphase resources in orthopaedics today is fiber, used in reinforced polymer composites. Carbon fiber is the most valuable reinforcement because it is stiff and strong in tension and compression. However, the materials usually used to make sockets, such as per loin fibers, carbon fabric, and PMMA polymer resin, are not sustainable and can release harmful gases and dust that require expensive safety equipment. Because natural fibers are easy to find and can be recycled, it is essential to learn how to use them. Recent studies have examined the use of plant fibers reinforced composites. To improve the modulus of elasticity and tensile strength, researchers changed the number of fiber layers and angles of direction. Laminations combining ramie, bamboo, carbon, and glass fibers have also been suggested to achieve physical and mechanical properties for the production of prosthetic sockets for the lower limbs below the knee [3].

The study aimed to determine the optimal method for constructing composite materials for lower limb prosthetic sockets by conducting a tensile test to assess their mechanical properties. The socket models were analyzed using the ANSYS finite element method to determine the equivalent Von Mises stress, equivalent Von Mises strain, safety factor distribution, and total deformation. This study showed the benefits of ramie and bamboo fiber mixed with polymethyl methacrylate (PMMA) resin composites, which haven't been tried before. The composites were mixed at different volume percentages and lamination layups.

Experimentation In This Research

Here, we will talk about how sockets are made, including the materials, tools, and preparation stages needed for suitable mechanical properties. We will also give details about the mechanical tests that can be used to judge how well the socket works.

Material Used in This Study

For their investigation into below-knee lamination sockets, researchers employed a composite material comprising a woven mat of ramie and bamboo fibers from Changzhou Doris Textile Co., Ltd, supplemented with woven carbon fiber, glass fiber, acrylic resin, polyvinyl alcohol (PVA) bag, hardening powder from Otto-Bock Healthcare, and gypsum for mold creation. Instrumentation included a 27x18x4 cm positive Jepson, a vacuum-forming system with a pump and associated tubing, sensitive weighing apparatus, and a digital vernier for precise measurement. The mechanical processing was conducted in the university's technical training and workshop department, utilizing CNC water jet cutting equipment to ensure exact dimensions of the materials, as detailed in Fig. 1.

Procedure of Sample Manufacturing

The production process for the samples entailed multiple steps. Firstly, a gypsum mold was created and connected to a vacuum system via pressure tubes. Then, the inner layer of the PVA bag was placed onto the gypsum molds, and the vacuum system was used to apply suction, avoiding bubble formation and bonding between the resin and gypsum mold. The lamination lay-up added Ramie fiber, bamboo fibers, carbon fiber, and person stockinet, as presented in Table I and illustrated in Fig. 1(b) and (c). Following this, an external layer of PVA was applied, and then a length of cotton thread was used to fix the PVC bag, keeping the smaller tip on the valve. A vacuum system was utilized to suck out the air between all bags, leaving the top end open for resin supply. The polymethyl methacrylate resin was mixed with the hardener according to the standard ratio of 2-3[4], and the resulting mixture was then placed inside the external (PVA) bag, where the matrix was distributed uniformly over the area of the lamination stockinet, as depicted in Fig.1 (d) The vacuum device was left running until the composite material solidified, after which the resulting lamination was removed from the gypsum mold. Six different types of laminated composite materials were produced, as demonstrated in Table 1.

Table 1: Presents a list of different laminated composite Materials used in this work.

Groups	Lamination	Lay-up symbol	The sum of Layers	No. of Lamination
Group 1	Lamination 1	242	8	(2Perlon +4Ramie fiber +2perlon)
	Lamination 2	22222	10	(2Perlon +2Ramie + 2Carbon +2Ramie + 2Perlon)
	Lamination 3	22222	10	(2Perlon +2Ramie + 2Glass +2Ramie +2Perlon)
Group 2	Lamination 4	242	8	(2Perlon +4Bamboo + 2perlon)
	Lamination 5	22222	10	(2Perlon +2 Bamboo + 2 Carbon+2 Bamboo +2Perlon)
	Lamination 6	22222	10	(2Perlon +2 Bamboo + 2Glass +2 Bamboo +2Perlon)

Fig. 1: Demonstrates the process of preparing test specimens for a knee prosthetic socket and depicts the various steps involved in the preparation procedure.

Materials Research Proceedings 68 (2026) 97-104 https://doi.org/10.21741/9781644904237-13

Testing For Modified Laminated Composite

The mechanical properties of a material refer to its response to externally applied forces. By evaluating a material's mechanical characteristics, including its predicted performance under specific conditions, it is possible to classify and identify the material according to established criteria.

Tensile Test

Tensile tests were done on the laminated composite samples to determine their mechanical properties. This involved creating a stress-strain curve for each sample; for making tensile laminated composite specimens, the ASTM standard (D638-14 Type I) is used as a guide [5]. The tensile test was done with a 5 mm/min feeding speed on an Instron universal testing machine until the specimen broke under load. The stress-strain curve was derived for each specimen through the utilization of the tensile test [6]. The specimens are subjected to this method to determine their tensile properties, such as ultimate tensile strength (UTS), elastic modulus (E), and percentage elongation at the fracture point. The instruments, a conventional tensile test sample, and the samples before and after testing are shown in Fig.2.

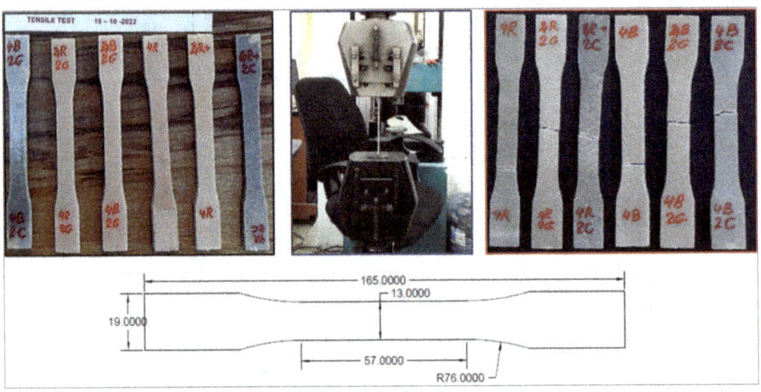

Fig. 2: (a) The samples before tensile testing, (b) Testing apparatus, (c) The samples after tensile testing, and (d) The standard sample dimensions.

Numerical Analysis for Modified Laminated Composite

To analyze the motion system of patients, a prosthetic socket was divided into four categories based on its weight-bearing properties (anterior, posterior, lateral, and medial), and interface pressures were measured between the below-knee socket and the residual lower limb during self-selected walking speed [7]. A finite element approach (ANSYS-22) was used to simulate the socket and evaluate the von Mises (stress and strain) and total deformation by applying boundary conditions. This involved measurement of the interface pressure distribution in various socket areas and fixing a support at the tip of the socket's bottom. For the model to give complete results, the mechanical properties of the prosthetic sockets of all groups of composite materials were added to the workbench ANSYS 22 data [8].

The finite element analysis was conducted using ANSYS Workbench 22 software on a prosthetic socket with a body mass of 76 kg. The process was completed using automatic meshing, resulting in 9351 elements and 18606 nodes. The meshing process was accurate, and the interface pressure was located correctly, as shown in Fig. 3. Table 2 summarizes the pressures on the plane of the socket, which were tested at the most dangerous point in the gait cycle when the heel hit the ground.

Table 2: Presents the location and average pressure data at the heel strike site observed in the prosthetic socket [9].

Location	Anterior	Lateral	Posterior	Medial
Interface pressure (MPa)	105	70	95	85

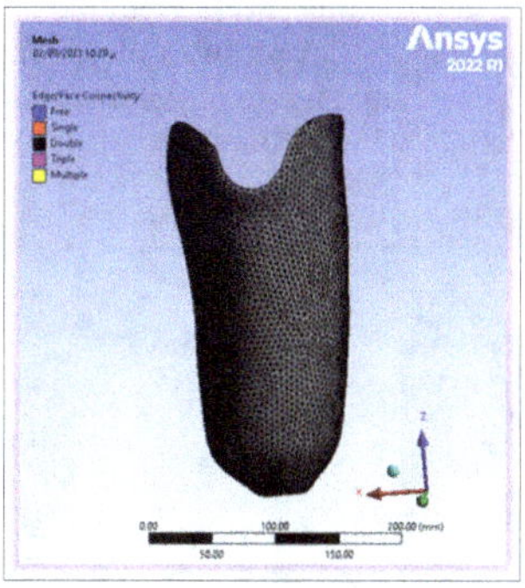

Fig. 3: Process of meshing sockets.

Experimental Result for Modified Laminated Composite

Tensile Strength of Modified Laminated Composites. Fig. (4) indicates that the ultimate tensile strength (UTS) of composite prosthetic sockets rises with an increase in the number of layers of ramie and bamboo. This implies incorporating more fiber layers can improve the laminated material's ultimate tensile strength (UTS). The different levels of ultimate tensile strength come from the different qualities of ramie, bamboo, carbon, and glass fibers. As you can also see, the laminates, including four layers of ramie fiber plus two layers of carbon fiber, had the greatest UTS values compared to the remanufactured eleventh sample. This is because, with carbon fibers included, stresses could be transferred more easily from the matrix to the fiber. Although adding bamboo fiber did not significantly enhance the properties of the matrix material, the matrix's characteristics were mainly responsible for the composites' overall tensile performance [9,10].

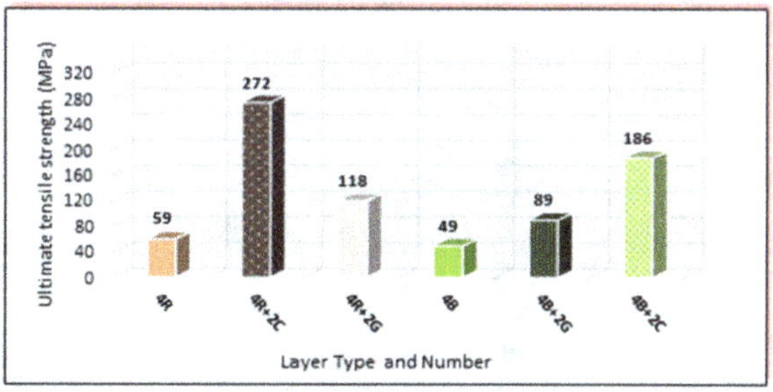

Fig. 4: Shows the influence of reinforced natural fibers (ramie and bamboo) plus synthetic fiber layers on the ultimate tensile strength.

Young's Modulus for modified laminated composite. The modulus of elasticity measures the rigidity of a component. Fig. (5) shows how the modulus of elasticity is affected by the amount and type of reinforcement fiber layers, such as ramie, bamboo, carbon, and glass fibers, added to the polymethyl methacrylate resin PMMA matrix. The results indicate that as the number of reinforcing layers increases, the elasticity modulus also increases. Adding two layers of carbon fiber with four layers of ramie and four layers of bamboo fiber led to a gradual increase in the modulus of elasticity due to the rising binding forces from the increased carbon fibre layers [11]. All twelve laminations, consisting of four layers of ramie fiber and two layers of carbon fiber, exhibited the highest modulus of elasticity among all laminations tested. Because of how carbon fibers are made, composite materials with carbon fiber have the highest modulus of elasticity [12,13].

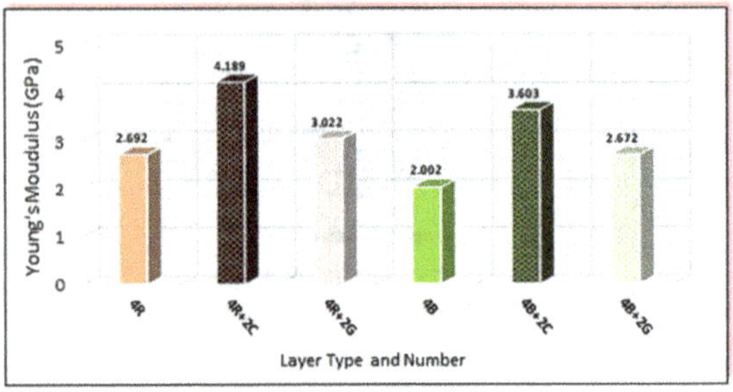

Fig. 5: Shows the influence of reinforced natural fibers (ramie and bamboo) plus synthetic fiber layers on Young's modulus.

Measure the elongation percentage at the failure to determine how the fiber layers affect the composite specimens' flexibility. The elongation percentage values decreased as the amount of ramie and bamboo used to reinforce the samples increased. Moreover, natural fiber-reinforced composites stretch less than natural plus synthetic fibre-reinforced composites. Also, it can be shown that adding these reinforcements to PMMA did not make it more flexible; the results, which are shown in Fig. (6), show that the elongation results from incorporating carbon fibers into ramie fibers and Perlon layers, decreasing from 7.6 % for (4ramie plus 2 Glass) to 6.5 % for (4ramie plus 2 Carbon). This indicates that Carbon fiber is more rigid than Glass fiber, but that rigidity also means it is not as durable, and Glass fiber has greater elongation-to-break.

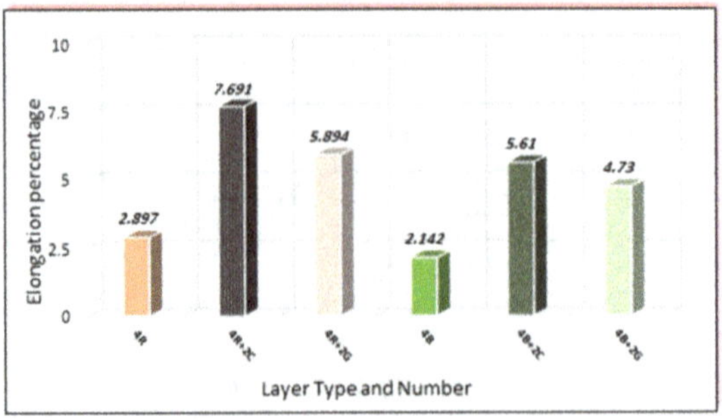

Fig. 6: Shows the influence of reinforced natural fibers (ramie and bamboo) plus synthetic fiber layers on the elongation percentage.

Conclusion

According to the study's findings, the laminated material prosthetic socket prototype serves as a safe and comfortable alternative for patients. Moreover, it is an environmentally friendly option that employs natural materials that are eco-friendly and recyclable. Among the tested materials, the hybrid has four layers of ramie plus two layers of glass fiber reinforcement and has the most significant density, thickness, and volume percentage. The ultimate tensile strength and elastic modulus of the below-knee socket changed significantly when the materials that made it up were changed. The stacking arrangement of four layers of Perlon plus four layers of ramie plus two layers of carbon fiber gave the highest tensile strength 272 MPa, and elastic modulus (4.46 GPa).

The most significant elongation percent value was 7.69%, from a mix of four ramies and two glass fibre layers. Last but not least, the bamboo plus carbon fiber reinforcement composites showed the most promising experimental and numerical results, making them a strong contender for enhancing the mechanical properties of below-knee prosthetic sockets.

Acknowledgment

The authors wish to acknowledge the support of Tikrit University and the Materials Engineering Department at the University of Technology for their provision of laboratory facilities and financial aid, which were essential to the completion of this study.

References

[1] M. J. Jweeg, Z. S. Hammoudi, and B. A Alwan, Optimised Analysis, Design, and Fabrication of Trans-Tibial Prosthetic Sockets, IOP Conf. Ser. Mater. Sci. Eng. 433 (2018) 012058. https://doi.org/10.1088/1757-899X/433/1/012058

[2] A. M. Bash, J. K. Oleiwi, and T. T. Othman, A Review of some Characteristics of a Composite Hybrid Socket for Prosthetics Derived from Plant Fibers. Key Eng. Mater. 937 (2022) 99-106. https://doi.org/10.4028/p-1554f7

[3] J. K. Oleiwi, Q. A. Hamad, and S. A. Abdulrahman, Flexural, impact and max. shear stress properties of fibers composite for prosthetic socket, Mater. Today: Proc. 56 (2022) 3121-3128. https://doi.org/10.1016/j.matpr.2021.12.368

[4] N. K. Faheed, J. K. Oleiwi, and Q. A. Hamad, Effect of different fiber reinforcements on some properties of prosthetic socket, Eng. Technol. J. 39 (2021) 1715-1726. https://doi.org/10.30684/etj.v39i11.2267

[5] H. R. Wasmi, J. S. Chiad, and A. A. Hamzah, Experimental study for materials prosthetic above knee socket under tensile or fatigue stress with varying temperatures effect, J. Eng. 20 (2014) 87-95. https://doi.org/10.31026/j.eng.2014.03.08

[6] H. Jena, M. K. Pandit, and A. K. Pradhan, Study the impact property of laminated bamboo-fibre composite filled with cenosphere, Int. J. Environ. Sci. Dev. 3 (2012) 456. https://doi.org/10.7763/IJESD.2012.V3.266

[7] N. K. Faheed, Q. A. Hamad, and J. K. Oleiwi, Enhancement of the flexural and impact properties laminated biocomposite by new natural fibres for artificial lower limb socket, Adv. Mater. Process. Technol. 8(4) (2022) 4347-4364. https://doi.org/10.1080/2374068X.2022.2073052

[8] M. B. Hunain, B. A. Abass, and J. M. Akhudair, Experimental and numerical studies of fatigue properties of carbon/glass fiber/epoxy hybrid composites enhanced with nano TiO2 powder, Diagnostyka, 22 (2021). https://doi.org/10.29354/diag/135146

[9] N. K. Faheed, Q. A. Hamad, and J. K. Oleiwi, Tensile and Stress Analysis of Hybrid Composite Prosthetic Socket Reinforced with Natural Fibers, J. Renew. Mater. 10(7) (2022) 1989. https://doi.org/10.32604/jrm.2022.017573

[10] H.A. Sharhan, Z.N. Rasheed, and J. K. Oleiwi, Effect of Polypropylene (PP) and Polyacrylonitrile (PAN) Fibers Reinforced Acrylic Resin on Compression, Hardness and Surface-Roughness for Denture Applications, Key Eng. Mater. 911 (2022) 9-16. https://doi.org/10.4028/p-197rys

[11] H.A. Sharhan, Z.N. Rasheed, and J.K. Oleiwi, Synthesis and Physical Characterization of PMMA/PP and PMMA/PAN Composites for Denture Applications J. Appl. Sci. Nanotechnol. (2021). https://doi.org/10.53293/jasn.2021.3759.1040

[12] J.K. Oleiwi, S.I. Salih, and H.S. Fadhil, Effect of Siwak and Bamboo Fibers on Tensile Properties of Self-Cure Acrylic Resin Used for Denture Applications, J. Mater. Sci. Eng. 06 (2017). https://doi.org/10.4172/2169-0022.1000370

[13] S.I. Salih, J.K. Oleiwi, and H.S. Fadhil, Preparation and Investigation of some Properties of Acrylic Resin Reinforced with Siwak Fiber Used for Denture Base Applications, Science 2 (2017) 309-314. https://doi.org/10.24017/science.2017.3.53

Physicochemical Characteristics of EPOLA/PEGDA as a 3D Scaffold for Tissue Engineering Application

SITI NURUL SYAFIKA Sheikh Ibrahim[1,a], MAT UZIR Wahit[1,b*], MARINA Talib[2,c*], NORHAYANI Othman[1,d], and FARAH FADZEHAH Hilmi[2,e]

[1]Faculty of Chemical and Energy Engineering, Universiti Teknologi Malaysia, 81310, UTM Johor Bahru, Johor, Malaysia

[2]Malaysian Nuclear Agency, Radiation Processing Technology Division, Bangi, 43000, Kajang, Selangor, Malaysia

[a]snsyafika@gmail.com, [b]r-uzir@utm.my, [c]marina@nm.gov.my, [d]norhayani@utm.my, [e]fadzehah@nm.gov.my

Keywords: 3D Scaffold, 3D Printing, SLA, Palm Oil, UV Radiation Crosslinking

Abstract. Ultraviolet (UV)-assisted curing has gained significant attention in tissue engineering due to its energy efficiency, reduced chemical consumption, and lower environmental impact. The development of 3D biomedical scaffolds requires bioresins that support essential properties such as biocompatibility, cell adhesion and proliferation, and mechanical strength comparable to native tissue. While synthetic polymers like PEGDA provide structural stability, they often lack biofunctionality; hence, natural polymers are incorporated to enhance biological performance. In this study, EPOLA, a natural-based polymer, was integrated into PEGDA to develop a novel EPOLA/PEGDA bioresin scaffold. Comprehensive characterization was conducted to validate successful UV-induced crosslinking, which is critical for achieving desired scaffold performance. Gel fraction analysis demonstrated a 27.48 % increase post-crosslinking, indicating enhanced polymer network formation. FTIR spectroscopy confirmed the presence of EPOLA functional groups, verifying chemical integration within the scaffold matrix. AFM results revealed a marked increase in surface roughness (*Rq* increased from 0.408 nm to 2.119 nm), suggesting a denser crosslinked structure. Additionally, compressive strength improved by 56.82 %, indicating that mechanical properties can be modulated through crosslinking density. These findings collectively confirm successful crosslinking of the EPOLA/PEGDA bioresin, establishing a foundation for subsequent evaluation of its biofunctionality in tissue engineering applications.

Introduction

3-Dimensional (3D) printing has emerged as a promising technology in tissue engineering development. This type of therapy is amongst the main therapies used in tissue engineering [1,2,3,4,5,6,7]. Stereolithography (SLA) 3D printing that uses ultraviolet (UV) radiation as the curing source to build tissue scaffold has become more popular since UV offers a fast, efficient, and environmentally friendly that requires less energy, generates less waste, and uses fewer chemicals [1, 8,9,10,11,12].

PEG based hydrogels have been a material of interest in tissue scaffold development [11, 19,20,21,22,23,24,25,26,27]. The flexibility of the synthetic polymer tailored according to the targeted end use [15,16,17] has made PEG based hydrogels one of the most used materials in tissue engineering studies. PEG material has proven to be advantageous for cell culture as it has characteristic properties such as high hydrophilic, good tissue biocompatibility, lack of toxicity and availability of reactive sites for chemical modification. However, the use of PEG alone will not offer the natural biomimetic properties offered by natural substances as PEG is a synthetic polymer, thus the addition of a natural based polymer is preferable to enhance scaffolds' properties

[4, 18,19,20,21] to make it more suitable for tissue engineering applications. One of the characteristics of a good scaffold is the high porosity of the scaffold to promote cell growth and proliferation. In this study, Polyethylene Glycol Diacrylate (PEGDA) is chosen to be the control scaffold material since acrylates monomers are known for their high reactivity, polymerise rapidly in the present of photogenerated free radicals that is suitable for crosslinking using UV radiation [11]. It produced strong crosslinked bonds between acrylate groups and showed promising biocompatibility from cellular tests. To enhance the biological functionality of PEGDA scaffold, Epoxidized Palm Oil Acrylate (EPOLA) is introduced into PEGDA photopolymer resin as a new bioresin for tissue scaffold development using stereolithography 3D printer through UV crosslinking process.

Epoxidized palm oil acrylate (EPOLA) is a modified form of palm oil that has undergone epoxidation, followed by a reaction with acrylic acid to introduce acrylate functional groups [13, 23,24]. This modification enhances the reactivity and properties of palm oil for various applications [13, 24,25,26,27,28]. The presence of acrylate functional groups can promote crosslinking reactions, leading to the formation of a 3D network structure [1]. This crosslinking can influence the porosity of the resulting scaffold [4, 28,29,30,31]. EPOLA has certain properties that could potentially be exploited in scaffold development, such as biocompatibility or the ability to be processed into various forms.

With the introduction of EPOLA into PEGDA, it is expected that there will be an increase in the crosslinking percentage of the scaffold which further increases the porosity of the scaffold. In this paper, the scaffold, EPOLA/PEGDA, is characterized accordingly to analyse the effects on its chemical, morphological, and mechanical properties.

Experimental

Design and Fabrication. A 3D scaffold of EPOLA/PEGDA was fabricated using stereolithography (SLA) 3D printer under FRGS/1/2019/TK05/UTM/02/15 (5F199) and IAEA Coordinate Research Project F23030. The scaffold model was fabricated using a mixture of 54.58 % PEGDA, 44.41 % EPOLA, with 1.0 % IR369 photoinitiator; UV crosslinked in SLA 3D Printer (ANYCUBIC LCD-based SLA 3D Printer, UV integrated light - wavelength 405nm) as in Fig.1. The resin was stirred for 24 hours at 700 rpm using a magnetic stirrer after mixing to ensure smooth printing process. The mechanism of UV crosslinking process using the SLA 3D printer is as illustrated in Fig.2. SLA 3D printer consists of a build platform, a resin tank, and an ultraviolet (UV) laser or light source. The build platform is positioned at the bottom of the resin tank, and the resin is poured into the tank. When the UV light hits the liquid resin, it causes a chemical reaction known as photopolymerisation. The UV light triggers the resin to solidify and harden, layer by layer, based on the pattern of the specific slice from the chosen scaffold model.

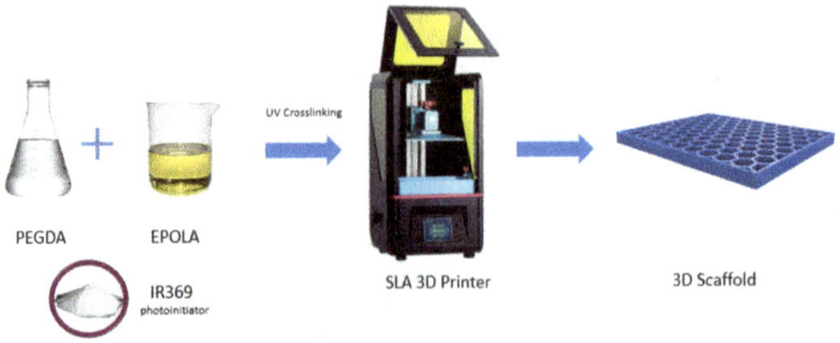

Fig 1 : Fabrication of 3D Scaffold using PEGDA and EPOLA bioresin with IR369 photoinitiator; UV crosslinked in SLA 3D Printer.

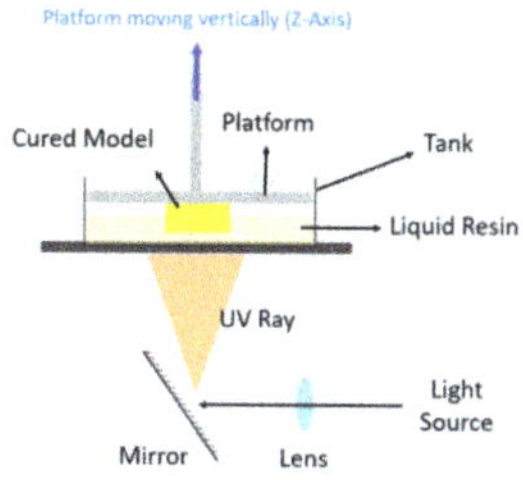

Fig. 2 : *Mechanism of UV crosslinking process using SLA 3D Printer.*

Characterisation. Following scaffold fabrication, a series of characterizations were conducted to evaluate the impact of EPOLA incorporation into PEGDA after the UV crosslinking process. These analyses aimed to confirm successful crosslinking by examining the chemical, morphological, and mechanical properties of the scaffold. The degree of crosslinking was first quantified using a gel fraction test, followed by further confirmation through FTIR spectroscopy, AFM surface analysis, and compressive testing using a universal testing machine.

Gel Fraction Test. The method to analyse the degree of crosslinking was adopted from Haryanto et.al [32]. The scaffolds were fully immersed in distilled water, each in a beaker, for 24 hours in a fume hood. They were then dried in the oven at 50 °C for the next 24 hours. After drying overnight, the printed samples were dried in vacuum oven at 50 °C until each of them obtained a constant weight. Their weights were recorded every 1-hour interval.

The gel fraction percentage is calculated as follows.

$$\text{Gel fraction } (\%) = (W_c/W_o)\times 100$$

Where W_c and W_o are the weight of dried scaffold and the initial weight of the scaffold, respectively.

Fourier Transform Infrared Spectroscopy (FTIR) Analysis. FTIR analysis (Tensor II, Bruker) was performed to verify the presence of EPOLA functional groups following EPOLA/PEGDA crosslinking. Liquid EPOLA and PEGDA resins were measured by placing a drop between salt plates, while the solid crosslinked sample was placed directly on the sample holder. Spectra were recorded in the 600–4000 cm^{-1} range and analyzed to identify characteristic absorption bands indicative of successful crosslinking.

Atomic Force Microscopy (AFM) Analysis. Characterisation of surface roughness and topography was analysed using tapping mode AFM (Shimadzu SPM-9500J2) to observe three-dimensional (3D) image and to compute surface roughness, R_q.

Compressive Strength Analysis. Compressive strength was determined and measured using SHIMADZU Autograph AGX-V Series. The sample was placed between compressive plates parallel to the surface and compressed at a uniform rate. The maximum load was recorded along with stress-strain data.

Results and Discussion

Gel Fraction Test. Determination of EPOLA/PEGDA scaffold gel fraction percentage after UV crosslinking is the basis of the successful crosslinking of the new bioresin formulation. Fig. 3 shows the comparison of gel fraction percentage of PEGDA and EPOLA/PEGDA.

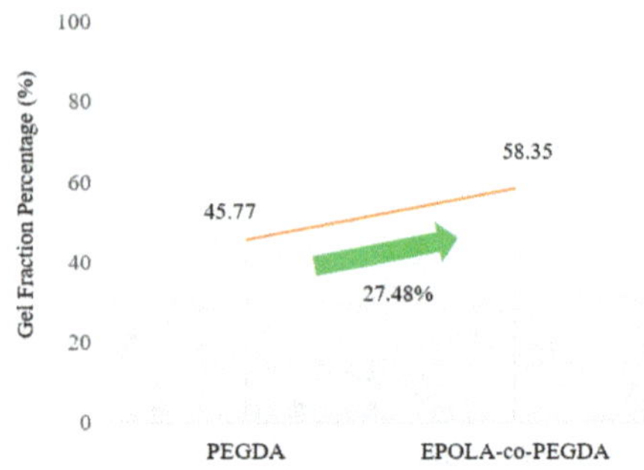

Fig.3 : Gel Fraction Percentage (%) of PEGDA and EPOLA/PEGDA.

From the gel fraction test conducted, gel fraction percentage of EPOLA/PEGDA was recorded at 58.35 %, an increment of 27.48 % from PEGDA, which recorded at 45.77 %. A higher gel fraction percentage suggests a higher degree of crosslinking in the polymer network as more polymer chains were bonded with covalent bonds during the UV crosslinking process. The formation of a more stable structure of the scaffold will result in improved mechanical strength, chemical resistance, and thermal stability.

Fourier Transform Infrared Spectroscopy (FTIR) Analysis. IR spectras provide valuable information about molecular structure and functional groups. Fig. 4 compares the FTIR spectral of PEGDA and EPOLA before crosslinking, with EPOLA/PEGDA after the UV crosslinking process.

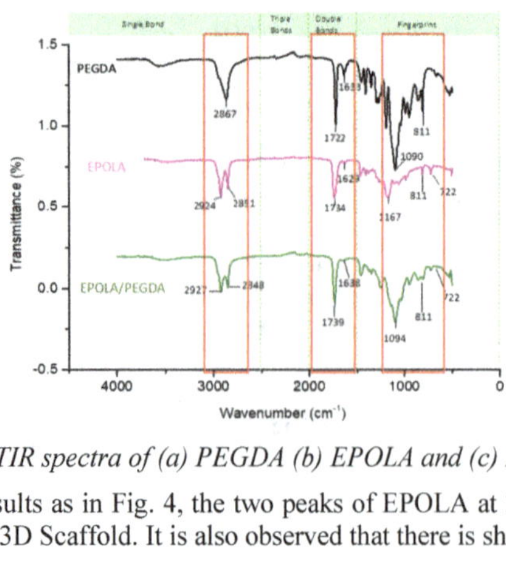

Fig. 4 : FTIR spectra of (a) PEGDA (b) EPOLA and (c) EPOLA/PEGDA.

Based on the FTIR results as in Fig. 4, the two peaks of EPOLA at 2924 cm^{-1} and 2851 cm^{-1} appear in the crosslinked 3D Scaffold. It is also observed that there is shifting in peaks from 2924

Materials Research Proceedings 68 (2026) 105-112 https://doi.org/10.21741/9781644904237-14

cm^{-1} to 2927 cm^{-1} and 2851 cm^{-1} to 2848 cm^{-1}. The intensity of the peaks at the area of 1638 cm^{-1} and 1739 cm^{-1} are also reduced, most probably due to the elimination of double bonds of C=C from PEGDA. Most peaks in the fingerprint region are also seen to shift and reduce in intensities in the area from 1500 cm^{-1} - 500 cm^{-1}. Obvious reductions can be seen at peak 1094 cm^{-1}, 811 cm^{-1}, and 722 cm^{-1}. The shifting and reduction in intensities suggest that there are changes in the molecular structure of EPOLA/PEGDA after UV crosslinking process.

Atomic Force Microscopy (AFM) Analysis. Topography changes can be observed after the crosslinking process. Fig. 5 shows the 3D surface topography image of (a) PEGDA and (b) EPOLA/PEGDA Scaffold.

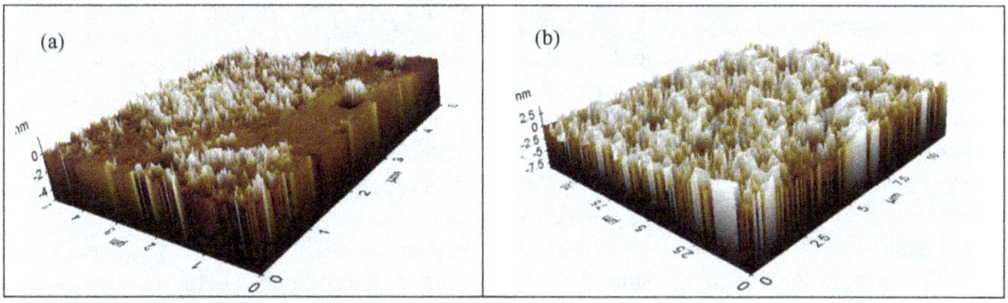

Fig. 5 : 3D Surface topography image of (a) PEGDA Scaffold and (b) EPOLA/PEGDA Scaffold.

The PEGDA scaffold exhibits a smooth and uniform surface with occasional bumps or ridges, whereas the EPOLA/PEGDA scaffold shows a slightly undulating texture attributed to random polymer chain orientation or arrangement. The root mean square (R_q) values for surface roughness were measured at 0.408 nm and 2.119 nm for PEGDA and EPOLA/PEGDA, respectively, indicating smoother and rougher surfaces, respectively. The higher R_q value of EPOLA/PEGDA suggests a highly crosslinked scaffold structure. This scaffold effectively inhibits nanoroughness, resulting from crosslinked polymer networks and UV-induced irregularities. Rough surfaces can enhance cell attachment and proliferation by offering more surface area and favorable topographical cues for cell anchorage. Studies demonstrate that cells adhere better to rough surfaces due to increased binding sites from surface irregularities and enhanced surface area provided by such scaffolds.

Compressive Strength Analysis. The compressive strength of PEGDA and EPOLA/PEGDA scaffolds determined how much load the samples can bear under compressive forces. Fig. 6 shows the comparison in compressive strength between both scaffolds.

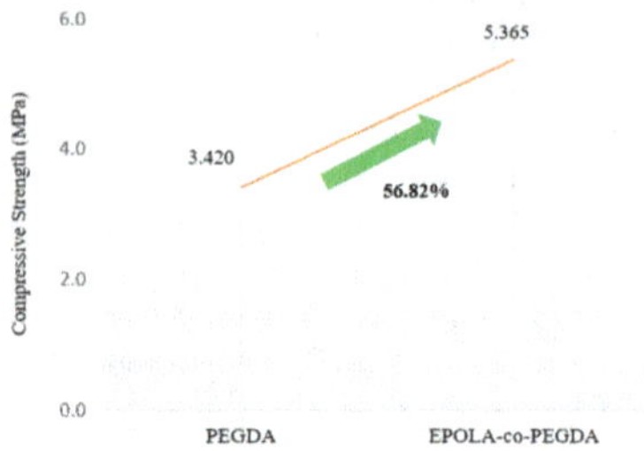

Fig. 6 : Compressive Strength of PEGDA and EPOLA/PEGDA.

The compressive strength is recorded at 5.365 MPa for EPOLA/PEGDA, an increase of 56.82 % as compared to PEGDA at 3.420 MPa. The compressive modulus is expected to be able to be tuned by varying the degree of crosslinking, which affects the density and stiffness of the polymer network. With the addition of EPOLA, the aim is to still have the flexibility of PEGDA scaffold while improving the ability to withstand natural loading conditions that is comparable to natural tissue.

Conclusion

The experimental data from the study proved a successful crosslinking between EPOLA and PEGDA, with gel fraction percentage increased by 27.48 %, FTIR results shown changes in the molecular structure of EPOLA/PEGDA, AFM showed a higher R_q value, and compressive strength increased by 56.82 %. All results support improvement of existing PEGDA resin served as resin for 3D scaffold development using UV crosslinking. Further research can be conducted to study the suitability of the new bioresin in terms of its biofunctionality to serve as a good tissue scaffold, potentially for bone or cartilage tissue scaffold.

Acknowledgement

This research was funded by a grant from Ministry of Higher Education of Malaysia FRGS/1/2019/TK05/UTM/02/15 (5F199) and IAEA Coordinate Research Project F23030. The authors gratefully acknowledge Universiti Teknologi Malaysia and Malaysian Nuclear Agency for materials and instruments support.

References

[1] Ibrahim SNSS, Wahit MU, Talib M, Othman N, Shukri NA, Hilmi FF, Arshad MA, UV Radiation Crosslinking of Acrylated Palm Olein (APO) Copolymer Resins for 3D Printing. Key Engineering Materials, Trans Tech Publ (2022). https://doi.org/10.4028/p-h1u335

[2] Masri S, Zawani M, Zulkiflee I, Salleh A, Fadilah NIM, Maarof M, Wen APY, Duman F, Tabata Y, Aziz IA, Bt Hj Idrus R, Fauzi MB, Cellular Interaction of Human Skin Cells towards Natural Bioink via 3D-Bioprinting Technologies for Chronic Wound: A Comprehensive Review, Int J Mol Sci (2022). https://doi.org/10.1016/j.ijbiomac.2024.131623

[3] Zhang F, King MW, Biodegradable Polymers as the Pivotal Player in the Design of Tissue Engineering Scaffolds, Adv Healthc Mater (2020). https://doi.org/10.1002/adhm.201901358

[4] Sivashankari PR, Prabaharan M, Three-dimensional porous scaffolds based on agarose/chitosan/graphene oxide composite for tissue engineering, Int J Biol Macromol (2020), pp.222-31. https://doi.org/10.1016/j.ijbiomac.2019.12.219

[5] Dai L, Cheng T, Duan C, Zhao W, Zhang W, Zou X, Aspler J, Ni Y, 3D printing using plant-derived cellulose and its derivatives: A review, Carbohydr Polym (2019), pp.71-86. https://doi.org/10.1016/j.carbpol.2018.09.027

[6] Corker A, Ng HC, Poole RJ, Garcia-Tunon E, 3D printing with 2D colloids: designing rheology protocols to predict 'printability' of soft-materials, Soft Matter (2019), pp.1444-56. https://doi.org/10.1039/C8SM01936C
[7] Alghamdi H, Nair SAO, Neithalath N, Insights into material design, extrusion rheology, and properties of 3D-printable alkali-activated fly ash-based binders, Materials & Design (2019).
[8] Bennett J, Measuring UV Curing Parameters of Commercial Photopolymers used in Additive Manufacturing, Addit Manuf (2017). pp.203-12. https://doi.org/10.1016/j.addma.2017.10.009
[9] Bryant SJ, Nuttelman CR, Anseth KS, Cytocompatibility of UV and visible light photoinitiating systems on cultured NIH/3T3 fibroblasts in vitro, J Biomater Sci Polym Ed (2000). pp.439-57. https://doi.org/10.1163/156856200743805.
[10] Park JW, Shim GS, Lee JG, Jang SW, Kim HJ, Choi JN, Evaluation of UV Curing Properties of Mixture Systems with Differently Sized Monomers, Materials (Basel) (2018). https://doi.org/10.3390/ma11040509
[11] Chen JY, Hwang JV, Ao-Ieong WS, Lin YC, Hsieh YK, Cheng YL, Wang J, Study of Physical and Degradation Properties of 3D-Printed Biodegradable, Photocurable Copolymers, PGSA-co-PEGDA and PGSA-co-PCLDA, Polymers (Basel) (2018). https://doi.org/10.3390/polym10111263
[12] Tajau R, Ibrahim MI, Yunus NM, Mahmood MH, Salleh MZ, Salleh NGN, Development of palm oil-based UV-curable epoxy acrylate and urethane acrylate resins for wood coating application, Advancing Nuclear Research and Energy Development (2014). pp.164-9. https://doi.org/10.1063/1.4866125
[13] Tang J, Zhang J, Lu J, Huang J, Zhang F, Hu Y, Liu C, An R, Miao H, Chen Y, Huang T, Zhou Y, Preparation and Properties of Plant-Oil-Based Epoxy Acrylate-Like Resins for UV-Curable Coatings, Polymers (Basel) (2020). https://doi.org/10.3390/polym12092165
[14] Wang Y, Ma M, Wang J, Zhang W, Lu W, Gao Y, Zhang B, Guo Y, Development of a Photo-Crosslinking, Biodegradable GelMA/PEGDA Hydrogel for Guided Bone Regeneration Materials, Materials (Basel) (2018). https://doi.org/10.3390/ma11081345
[15] Stillman ZS, Jarai BM, Raman N, Patel P, Fromen CA, Degradation Profiles of Poly(ethylene glycol) diacrylate (PEGDA)-based hydrogel nanoparticles, Polym Chem (2020). pp.568-80. https://doi.org/10.1039/C9PY01206K
[16] Burke G, Cao Z, Devine DM, Major I, Preparation of Biodegradable Polyethylene Glycol Dimethacrylate Hydrogels via Thiol-ene Chemistry, Polymers (Basel) (2019). https://doi.org/10.3390/polym11081339
[17] Urrios A, Parra-Cabrera C, Bhattacharjee N, Gonzalez-Suarez AM, Rigat-Brugarolas LG, Nallapatti U, Samitier J, DeForest CA, Posas F, Garcia-Cordero JL, Folch A, 3D-printing of transparent bio-microfluidic devices in PEG-DA, Lab Chip (2016). pp.2287-94. https://doi.org/10.1039/c6lc00153j
[18] Ullah S, Zainol I, Chowdhury SR, Fauzi MB, Development of various composition multicomponent chitosan/fish collagen/glycerin 3D porous scaffolds: Effect on morphology, mechanical strength, biostability and cytocompatibility, Int J Biol Macromol (2018). pp.158-68. https://doi.org/10.1016/j.ijbiomac.2017.12.136
[19] Saber-Samandari S, Saber-Samandari S, Biocompatible nanocomposite scaffolds based on copolymer-grafted chitosan for bone tissue engineering with drug delivery capability, Mater Sci Eng C Mater Biol Appl (2017). pp.721-32. https://doi.org/ 10.1016/j.msec.2017.02.112
[20] Lei Y, Xu Z, Ke Q, Yin W, Chen Y, Zhang C, Guo Y, Strontium hydroxyapatite/chitosan nanohybrid scaffolds with enhanced osteoinductivity for bone tissue engineering, Mater Sci Eng C Mater Biol Appl (2017). pp.134-42. https://doi.org/10.1016/j.msec.2016.11.063

[21] Shamekhi MA, Rabiee A, Mirzadeh H, Mahdavi H, Mohebbi-Kalhori D, Baghaban Eslaminejad M, Fabrication and characterization of hydrothermal cross-linked chitosan porous scaffolds for cartilage tissue engineering applications, Mater Sci Eng C Mater Biol Appl (2017).pp:532-42. https://doi.org/10.1016/j.msec.2017.03.194
[22] Tormos CJ, Madihally SV. Chitosan for cardiac tissue engineering and regeneration. Chitosan Based Biomaterials Volume 2 (2017). p.115-43. https://doi.org/10.1016/B978-0-08-100228-5.00005-5
[23] Tajau R, Rohani R, Salleh MZ, Physicochemical and Thermal Properties of Acrylated Palm Olein as a Promising Biopolymer, Journal of Polymers and the Environment (2020). pp.2734-48. https://doi.org/10.1007/s10924-020-01797-6
[24] Yeoh FH, Lee CS, Kang YB, Wong SF, Cheng SF, Ng WS. Production of Biodegradable Palm Oil-Based Polyurethane as Potential Biomaterial for Biomedical Applications, Polymers (Basel) (2020). https://doi.org/10.3390/polym12081842
[25] Tajau R, Rohani R, Wan Isahak WNR, Salleh MZ, Ghazali Z, Development of new bio-based polyol ester from palm oil for potential polymeric drug carrier, Advances in Polymer Technology (2018). pp.3552-60. https://doi.org/10.1002/adv.22139
[26] Lee D, Song SH, Investigation of Epoxidized Palm Oils as Green Processing Aids and Activators in Rubber Composites, International Journal of Polymer Science (2019). pp.1-7. https://doi.org/10.1155/2019/2152408
[27] Tajau R, Rohani R, Wan Isahak W, Salleh MZ, Palm oil-based precursors for development of polymeric delivery system, Malays J Anal Sci (2017). pp.496-511. https://doi.org/ 10.17576/mjas-2017-2102-25
[28] Ng WS, Lee CS, Chuah CH, Cheng S-F, Preparation and modification of water-blown porous biodegradable polyurethane foams with palm oil-based polyester polyol, Industrial Crops and Products (2017). pp.65-78. https://doi.org/10.1016/j.indcrop.2016.11.066
[29] Fakhari A, Rahmat AR, Wahit MU, Arjmandi R, Bio-Based Hybrid Polymers from Vinyl Ester Resin and Modified Palm Oil: Synthesis and Characterization, International Polymer Processing (2017). pp.20-5. https://doi.org/10.3139/217.3166
[30] Park J-W, Lee J-G, Shim G-S, Kim H-J, Kim Y-K, Moon S-E, No D-H, Evaluation of the Ultraviolet-Curing Kinetics of Ultraviolet-Polymerized Oligomers Cured Using Poly (Ethylene Glycol) Dimethacrylate, Coatings (2018). https://doi.org/10.3390/coatings8030099
[31] Ramanathan G, Singaravelu S, Muthukumar T, Thyagarajan S, Perumal PT, Sivagnanam UT. Design and characterization of 3D hybrid collagen matrixes as a dermal substitute in skin tissue engineering, Mater Sci Eng C Mater Biol Appl (2017). pp.359-70. https://doi.org/10.1016/j.msec.2016.11.095
[32] Ullah S, Zainol I, Idrus RH, Incorporation of zinc oxide nanoparticles into chitosan-collagen 3D porous scaffolds: Effect on morphology, mechanical properties and cytocompatibility of 3D porous scaffolds, Int J Biol Macromol (2017). pp.1020-9. https://doi.org/10.1016/j.ijbiomac.2017.06.080
[33] Haryanto, Kim S, Kim JH, Kim JO, Ku S, Cho H, Han DH, Huh P, Fabrication of poly(ethylene oxide) hydrogels for wound dressing application using E-beam, Macromolecular Research (2013). pp.131-8. https://doi.org/10.1007/s13233-014-2023-z

Optimization of EPOLA/PEGDA Formulation for 3D Scaffold Development using Design of Experiment (DOE - Mixture)

SITI NURUL SYAFIKA Sheikh Ibrahim[1,a], MAT UZIR Wahit[1,b*], MARINA Talib[2,c], NORHAYANI Othman[1,d], and FARAH FADZEHAH Hilmi[2,e]

[1]Faculty of Chemical and Energy Engineering, Universiti Teknologi Malaysia, 81310, UTM Johor Bahru, Johor, Malaysia

[2]Malaysian Nuclear Agency, Radiation Processing Technology Division, Bangi, 43000, Kajang, Selangor, Malaysia

[a]snsyafika@gmail.com, [b]r-uzir@utm.my, [c]marina@nm.gov.my, [d]norhayani@utm.my, [e]fadzehah@nm.gov.my

Keywords: Design of Experiment, Design Expert, DOE – Mixture, 3D Printing, SLA

Abstract. This paper explores the optimization of Epoxidized Palm Oil Acrylate (EPOLA) and Polyethylene Glycol Diacrylate (PEGDA) formulations for developing 3D scaffolds via stereolithography (SLA) 3D printing. By leveraging Design-Expert software, experiments were designed to enhance the PEGDA and EPOLA mixture with the addition of the photoinitiator Irgacure-369 (IR 369). EPOLA, a natural-based polymer, complements PEGDA, a synthetic polymer, addressing the scaffold's mechanical robustness necessary for natural tissue-mimicking properties under loading conditions, while facilitating biofunctionality crucial for cell attachment and proliferation. The study focuses on crosslinking the bioresin mixture in an SLA environment, utilizing ultraviolet (UV) light as the crosslinking agent within the 3D printer to fabricate the scaffolds. Employing Design of Experiment – Mixture (DOE-Mixture) methodology in Design-Expert software version 13 enabled systematic optimization. The findings reveal that the highest crosslinking percentage reached 58.349%, achieved with a formulation composed of 54.587% PEGDA, 44.413% EPOLA, and 1.0% IR369, yielding a desirability score of 0.921.This statistical experimental design approach facilitated the systematic determination of the optimal formulation, showcasing the efficacy of integrating natural-based and synthetic polymers for advanced 3D scaffold development via SLA printing technology.

Introduction

The use of software in designing experiments is one of the remarkable improvements in the research world. Using software such as JMP, Design-Expert, Minitab, MATREX, CARD, or R packages like RSM or Mixexp is essential in conducting experiments in the current research approach [1,2,3,45]. These tools are useful in designing and analyzing experiments as they can help to streamline the process and provide statistical support for (Design of Experiment) DOE analysis. It has become necessary for researchers to use these tools to help them eliminate or reduce complexity when designing experiments. Applications of DOE have been growing rapidly in manufacturing and non-manufacturing industries [1,2,3,4,5,6,7]. The systematic, efficient approach enables scientists and engineers to study the relationship between multiple factors and responses. It is a structured approach for collecting data and making discoveries [1,2,3,4,5, 8].

Mixture designs are used when factors are interdependent and when each component in a mixture depends on other component settings [1,9,10]. When dealing with mixtures, the optimization should consider a few methods depending on the factors that act as the manipulated process inputs to cause a change in its output or response. There are three (3) common designs used for mixtures: simplex lattice or simplex centroid, screening, and optimal. A simplex lattice or centroid is used when all factor ranges are identical. Screening is used when the estimation is made on linear models, while optimal is used when the factors have different ranges.

Polyethylene Glycol Diacrylate (PEGDA) scaffolds are pivotal in tissue engineering for their biocompatibility and customizable properties. Recent advancements, such as stereolithography-based fabrication, enhance scaffold architecture and mechanical integrity, crucial for tissue regeneration. This study innovatively combines PEGDA with Epoxidized Palm Oil Acrylate (EPOLA) to address shortcomings of traditional natural polymers like collagen and chitosan, which often lack sufficient mechanical strength and consistent degradation rates [11,12]. These hybrid scaffolds promise improved efficacy in diverse biomedical applications. In this study, as the ranges for PEGDA, EPOLA, and IR369 photoinitiator are not the same, and the output model is expected not to be linear, the optimal design is used as the most suitable method for the optimization process [1,10].

Experimental

Design. The range of each mixture must be known to enable DOE–Mixture to design the experiment [10, 13, 14]. In this study, a preliminary experiment using the One Factor at a Time (OFAT) method was conducted to determine the range of each mixture, and the results were used in the DOE approach. The determined range of 40 – 60 wt% was inserted in the design step of DOE for both EPOLA and PEGDA, with 0 – 1 wt% for the IR369 photoinitiator. A constraint of A+B+C = 100 was also inserted as the total mixture of the formulation should be 100 %. DOE will then automatically help to adjust the range accordingly with the specific constraints for each element. The range of each element configured in DOE is tabulated in Table 1.

Table 1 : Design of Experiment – Mixture for A:PEGDA and B: EPOLA bioresin with C:IR369 photoinitiator; design limit and constraint.

Low Limit		*Constraint*		*High Limit*
40.000	≤	A:PEGDA	≤	60
40.000	≤	B:EPOLA	≤	60
0.000	≤	C:IR369	≤	1.0
		A+B+C	=	100.0

Analysis. Analysis of Variance (ANOVA) was used to analyze the input, as in Table 1. Table 2 shows the design arrangements tabulated by the software. Sixteen runs were run accordingly to achieve the optimized results. The response, crosslinking percentages (%), were recorded accordingly. Model validation represented by coded equations will be generated after suitable models identified by the software are analyzed, and the most suitable model is chosen [9, 15].

Table 2: Design of Experiment – Mixture for A:PEGDA and B:EPOLA bioresin with C:IR369 photoinitiator with crosslinking (%) as a response.

RUN	A:PEGDA	B:EPOLA	C:IR369	RUN	A:PEGDA	B:EPOLA	C:IR369
1	40.800	59.200	0.000	9	49.731	49.269	1.000
2	59.239	40.000	0.761	10	57.971	42.030	0.000
3	48.536	51.064	0.400	11	48.536	51.064	0.400
4	53.387	45.613	1.000	12	42.789	56.211	1.000
5	40.800	59.200	0.000	13	59.239	40.000	0.761
6	55.932	44.068	0.000	14	52.059	47.941	0.000
7	48.536	51.064	0.400	15	45.700	53.800	0.500
8	40.000	59.244	0.756	16	40.000	59.244	0.756

Optimization. After the analyses using ANOVA and selecting the most suitable model, several optimization solutions were suggested and represented by ramps analysis. The highest desirability was then selected from the suggested solutions [14], and the plots to the optimization for the specific response, crosslinking percentage (%), will be graphically presented by a contour plot and a 3-Dimensional (3D) perspective.

Post Analysis. Model validation was carried out to validate the results from DOE. The experiments were conducted three (3) times with the same optimized data, and the responses were keyed into the software. Confirmation data will analyze the sample average by comparing the prediction interval of the model [13,14]. The model is 'confirmed' if the average of the sample is inside the prediction interval. The Coefficients Table shows the coefficient magnitudes and significance for all analyzed responses.

Results and Discussion

Design. The determination of EPOLA/PEGDA scaffold crosslinking percentage after UV crosslinking is the basis of the successful crosslinking of the new bioresin formulation. A higher crosslinking percentage suggests a higher density of gel components in the interconnected polymer chains network due to the presence of more linkages per length of polymer chain [16-19]. Table 1 shows the design arrangement of the mixture (A:PEGDA, B:EPOLA, and C:IR369) created by the software and the response (crosslinking %) resulting from the experiments.

Table 3: Crosslinking (%) response for Design of Experiment – Mixture for A:PEGDA and B:EPOLA bioresin with C:IR369 photoinitiator.

Run	Crosslinking %	Run	Crosslinking %
1	44.60	9	59.58
2	55.12	10	55.70
3	53.65	11	54.76
4	58.33	12	44.90
5	44.86	13	55.15
6	56.00	14	56.01
7	54.30	15	50.80
8	44.03	16	44.30

Analysis. ANOVA provides statistical information on suitable models that could fit the response. Table 4 shows the results from ANOVA with 4 (a) Sequential Model Sum of Squares, 4 (b) Model Summary Statistics, and 4 (c) ANOVA Table for Quadratic Model. From the analysis of the Sequential Model Sum of Squares as in Table 4 (a), two (2) models were suggested, a Quadratic Model and a Special Quartic Model, by considering the highest order polynomial where the additional terms are significant, and the model is not aliased. After further statistical analysis, the Quadratic Model is identified as the most suitable model, as in Table 4 (b) - Model Summary Statistics. In the Special Quartic Model, the predicted R^2 is substantially less than R^2, which may indicate that the model is over-fit; thus, it is not in this case. With the Quadratic Model, the Predicted R^2 of 0.83 is in reasonable agreement with the Adjusted R^2 of 0.91, as the difference is less than 0.2.

Referring to the ANOVA Table in Table 4 (c), the Model F-value of 31.05 implies the model is significant. There is only a 0.01 % chance that an F-value this large could occur due to noise. The correlation between the mixture components AB, AC, and BC shows that only A and B depend on each other, as the results show a significant p-value of AB. The results also validate the use of a photoinitiator which is only for initiating the free radical formation without affecting the chemical structure of the mixture [20]. The Lack of Fit F-value of 39.09 implies the Lack of Fit is significant. There is only a 0.01% chance that a Lack of Fit F-value this large could occur due to noise. P-values less than 0.0500 indicate that the model terms are significant. Values greater than 0.1000 indicate that the model terms are insignificant [21].

Table 4: ANOVA Analyses to determine the most suitable model using (a) Sequential Model Sum of Squares; (b) Model Summary Statistics; and (c) ANOVA Table for Quadratic Model.

(a) Sequential Model Sum of Squares

Source	*Sum of Squares*	*df*	*Mean Square*	*F-value*	*p-value*	
Mean vs Total	43273.36	1	43273.36			
Linear vs Mean	319.80	2	159.90	14.93	0.0004	
Quadratic vs Linear	*111.41*	*3*	*37.14*	*13.37*	*0.0008*	*Suggested*
Sp Cubic vs Quadratic	2.40	1	2.40	0.8517	0.3802	
Cubic vs Sp Cubic	22.37	3	7.46	14.92	0.0034	
Quartic vs Cubic	2.31	1	2.31	16.65	0.0095	Aliased
Residual	0.6928	5	0.1386			
Sp Quartic vs Quadratic	*23.97*	*3*	*7.99*	*14.70*	*0.0021*	*Suggested*
Quartic vs Sp Quartic	3.11	2	1.56	11.23	0.0142	Aliased
Residual	0.6928	5	0.1386			
Total	43732.34	16	2733.27			

(b) Model Summary Statistics

Source	*Std. Dev.*	R^2	*Adjusted* R^2	*Predicted* R^2	*PRESS*	
Linear	3.27	0.6968	0.6501	0.5307	215.40	
Quadratic	*1.67*	*0.9395*	*0.9092*	*0.8246*	*80.52*	*Suggested*
Special Cubic	1.68	0.9447	0.9079	0.7805	100.77	
Cubic	0.7071	0.9935	0.9837	-0.6745	768.58	
Special Quartic	*0.7372*	*0.9917*	*0.9822*	*-12.4973*	*6195.04*	*Suggested*
Quartic	0.3722	0.9985	0.9955		*	Aliased

(c) ANOVA Table for Quadratic Model

Source	*Sum of Squares*	*df*	*Mean Square*	*F-value*	*p-value*	
Model	*431.21*	*5*	*86.24*	*31.05*	*< 0.0001*	*significant*
(1)*Linear Mixture*	319.80	2	159.90	57.57	< 0.0001	
AB	111.22	1	111.22	40.04	< 0.0001	
AC	1.26	1	1.26	0.4536	0.5159	
BC	1.46	1	1.46	0.5260	0.4849	
Residual	27.77	10	2.78			
Lack of Fit	27.08	5	5.42	39.09	0.1100	Not significant
Pure Error	0.6928	5	0.1386			

Model Validation. The chosen Quadratic Model is computed according to Scheffe's polynomial equation for quadratic models. The equation in terms of coded factors is shown in Table 5. This data can be used to predict the response for given levels of each factor. The equation in terms of coded factors is useful for identifying the relative impact of the factors by comparing the factor coefficients [22].

Table 5 : The equation in terms of Coded Factors

Crosslinking	=
+53.50	A
+43.34	B
+1135.20	C
+28.21	AB
-1065.39	AC
-1149.24	BC

In this case, the individual components of A (PEGDA), B (EPOLA), and C (IR369) impact the crosslinking percentage (%), while the only dependent components are between A and B.

Optimization. Table 6 shows the four (4) solutions that were suggested from the ramp analysis that can fit into the model. Based on the highest desirability, solution one (1) was selected and represented by a graphical ramp report as in Fig. 1. The results are illustrated using a contour plot and a 3-D perspective as in Fig. 2 (a) and 2 (b).

Table 6: Summary of solutions suggested by the software.

Number	*A:PEGDA*	*B:EPOLA*	*C:IR369*	*Crosslinking*	*Desirability*	
1	*54.587*	*44.413*	*1.000*	*58.349*	*0.921*	*Selected*
2	50.797	48.203	1.000	57.337	0.856	
3	50.170	48.830	1.000	56.974	0.832	
4	49.879	49.121	1.000	56.787	0.820	

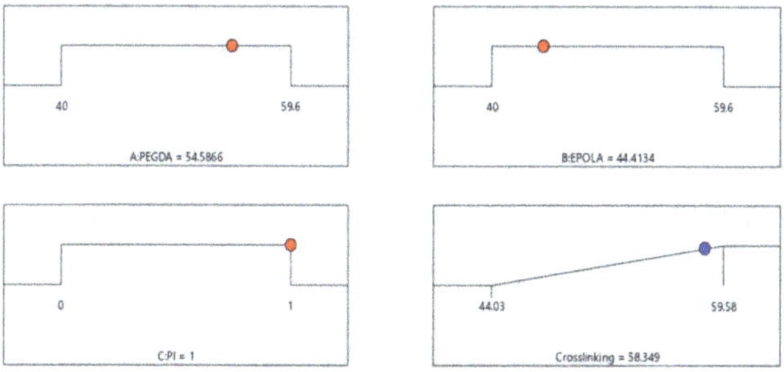

Fig. 1 : Solution (1) from Ramp analysis for optimum crosslinking % with the optimized resin formulation.

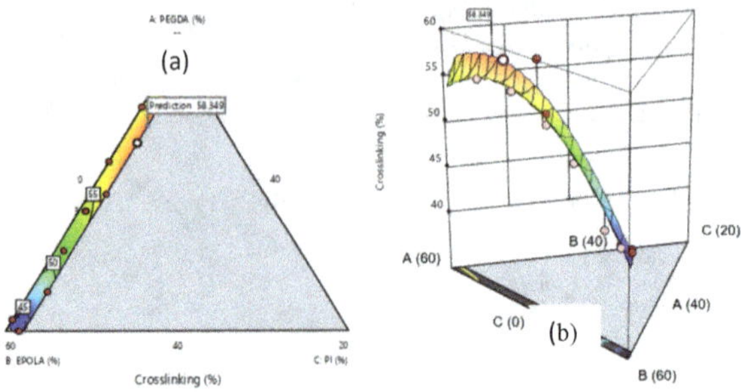

Fig. 2 : Graphical analysis for optimum crosslinking % with optimized resin formulation; (a) contour plot and (b) 3-dimensional perspective based on solution (1) from Ramp analysis.

The colour contours used in the graphical analysis represent a better perspective of the optimized result generated based on the solution by ramp analysis—the highest response, which is the crosslinking %, is represented in red. The dependent variables can also be seen clearly with the graphical presentations, where in this case, AB has the highest dependency compared to AC and BC. Only 1.0 % of C – the photoinitiator is needed in the formulation; thus, only a small portion of the triangle in Fig. 2 (a) is contoured. Also, as represented by the 3D perspective in Fig. 2 (b), only a narrow 'umbrella-like' image represents the formulation of (A) 54.587 % PEGDA, (B) 44.413 % EPOLA, and (C) 1.0 % IR369.

Post Analysis. After conducting experiments another three times with the same optimized data for the confirmation, the actual mean was calculated at 58.530. The results of the predicted mean are shown in the Confirmation Table as in Table 7 (a) and validated using the Coefficients Table as in Table 7 (b).

Table 7: Model confirmation and validation by (a) Confirmation and (b) Coefficients Table.

(a) Confirmation Table

Analysis	*Predicted Mean*	*Predicted Median*	*Std Dev*	*n*	*SE Pred*	*95% PI low*	*Data Mean*	*95% PI high*
Crosslinking	58.349	58.349	1.66658	3	1.46672	55.0809	58.53	61.617

(b) Coefficients Table

	A	*B*	*C*	*AB*	*AC*	*BC*
Crosslinking	53.4954	43.3416	1135.2	28.2109	-1065.39	-1149.24
p-values				< 0.0001	0.5159	0.4849

In Table 7, the Prediction Interval (PI) of 95% confidence is kept between 55.3755 and 61.3225. After the software configuration, if the Data Mean reading is outside the PI range, it will be shown in red and bold, indicating a failure to confirm the model. In this case, the mean reading shown in Table 7 is 58.53 and is neither in red colour nor bold after entering the responses, which indicates that the model is in conformance with the response. Relative error percentage of 0.181 between the actual (58.530) and predicted (58.349) is within the acceptable range (<5%).

Conclusion

The results obtained from the study provide important information on the suitability of EPOLA/PEGDA as a bioresin mixture in the development of a tissue scaffold using the UV crosslinking process. From the complete analysis using DOE, the optimum formulation of the mixture is agreed to be at PEGDA 54.587 %, EPOLA at 44.413 %, and IR369 PI at 1.0 %. Successful crosslinking is achieved at 58.349 % with a desirability of 0.921 when using the optimized formulation, supporting the improvement of existing PEGDA resin served as resin for 3D scaffold development using UV crosslinking. Validation using the coded quadratic model also proved that the model was confirmed and achieved after the added experiment was conducted for validation. All results can be further used to study the ability of the developed 3D scaffold to serve and function as a promising tissue scaffold. Using software such as Design Expert as a tool for optimization has been proven to streamline the process by reducing the complexity of an experiment, especially in the number of experiments needed to be run. The tool provides statistical support that is crucial in experiments requiring precise results. These can further reduce time, energy, money, and resources.

Acknowledgement

This research was funded by a grant from the Ministry of Higher Education of Malaysia FRGS/1/2019/TK05/UTM/02/15 (5F199) and IAEA Coordinate Research Project F23030. The authors gratefully acknowledge Universiti Teknologi Malaysia and the Malaysian Nuclear Agency for the support of materials and instruments.

References

[1] Sopyan I, Gozali D, Sriwidodo, Guntina RK.Design-Expert Software (Doe): An Application Tool for Optimization in Pharmaceutical Preparations Formulation, International Journal of Applied Pharmaceutics (2022) 55-63. https://dx.doi.org/10.22159/ijap.2022v14i4.45144

[2] Kasemiire A, Avohou HT, De Bleye C, Sacre PY, Dumont E, Hubert P, Ziemons E.Design of experiments and design space approaches in the pharmaceutical bioprocess optimization, Eur J Pharm Biopharm (2021) 144-54. https://doi.org/10.1016/j.ejpb.2021.06.004

[3] Tavares Luiz M, Santos Rosa Viegas J, Palma Abriata J, Viegas F, Testa Moura de Carvalho Vicentini F, Lopes Badra Bentley MV, Chorilli M, Maldonado Marchetti J, Tapia-Blacido DR.Design of experiments (DoE) to develop and to optimize nanoparticles as drug delivery systems, Eur J Pharm Biopharm (2021) 127-48. https://doi.org/10.1016/j.ejpb.2021.05.011

[4] Bowden GD, Pichler BJ, Maurer A.A Design of Experiments (DoE) Approach Accelerates the Optimization of Copper-Mediated (18)F-Fluorination Reactions of Arylstannanes, Sci Rep (2019) 11370. https://doi.org/10.1038/s41598-019-47846-6

[5] Beg S, Swain S, Rahman M, Hasnain MS, Imam SS.Application of Design of Experiments (DoE) in Pharmaceutical Product and Process Optimization, Pharmaceutical Quality by Design (2019) 43-64. https://doi.org/10.1016/B978-0-12-815799-2.00003-4

[6] Joas S, Tovar GEM, Celik O, Bonten C, Southan A.Extrusion-Based 3D Printing of Poly(ethylene glycol) Diacrylate Hydrogels Containing Positively and Negatively Charged Groups, Gels (2018). https://doi.org/10.3390/gels4030069

[7] Alagumurthi N, Palaniradja K, Soundararajan V.Optimization of grinding process through design of experiment (DOE)—A comparative study, Materials and manufacturing processes (2006) 19-21. https://doi.org/10.1080/AMP-200060605

[8] Trivedi D KV, Spandana AK, Kuppusam G.Design of Experiments: Optimization and applications in pharmaceutical nanotechnology, Chem Sci Rev Lett (2015) 109-20. https://doi.org/10.4103/japtr.japtr_690_22

[9] Durakovic B.Design of experiments application, concepts, examples: State of the art, Periodicals of Engineering and Natural Sciences (PEN) (2017). https://doi.org/10.21533/pen.v5i3.145

[10] Snee RD.Design and Analysis of Mixture Experiments, Journal of Quality Technology (2018) 159-69. ttps://doi.org/10.1080/00224065.1971.11980489

[11] Chua CK, Leong KF. 3D Printing and Additive Manufacturing: Principles and Applications, Singapore World Scientific Publishing Company Pte Limited; 2014. https://doi.org/10.1142/9008

[12] Cui H, Nowicki M, Fisher JP, Zhang LG.3D Bioprinting for Organ Regeneration, Adv Healthc Mater (2017). https://doi.org/10.1002/adhm.201601118

[13] Davidov-Pardo G, Arozarena I, Marín-Arroyo MR.Optimization of a wall material formulation to microencapsulate a grape seed extract using a mixture design of experiments, Food and Bioprocess Technology (2013) 941-51. https://doi.org/10.1007/s11947-012-0848-z

[14] Jeirani Z, Jan BM, Ali BS, Noor IM, Hwa SC, Saphanuchart W.The optimal mixture design of experiments: Alternative method in optimizing the aqueous phase composition of a microemulsion, Chemometrics and Intelligent Laboratory Systems (2012) 1-7. https://doi.org/10.1016/j.chemolab.2011.10.008

[15] St L, Wold S.Analysis of variance (ANOVA), Chemometrics and intelligent laboratory systems (1989) 259-72. https://doi.org/10.1016/0169-7439(89)80095-4

[16] Naghieh S, Karamooz-Ravari MR, Sarker MD, Karki E, Chen X.Influence of crosslinking on the mechanical behavior of 3D printed alginate scaffolds: Experimental and numerical approaches, J Mech Behav Biomed Mater (2018) 111-8. https://doi.org/10.1016/j.jmbbm.2018.01.034

[17] Irvine SA, Agrawal A, Lee BH, Chua HY, Low KY, Lau BC, Machluf M, Venkatraman S.Printing cell-laden gelatin constructs by free-form fabrication and enzymatic protein crosslinking, Biomed Microdevices (2015). https://doi.org/10.1007/s10544-014-9915-8

[18] Irvine SA, Agrawal A, Lee BH, Chua HY, Low KY, Lau BC, Machluf M, Venkatraman S.Printing cell-laden gelatin constructs by free-form fabrication and enzymatic protein crosslinking, Biomed Microdevices (2015) 16. https://doi.org/10.1007/s10544-014-9915-8

[19] Awang MA, Firdaus MA, Busra MB, Chowdhury SR, Fadilah NR, Wan Hamirul WK, Reusmaazran MY, Aminuddin MY, Ruszymah BH.Cytotoxic evaluation of biomechanically improved crosslinked ovine collagen on human dermal fibroblasts, Biomed Mater Eng (2014) 1715-24. https://doi.org/10.3233/BME-140983

[20] Czech Z, Kabatc J, Bartkowiak M, Licbarski A, Mozelewska K, Kwiatkowska D.Novel Photoreactive Pressure-Sensitive Adhesives (PSA) Based on Acrylics Containing Additionable Photoinitiators, Materials (Basel) (2020). https://doi.org/10.3390/ma13225151

[21] Wiens DP.Maximin power designs in testing lack of fit, Journal of Statistical Planning and Inference (2019) 311-7. https://doi.org/10.1016/j.jspi.2018.07.007

[22] Malaise S, Rami L, Montembault A, Alcouffe P, Burdin B, Bordenave L, Delmond S, David L.Bioresorption mechanisms of chitosan physical hydrogels: a scanning electron microscopy study, Mater Sci Eng C Mater Biol Appl (2014) 374-84. https://doi.org/10.1016/j.msec.2014.04.060

Exploring the Impact of Carbon Fiber Incorporation on the Mechanical Properties of PLA Composites

Yusrina MAT DAUD[1,2,a*], Hazwani ZAIDOL[2,b], Mohammad Firdaus ABU HASHIM[1,3,c], Luqman MUSA[1,2,d], Mohamad Syahmie MOHAMAD RASIDI[1,2,e], Ahmad Azrem AZMI[1,2,f], Lokman Hakim IBRAHIM[1,2,g], Muhammad Al Fatih HENDRAWAN[3,h], Kholqillah Ardhian ILMAN[3,i], Agus Dwi ANGGONO[3,j], and Bambang Waluyo FEBRIANTOKO[3,k]

[1]Green Technology (CEGeoGTech), School of Materials Engineering, Universiti Malaysia Perlis (UniMAP), P.O. Box 77, d/a Pejabat Pos Besar, 01000 Kangar, Perlis, Malaysia

[2]Faculty of Chemical Engineering & Technology, Universiti Malaysia Perlis (UniMAP), Kompleks Pusat Pengajian Jejawi 2, 02600 Arau, Perlis, Malaysia

[3]Faculty of Mechanical Engineering & Technology, Universiti Malaysia Perlis (UniMAP), Kampus Tetap Pauh Putra, 02600 Arau, Perlis, Malaysia

[4]Department of Mechanical Engineering, Faculty of Engineering, Universitas Muhammadiyah Surakarta, Indonesia

[a]yusrina@unimap.edu.my, [b]hazwanizaidol@gmail.com, [c]firdaushashim@unimap.edu.my, [d]luqman@unimap.edu.my, [e]syahmie@unimap.edu.my, [f]ahmadazrem@unimap.edu.my, [g]lokmanh@unimap.edu.my, [h]mah246@ums.id, [i]kai252@ums.ac.id, [j]ada126@ums.ac.id, [k]bw276@ums.ac.id

Keywords: PLA, Carbon Fiber, Coconut Husk

Abstract. This research investigates the influence of carbon fiber loading on PLA/Carbon Fiber composites utilizing coir coconut husk (CCH) as a sustainable resource. The objective is to evaluate the mechanical and physical characteristics of PLA with varying carbon fiber loadings (0%, 5%, 10%, 15%, and 20%). Fiber is extracted from CCH, carbonized to generate carbon fiber, and compounded with PLA. Test results indicate that a 10% carbon fiber loading yields the most favorable mechanical properties. The study concludes that introducing carbon fiber at its optimum value is highly recommended to attain exceptional properties in PLA.

Introduction

In recent years, the utilization of polylactic acid (PLA) in plastic applications has surged due to its favorable properties. Researchers have found that incorporating natural fibers as fillers in PLA composites enhances the materials' strength and rigidity [1]. Furthermore, the addition of carbon fibers to PLA composites produces high-performance materials that are notably lighter than aluminum and steel. Carbon fibers, renowned for their exceptional thermal characteristics and tensile strength, emerge as versatile and compelling reinforcement materials. The focus has shifted towards developing renewable and environmentally friendly carbon fibers from natural resources. PLA/carbon fiber composites offer biodegradable and reusable alternatives for various applications. In the realm of 3D printing, where PLA is prevalent, the inclusion of carbon fibers has shown promise in enhancing the mechanical properties and 3D printing suitability of PLA composites [2]. In summary, integrating natural fibers and carbon fibers into PLA composites presents opportunities to enhance mechanical properties, reduce weight, and expand applications while upholding their renewable and environmentally friendly status.

Methodology

Preparation of raw materials. PLA pellets sourced from TT Biotechnologies Sdn Bhd were paired with carbon fiber extracted from lignin sourced from coconut coir husk (CCH). The extraction process involved utilizing ethanol, a 10% NaOH solution, sulfuric acid, and distilled water.

Sample preparation. The process commences with an initial alkaline treatment of coconut coir husk (CCH), followed by carbonization to extract carbon constituents. These carbon fibers are then blended with a PLA matrix utilizing a heated two-roll mill, in accordance with the formulation outlined in Table 1. The compounded blend undergoes a drying phase, is subsequently placed in a preheated mold, and subjected to compression through the application of a hot press. The resultant PLA/CF composites are subsequently demolded and are poised for subsequent testing and characterization.

Table 1. Formulation of PLA/Carbon Fiber by 100g.

Composition (%)	PLA (g)	Carbon Fiber (g)
0	100	0
5	95	5
10	90	10
15	85	15
20	80	20

Testing and characterization

Fourier Transform Infrared Spectroscopy. FTIR spectroscopy with a wavenumber range of 400 cm^{-1} to 4000 cm^{-1} was used to analyze the functional groups of treated and untreated extracted fiber.

Tensile Test. The experiment adhered to the ASTM D 638 standard, outlining the procedure for assessing the tensile properties of plastics. A gauge length of 33 mm was demarcated on each specimen, with the gauge line positioned 16.5 mm from the center on both sides. The crosshead speed for the test was established at 10 mm/min.

Flexural Test. The flexural test, commonly referred to as the 3-point bend test, was conducted in accordance with ASTM D 790 standards. The crosshead speed was maintained at 10 mm/min, and the span length for each specimen was established at 90 mm.

Results and Discussion

Functional group identification of treated and untreated fiber. FTIR analysis was conducted to assess the impact of alkaline treatment on the functional groups of extracted fiber. A comparison of the results with wave numbers reported in prior studies indicated a relatively consistent pattern, affirming the presence of lignin in the extract. The FTIR spectra presented in Fig. 1 displayed absorbance peaks characteristic of fiber lignin, including aromatic ring stretching (C=C) and aliphatic ring stretching (C-H). Contrasting samples treated with 10% sodium hydroxide against untreated samples revealed variations in peak intensities and wave number shifts. The treatment process induced changes in the chemical composition by reducing the number of hydroxyl groups and altering hydrogen bonding interactions. The analysis demonstrated that the extraction procedure induced significant chemical modifications in fiber, providing confirmation of its

Materials Research Proceedings 68 (2026) 121-125 https://doi.org/10.21741/9781644904237-16

presence in the extracted samples. The analysis will contribute to the enhancement interfacial adhesion and, consequently, the strength of the composite.

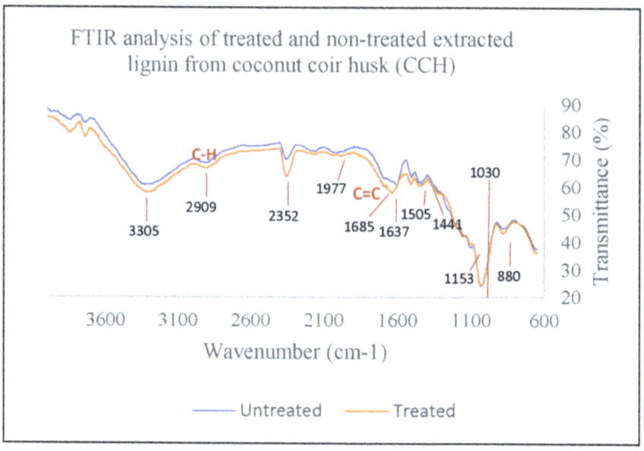

Fig. 1: FTIR spectrogram of untreated and treated extracted lignin.

Effect of carbon fiber loading on tensile strength. The investigation into the tensile strength of PLA composites involved varying carbon fiber loadings. As illustrated in Fig. 2, the optimal value of 20.01 MPa was achieved at a 10% carbon fiber loading. Consistent with prior studies, the incorporation of carbon fiber as reinforcement enhances the tensile strength compared to the PLA matrix alone.

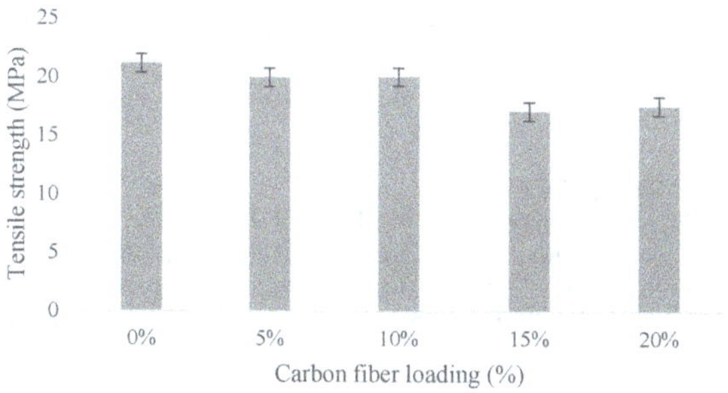

Fig. 2: Effect of carbon fiber loading on the tensile strength of PLA composites.

However, beyond a 10% filler loading, there is a reduction in the ductility of the composites. This aligns with the findings, who noted a decrease in sample ductility with increased carbon fiber concentrations [3]. The decline in tensile strength beyond 10% can be attributed to diminished fiber-matrix contact due to heightened fiber-to-fiber interfacial adhesion [4]. Notably, there is a discernible trend of decreasing tensile strength at 15% and 20% carbon fiber by weight, potentially linked to issues such as nozzle obstruction.

Effect of carbon fiber loading on flexural strength. Flexural strength, indicating a material's ability to withstand bending forces until specimen fracture, is a critical parameter. As illustrated in Figure 3, the flexural strength of PLA/carbon fiber composites at 5% is lower than other carbon fiber loadings, attributed to the filler's inability to withstand stress from the matrix.

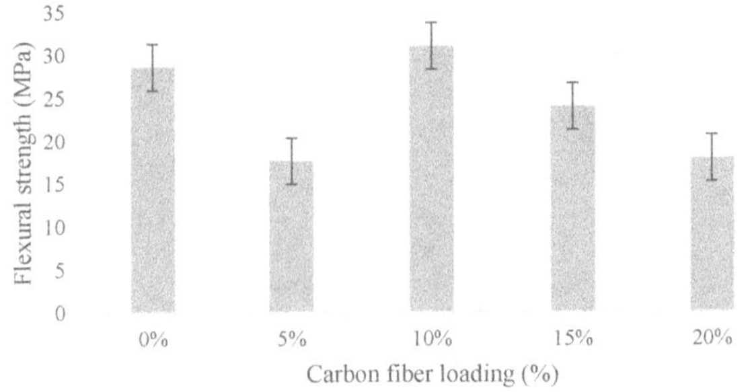

Fig. 3: Effect of carbon fiber loading on the flexural strength of PLA composites.

Remarkably, at a 10% carbon fiber loading, the sample exhibits the highest flexural strength, signifying improved compatibility and stress distribution between the PLA matrix and carbon fibers. The interfacial strength between the matrix and reinforcement is crucial for the three-point bending test and flexural properties. The transfer of burden from resin to fiber enhances the load-bearing capacity of carbon fibers post-fracture [5]. Despite this, flexural strength diminishes at 15% and 20% filler loadings, aligning with previous research attributing this decline to the filler's inability to withstand stresses and weak interfacial bonding. The reduction in flexural strength is further influenced by particle aggregation and the presence of defects [6,7].

Summary

The investigation effectively fabricated PLA composites with diverse weight percentages of carbon fiber as a filler. The analysis utilizing lignin extracted from coconut coir fiber yielded favorable outcomes. Consequently, the study unveiled that the mechanical properties reached an optimum at 10% carbon fiber content, with varying strengths at higher concentrations attributed to the influence of fiber concentration on fiber distribution and interfacial bonding.

Acknowledgement

The author would like to acknowledge the financial support in the form of publication incentive grant from Universiti Malaysia Perlis (UniMAP).

References

[1] R. Siakeng, M. Jawaid, H. Ariffin, S.M. Sapuan, M. Asim, and N. Saba, Natural fiber reinforced polylactic acid composites : A review, Polym. Compos. 40 (2018) 446-463. https://doi.org/10.1002/pc.24747.

[2] S.H. Sanei and D. Popescu, 3D printed carbon fiber reinforced polymers composites : A systemic review, J. Compos. Sci.4 (2020) 98. https://doi.org/10.3390/jcs4030098

[3] T. Yu, Z. Zhang, S. Song, Y. Bai, and D. Wu, Effect of fiber-layer position on mechanical properties of carbon fiber reinforced materials manufactured by fused deposition modelling, Compos. Struct. 225 (2019) 111147. https://doi.org/10.17222/mit.2022.419

[4] S. Wasti, E. Triggs, R. Farag, M. Auad, S. Adhikari, D. Bajwa, M. Li, and A.J. Ragauskas, Comparison of additive manufacturing and injection molding of biocomposites reinforced with alkali-treated wood flour derived from recycled wooden pallets. Composites Part B: Engineering. 205 (2021) 108483. https://doi.org/10.3390/polym17152004

[5] H.P.S. Abdul Khalil, H.M. Fizree, A.H. Bhat, M. Jawaid, and C.K. Abdullah, Development and characterization of epoxy nanocomposites based on nano-structured oil palm ash, Compos. B Eng. 53 (2013) 324-333. https://doi.org/10.1016/j.compositesb.2013.04.013

[6] I.N. Nasidi, L.H. Ismail, and E.M. Samsudin, Pertanika Journal of Science and Technology. 29 (2021) 693-706. https://doi.org/10.1038/s41598-025-10825-1

[7] V. Thakur, R. Kumar, R. Kumar, R. Singh, and V. Kumar, Recent advances in fused deposition modeling based additive manufacturing of thermoplastic composite structures: A review. J. Thermoplas. Compos. Mater. 36 (2023) 3094-3132.

Materials Research Proceedings 68 (2026) 126-133 https://doi.org/10.21741/9781644904237-17

The Effect of Kernel Shell Oil Palm Empty Fruit Bunches (KS-OPEFB) and Rice Husks (RH) Hybrid Fillers on Thermoplastic Starch Films

Sinar Arzuria ADNAN[1,2,a*], Azlin Fazlina OSMAN[1,2], Nur Hidayah AHMAD ZAIDI[1], Muhammad Hanis ROSLI[1], Yahmunaa JAVERSEETHARAMAN[1], Ameirul Qayyum Eiman RAMLI[1], Haiza HAROON[1,2], Syazwani MAHMAD PUZI[1] and Sharifah Hanis Yasmin SAYID ABDULLAH[3]

[1]Faculty of Chemical Engineering & Technology, Universiti Malaysia Perlis (UniMAP), Arau, 02600, Perlis, MALAYSIA

[2]Biomedical and Nanotechnology Research Group, Centre of Excellence Geopolymer and Green Technology (CeGeoGTech), Universiti Malaysia Perlis (UniMAP), Arau, 02600, Perlis, MALAYSIA

[3]Faculty of Innovative Design & Technology, Universiti Sultan Zainal Abidin, Kuala Terengganu, 21030, MALAYSIA

[a]sinar@unimap.edu.my

Keywords: Kernel Shell, Oil Palm Fruit Bunches, Rice Husks, Hybrid Fillers, Thermoplastic Starch Film

Abstract: Thermoplastic starch (TPS) films have attracted increasing attention as sustainable alternatives to conventional petroleum-based plastics; however, their widespread application remains limited by poor mechanical strength and high moisture sensitivity. In this study, agricultural waste-derived hybrid fillers from oil palm kernel shell empty fruit bunches (KS-OPEFB) and rice husk (RH) were incorporated into TPS films to enhance their mechanical performance and biodegradation behaviour. TPS composite films were prepared via a solution casting method using corn starch and glycerol, with varying ratios of KS-OPEFB and RH fillers. The effects of hybrid filler loading on tensile properties, structural characteristics, water absorption, and soil biodegradability were systematically investigated. The mechanical testing results showed that the addition of 16 wt.% KS-OPEFB and 4 wt.% RH improved the tensile strength of the TPS films, with an increase of up to 4.53 MPa compared to the neat TPS. The results of the Fourier transform infrared analysis indicated improved interfacial interactions between the starch matrix and the hybrid fillers. The results of the soil burial test showed improved biodegradability of the hybrid filler-filled TPS films, particularly for the films containing higher amounts of KS-OPEFB. The results showed improved biodegradability due to the increased accessibility of the films for biodegradation. The results also showed improved biodegradability due to the hydrophilic nature of the fillers. The results showed that the incorporation of the hybrid filler had a substantial effect on the water absorption of the TPS films. The results showed that the porous nature of the lignocellulosic fillers improved the water absorption of the films. The results showed that the hydrophilic nature of the fillers improved the water absorption of the films.

Introduction

The environmental issues happening globally because of the use of petroleum-based plastics has increased the global initiatives trying to find sustainable and biodegradable substitutes for use in packaging. The qualities of traditional plastics make them extremely durable, resistant to being broken down, and creates an ongoing accumulation of waste in nature, which has long-lasting negative consequences. Biodegradable polymers are being constructed from renewable resources with increasing amounts of attention. Thermoplastic starch (TPS) is a popular source due to its relatively low cost, easy accessibility, along with its inherent nature of being biodegradable [5]. While TPS has a lot of advantages, the use of TPS films still presents several inherent disadvantages that impede the practical use of this new product: poor mechanical strength, high moisture sensitivity and low resistance to water absorption [6].

Many reinforcement strategies have been studied to mitigate some of these disadvantages; one being the addition of natural fibres and fillers into the TPS matrix has shown to be the most favourable method. Lignocellulosic materials created from agricultural by-products are excellent considerations because they are: renewable, biodegradable and they possess favourable mechanical characteristics. Empty fruit bunches (KS-OPEFB) from palm oil and rice husks are two very abundant agroindustrial waste material in many agricultural areas. KS-OPEFB contains a large amount of cellulose and lignin, which give it rigidity and the ability to serve as a reinforcer, while RH contains a unique combination of both cellulose and silica which adds to the stiffness of the final combined materials. The synergistic interaction between different lignocellulosic fillers can enhance stress transfer efficiency, interfacial bonding, and overall composite performance [2]. Nevertheless, systematic investigations on the combined use of oil palm kernel shell and rice husk as hybrid fillers in thermoplastic starch films remain limited, particularly in relation to their simultaneous effects on mechanical performance, water absorption behaviour, and biodegradability.

Therefore, the objective of this study is to investigate the influence of KS-OPEFB and RH hybrid fillers on the mechanical, structural, water absorption, and soil biodegradation properties of thermoplastic starch films. Composite TPS films were prepared using the solution casting technique with different KS-OPEFB and RH ratios. The films were mechanically tested for tensile strength, FTIR analysis, water absorption measured, and evaluated for biodegradation in soil through burial. This study's objective is to use two agricultural waste fill materials that have complementary properties to create a sustainable design method to improve the functionality of starch-based biodegradable films as a viable alternative to traditional plastic packaging [5,9].

Experimental

Material. TPS films were prepared using corn starch as the base polymer matrix. For the purpose of creating a plasticizer, glycerol was added and distilled water was used as the solvent to produce solutions. Hybrid fillers were created from both Oil Palm Kernel Shell (KS-OPEFB) and Rice Husk (RH). Filler treatment comprised of the use of hydrogen peroxide (H2O2) and acetic acid (C6H4O2). All materials (e.g., corn starch, glycerol, distilled water DK) were received and utilized without any additional processing/cleaning or purification. A table containing all the formulations can be found in Table 1.

Table 1: The Formulation KS-OPEFB-RH/TPS Film Composites.

Sample	*TPS (g)*	*KS-OPEFB (wt.%)*	*RH (wt.%)*
Pure TPS	5	0	0
KS-OPEFB/TPS	5	20	0
RH/TPS	5	0	20
KS-OPEFB-RH/TPS	5	16	4
KS-OPEFB-RH/TPS	5	12	8
KS-OPEFB-RH/TPS	5	8	12
KS-OPEFB-RH/TPS	5	4	16

Preparation of KS-OPEFB-RH/TPS Film Composites. The fillers KS-OPEFB and RH were first cleaned by removing their foreign materials, then dried and ground into fine pieces. After that a bleaching treatment was performed using hydrogen peroxide with the addition of acetic acid to help improve its cleanliness, flexibility, and exposure of the cellulose fiber prior to an application. After being treated, the fillers were cleaned thoroughly with distilled water and then dried before they were used. This treatment process aimed to enhance filler matrix compatibility by reducing surface impurities and increasing hydroxyl group availability [12].

Thermoplastic starch films were made using a solution casting method. Corn starch (5 g) was mixed with glycerol (2 g) in 80 mL of distilled water while stirring continuously. The mixture was heated to 75–80 °C until the starch completely gelatinized. In a separate container, the hybrid fillers were mixed with distilled water and ultrasonicated for 1 hour at a frequency of 53 Hz, 100% power, and about 24% amplitude to ensure they were well-dispersed. After that, the filler suspension was added to the gelatinized TPS solution and stirred for 5–10 minutes to achieve a uniform mixture. The resulting solution was cast into an 8-inch mould and dried in an oven at 40 °C for 18 h. After drying, the films were removed and conditioned at ambient conditions for 10–15 min prior to testing.

Testing and Characterization. This research examines the effect of hybrid fillers made from rice husk (RH) and palm kernel shell empty bunches (KS-OPEFB) on KS-OPEFB-RH/TPS Film Composites. Composite mechanical strength and chemical and degradation properties were examined using an assortment of testing and characterizing methods. To determine the tensile properties of TPS composite films, universal testing apparatus was used. Test specimens were produced in a "dumbbell" shape for measuring the tensile strength, Young's modulus and elongation at break utilizing the stress-strain curves obtained from the universal testing results. These tests were carried out to investigate how hybrid filler loading affected the TPS film's mechanical performance [11].

To evaluate the functional groups and interfacial bonding between the TPS matrix and hybrid fillers, FTIR spectroscopy was used to generate the FTIR spectra for both neat TPS films, TP films containing hybrid fillers and treated and untreated KS-OPEFB fillers. A comparison of FTIR by way of examining shifts of characteristic absorptions for hydroxyls and ethers showed potential inter-filling bonding types between the filler and starch matrix [4]. Water absorption behaviour of the TPS composite films was tested as per ASTM D570. Film pieces were dried until a constant weight was attained and subsequently, they were plunged into distilled water. The specimens were taken out at regular intervals, surface dried, and weighed. Water absorption was expressed as the increment of weight percentage change related to the first dry weight, thus, giving the moisture sensitivity of the composite films.

Soil burial biodegradation experiments were performed following EN ISO 846:1997 guidelines. Film samples were buried in the natural soil and taken out after certain time intervals up to 20 days. After cleaning and drying, the samples were weighed to calculate the percentage weight loss that correlates with the burial time. The degradation rate served as a measure to assess the biodegradability of the TPS composite films and the effect of the hybrid filler composition on the microbial degradation behaviour [9].

Result and Discussion

Tensile Strength. The tensile properties of the hybrid filler ratios of the KS OPEFB RH TPS composite films are displayed in Table 2. In order to determine the strengthening capacity of the oil palm kernel shell from the empty fruit bunches (KS OPEFB) and the rice husk (RH) within the thermoplastic starch (TPS) matrix, tensile tests were performed. From Table 2, it can be seen that the neat TPS had the least tensile strength (3.20 MPa), which represents the poor mechanical performance expected from starch-based composites. When the single filler's incorporated into the TPS matrix, the resulting materials produced a moderate increase in the composite's tensile strength, with 3.95 MPa observed (TPS KS OPEFB) as compared to TPS RH's value of 3.68 MPa.

This can be attributed to the greater content of lignocellulosic material and lignin found in the KS OPEFB, contributing to the rigidity and strength of the TPS matrix [12].

Table 2: Tensile Properties of KS-OPEFB-RH/TPS Film Composites.

Sample	Tensile Strength (MPa)	Young Modulus (MPa)	Elongation at Break (%)
Pure TPS	3.2	11.7	8.85
TPS-RH	3.68	16.62	6.76
TPS-KS-OPEFB	3.95	24.21	13.52
16 WT% KS-OPEFB 4 WT% RH	4.53	22.75	12.58
4WT% KS-OPEFB 16 WT% RH	3.51	25.20	9.48
12 WT% KS-OPEFB 8WT% RH	3.92	13.25	7.18
8 WT% KS-OPEFB 12 WT% RH	3.39	24.55	4.18

The further increase was more substantial for the hybrid, filled composites. The TPS film with 16 wt. % KS, OPEFB and 4 wt. % RH in the mixture showed the strongest tensile strength of 4. 53 MPa, which means that the reinforcing effect of the two fillers is synergistic. The best composition here reveals that combined stress transfer efficiency is higher and stress concentration is lower because of good dispersion of the filler and interfacial bonding in the TPS matrix [11].

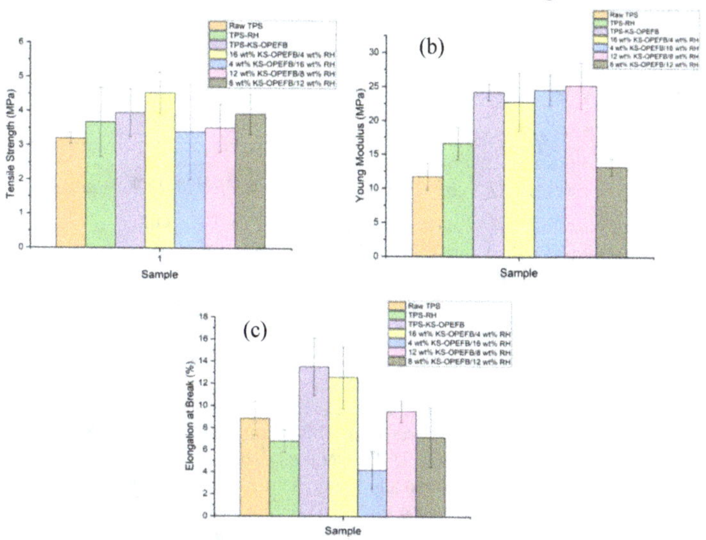

Fig. 1: Tensile results (a) Tensile Strength, (b) Young Modulus, and (c) Elongation at Break for KS-OPEFB-RH/TPS Film Composites.

The relationship between tensile strength, Young's modulus and elongation at break of the composites with the hybrid filler loading is shown in Fig.1 (a) through (c). As seen, in Fig. 1(a) the tensile strength of the composites increased with increasing amounts of KS-OPEFB to a certain point, which then resulted in decreased tensile strength for increasing additional RH. Young's modulus of the composites also increased with increasing amounts of KS-OPEFB, as illustrated in Fig. 1(b), indicating increased stiffness of the composite due to decreased mobility of the polymer chains. Fig. 1(c) illustrates a decrease in elongation at break of the composites containing higher amounts of KS-OPEFB; this is a common dilemma observed with natural fibre-reinforced polymers that is, increased stiffness of the composite versus increased flexibility [6].

Fourier Transform Infrared Spectroscopy (FTIR) Analysis. Fig. 2(a) presents FTIR spectra for TPS composite films composed of several different types of hybrid filler compositions and Fig. 2(b) shows a comparison of the FTIR spectra for treated vs. untreated KS-OPEFB fillers. All TPS based films containing characteristic absorption bands due to starch structure, such as a very broad hydroxyl (–OH) stretching band in the range of 3200 to 3500 cm-^{1}, and a C–O stretching band approximately between 1000 and 1100 cm-^{1}. The C-O stretching vibrations noted in Fig. 2(a) near 1007 to 1010 cm-^{1} represent the C-O stretching due to the presence of polysaccharide structures coming from both the TPS matrix itself as well as from the lignocellulosic fillers. Minor peak shifts and intensity variations suggest enhanced intermolecular interactions between starch and hybrid fillers, primarily through hydrogen bonding.

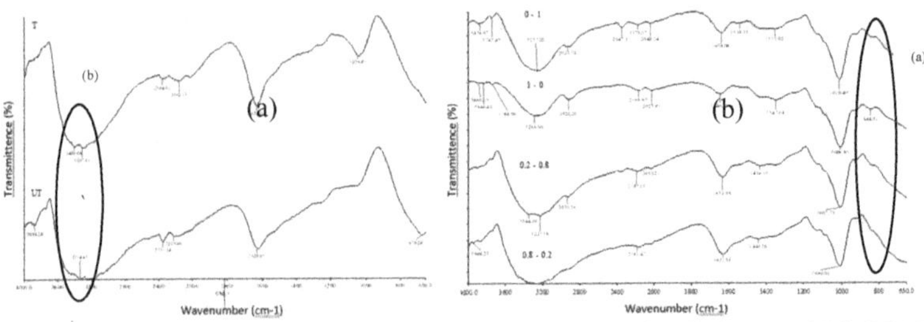

Fig. 2: FTIR spectra of (a) TPS composite film and (b) treated and Untreated KS-OPEFB.

The FTIR spectra in Fig. 2(b) reveal that treated KS-OPEFB exhibits a broader and more intense –OH absorption band at approximately 3330–3400 cm^{-1} compared to untreated KS-OPEFB. This observation indicates increased exposure of hydroxyl groups following the bleaching treatment, which enhances filler–matrix compatibility by promoting hydrogen bonding with the starch matrix [4]. Improved interfacial interactions contribute directly to the enhanced tensile performance observed in Section 3.1, consistent with previous studies on lignocellulosic-reinforced starch composites [2].

Soil Biodegradability Test. The soil burial biodegradation behaviour of KS-OPEFB–RH/TPS film composites is summarised in Table 3. All samples exhibited progressive weight loss with increasing burial time, confirming their biodegradable nature. As shown in Table 3, the composite film containing 16 wt.% KS-OPEFB and 4 wt.% RH demonstrated the highest degradation rate, achieving 58.76% weight loss after 20 days. This enhanced biodegradation can be attributed to the hydrophilic and porous structure of KS-OPEFB, which facilitates moisture penetration and microbial colonisation within the composite matrix [9].

Table 3: Weight loss (%) of the Soil biodegradability of KS-OPEFB-RH/TPS Film Composites.

Sample	0 Days	5 Days	10 Days	15 Days	20 Days
TPS-RH	0 %	20.72 %	25.23 %	30.12 %	32.33 %
TPS-KS-OPEFB	0 %	5.23 %	6.77 %	8.32 %	10.46 %
16 wt% KS-OPEFB/4 wt% RH	0 %	30.25 %	45.87 %	54.29 %	58.76 %
12 wt% KS-OPEFB/8 wt% RH	0 %	14.34 %	16.27 %	20.48 %	21.73 %
8 wt% KS-OPEFB/12 wt% RH	0 %	18.29 %	22.35 %	28.47 %	29.75 %
4 wt% KS-OPEFB/16 wt% RH	0 %	18.23 %	21.34 %	26.48 %	26.66 %

Composites with higher RH content exhibited comparatively lower degradation rates, likely due to the presence of silica in rice husk, which may partially hinder microbial access. These results indicate that hybrid filler composition plays a critical role in tailoring the biodegradation behaviour of TPS films.

Water Absorption Test. The water absorption behaviour of KS-OPEFB–RH/TPS film composites is presented in Table 4. All samples showed increasing water uptake with immersion time due to the hydrophilic nature of starch and lignocellulosic fillers. The composite film containing 16 wt.% KS-OPEFB and 4 wt.% RH exhibited the highest water absorption, reaching 225.85% after 20 days, as shown in Table 4. This behaviour is associated with the high porosity and hydroxyl group density of KS-OPEFB, which promotes water diffusion into the composite structure [1]. On the other hand, higher relative humidity content meant less absorption of water likely because of the presence of silica in rice husk which inhibited moisture from penetrating through. This study illustrates how hybrid designs of filler can provide a way to balance the dual requirements of water absorption and biodegradability based on their specific application.

Table 4: Water Absorption Content (%) from Water Absorption Test of KS-OPEFB-RH/TPS Film Composites.

Sample	0 Days	5 Days	10 Days	15 Days	20 Days
TPS-RH	0 %	20.72 %	30.29 %	41.29 %	45.93 %
TPS-KS-OPEFB	0 %	20.14 %	32.78 %	60.46 %	97.15 %
16 wt% KS-OPEFB/4 wt% RH	0 %	52.39 %	120.21 %	180.34 %	225.85 %
12 wt% KS-OPEFB/8 wt% RH	0 %	14.34 %	36.29 %	52.22 %	70.20 %
8 wt% KS-OPEFB/12 wt% RH	0 %	18.29 %	34.23 %	58.21 %	70.61 %
4 wt% KS-OPEFB/16 wt% RH	0 %	18.23 %	33.23 %	55.23 %	78.95 %

Conclusion

This study demonstrates that hybrid agricultural waste fillers derived from oil palm kernel shell empty fruit bunch (KS-OPEFB) and rice husk (RH) are effective in enhancing the performance of thermoplastic potato starch (TPS) films. The incorporation of hybrid fillers significantly improved the mechanical properties of TPS compared to neat films. The optimum formulation, consisting of 16 wt% KS-OPEFB and 4 wt% RH, achieved the highest tensile strength of 4.53 MPa. This enhancement is attributed to the presence of rigid lignocellulosic fibres within the starch matrix, which promotes efficient stress transfer and stronger interfacial bonding between the filler and polymer phases.

Materials Research Proceedings 68 (2026) 126-133 https://doi.org/10.21741/9781644904237-17

FTIR analysis confirmed improved compatibility between the starch matrix and the hybrid fillers through hydrogen bonding interactions between hydroxyl (–OH) groups on the starch chains and the treated fibre surfaces. These interactions contributed to the formation of a more cohesive and structurally stable composite film network. In addition, soil burial tests revealed that hybrid-filled TPS films exhibited accelerated biodegradation compared to neat TPS, achieving a maximum average weight loss of 58.76% after 20 days. The enhanced degradation behaviour is primarily associated with the hydrophilic nature, porous structure, and lignocellulosic composition of KS-OPEFB, which facilitate microbial penetration and activity.

The hybrid fillers used in the study affected the water absorption behavior of the materials being tested. The composite films showed increased moisture absorption because higher levels of KS-OPEFB content created more porous structures and the filler materials contained numerous hydroxyl groups. The researchers discovered that controlled hybrid filler loading enables precise adjustment of mechanical properties and degradation rate and moisture absorption characteristics in TPS films. The combination of KS-OPEFB and RH produces an environmentally friendly method to enhance thermoplastic starch films through agricultural waste valorization. The hybrid TPS films developed by the team demonstrate high potential for use in biodegradable packaging because they reduce dependence on petroleum-based plastics and support biodegradation. The upcoming research needs to improve fiber surface treatment methods and develop barrier coating solutions, which will strengthen waterproof capabilities and increase the lifespan of eco-friendly composite films.

References

[1] Z.C. Berry, N.C. Emery, S.G. Gotsch, G.R. Goldsmith, Foliar water uptake: Processes, pathways, and integration into plant water budgets, Plant, Cell & Environment 42 (2019) 410–423. https://doi.org/10.1111/pce.13439

[2] A.M. Díez-Pascual, A.L. Díez-Vicente, ZnO-reinforced poly(3-hydroxybutyrate-co-3-hydroxyvalerate) bionanocomposites with antimicrobial function for food packaging, ACS Applied Materials & Interfaces 6 (2014) 9822–9834. https://doi.org/10.1021/am501523c

[3] S. Fiedler, M. Graw, Decomposition of buried corpses, with special reference to the formation of adipocere, Naturwissenschaften 90 (2003) 291–300. https://doi.org/10.1007/s00114-003-0437-0

[4] M.L. LaBerge, The heart of the madder: An important prehistoric pigment and its botanical and cultural roots, PhD Dissertation, University of Wisconsin–Milwaukee, 2018.

[5] H. Lang, X. Chen, J. Tian, J. Chen, M. Zhou, F. Lu, S. Qian, Effect of microcrystalline cellulose on the properties of PBAT/thermoplastic starch biodegradable film with chain extender, Polymers 14 (2022) 4517. https://doi.org/10.3390/polym14214517

[6] W.G. Perkins, Polymer toughness and impact resistance, Polymer Engineering & Science 39 (1999) 2445–2460. https://doi.org/10.1002/pen.11614

[7] N.S. Rosli, S. Harun, J.M. Jahim, R. Othaman, Chemical and physical characterization of oil palm empty fruit bunch, Malay J. Anal. Sci. 21 (2017) 188–196.

[8] M.S. Umikalsom, A.B. Ariff, H.S. Zulkifli, C.C. Tong, M.A. Hassan, M.I.A. Karim, The treatment of oil palm empty fruit bunch fibre for subsequent use as substrate for cellulase production by Chaetomium globosum Kunze, Bioresour. Technol. 62 (1997) 1–9. https://doi.org/10.1016/S0960-8524(97)00114-0

[9] S. Venkatesh, S. Mahboob, M. Govindarajan, K.A. Al-Ghanim, Z. Ahmed, N. Al-Mulhm, S. Vijayalakshmi, Microbial degradation of plastics: Sustainable approach to tackling

environmental threats facing big cities of the future, J. King Saud Univ. Sci. 33 (2021) 101362. https://doi.org/10.1016/j.jksus.2021.101362

[10] H.M. Wang, T.Q. Yuan, G.Y. Song, R.C. Sun, Advanced and versatile lignin-derived biodegradable composite film materials toward a sustainable world, Green Chem. 23 (2021) 3790–3817. https://doi.org/10.1039/D0GC04145A

[11] M.R. Wisnom, D. Green, Tensile failure due to interaction between fibre breaks, Composites 26 (1995) 499–508. https://doi.org/10.1016/0010-4361(95)96856-5

[12] N.L.I. Zailuddin, S. Husseinsyah, Tensile properties and morphology of oil palm empty fruit bunch regenerated cellulose biocomposite films, Procedia Chem. 19 (2016) 366–372. https://doi.org/10.1016/j.proche.2016.03.040

Acid and Alkali Treatment on Natural Fibers for Better Tensile Strength of Epoxy Composites

Nor Azwin AHAD[1,a*] and Rama Murthi A/L Sooria MOORTHI[1]

[1]Faculty of Chemical and Engineering Technology, Universiti Malaysia Perlis, Kompleks Pengajian Jejawi 2, 02600 Jejawi, Perlis, Malaysia

[a]norazwin@unimap.edu.my

Keywords: Natural Fiber, Acid Treatment, Alkali Treatment, Tensile Strength

Abstract. Natural fiber which known are from agricultural waste, fruit skin, stalks, and so on is considered recyclable also creating great demand as they come at a very low cost, are biodegradable, and have low density. However, natural fibers are hydrophilic in nature, so they tend to absorb water. Chemical treatments can usually overcome this problem. Usually, acid and alkali treatments are very popular among researchers. In this study, we focused on seeing the differences in the natural treatment of acidic and alkaline using citric acid (vinegar) and sodium bicarbonate (soda) of certain three types of selected fibers. The 24-hour treatments were studied on coconut coir, banana stem fiber, sugarcane bagasse, and corn stalk fiber. All the fibers acted as fillers in epoxy composites. The tensile strength has been investigated and compared between untreated and treated as well. All fibers were observed highest tensile strength after being treated with alkali compared to acid. Treatment with alkali is more suitable for natural fiber to clean the fiber surface, thus helping to get better fiber-matrix interaction. These are the reasons for the highest tensile strength.

Introduction

Natural fibers are the one of popular resources that are used as reinforcement or filler in polymer matrices. According to the previous study [1], coconut coir, banana stem fibers, sugarcane bagasse, and corn stalk fiber are some of the sources of natural fibers that have the potential to be used as fillers in the polymer. In contrast, the main problem with the use of natural fibers is the hydrophilic nature of all the cellulose fiber for reinforcing or filler in polymer [2]. This property is not compatible with the hydrophobic nature of the matrix material [3]. Due to this major issue, natural fibers need to be treated before use as filler, to reduce the hydrophilic properties and to optimize the interface between fiber and matrix.

The properties of compatibility of fiber-matrix as well as their interface or interfacial adhesion are of utmost crucial for the mechanical properties of composite materials [3]. According to Rajala et al. [4], the biggest weakness of these natural fiber-reinforced polymer composites is the poor interfacial adhesion between the fiber and the matrix, then the high moisture absorption ability of natural fiber due to their hydrophilic nature. This can result in ineffective stress transfer between fiber and matrix which leads to unsatisfactory mechanical properties.

All these drawbacks could be improved by modifying the fiber surface with chemical treatment [4]. The chemical method for the modification of natural fibers was much discussed. It is applied depending on what type of chemical we want to use and since natural fiber has different climate conditions, the need to find a suitable chemical method of modification cannot be overemphasized. [3]. The chemical treatment process is cheap and easy to operate. It involves the immersion of natural fibers in a concentrated chemical solution for some time. The chemical solution will remove certain unwanted portions of the unwanted elements from the external surface of the natural fibers.

In this study, the use of chemicals is simply using vinegar and soda. These two materials were chosen because we wanted to understand their potential as chemical treatment materials for natural fibers. Previous researchers tend to treat with natrium hydroxide, hydrochloric acid, kalian hydroxide, peroxide, and others. All of these chemicals have been shown to have the potential to be good treatment materials for natural fibers. We aim to evaluate the effect of an eco-friendly and cost- effective treatment method based on the use of commercial baking soda (alkali treatment) and vinegar (acid treatment) on these three types of natural fibers.

Alkali treatment is the most used chemical treatment of natural fiber, but fewer reported on acid treatment, which gave severe reactions. Juliana et al. presented that alkali treatment on sugarcane bagasse produced the best reinforcement for composites due to excellent fiber-resin bonding. This showed that alkaline treatment is highly proven to improve the properties of the fibers. Bakri et al. [5] concluded that, better adhesion of fiber-matrix in banana stem fiber/epoxy composites. Vijay et al. [6] agreed that acid treatments have poor strength due to more removal of cellulose, and destructing both the fiber structure. Mostapha et al. [7] proved that the acid treatment significantly decreased the hydrophilicity of the fibers. To determine the difference between these two types of treatment, this study was done on four different types of natural fibers. All of these fibers will be used as filler for composite epoxy, whether treated or untreated. The tensile strength of the epoxy/natural fiber has been investigated and compared.

Experimental

Materials. Epoxy resin and hardener (EPOX-IT 80) were supplied from HH Saintifik Enterprise. All natural fibers (coconut coir, banana stem, sugarcane bagasse, and corn stalk) were collected from markets, stalls, and villages. They were washed and dried under direct sunlight for days to evaporate all the moisture. The dried fibers were shredded and separated into each fiber, then were cut to 1.5 cm, manually.

Fiber treatments and preparation. The four types of natural fibers (coconut coir, banana stem fiber, sugarcane bagasse, and corn stalk fiber) were immersed in vinegar and soda solution for 24 hours. Before that, the vinegar and soda solution were tested the pH value using a pH meter. Vinegar was on the pH 3 scale and soda on the pH 8 scale. After 24 hours of immersion, the treated fibers were washed with water until the pH value became 7, which is neutral. The epoxy composite was prepared by hand mixing. Mixing was carried out at room temperature with the specification of mix ratio 1 : 1 work time of 40 min and cure time of 24 hours. The mixture was cast in a dumbbell shape silicon mold, until hardened for about 24 hours.

Tensile testing. The hardened samples of epoxy composited which filled with untreated and treated coconut coir, banana stem fiber, sugarcane bagasse, and corn stalk fiber were underwent tensile testing. Tensile testing is to measure the tensile properties and was done using an Instron Testing Machine. The dumbbell shape of the sample was about 50 mm according to ASTM 638.

Results and Discussion

The results which were presented here are on the tensile strength of the composites that have been studied. A comparison of the tensile strength results was made against four types of natural fibers that are untreated and that have been treated with vinegar and soda bicarbonate. The tensile strength is inferred in Fig. 1 for both treatments. The neat composite was used as a control unit with various tensile measured values in the range of 18 to 25 MPa. This indicated lower than the epoxy composites which are filled with treated natural fibers. Rajala et. al [4] proved through their study that alkali-treated natural fiber has better tensile strength than the untreated one.

This is supported by the statement of most research studies have found and reported that surface modification by alkali treatment is effectiveness in increasing the adhesion between the fiber surface and the matrix, thereby enhancing the overall performance of the natural fiber-reinforced composites [8]. Then, if we relate to the nature of natural fiber which contains lignin, hemicellulose, and others, the alkali treatment tends to remove the lignin and hemicellulose thus affecting the tensile properties of the fibers [9]. After the hemicellulose is removed, the fibrils are more capable of rearranging themselves along the direction of tensile deformation. Alkali treatment with appropriate concentration also has a significant impact on tensile properties. It will make the surface become rough and will promote the mechanical anchoring between fiber and matrix [10]. All this will help in the improvement of tensile strength.

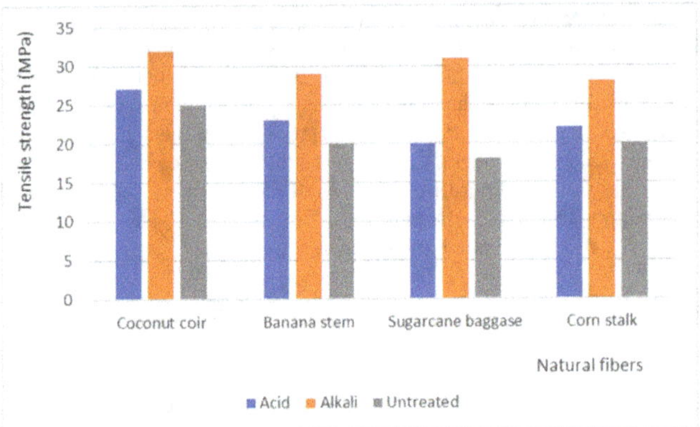

Fig. 1: Tensile strength of untreated and treated natural fiber as filler in epoxy composites.

This figure illustrated the evolution of the tensile strength and the marked highest was observed at coconut coir, after alkali treatment and also highest at treated with acid. By looking only at the four different types of natural fiber, it can be seen that the tensile strength is affected by filler type and their surface area [11]. By comparing both treatments for each natural fiber; coconut coir for example, it showed the tensile strength of the epoxy composites is higher for alkaline treated than acid treated. This is followed by sugarcane bagasse, banana stem fiber, and lastly corn stalk fiber, respectively. For acid treatment, after coconut coir, resulted from banana stem fiber, corn stalk, and sugarcane bagasse.

Alkali treatments are significantly effective in cleaning the fiber surface more than acid treatment. This may explain why most of the researchers tend to choose alkaline treatment instead of acidic treatment. Alkaline treatment will give a better and more desirable result. The acid treatment made surface damage on the surface and this can impart some hindrance causing the composite's strength reduction [6]. The cleanliness and roughness in the fiber surface due to the fiber delignification process were observed to increase by a different type of treatment between acid and alkali [12] and the effect of a different type of fiber as well. Most of the properties showed higher tensile strength readings for treated fiber and its composites rather than untreated. This is attributed to the fibers tending to get better fiber-matrix interaction after treatment. Untreated fibers have low tensile strength because they are hydrophilic and incompatible with the matrix [13].

Most of the researchers conclude improvements in tensile strength after treatments and proved by morphological examination. The treatment will improve adhesion between the hydrophilic fibers and hydrophobic epoxy. The weak filler-matrix adhesion was due to low compatibility between both and thus led to poor strength properties of the epoxy composites filled with untreated natural fibers. Treatment also showed an increase in surface roughness and quality in terms of the reduction of impurities. Most of the findings found the same reason and stated that the roughening of the surface of the fiber is due to the removal of non-cellulose with acid treatment. The removal of certain portions of hemicellulose, lignin, waxy substances, and natural oils covering the surface of the fiber surface.

Besides, as we know, every natural fiber has its surface morphology which decides the interfacial fiber-matrix adhesion and these will affect the mechanical/ physical properties [3]. This explains why a different type of natural fiber will give different highest to lowest tensile strength values, even have been treated with the same medium, soaking time, and concentration. Furthermore, the untreated fiber which is filled in other matrices also will give the same trend, as revealed in our previous research [1].

Conclusion

It is well known that; cellulose fiber of natural fibers is inherently compatible with hydrophobic matrices like epoxy resin. Chemical treatment is significant for modifications to improve interfacial adhesion. Four different natural fibers from agricultural resources; coconut coir, banana stem fiber, sugarcane bagasse, and corn stalk fiber, and two types of chemical treatments have been compared among alkali treatment of baking soda and acid treatment of vinegar. All fibers were observed highest tensile strength after being treated with alkali compared to acid. Alkali treatment results in improving bonding between hydrophilic fibers and hydrophobic polymer matrix. Among them, coconut coir showed the highest tensile strength either after alkali treatment or acid treatment.

References

[1] N.A Ahad, N.F.H. Radzi, N, S Sadli, and J. Tan, Comparing Various Types of Natural Fibers as Filler in TPU: Mechanical Properties, Morphological and Oil Absorption Behavior, MATEC Web Conf. 97 (2017) 01062. https://doi.org/10.1051/matecconf/20179701062

[2] M. Ravi, R.D. Rajnish, Effect of surface treatment on natural fibers composites, IOP Conf. Ser. : Mater. Sci. Eng. 376 (2018) 012053. https://doi.org/10.1088/1757-899X/376/1/012053

[3] F.A. Koyade, Surface Treatments of Natural Fibres - A Review: Part 1, Open J. Polym. Chem. 5 (2015) 41-46. https://doi.org/10.4236/ojpchem.2015.53005

[4] A. Rajala, K. Yoganandan, and V. Mohanavel, Effect of chemical treatment on natural fiber for use in fiber reinforced composites - Review, Mater. Today Proc. 33 (7) (2020) 2996-2999. https://doi.org/10.1016/j.matpr.2020.02.982

[5] M. K. Bakri, E. Jayamani, and S. Hamdan, Processing and Characterization of Banana Fiber/Epoxy Composites: Effect of Alkaline Treatment, Mater. Today Proc. 4(2) (2017) 2871-2878. https://doi.org/10.1016/j.matpr.2017.02.167

[6] R. Vijay, & A. Vinod, Characterization of chemical treated and untreated natural fibers from Pennisetum orientale grass - A potential reinforcement for lightweight polymeric applications, Int. J. Lightweight Mater. Manuf. 4 (1) (2021) 43-49. https://doi.org/10.1016/j.ijlmm.2020.06.008

[7] Mustapha Nouri, Mahfoud Tahlaiti & Fréderic Grondin, Effect of Chemical and Physical Treatments on Mechanical Properties of Diss Fibers-Based Biocomposites Materials, J. Nat. Fibers 20 (2021) 2128148. https://doi.org/10.1080/15440478.2022.2128148

[8] K. Mujahid, R. Sayyadh, Effect of NaOH treatment on the mechanical strength of banana/epoxy laminates, Polym. Renew. Resources, 10 (1-3) (2019) 19-26. https://doi.org/10.1177/2041247919863626

[9] J. Flandez, I. González, J. B. Resplandis, N. E. El Mansouri, F. Vilaseca, and P. Mutjé, Management of corn stalk waste as reinforcement for polypropylene injection moulded

composites, BioResources. 7(2) (2012) 1836-1849. https://doi.org/10.15376/biores.7.2.1836-1849

[10] A. A. Kibret (Yeshurun), Effect of Chemical Treatment of Fibres on the Mechanical Properties of Natural Fibre Reinforced Polymer Composites, IJSER 9 (2018) 73.

[11] K. Myrath, H. Onggo, and A. Syampurwadi, Physical and Mechanical Properties of Natural Fibers Filled Polypropylene Composites and Its Recycle, J. Biol. Sci. 7 (2) (2007) 393-396. https://doi.org/10.3923/jbs.2007.393.396

[12] M.Y. Hashim, A.M. Amin, O.M.F, Marwah, M.H. Othman, M.R.M. Yunus, and N. C. Huat, The effect of alkali treatment under various conditions on physical properties of kenaf fiber, J. Phys. Conf. Ser. 914 (1) (2017) 1-15. https://doi.org/10.1088/1742-6596/914/1/012030

[13] P. Ramadevi, and B. Basavaraju, Effect of Chemical Treatments on Tensile Strength of Areca Fiber Reinforced Natural Rubber Composites, IOSR Journal of Applied Chemistry (IOSR-JAC) 8 (2015) 43-52.

Materials Research Proceedings 68 (2026) 139-147 https://doi.org/10.21741/9781644904237-19

Effect of Curing Time and Moisture Exposure on the Mechanical Response of Polyurethane with Varying Polyol-to-Isocyanate Ratios

Ahmad Syafiq Iskandar AHMAD NAZRI[1,b], Diana CHE LAT[1,a], Ab Aziz MOHD YUSOF[2,c], Ismacahyadi Bagus MOHAMED JAIS[3,d], Doris Asmani MAT YUSOF[1,e], Mohamed Azizi MD ALI[4,f], and Roslizayati RAZALI[1,g]

[1]Faculty of Civil Engineering, Universiti Teknologi MARA Kampus Pasir Gudang, 81750, Masai, Johor, Malaysia

[2]Faculty of Mechanical Engineering, Universiti Teknologi MARA Kampus Pasir Gudang, 81750, Masai, Johor, Malaysia

[3]Faculty of Civil Engineering, Universiti Teknologi MARA Shah Alam, 40450, Selangor, Malaysia

[4]Geocon (M) Sdn Bhd, Shah Alam, 40400, Selangor

[a]dianacl@uitm.edu.my, [b]2024610478@student.uitm.edu.my, [c]abaziz86@uitm.edu.my, [d]ismac821@uitm.edu.my, [e]dorisasmani@uitm.edu.my, [f]office@geoconmsb.com, [g]roslizayati440@uitm.edu.my

Keywords: Polyurethane, Polyol-to-Isocyanate Ratios, Curing Time, Moisture Exposure

Abstract. Pavement maintenance and rehabilitation are critical components of modern transportation infrastructure management. Recent advancements focus on enhancing its performance through various additives and modifications, such as the incorporation of polyurethane, epoxy resins and recycled materials. The unique structure and tunable formulation of PU have made it attractive for asphalt modification, significantly improving asphalt performance. In this study, the tensile strength test is conducted using a Universal Testing Machine (UTM) to determine the ability of the PU to resist tensile (pulling) forces. This study investigates the mechanical performance of polyurethane (PU)-modified cold mix asphalt under varying polyol-to-isocyanate ratios (1:1 and 2:1), curing durations (1, 3, 7, and 15 days), and environmental conditions (dry and wet). The findings indicate that the 2:1 ratio exhibits superior early-stage flexibility and higher strain values, suggesting enhanced ductility and suitability for rapid-setting applications. However, as curing time increases, the 1:1 stoichiometric ratio outperforms the 2:1 mixture in terms of maximum stress, particularly on Day 15 under both dry (12.345 MPa) and wet (9.412 MPa) conditions. This highlights the 1:1 ratio's greater long-term strength and durability, making it more appropriate for applications requiring sustained load-bearing capacity and environmental resistance. Overall, the results suggest that the selection of PU ratio should be application-specific: the 2:1 ratio favors early-stage flexibility, while the 1:1 ratio ensures long-term mechanical stability.

Introduction

Pavement maintenance and rehabilitation are critical components of modern transportation infrastructure management. Among the various pavement maintenance techniques, pavement patching has emerged as a cost-effective and practical solution for addressing localized pavement defects [1]. Pavement patching is a critical maintenance activity aimed at extending the lifespan of road surfaces and ensuring safety for users. The process involves various materials and methods, each with distinct characteristics and applications. Cold mix asphalt (CMA) is a type of asphalt that is mixed and applied at ambient temperatures, making it particularly suitable for road maintenance and patching applications. Its key characteristics include ease of preparation, lower energy requirements, and the ability to be transported over long distances without heating [14]. Historically, CMA has been utilized primarily for pothole repairs and the rehabilitation of lightly trafficked roads. Recent advancements focus on enhancing its performance through various additives and modifications, such as the incorporation of polyurethane, epoxy resins and recycled materials [3].

Polyurethane (PU) is a flexible polymer widely used due to its unique properties. It is formed by reacting polyols (alcohols with multiple hydroxyl groups) with isocyanates (compounds with -N=C=O groups) [15]. PU consists of two main parts: the hard segment, which gives strength and rigidity, and the soft segment, which provides flexibility and elasticity. By adjusting the ratio between these segments, the mechanical properties of PU can be customized for different applications [5]. Isocyanates used in PU can be aromatic or aliphatic. Aromatic types give higher strength and stiffness, while aliphatic ones offer better flexibility and weather resistance [6]. Common isocyanates include toluene diisocyanate (TDI), methylene diphenyl diisocyanate (MDI), and polyaryl polymethylene isocyanate (PAPI), as shown in Fig. 1.

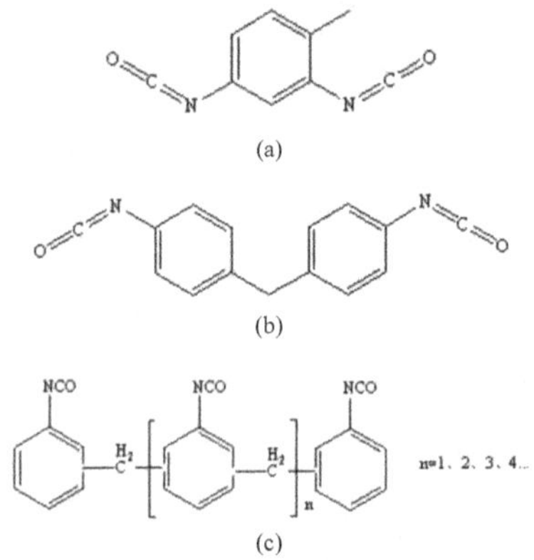

Fig. 1: Isocyanate molecular structure formula, (a) 2,4 Toluene diisocyanate (TDI), (b) 4,4 Methylene diphenyl diisocyanates (MDI), (c) Polyaryl polymethylene isocyanate (PAPI).

Polyurethane (PU) is a block polymer widely used in industries like automotive, civil engineering, and medical equipment, primarily as coatings, adhesives, and foam [11]. It consists of alternating hard and soft segments. Hard segments (from isocyanates and chain extenders) give high-temperature resistance, while soft segments (from polyols) offer flexibility and better low-temperature properties [7]. The unique structure and tunable formulation of PU have made it attractive for asphalt modification, significantly improving asphalt performance [10].

Polyurethane (PU) asphalt offers notable environmental benefits compared to traditional asphalt, which is energy-intensive and emits significant greenhouse gases. PU asphalt can incorporate recycled materials, such as waste polyurethane, promoting circular economy practices. Additionally, PU formulations allow for lower mixing temperatures, reducing energy consumption [12]. Using bio-based polyols further enhances sustainability by decreasing dependence on fossil fuels. PU asphalt is well-suited for road patching, cold mix applications, and sustainable infrastructure due to its mechanical strength and durability. Its rapid curing and high performance make it ideal for pothole repairs, while cold mix usage simplifies application without heating. These properties align with the growing demand for eco-friendly and resilient infrastructure [4].

Materials Research Proceedings 68 (2026) 139-147 https://doi.org/10.21741/9781644904237-19

One of the key advantages of PU-modified asphalt is its superior resistance to aging, fatigue, and thermal deformation, improving pavement longevity [17]. Thermoplastic polyurethane (TPU), for example, enhances performance across a wide temperature range, minimizing rutting and cracking risks. PU also improves adhesion with aggregates, reducing moisture susceptibility [13]. As a binder, PU enhances elasticity, bonding, and durability. This elasticity helps absorb traffic and environmental stresses, preventing cracking and deformation. PU's chemical structure fosters strong binder-aggregate interaction, creating a cohesive mix that withstands environmental challenges [16]. PU's compatibility with aggregates ensures improved bonding strength, critical under heavy loads. Strong chemical interactions between PU and aggregate surfaces reduce stripping and water damage [7]. Its hydrophobic nature further limits water infiltration, extending pavement life, particularly in wet climates [13]. PU-modified asphalt also shows excellent thermal stability, maintaining flexibility at low temperatures and rutting resistance at high temperatures. Its durability surpasses traditional asphalt, minimizing repairs and preserving long-term performance. For cold mix applications, PU is particularly effective. CMA with PU offers improved adhesion and flexibility, critical for long-lasting pothole repairs [14]. Its moisture resistance prevents early deterioration, ensuring reliability in challenging environments [13].

While polyurethane (PU) is used to enhance the performance of the asphalt, problems can still occur because of uneven settling (subsidence) on either side of the road joint and the continuous stress from traffic loads. These conditions cause the interface between the polyurethane and the original pavement material to experience tensile stress and shear stress. Fig. 2 shows the stress analysis diagram of splicing pavement [17]. This interface zone is critical because it involves two different materials coming together, and such transition zones are usually the weakest part of any structure. If this bond is not strong enough, the repair may fail prematurely.

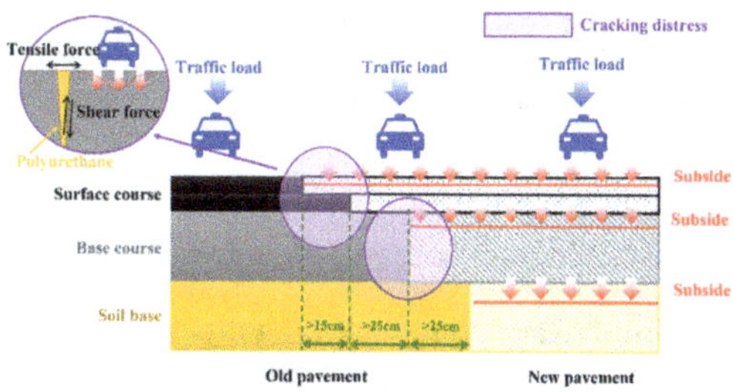

Fig. 2: Stress Analysis Diagram of Splicing Pavement [17].

Therefore, studying the mechanical properties such as tensile strength and durability of the interface between polyurethane and pavement material is very important. It helps improve the quality, durability, and performance of longitudinal joint repairs in asphalt pavements.

In this study, tests are done to measure PU in terms of strength, durability, and how well it sticks to surfaces. These tests help understand how well the material can resist cracking, bending, and damage from water before added to asphalt as an additive, which are important for the long-lasting quality of roads. The study also compares four durations of curing of PU, which are 1, 3, 7 and 15 days. These variety in curing period can help in observing the strength achieved of PU according to the timeline. Additionally, the test will include the comparison of presence of moisture damage. In order to formulate the optimum bonding of PU-asphalt composite, the ratio of the PU added to asphalt will be tested and compared.

Materials Research Proceedings 68 (2026) 139-147 https://doi.org/10.21741/9781644904237-19

Methodology

In this study, the tensile strength test is conducted using a Universal Testing Machine (UTM) as shown in Fig. 3, to determine the ability of the PU to resist tensile (pulling) forces. The objective of this test is to determine the maximum stress the material can withstand before breaking. Polyols and isocyanate are prepared in separate cups to facilitate easier mixing. The polyol and isocyanate are mixed in a 2:1 ratio in a separate cup. Using a syringe, 26ml of the mixed polyurethane is slowly with-drawn, ensuring bubbles are avoided inside the syringe. The polyurethane is then injected into a cuboid silicone mold (100mm x 20mm x 10mm), and the process is repeated to produce 8 samples. The samples are allowed to rest and set for one day before being carefully demolded. The samples are cured by air curing. The curing period are one (1), three (3), seven (7), and fifteen (15) days, respectively for each two samples. After the end of each curing period, one of the samples is taken and submerged into water for one day. This is to introduce the samples to moisture damage. Then, the samples are taken out of the water and let to rest for two hours before conducting the test. The process is repeated using a 1:1 ratio of polyol and isocyanate to create different polyurethane variables. The samples are inspected for defects, such as air bubbles or cracks, that could affect the test results, and defective samples are discarded.

The Universal Testing Machine (UTM) is calibrated according to the manufacturer's guidelines to ensure accurate results. Wedge grips are installed onto the UTM, and the testing software is configured to record the applied load, extension, and stress-strain curve. The PU sample is carefully clamped between the upper and lower grips of the UTM, ensuring vertical alignment and avoiding pre-loading or bending forces. The grip pressure is adjusted to securely hold the sample without causing damage or deformation. The test is initiated by applying a uniaxial tensile load to the sample at a constant crosshead speed, as specified in the ASTM D638 standard. As the load increases, the UTM records the stress (force per unit area) and strain (elongation per unit length) until the sample fails. The machine continuously records data, producing a stress-strain curve that represents the sample's mechanical behavior. The test is repeated for a minimum of three samples per condition, such as 2:1 and 1:1 polyol-to-isocyanate ratios, to ensure statistical reliability and reproducibility.

PU consists of two main parts which are the hard segment that governs its strength and rigidity, and the soft segment, which provides flexibility and elasticity. Hard segment of PU comes from isocyanate while soft segment of PU is given from polyol. In common practice, 1:1 of polyol-to-isocyanate ratio is considered stoichiometric balance. However, ratio 2:1 of polyol-to-isocyanate ratio is also selected to be tested so that the flexibility of PU can be compared. These variation in polyol-to-isocyanate ratios also gives and insight of how the flexibility of PU affects the durability of the material.

Fig. 3 : Universal Testing Machine (UTM).

Materials Research Proceedings 68 (2026) 139-147 https://doi.org/10.21741/9781644904237-19

Results and discussion

Tensile Strength of Polyurethane. The stress-strain curves of polyurethane (PU) samples cured over 1, 3, 7, and 15 days, as shown in Fig. 4,5,6 and 7 respectively, under varying polyol-to-isocyanate ratios of 1:1 and 2:1 and environmental conditions, dry and wet reveal critical insights into the development of mechanical properties over time. The data demonstrates how both curing duration and formulation influence the strength and ductility of PU.

Fig. 4 illustrates the strength of PU at early stage, where all samples exhibit relatively low stress values, indicating incomplete curing and weak internal bonding. The 2:1 Wet sample shows the highest peak stress and largest strain capacity among the four conditions. This suggests that the presence of excess polyol, combined with moisture exposure, may slow down hardening and cross-linking, allowing for greater flexibility and ductile response. In contrast, both 1:1 Dry and 2:1 Dry samples show lower stress and limited strain, reflecting a stiff but underdeveloped matrix due to insufficient curing time.

The improvement of mechanical properties for all samples can be seen in Fig. 5. Especially, the 2:1 Wet sample maintains the highest strength and strain values, reflecting a more developed yet still ductile structure. The 2:1 Dry sample also shows increased stress resistance, though it fails at a lower strain, indicating a stiffer and more brittle behavior. The 1:1 Dry formulation remains the weakest, exhibiting early failure and minimal deformation capacity, which is likely due to its limited toughness and early-stage brittleness.

At seven days of curing, substantial differences in behavior develop according to Fig. 6. Both 2:1 Wet and 2:1 Dry samples show significantly improved performance. The 2:1 Dry sample demonstrates a well-defined stress peak and moderate ductility, indicating a more balanced mechanical profile. Conversely, the 1:1 Dry sample shows a sharp increase in stress followed by sudden failure, revealing that although its stiffness has improved, it remains brittle. The 2:1 Wet sample continues to outperform in terms of strain, suggesting that moisture exposure could contributes to greater deformability even as strength increases.

Fig. 7 shows the strength of PU after fifteen days. The 1:1 Dry sample achieves the highest peak stress, suggesting that, with sufficient curing time, optimal PU molar ratio of 1:1 can attain a highly cross-linked and strong matrix. However, the curve shows a sharp drop after peak stress, indicating brittle fracture. The 2:1 Dry sample records moderate stress values and exhibits improved ductility, pointing to a more flexible matrix with enhanced toughness. Both wet-conditioned samples 1:1 and 2:1 Wet show lower stress levels compared to their dry counterparts. This suggests that even when moisture exposure introduced only on the final day of curing, it negatively affects the ultimate strength development of PU. However, their stress-strain curves display more gradual slopes and extended deformation, indicating improved elasticity and ductility.

The 1:1 polyol-to-isocyanate ratio have the best quality in terms of tensile strength, while higher polyol content in 2:1 polyol-to-isocyanate ratio gives more flexibility to the material, this pattern is compatible with results from Mohamed (2021).

Materials Research Proceedings 68 (2026) 139-147 https://doi.org/10.21741/9781644904237-19

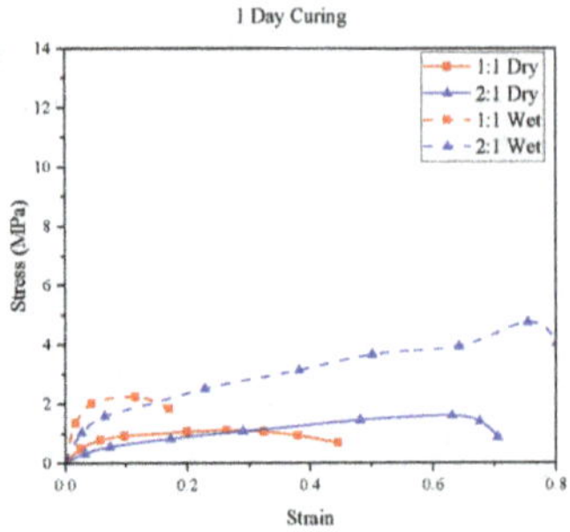

Fig 4: Stress-Strain Curve for 1 Day Curing.

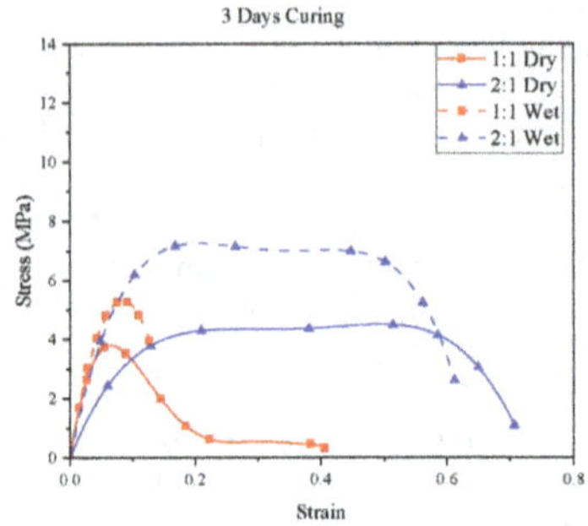

Fig 5 : Stress-Strain Curve for 3 Days Curing.

According to Table 1, the 2:1 polyol-to-isocyanate ratio consistently demonstrates higher average maximum stress in the early curing periods compared to the 1:1 ratio, under both dry and wet conditions. For example, on Day 1 under dry conditions, the 2:1 ratio achieved a maximum stress of 1.594 MPa, which is approximately 45% higher than the 1.099 MPa recorded for the 1:1 ratio. A similar trend is seen on Day 3 and Day 7. This suggests that the excess polyol content in the 2:1 mixture may lead to better initial bonding and flexibility, due to more available hydroxyl groups for reaction and improved adhesion to the asphalt matrix [2].

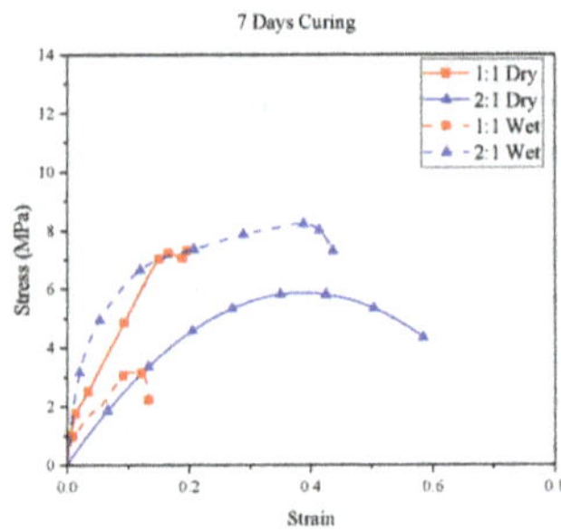

Fig. 6 : Stress-Strain Curve for 7 Days Curing.

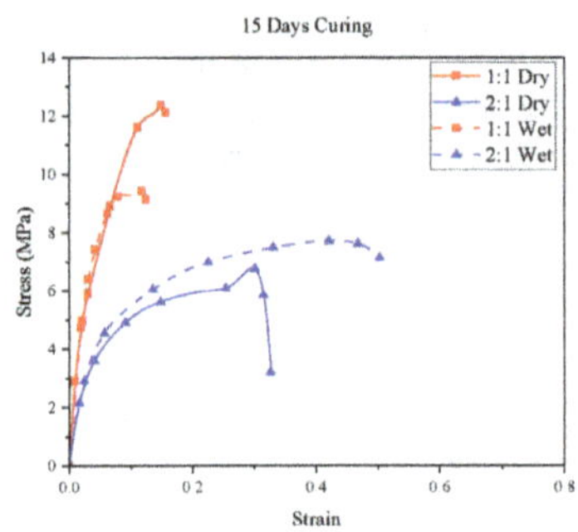

Fig. 7 : Stress-Strain Curve for 15 Days Curing.

However, as curing progresses, the performance of the 1:1 ratio improves significantly. By Day 15 under dry conditions, the 1:1 mixture recorded the highest stress value of 12.345 MPa, surpassing the 2:1 ratio, which only reached 6.762 MPa. This indicates that a balanced stoichiometric (optimum cross-linking ratio), 1:1 ratio, composition benefits from extended curing, allowing full polymerization and crosslinking to take place. A similar trend is observed under wet conditions, where the 1:1 ratio also exhibits higher stress on Day 15 (9.412 MPa) compared to the 2:1 ratio (7.705 MPa), suggesting better long-term moisture resistance and network stability for the stoichiometric blend.

In terms of strain behavior, the 2:1 ratio consistently shows higher strain at maximum stress across all curing periods. For instance, on Day 1 under dry conditions, the strain for 2:1 was 0.619, compared to 0.265 for the 1:1 ratio. This indicates that the 2:1 mixture is more ductile and capable of withstanding deformation before failure. The higher strain values associated with the 2:1 ratio suggest a more flexible and elastic material structure, which is advantageous for early-stage applications such as patching, where conformability is important. On the other hand, the 1:1 ratio, while initially more brittle, gains stiffness and strength over time, reflecting improved structural integrity and resistance to deformation.

Table 1 : Average Maximum Stress and Strain at Maximum Stress for All Variable.

Curing Period (Days)	Ratio	Condition	Avg Max Stress (MPa)	Strain at Max Stress
1	1:1	Dry	1.099	0.265
1	2:1	Dry	1.594	0.619
1	1:1	Wet	2.229	0.107
1	2:1	Wet	4.730	0.757
3	1:1	Dry	3.791	0.066
3	2:1	Dry	4.485	0.494
3	1:1	Wet	3.163	0.074
3	2:1	Wet	7.246	0.205
7	1:1	Dry	7.303	0.196
7	2:1	Dry	5.852	0.386
7	1:1	Wet	5.096	0.113
7	2:1	Wet	8.233	0.384
15	1:1	Dry	12.345	0.150
15	2:1	Dry	6.762	0.297
15	1:1	Wet	9.412	0.117
15	2:1	Wet	7.705	0.427

Fig. 8 shows the trend of maximum stress in polyurethane (PU) samples over different curing periods (1, 3, 7, and 15 days) under varying material ratios and exposure conditions. The 2:1 wet samples consistently demonstrate the highest stress values at all curing durations, indicating superior mechanical performance over time. Although this ratio represents an excess of polyol compared to isocyanate, the improved strength is not necessarily due to increased cross-linking [9]. Instead, it suggests that the higher polyol content contributes to enhanced toughness and flexibility in the PU matrix, allowing the material to better withstand applied stress without premature failure. In contrast, the 1:1 Wet samples exhibit the lowest strength across all curing periods, highlighting the detrimental impact of moisture exposure, especially when the formulation already has limited toughness. While all samples show a general increase in strength with curing time, indicating continued material development, the wet-conditioned samples consistently underperform. This is likely due to water absorption interfering with the integrity of the PU structure, even after surface drying. Overall, the results suggest that longer curing durations enhance PU strength, but the combination of higher polyol content and dry curing yields the most favorable mechanical properties, possibly due to a more flexible and resilient polymer network rather than a more cross-linked one.

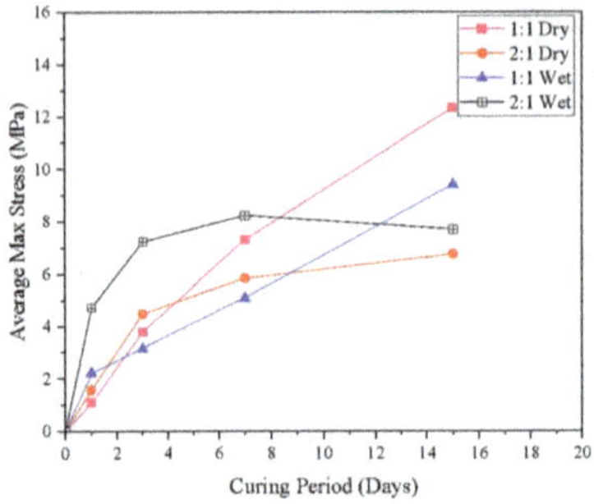

Fig 8: Average Max Stress vs Curing Period Curve.

Conclusion

The results indicate that curing time plays a critical role in enhancing both the strength and stiffness of PU. The 1:1 Dry sample achieves the highest ultimate strength at 15 days, although at the cost of brittleness. On the other hand, the 2:1 Wet samples consistently demonstrate higher strain capacity across curing periods, indicating improved ductility and toughness. While wet conditions generally reduce strength, they may contribute to greater flexibility, especially in polyol-rich formulations. These findings highlight the need to balance strength and toughness in PU systems based on application-specific requirements, with careful consideration of curing conditions and formulation ratios.

Acknowlegement

The project is financially supported by the Fundamental Research Grant Scheme (FRGS/1/2024/TK01/UITM/02/10).

References

[1] Chen, D. H., Lin, H. H., & Sun, R. (2011). Field performance evaluations of partial-depth repairs. Construction and Building Materials, 25(3), 1369–1378. https://doi.org/10.1016/j.conbuildmat.2010.09.007

[2] Desai, S. D., Patel, J. V, & Sinha, V. K. (2003). Polyurethane adhesive system from biomaterial-based polyol for bonding wood. International Journal of Adhesion and Adhesives, 23(5), 393–399. https://doi.org/https://doi.org/10.1016/S0143-7496(03)00070-8

[3] Fan, J., Liang, H., Sun, Z., & Chen, Z. (2023). Investigation on strength characteristics and mechanism of cold mix based on geopolymerization reaction. Frontiers in Built Environment, 9(December), 1–12. https://doi.org/10.3389/fbuil.2023.1301126

[4] Hong, B., Lu, G., Gao, J., Dong, S., & Wang, D. (2021). Green tunnel pavement: Polyurethane ultra-thin friction course and its performance characterization. Journal of Cleaner Production, 289, 125131. https://doi.org/10.1016/j.jclepro.2020.125131

[5] Hussain, N. I. A. M., Bonnia, N. N., Asli, N. A., Ali, E. S., Hassan, N. A., & Affandi, N. D. N. (2022). Mechanical Properties of Green and Petroleum based Polyurethane as Grouting Materials. Journal of Mechanical Engineering, 19(1), 71–82. https://doi.org/10.24191/JMECHE.V19I1.19687

[6] Kemona, A., & Piotrowska, M. (2020). Polyurethane recycling and disposal: Methods and prospects. Polymers, 12(8). https://doi.org/10.3390/POLYM12081752

[7] Li, X., Li, J., Wang, J., Yuan, J., Jiang, F., Yu, X., & Xiao, F. (2021). Recent applications and developments of Polyurethane materials in pavement engineering. Construction and Building Materials, 304(June), 124639. https://doi.org/10.1016/j.conbuildmat.2021.124639

[8] Mohamed, D. J., Hadi, N. J., & Alobad, Z. K. (2021). Investigation of the Polyol Types and Isocyanate Concentrations on the Rheological, Morphological and Mechanical Properties of Polyurethane Foams. IOP Conference Series: Materials Science and Engineering, 1094(1), 12157. https://doi.org/10.1088/1757-899X/1094/1/012157

[9] Mohd Fauzi, F. N. A., Bonnia, N. N., Setik, R., Anuar, A. U., Rahman, Z. A., Hizrin, R. S. F. N., & Affandi, N. D. N. (2024). Physical and Morphological Analysis of Polyurethane with Different Ratios of Polyol & Isocyanate for Grouting Applications. Malaysian Journal of Chemistry, 26(6), 46–57. https://doi.org/10.55373/mjchem.v26i6.46

[10] Sheng, X., Wang, M., Xu, T., & Chen, J. (2018). Preparation, properties and modification mechanism of polyurethane modified emulsified asphalt. Construction and Building Materials, 189, 375–383. https://doi.org/10.1016/j.conbuildmat.2018.08.177

[11] Shi, S., Xu, T., Wang, D., & Oeser, M. (2020). The Difference in Molecular Orientation and Interphase Structure of SiO2/Shape Memory Polyurethane in Original, Programmed and Recovered States during Shape Memory Process. In Polymers (Vol. 12, Issue 9). https://doi.org/10.3390/polym12091994

[12] Wang, X., Yu, T., Wu, Y., Sheng, Y., Wang, Y., & Hang, Y. (2023). Study on Mechanical Properties of Two-Component Polyurethane Based on Multi-Scale Molecular Simulation. Materials, 16(3). https://doi.org/10.3390/ma16031006

[13] Wu, S., Xu, W., Zhang, F., & Wu, H. (2022). Effect of Polyurethane on High-and Low-Temperature Performance of Graphene Oxide-Modified Asphalt and Analysis of the Mechanism Based on Infrared Spectrum. Coatings, 12(5). https://doi.org/10.3390/coatings12050590

[14] Wulandari, P. S., & Tjandra, D. (2023). Properties evaluation of cold mix asphalt based on compaction energy and mixture gradation Properties evaluation of cold mix asphalt based on compaction energy and mixture gradation. https://doi.org/10.1088/1755-1315/1195/1/012024

[15] Zahedifar, P., Pazdur, L., Vande Veld, C. M. L., & Billen, P. (2021). Multistage chemical recycling of polyurethanes and dicarbamates: A glycolysis–hydrolysis demonstration. Sustainability (Switzerland), 13(6). https://doi.org/10.3390/su13063583

[16] Zhao, W., Sun, X., Ou, Z., Li, Z., Liu, Z., & Qin, X. (2023). Modifying Effect and Mechanism of Polymer Powder on the Properties of Asphalt Binder for Engineering Application. Polymers, 15(24). https://doi.org/10.3390/polym15244659

[17] Zhong, Y., Xu, S., Chi, J., Zhang, B., Shen, G., Yang, Z., Cheng, H., & Wang, B. (2023). Experimental study on the interface bonding characteristic of polyurethane and pavement materials. Construction and Building Materials, 397(July), 132390. https://doi.org/10.1016/j.conbuildmat.2023.132390

Characterisation of Moisture Retention Properties in Commercial Hydrogel Dressings for Sustained Wound Hydration

Diviya Devi PARAMASIVAM[1,2,a], Raimi DEWAN[1,2,3,4,b*], Nadia ADRUS[3,5,c], Kok Yeow YOU[2,d], and Faishal Adilah SURYANATA[1,2,e]

[1]Advanced Radio Frequency & Microwave Research Group, Faculty of Electrical Engineering, Universiti Teknologi Malaysia, 81310 UTM Johor Bahru, Malaysia

[2]Faculty of Electrical Engineering, Universiti Teknologi Malaysia, 81310 UTM Johor Bahru, Johor, Malaysia

[3]IJN-UTM Cardiovascular Engineering Centre, Institute of Human Centered Engineering, Universiti Teknologi Malaysia, 81310 UTM Johor Bahru, Johor, Malaysia

[4]Department of Biomedical Engineering and Health Sciences, Faculty of Electrical Engineering, Universiti Teknologi Malaysia, 81310 UTM Johor Bahru, Johor, Malaysia

[5]Biopolymer Research Group, Faculty of Chemical and Energy Engineering, Universiti Teknologi Malaysia, 81310 UTM Johor Bahru, Malaysia

[a]diviyadevi@graduate.utm.my, [b]raimi.dar@utm.my, [c]nadia@utm.my, [d]ykyeow@utm.my, [e]asfaishal@graduate.utm.my

Keywords: Hydrogel Dressing, Moisture Retention, Wound Hydration

Abstract. Wound care is a global concern due to the rising number of acute and chronic wound cases. Improper wound moisture management hinders healing; low moisture delays it, while excess causes maceration. Therefore, maintaining appropriate moisture at the wound site is crucial and can be achieved using moisture-balancing dressings, such as hydrogel. However, the moisture retention property of the hydrogel was not evaluated with varying amounts of simulated wound fluid amount. Therefore, this study investigated a commercial hydrogel dressings' moisture retention property with 0, 0.5, 1, 2, 3, and 4 mL of phosphate buffer solution to simulate the fluctuating amount of wound fluid. Prior to that, the swelling degree of hydrogel was assessed. The results showed that the hydrogel could swell up to 1234% of its original weight and achieve equilibrium of moisture retention after 6 hours. When the swollen hydrogel's moisture retention was evaluated, the findings suggested that sample A with 0 mL of phosphate-buffered saline (PBS) addition exhibited minimal moisture retention of 3.93%. Conversely, samples B, C, D, E, and F, with the addition of 0.5, 1, 2, 3, and 4 mL of PBS, respectively, exhibited moisture retention of 12.03%, 17.00%, 44.00%, 48.00%, and 62.90%, respectively. Notably, the moisture retention was reduced from sample F to B as the volume of PBS was reduced. This emphasises that the amount of wound fluid significantly affects the moisture retention properties of the hydrogel. Hence, a thorough evaluation of moisture retention can contribute to developing hydrogel with enhanced retention properties, which is crucial for long-term wearable wound dressings.

Introduction

A wound is a damage or injury to the skin tissue caused by diverse factors such as diseases, traumatic events, prolonged exposure to excessive pressure, and unintended physical damage such as cuts and abrasions [1]. The injured skin tissues will undergo a sophisticated process of restoration involving several phases such as haemostasis, inflammation, proliferation, and remodeling, and this process is known as wound healing. If the normal progression of a wound is disturbed, the inflammatory phase will be prolonged and eventually lead to a chronic wound [2]. In the case of chronic wounds, improper care will lead to repeated infections or even amputation

[3]. Therefore, it is vital to monitor the wound healing status in terms of the physical parameters of the wound site in a timely and effective manner [3]. This is because the optimum physical characteristics, such as temperature, moisture, and pressure, are essential for wound healing, and these parameters are significant and can be utilised to detect the status of wound progression [3].

Among these parameters, the optimum local moisture level of the wound site is crucial for efficient wound healing [3, 4]. This is because an adequate level of moisture is proven to be significant for initiating re-epithelialisation and granulation of tissues [4]. In addition, moisture stands out as a key parameter as it can be readily managed through the selection of appropriate dressings, unlike other factors that often require systemic treatment or advanced monitoring technologies [5]. Besides that, when comparing dry and moist environments for healing, a moist wound environment is beneficial as it reduces eschar formation, dead tissue, and necrotic tissue formation by 40%, resulting in faster wound healing [4]. However, the excess moisture at the wound bed causes maceration, while insufficient moisture dries out the wound [6]. Therefore, maintaining optimal moisture levels is vital to ensure normal wound healing and prevent complications associated with overhydration or dehydration.

Several types of wound dressings have been widely used, such as traditional wound dressings comprising gauze and bandages and modern dressings comprising films, foams, hydrocolloids, and hydrogel dressings. When comparing these dressings, hydrogel dressings have outstanding characteristics that enhance their effectiveness in wound healing, primarily due to three dominant factors. First, hydrogels closely resemble the extracellular matrix (ECM), which makes hydrogels biocompatible and non-toxic, minimising irritation at the wound site [7]. Second, hydrogels have a three-dimensional (3D) porous structure that facilitates gas exchange by allowing oxygen and water vapor to pass between the wound and the external environment . This ability is essential for maintaining a moist environment, which is crucial for efficient re-epithelisation and granulation. Third, the hydrophilic 3D polymeric network of hydrogels contributes to their water-absorbing capacity, which the hydrogels can absorb low-to- moderate-wound exudate [8]. Not only that, hydrogels can retain [1] and donate moisture. Thus, it creates a moist environment for the wound, which is optimal for re-epithelisation and granulation. Hence, all these characteristics make hydrogels create a favourable environment for optimum tissue regeneration [9].

Based on the above-discussed characteristics and availability of distinct options for hydrogel classification in terms of materials, cross-linking method, configuration, and physical appearance that can be tailored to satisfy specific criteria, such as particular wound size, location, and severity, have led to extensive research on developing hydrogels with various functions for wound healing [10]. In addition, this caused the introduction of hydrogel wound dressings for commercial use. Off-the-shelf hydrogel dressings have been available in different forms, such as sheets, amorphous gels, and those impregnated into gauze [4]. These dressings are used for various types of wounds, from dry to wet and acute to chronic. However, it must be noted that the moisture retention property of this dressing has not been investigated with varying amounts of exudate conditions over time, as the reported studies evaluated utilising the limited varying quantity of 1 mL and 0.5 mL of phosphate-buffered saline (PBS) solution [11] or in a dry state [12]. Hence, the restriction in simulating the dynamic nature of real wounds lacks the determination of the moisture retention capability of the dressing over a period. Thus, in this study, the moisture retention properties of the hydrogel dressing will be evaluated with varying amounts of PBS to simulate real-world situations that change over time.

Methodology

This section provides a detailed overview of the steps involved in evaluating commercial hydrogel dressing for moisture retention. Fig. 1 illustrates the overall workflow of this study, representing the step-by-step experimental processes.

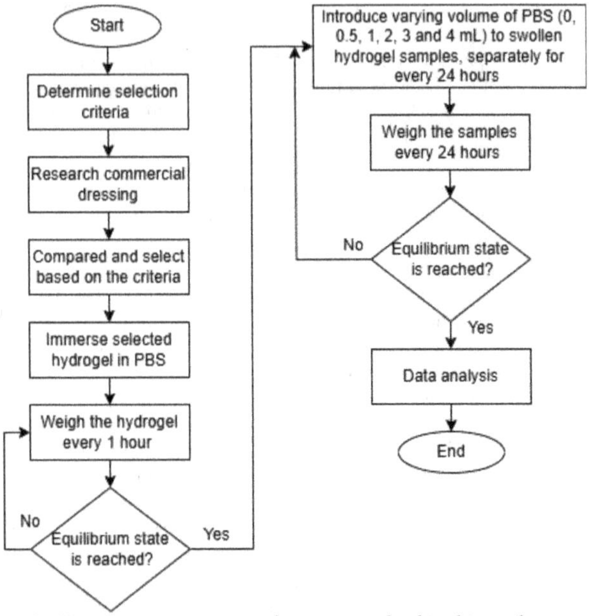

Fig. 1. Flowchart summarising the steps involved in this study.

Selection of Hydrogel Dressings. This study selected commercial hydrogel dressing based on its key characteristics, as presented in Table 1. The available dressings were compared to identify the dressing that satisfies the predefined selection criteria, including widespread clinical use. Among the analysed options, the dressing best suited to these criteria was selected for moisture retention evaluation.

Table 1: Criteria used for selecting the commercial dressing.

Criteria	Description
Biocompatibility	Dressing must be non-toxic, non-irritating, and safe for prolonged contact with the skin.
Regulatory Approval	Dressing must receive regulatory approval for clinical use to ensure safety.
Moisture retaining ability	Dressing must be able to retain moisture over a period.
Moisture donating ability	Dressing must be able to donate moisture to reduce the risk of desiccation.
Ease of removal	The dressing must be able to be removed without causing secondary damage to the wound site.

Swelling Capacity. Prior to the moisture retention testing, the swelling capacity of the hydrogel was assessed to ensure it reached a fully swollen state, which is necessary for the evaluation of retention properties. The hydrogel was dried at room temperature to completely remove the moisture, followed by weighing. The dried hydrogel with the absence of moisture was fully immersed in the PBS and left at room temperature. Subsequently, the hydrogel was removed from the solution periodically, and the surface was wiped with tissue paper to remove excess solution. The swollen weight of the hydrogel was recorded every 1 hour until there was no significant increase in the weight. Finally, the swelling ratio was calculated as shown in Eq. 1 [13]:

$$\text{Swelling degree (\%)} = \frac{W_s}{W_d} \times 100 \tag{1}$$

where W_d is the dry weight of the hydrogel and W_s is the swollen weight of the hydrogel. All measurements were performed in triplicate (n = 3) and data are presented as mean ± standard deviation (SD).

Moisture Retention Test. The hydrogels immersed in PBS solution from the swelling test were used and their weight was recorded as the initial weight. Then, the fully swollen hydrogels were placed in the petri dish separately and labeled as in Table 2. Six samples were subjected to varying amounts of PBS to simulate the real-wound exuding condition. For this study, the highest amount of PBS was limited to 4 mL, owing to the suitability of the selected dressing, which is for the minimally exuding wound that is expected to produce less than 5 mL of wound fluid per 24 hours [14]. Meanwhile, a minimum of 0.5 mL was determined based on the study conducted by Lu et al.[11] for a minimally exuding wound.

Table 2: The volume of PBS for each sample.

Samples	Volume of PBS introduced every 24 hours (mL)
Sample A	0.0
Sample B	0.5
Sample C	1.0
Sample D	2.0
Sample E	3.0
Sample F	4.0

Sample A was served as a controlled group, as there would be no addition of PBS, whilst other samples were introduced to the designated volume of PBS every 24 hours until it reached equilibrium. The weight of the samples was recorded every 24 hours until no changes in the weight were observed. The moisture retention was calculated using Eq. 2 [15]:

$$\text{Moisture retention (\%)} = \frac{W_t}{W_i} \times 100 \tag{2}$$

where W_i is the initial weight of the swollen hydrogel, and W_t is the weight of the hydrogel at a certain period. All measurements were performed in triplicate (n = 3) and data are presented as mean ± standard deviation (SD).

Results and Discussion

In this section, the selection of the commercial hydrogel dressing that meets the determined criteria was discussed thoroughly. Then, the swelling capacity of the selected hydrogel was analysed. Finally, the moisture retention results of the swollen hydrogel were presented.

Selection of Commercial Hydrogel Dressing. Three commercial hydrogel dressing brands were selected based on biocompatibility and regulatory approval. The dressings were then further compared based on the criteria outlined in Table 1 to ensure the most suitable product was chosen.

Brand A hydrogel sheet is a primary dressing comprising hybrid polymers such as polyethylene glycol (PEG), polyvinylpyrrolidone (PVP), and agar. These hydrogels exhibit good mechanical strength owing to their cross-linking method, which is through an electron beam. This dressing comprises 90% water and can absorb and donate water to the wound site. However, this hydrogel sheet lacks adhesion properties. Thus, it requires secondary dressings such as a bandage or net. In addition, this dressing requires a poultice to restore its elasticity and moisture, as it will be dried faster. Lastly, soaking is also necessary before removing the dressing to avoid damage to the granulating wound.

On the other hand, Brand B hydrogel dressing is suitable for moderately exuding wounds. Unlike other dressings, the wear time and frequency of replacement are influenced by the rate at which the dressing becomes periodically saturated with wound exudate. However, if this dressing is applied to dry wounds, it needs to be frequently monitored to prevent it from drying out and adhering to the wound site, which will disrupt the healing when removed. Thus, the dried dressing requires rehydration prior to removal.

Brand C moisturises, soothes, and facilitates the osmotic and autolytic debridement of necrotic tissue, providing a microenvironment for ideal healing for lightly exuding and sloughy wounds. This hydrogel dressing is made up of two layers. The top layer, polyurethane (PU), is flexible, fluid-repellent, and highly breathable, which acts as an external protective layer, impermeable to external bacteria. However, it is permeable to moisture. The bottom layer is a hydrogel, where the composition information is not classified in the product sheet, and it will be in contact with the wound site to interact with the wound exudate. The absence of an external adhesive layer prevents the strong adhesion of the dressing to the wound on removal by maintaining the gel-like consistency. Hence, it reduces pain during the dressing changes. Most importantly, this dressing reduces the risk of desiccation or dehydration of the wound by donating moisture, thus increasing the granulation tissue of chronic wounds. Besides that, this dressing also prevents maceration of the peri-wound area by absorbing the excess wound fluid. This showed that the Brand C hydrogel can balance the moisture of the wound. Therefore, in this study, the Brand C dressing was selected due to the good moisture balance and the fact that they do not require rehydration prior to removal, even though it requires a secondary dressing to hold them in place after applying them to the wound. The Brand C hydrogel was used for the rest of the study.

Swelling Capacity of the Hydrogel. The selected hydrogel's swelling capacity is illustrated in Fig. 2, providing insight into the increase in the percentage of hydrogel weight, which is significant for determining moisture retention properties. As shown in Figure 2, the hydrogel swelled rapidly from 0 to 2 hours, reaching swelling degrees of 364% and 540%, respectively, indicating high water absorption due to its porous structure that allows fast diffusion of water into the polymeric network. The increment in water absorption rate remains steady from 2 to 5 hours, with swelling degrees of 827% and 1079%, respectively. However, the water absorption rate began to slow after 6 hours. This shows that the hydrogel was gradually reaching equilibrium as water occupied all the available sites in the polymeric structure. By the 9th hour, the swelling curve plateaued at a swelling degree of 1234%, indicating that the hydrogel network had reached its maximum capacity. Thus, the swollen hydrogel was then utilised for moisture retention testing.

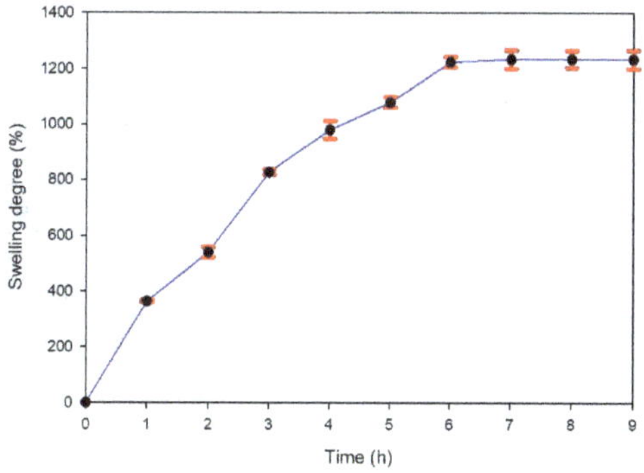

Fig. 2: Swelling degree of hydrogel. Data points represent the mean values of triplicate measurements. Error bars indicate the standard deviation (SD) among replicates.

Analysis of Moisture Retention. The practical application of moisture retention of commercial hydrogel dressing was simulated by introducing various amounts of PBS to the samples. Fig. 3 shows the moisture retention curve of the samples over 8 days. Notably, the control sample A lost water quickly from 0 to 2 days compared to other samples. After 2 days, sample A gradually achieved equilibrium as the curve plateaued. The sample A managed to retain 3.93% of water, which is the lowest among the samples.

Meanwhile, the moisture retention of hydrogel samples B to F was reduced over time despite the addition of PBS every 24 hours, preventing excessive fluid accumulation. The reduction indicates that the hydrogel can regulate its moisture by maintaining a balance between absorption and evaporation, thus preventing over-hydration of the wound that might delay the healing. On the other hand, this shows that the amount of wound fluid present influences moisture retention of the dressing at the wound site. For instance, the lower the wound fluids amount, the lower the capability of the hydrogel to retain moisture.

Conversely, Sample C exhibited a slight increase in moisture retention after Day 3, owing to localised variation in surface hydration or delayed absorption dynamics. Hydrogels often exhibit transient swelling, especially under dynamic fluid environments, where delayed polymer chain relaxation or non-uniform fluid distribution can result in temporary moisture overshoots [16]. By Day 4, the subsequent decrease suggests a return to equilibrium as the hydrogel released excess water through evaporation or internal redistribution.

Furthermore, samples D, E, and F exhibited similar moisture retention percentages, ranging from 70.34% to 72.81% on day 3, suggesting that these samples have reached the point of equilibrium at the same time. On day 8, all the samples achieved an equilibrium state, where the moisture retention of samples B, C, D, E and F were 12.03%, 17.00%, 44.00%, 48.00%, and 62.90%, respectively. Therefore, investigating the moisture retention properties of the hydrogel dressing with various amounts of PBS provides detailed insights into the ability of the commercial hydrogel dressing to maintain moisture in practical applications.

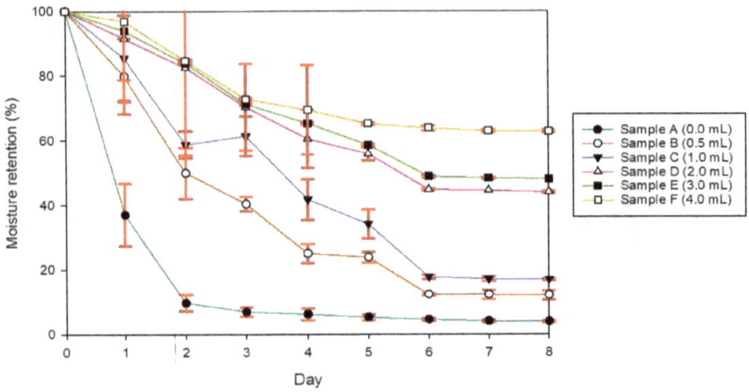

Fig. 3 : Moisture retention of samples with varying amounts of PBS was added every 24 hours until an equilibrium state was obtained. Data points represent the mean values of triplicate measurements. Error bars indicate the standard deviation (SD) among replicates.

Conclusion

Wound moisture is a crucial parameter for optimum tissue regeneration during wound healing. Therefore, assessing the moisture retention properties of wound dressing, particularly hydrogel dressing, at various stages of wound healing, provides an overview of the capability of the dressing to maintain moisture. This study assessed the moisture retention of selected commercial hydrogel dressing by introducing 0, 0.5, 1, 2, 3, and 4 mL of PBS every 24 hours until an equilibrium state was achieved. The hydrogel demonstrated a swelling capacity of up to 1234% of its original weight and reached moisture retention equilibrium within 8 hours. In addition, the findings revealed that the swollen hydrogel's moisture retention increased with higher volumes of PBS, with sample A (0 mL PBS) retaining only 3.93% moisture, while samples B to F retained 12.03%, 17.00%, 44.00%, 48.00%, and 62.90%, respectively, confirming that wound fluid volume significantly influences the hydrogel's retention capability. Therefore, the simulation of a practical application where the wound fluid level varies allows for a comprehensive understanding of how effectively hydrogel dressing can retain moisture over an extended period. Thus, these insights can guide the development of hydrogel dressings with enhanced retention properties, suitable for extended wear in real-world wound care applications.

Acknowledgement

The authors would like to thank the Ministry of Higher Education (MOHE), School of Postgraduate Studies (SPS), Research Management Centre, Advanced RF and Microwave research group, IJN-UTM Cardiovascular Engineering Centre and Universiti Teknologi Malaysia (UTM) Johor Bahru, for the support of the research under UTM Fundamental Research Q.J130000.3823.24H07 and Fundamental Research Grant Scheme titled FRGS/1/2023/TK07/UTM/02/23.

References

[1] J. Su, J. Li, J. Liang, K. Zhang, and J. Li, Hydrogel Preparation Methods and Biomaterials for Wound Dressing, Life (Basel). 11 (2021) 'https://doi.org' 10.3390/life11101016

[2] P. Schilrreff and U. Alexiev, Chronic Inflammation in Non-Healing Skin Wounds and Promising Natural Bioactive Compounds Treatment, Int J Mol Sci. 23 (2022) 'https://doi.org' 10.3390/ijms23094928

[3] L. Shao-Hao *et al.*, Multimodal sensing and therapeutic systems for wound healing and management: A review, Sensors and Actuators Reports. 4 (2022) 100075. 'https://doi.org' https://doi.org/10.1016/j.snr.2022.100075

[4] K. Nuutila and E. Eriksson, Moist Wound Healing with Commonly Available Dressings, Adv Wound Care (New Rochelle). 10 (2021) 685-698. 'https://doi.org/10.1089/wound.2020.1232

[5] N. Tang *et al.*, Wearable Sensors and Systems for Wound Healing-Related pH and Temperature Detection, Micromachines (Basel). 12 (2021) 'https://doi.org' 10.3390/mi12040430

[6] E. Crowe, C. Scott, S. Cameron, J. H. Cundell, and J. Davis, Developing Wound Moisture Sensors: Opportunities and Challenges for Laser-Induced Graphene-Based Materials, Journal of Composites Science. 6 (2022) 176. [Online]. Available: https://www.mdpi.com/2504-477X/6/6/176

[7] C. Sara and L. Gaetano, Hydrogel-based commercial products for biomedical applications: A review, International Journal of Pharmaceutics. 573 (2020) 118803. 'https://doi.org' https://doi.org/10.1016/j.ijpharm.2019.118803

[8] W. V. Sathyaraj, L. Prabakaran, J. Bhoopathy, S. Dharmalingam, R. Karthikeyan, and R. Atchudan, Therapeutic Efficacy of Polymeric Biomaterials in Treating Diabetic Wounds—An Upcoming Wound Healing Technology, Polymers. 15 (2023) 1205. [Online]. Available: https://www.mdpi.com/2073-4360/15/5/1205

[9] T. Subramaniam, M. B. Fauzi, Y. Lokanathan, and J. X. Law, The Role of Calcium in Wound Healing, Int J Mol Sci. 22 (2021) 'https://doi.org' 10.3390/ijms22126486

[10] P. Viaña-Mendieta, M. L. Sánchez, and J. Benavides, Rational selection of bioactive principles for wound healing applications: Growth factors and antioxidants, Int Wound J. 19 (2022) 100-113. 'https://doi.org' 10.1111/iwj.13602

[11] J. Lu *et al.*, Construction and function of robust and moist bilayer chitosan-based hydrogel wound dressing, Materials & Design. 226 (2023) 111604. 'https://doi.org' https://doi.org/10.1016/j.matdes.2023.111604

[12] Z. Z. Zhang *et al.*, Water-retaining and separable adhesive hydrogel dressing for wound healing without secondary damage, SCIENCE CHINA-MATERIALS. 66 (2023) 3337-3346. 'https://doi.org' 10.1007/s40843-022-2466-7

[13] H. Chopra, S. Bibi, S. Kumar, M. S. Khan, P. Kumar, and I. Singh, "Preparation and Evaluation of Chitosan/PVA Based Hydrogel Films Loaded with Honey for Wound Healing Application," *Gels*, vol. 8, no. 2, doi: 10.3390/gels8020111

[14] J. B. Lutz, C. L. Zehrer, S. E. Solfest, and S. A. Walters, A new in vivo test method to compare wound dressing fluid handling characteristics and wear time, Ostomy Wound Manage. 57 (2011) 28-36.

[15] A. Rezaei, H. Ehtesabi, and S. Ebrahimi, Incorporation of Saqez essential oil into polyvinyl alcohol/chitosan bilayer hydrogel as a potent wound dressing material, International Journal of Biological Macromolecules. 226 (2023) 383-396. 'https://doi.org' https://doi.org/10.1016/j.ijbiomac.2022.12.036

[16] N. A. Peppas, P. Bures, W. Leobandung, and H. Ichikawa, Hydrogels in pharmaceutical formulations, European Journal of Pharmaceutics and Biopharmaceutics. 50 (2000) 27-46. 'https://doi.org' https://doi.org/10.1016/S0939-6411(00)00090-4

Study on the Process of Culantro Teabag (*Eryngium foetidum*)

Nguyen THI HIEN[1,a*], Nguyen KIM PHUNG[1,b]

[1]School of Agriculture and Aquaculture, Tra Vinh University, Tra Vinh City, Vietnam

[a]hiennguyen@tvu.edu.vn, [b]nphung@tvu.edu.vn

Keywords: Culantro, Tea Bag, Total Polyphenol Content, Bioactive Compounds

Abstract. *Eryngium foetidum*, commonly known as Culantro or long coriander, is a widely used herbaceous plant, both as a food and as a food ingredient, that contains numerous health-beneficial nutrients but has relatively low economic value. Phytochemical analysis revealed the substantial concentrations of phenolics, flavonoids, tannins, alkaloids and saponins. The study aims to combine culantro leaves, green tea and Chrysanthemum to create new products that are good for health. The research focuses on (i) The drying temperature was one of the important factors influencing the final quality of the products. Different drying temperatures from 50 to 90°C were evaluated. Convective hot-air drying, a traditional drying method was employed, in which heated air is circulated over the materials to remove moisture through evaporation. (ii) Then the ingredients mixed differently in proportions culantro, green tea, and chrysanthemum. The results indicate that drying at 70oC for 210 minutes, the total polyphenol content is 85,402 ± 0,084 mg GAE/100g dry weight, moisture content of 9%. The mixing ratio of 60% culantro: 30% green tea: 10% Chrysanthemum, produced a product with high sensory scores for color (5.05 ± 0.19), aroma (5.40 ± 0.11), and taste (5.95 ± 0.10). The total polyphenol content of the product was found to be 936.05 ± 0.716 mg GAE/100g dry weight. This study demonstrates the potential of culantro in producing tea bags with high nutritional value and favorable properties.

Introduction

Tea contains bioactive compounds including polyphenols, catechins, and polysaccharides. Polyphenols have demonstrated anti-inflammatory properties that may contribute to improved vascular health, reduced blood pressure, lipid and cholesterol levels. These beneficial effects are attributed to the antioxidant, antibacterial, antifungal, and antiviral activities of polyphenols [1]. Catechins and polysaccharides are recognized for their ability to lower blood sugar levels. Phenolic antioxidants have been extensively studied for their potential role as dietary components in managing diabetes, cardiovascular diseases, arthritis, cognitive decline, and cancer [2]. The phenolic compounds present in tea encompass flavanols, various other flavonoids, and phenolic acids. The antioxidant and antibacterial properties of polyphenol-catechins can be harnessed as natural, non-toxic preservatives or employed in the formulation of functional products. There is a notable upward trend in the expansion of the functional food market, as herbal teas not only function as refreshing beverages but also provide supplementary health benefits. These benefits are attributed to the bioactive compounds present in the teas, which exhibit strong antioxidant properties that contribute to overall health.

Eryngium foetidum L., commonly known as culantro, is a plant species of significant nutritional and pharmacological value. It has various applications in disease treatment but is currently used primarily as a common vegetable for culinary purposes. Culantro contains bioactive compounds such as phenols, flavonoids, tannins, and polyphenols. The bioactive properties of culantro include antibacterial, antioxidant, and anti-inflammatory activities [3]. Its medicinal potential is also highly recognized [4]. Culantro is a rich source of nutrients including calcium, iron, carotene, riboflavin, protein, and vitamins, contributing beneficially to health [5]. The herb's content of essential oils and bioactive compounds has attracted considerable research attention [6,7]. Studies have revealed a diverse array of chemical constituents in Culantro. The chemical profile of the herb is notably characterized by the presence of aromatic aldehydes, with (2E)-2-dodecenal being predominant in the leaves. Other compounds identified include carotenoids, flavonoids, and phenols. Marina Silalahi (2021) reported that the essential oils extracted from *E. foetidum* leaves exhibit significant antibacterial and anti-inflammatory properties [8]. Furthermore, Leitão *et al.* (2020) confirmed the herb's bioactive and antioxidant potential, highlighting the role of polyphenolic compounds in various biological activities such as anti-inflammatory, hepatoprotective, antioxidant, antithrombotic, anticancer, free radical scavenging, antibacterial, antimicrobial, and hypoglycemic effects [9]. The leaves of *E. foetidum* are extensively employed in the treatment of gastrointestinal disorders and possess antibacterial, analgesic, anti-inflammatory, anthelmintic, anticonvulsant, and anticancer properties [10,12]. The notable nutritional and medicinal value of *Eryngium foetidum* is attributed to its diverse array of bioactive metabolites, including aldehydes, carotenoids, phenolic compounds, and anthraquinones [13]. However, current technologies in food research and development face significant challenges in preserving the polyphenol content in food products. Therefore, the study titled "Research on the production process of culantro teabags" was conducted with the objective of developing a new, convenient product that is beneficial for health and contributes to economic value enhancement.

Materials and methods

Material, chemicals, and equipment. The tea bags contained a formulated blend of culantro, green tea, and *Chrysanthemum indicum*. Fresh and healthy culantro leaves were harvested from Tra Vinh province, Vietnam. The remaining ingredients were produced from a local supermarket in the same region. The materials utilized in the bioactivity analysis comprise Folin - Ciocalteu reagent (Novozyme, Denmark), methanol (Merck, Germany), ethanol (Merck, Germany), acid gallic (Merck, Germany), Na_2CO_3 (Merck, Germany). The equipment employed in this study included a food dehydrator, 4-digit analytical balance, dry blender, UV-VIS spectrophotometer (Genesys 6. Thermo spectroic USA), Vortex, portable colorimeter.

Research methods:

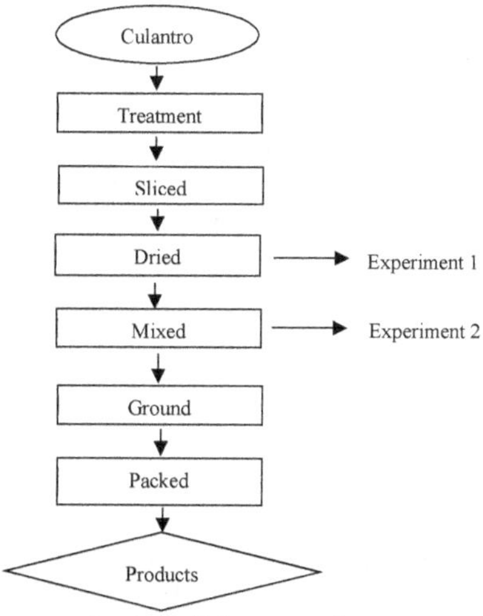

Fig 1: Process of making Culantro tea bags.

The proposed technological process to produce culantro-based tea bags is illustrated in Fig. 1. Fresh culantro leaves were first washed, drained, and then cut into slices approximately 0.5 cm in length. The sliced leaves were subsequently dried at various temperatures to reduce their moisture content to below 10%. Following dehydration, culantro was blended with green tea and *Chrysanthemum indicum* in different mixing ratios to formulate experimental samples. The blended materials were then ground to an appropriate particle size and packed into tea bags.
The factors affecting the quality of Culantro tea bags were investigated through experimental studies:

a) Experiment 1: Analysis of basic chemical composition

Culantro was washed, drained, and chopped into 0.5 cm pieces. The material was analyzed to determine moisture, total polyphenol content, and ash.

b) Experiment 2: Effect of drying temperature on the quality of Culantro

Samples were investigated at various temperatures of 60°C, 65°C, 70°C, 75°C, and 80°C; through the drying process, the equilibrium moisture content was reduced to 10%. The material was analyzed to determine moisture and total polyphenol content.

c) Experiment 3: Effect of the mixing ratio of raw materials to the quality of Culantro teabags

The experiment was designed with the variable factor being the mixing ratio, consisting of three components: culantro, green tea, and chrysanthemum. After mixing (table 1), the samples were stored in teabags and analyzed for sensory evaluation and total polyphenol content.

Table 1: The mixing ratio of materials.

Ratio	Culantro (%)	Green tea (%)	chrysanthemum (%)
1	60	30	10
2	65	25	10
3	70	20	10
4	75	15	10
5	80	10	10

Analytical Methods. Moisture content: moisture is determined by the drying method, in which the weight was perceived at a constant level (based on Vietnam Standard No 4415:1987) [14].
Sensory evaluation. Sensory evaluation was performed for overall acceptance. The sensory evaluation committee consists of 90 members with expertise in food quality assessment, according to the level of preference from 1 to 7 on the basis of 9-point Hedonic scale [15].
Determine the total polyphenol content (TPC) [16]. The Folin-Ciocalteu's reagent was used for total phenol content determination. 3 ml of gallic acid solutions were added to 0.5 ml of Folin-Ciocalteu's reagent and 2 ml of 20 % sodium carbonate. After vortexing, the samples stabilised for 1 min at 100oC, and the total phenol content was determined by using a spectrophotometer at a wavelength of 765 nm. The standard curve was prepared using solutions of gallic acid (GA) with concentrations of 0, 10, 20, 30, 40, 50μg/mL and then total phenol content is expressed as mg GA equivalents in the sample. The extracted samples were processed similarly. TPC is calculated by the formula:

$$P=(a^* V^* K)/(1000^*m^*(1-w)) \quad (1)$$

In which: P is TPC (mgGAE/g dry matter); a is Concentration of gallic acid from the standard curve with gallic acid (μgGAE/mL); V is the volume of the extract solution (mL); m is the sample mass (g); w is the moisture content of the sample; K is the dilution.
Determine the ash content by the total ash method based on Vietnam Standard No 5611:2007.

Statistical analysis. The experiments were repeated three times. The results presented are the mean ± standard deviation (SD). Data was examined using Statgraphics Centurion XVI and Microsoft Excel software.

Results and discussion

The basic chemical composition of culantro. In the field of production, the selection of appropriate raw materials not only helps in controlling the quality but also minimizes undesirable changes in the product during processing. Therefore, the determination of the chemical constituents of culantro is essential as a foundation for the research and development of culantro tea. The basic components of the raw material are meticulously analyzed, providing critical data for subsequent experiments.

Table 2: The basic chemical composition of culantro.

Components	Contents
Moisture (%)	87,74 ± 0,391
Ash (%)	1,7 ± 0,038
Total phenolic content (mg GAE/g dry matter)	136,477 ± 0,084

The identification of the basic chemical composition of culantro in Tra Vinh shows results, as presented in Table 2. The moisture content in culantro is 87.74 ± 0.391%, which is a crucial parameter affecting preservation and influencing the drying process. The ash content in the present study ranged from 1.1% to 3.3%, consistent with the findings reported by Lepcha *et al.* (2018) [17]. The relatively low ash values observed may be attributed to environmental factors and seasonal variation. Furthermore, differences in moisture and ash content of culantro samples collected in Tra Vinh may result from variations in harvesting time, drying techniques, as well as storage and handling conditions [17]. The TPC in the extract is based on the absorbance of the sample and the Folin-Ciocalteu reagent mixture at 765 nm, which was determined to be 136.477 ± 0.084 (mg GAE/g dry matter). Phenolic compounds play a significant role in antioxidant activities, highlighting the necessity for further investigation [18].

Effect of drying temperature on moisture changes over time. The research results indicate the impact of drying temperature on the quality changes of culantro as illustrated in Fig. 2. The moisture content of the dried culantro samples is below 10% in accordance with TCVN 7975:2008 [19]. An increase in drying temperature leads to a faster drying rate and a reduction in drying time. As shown in Fig. 2, the slope of the drying curve increases progressively with rising temperature, with the smallest slope observed at 60°C and the largest at 80°C.

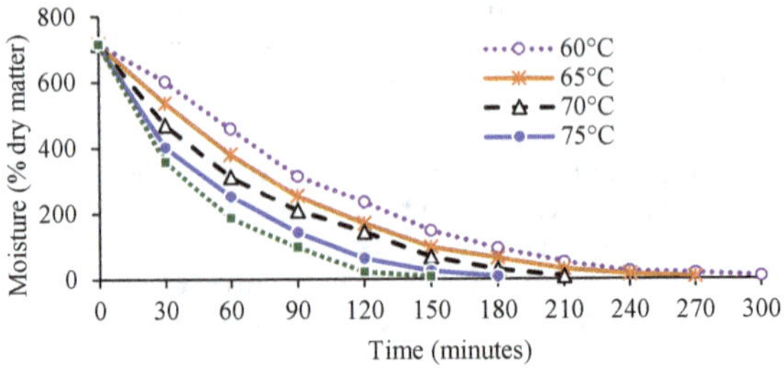

Fig. 2: The drying curve of culantro at different temperatures.

At lower drying temperatures, the moisture content in the material decreases more slowly due to the reduced pressure difference between the material surface and the partial vapor pressure in the surrounding air. At 60°C, the drying time reached a maximum of 300 minutes. As the drying temperature increases, the drying rate correspondingly rises due to enhanced heat transfer to the material, promoting moisture diffusion toward the surface, reaching approximately 150 minutes at 80°C. High drying temperatures cause rapid moisture diffusion initially, but eventually form a hard layer on the surface that prevents moisture from moving outward. Conversely, at excessively low temperatures, the slow drying rate allows microbial activity, which can degrade product quality.

Effect of drying temperature on TPC. The TPC in culantro was found to vary significantly when dried at different temperatures ($p < 0.05\%$). Figure 3 shows that the highest TPC was recorded in the sample dried at 70°C, reaching 85.40 ± 0.084 mg GAE/g dry matter. A trend of increasing polyphenol content was observed as the drying temperature increased from 60°C to 70°C, rising from 51.24 ± 0.049 mg GAE/100g dry matter to 85.40 ± 0.079 mg GAE/100g dry matter, and then decreasing at 75°C to 80.04 ± 0.097 mg GAE/100g dry matter. This result is consistent with the study by Hien *et al.* (2020), which found that drying mangrove-apple at 70°C yielded the highest TPC of 275.52 mg GAE/100g dry matter among the tested temperature ranges [20]. Similarly, the study on dried filter tea made from Wood apple at 70°C for 150 minutes reported a TPC of 30.87 mg GAE/100g dry matter [20]. According to the author Abhay *et al.* (2016), thermal degradation plays a more significant role in the reduction of polyphenols when drying at high temperatures ranging from 75°C to 90°C, whereas at temperatures lower than 70°C, enzymatic degradation predominates during the drying process [21]. Studies conducted by Truong Quoc Tat *et al.* (2021) [22] and Abhay *et al.* (2015) [21] on the effect of drying temperature on the polyphenol content, demonstrated that drying at 70 °C resulted in the highest total phenolic content compared to samples dried at other temperatures.

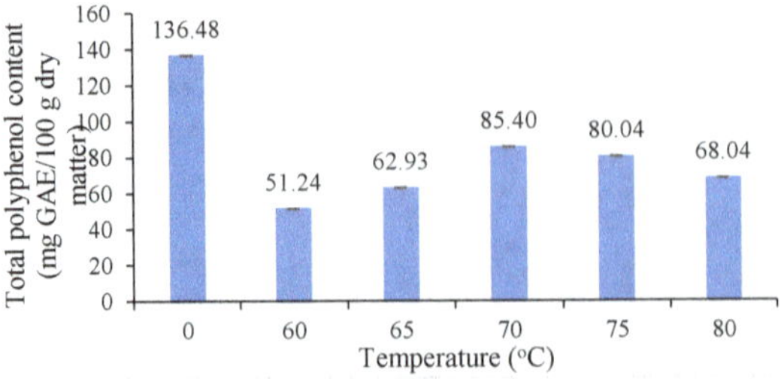

Fig. 3 : The impact of drying temperature on TPC.

Effect of drying temperature on sensory value of the product. Culantro contains a high content of oils, are compounds playing a crucial role in aroma formation [12]. This is an important parameter to control during the drying process to ensure the production of a product with a well-balanced aroma. Additionally, the color of culantro affects the sensory quality of the final product. The drying temperature for culantro was surveyed in the range of 60°C to 80°C, which was evaluated using a preference scale ranging from 1 to 7.

Table 3: The value of sensory at different drying temperatures.

Drying temperature	The value of sensory	
	Color	Aroma
60ºC	3.76 ± 0.236^{c}	3.64 ± 0.177^{c}
65ºC	4.19 ± 0.246^{c}	4.13 ± 0.185^{b}
70ºC	5.35 ± 0.109^{ab}	5.40 ± 0.165^{a}
75ºC	5.69 ± 0.058^{a}	5.16 ± 0.189^{a}
80ºC	5.17 ± 0.065^{b}	5.25 ± 0.221^{a}

(Note: Mean of three replicas ± standard deviation. Different letters in a column indicated significant difference ($p < 0.05$))

The results of the study affect the color and smell of Culantro, it is shown on table 3. At the sample dried at 70ºC, the sensory value for aroma is 5.40 ± 0.165; color is 5.35 ± 0.109; the color and aroma products are characteristic and harmonious. At the sample dried higher or less temperature, the color and the aroma were not characteristic. To achieve a product with a harmonious aroma, suitable quality, and optimal polyphenol content, the culantro sample dried at 70°C was chosen for the subsequent mixing experiment.

Effect of mixing ratio on total phenolic content. Choosing the mixing ratio to create new products, additional ingredients not only help to flavor but also bring nutritional value to Culantro tea products. The percentage of ingredients includes Culantro accounting for 60% to 80%, green tea from (10 - 30%) and fixed chrysanthemum 10%; put into filter bags, packaged at different concentrations to come up with the most suitable formula.

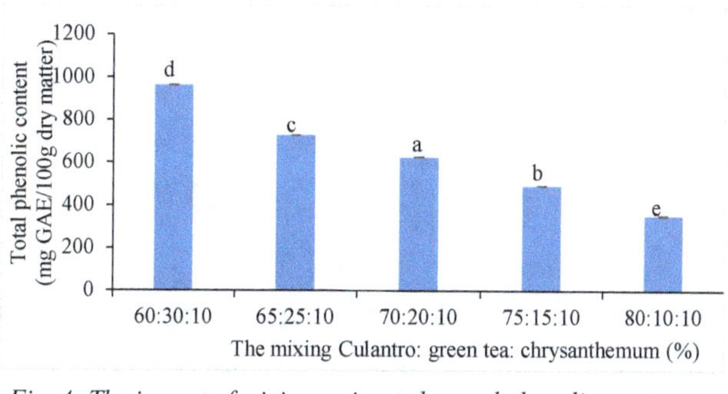

Fig. 4: The impact of mixing ratio on the total phenolic content. (Non-similar letters indicate a significant difference between treatments)

The mixed ingredients were extracted with hot water at 100°C for 10 minutes, the tea samples were taken to evaluate the total phenolic content in each blended sample. The results showed that the total phenolic content tended to decrease with the increasing percentage of Culantro addition (Fig 4). The highest and lowest total phenolic content at the Culantro addition ratio of 60% and 80% with the content of 963.047± 0.716 (mg GAE/100g dry matter), 351.486 ± 1.149 (mg GAE/100g dry matter), respectively. Under the same implementation conditions, the two raw materials of chrysanthemum and green tea were also analyzed for TPC. The results showed that the total phenolic content of chrysanthemum was 414.53 ± 0.364 (mg GAE/100g dry matter); green tea was 1020.36 ± 0.716 (mg GAE/100g dry matter). This also explains that increasing the mixing ratio of green tea also affects the phenolic content of the Culantro tea sample. Through the research process of mangrove apple teabag products, it was also shown that the total phenolic content was affected by the mixing ratio, the total phenolic content of the product increased when increasing the ratio of mangrove apple [23]. Acording to Giuliana Vinci *et al.* (2020) the total phenolic content of green tea was 916.12–1169.81 mg GAE/g [24]. The total phenolic content of green tea was lower than the research results by about 73.03 (mg GAE/100g dry matter). Compared with the experimental results, the total phenolic content of chrysanthemum for blending was measured at 414.53 ± 0.364 (mg GAE/100g dry matter). This result was higher than the research results of Ruenwai, (2020), the total phenolic content in yellow chrysanthemum was 338.31 (mg GAE/100g dry matter) [25]. The total phenolic content of the Culantro teabag of 60% Culantro: 30% green tea: 10% chrysanthemum was the highest compared to the remaining samples.

Effect of mixing ratio on sensory value of the product. Sensory evaluation is an important factor contributing to the assessment of product quality. The color, smell, clarity and taste test are affected by the mixing ratio because to create a quality new tea product. The mixing of ingredients has the effect of creating aroma, increasing the sensory value of the product to make it more attractive. The level of preference of the mixed samples has a difference between the mixing ratios in terms of color, aroma, and taste.

In Fig. 5 and Fig. 6, it shows the effect of the mixing ratio of Culantro: green tea: chrysanthemum (%) on the sensory quality of the product, especially the taste of the product. The results show that at a mixing ratio of 60% Culantro, the highest sensory score is for clarity of 5.01 ± 0.08; smell of 5.40 ± 0.11; taste of 5.95 ± 0.10 and color of 5.05 ± 0.19; the sample has a bright yellow color, a slightly sweet taste, and a slightly sweet. On the contrary, with the Culantro blending ratio of 80%, the sensory value of the sample is the smallest because it does not create a harmonious taste between the ingredients. The sensory score for color is only 4.41 ± 0.15, smell 4.06 ± 0.18, taste 4.09 ± 0.21 and clarity 4.54 ± 0.13. When the Culantro ratio is increased, the product shows the over characteristic smell of Culantro.

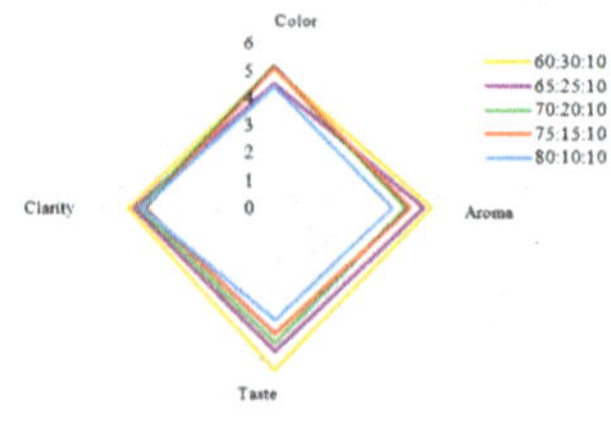

Fig. 5: Effect of raw material mixing ratios (% Culantro:% green tea:% chrysanthemum) on sensory value.

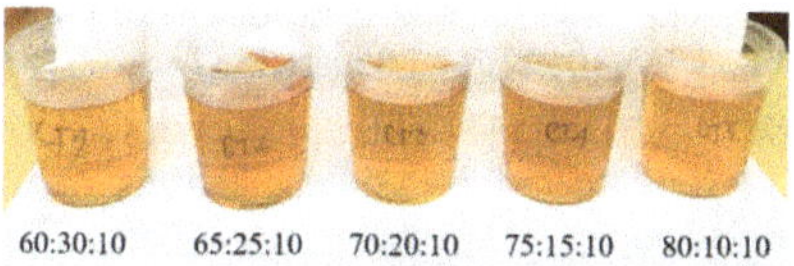

Fig. 6: Samples of Culantro teabas at different mixing ratios.

Conclusion

The research results show that the drying time at 70oC for 210 minutes, with a mixing ratio of 60% Culantro: 30% green tea: 10% chrysanthemum, the Culantro tea product achieved the highest sensory value; the TPC was 963.017 mg GAE/100g dry matter. This research result will contribute to perfecting the technical process of producing Culantro tea.

Acknowledgement

The authors would like to thank Tra Vinh University.

References

[1] Lin J.K., Lin-Shiau S.Y., Mechanisms of hypolipidemic and anti-obesity effects of tea and tea polyphenols. Molecular Nutrition & Food Research, 50(2) (2006) 211-217. https://doi.org/10.1002/mnfr.200500138

[2] Friedman M., Overview of antibacterial, antitoxin, antiviral, and antifungal activities of tea flavonoids and teas. Molecular Nutrition & Food Research, 51 (2006) 116 - 134. https://doi.org/10.1002/mnfr.200600173

[3] Tabarak M., Devendra K.P., Priyanka R., Annie O., Evaluation of Phytochemicals, Antioxidant, Antibacterial and Antidiabetic Potential of Alpinia galanga and *Eryngium foetidum* Plants of Manipur (India). Pharmacognosy Journal, 8(5) (2016). https://doi.org/ 10.5530/pj.2016.5.8

[4] Thomas P.S., Essien E.E., Ntuk S.J., Choudhary M.I., *Eryngium foetidum* L. essential oils: chemical composition and antioxidant capacity. Medicines, 4(2) (2017) 24. https://doi.org/ 10.3390/medicines4020024

[5] Aly A., Biosynthesis of phenolic compounds and water soluble vitamins in culantro (*Eryngium foetidum* L.) plantlets as affected by low doses of gamma irradiation. Analele Universitatii din Oradea, Fascicula Biologie, 17 (2010). http://www.bioresearch.ro/bioresearch/2010-2/356-361%20-%20ALY%20A.A.%20-%20N.C.R.R.T.Cairo.Eg.%20-%20Biosynthesys%20of%20phenolic.pdf

[6] Bhavana G.P., Saraswathi K.J., Quantitative determination of secondary compounds in populations of *Eryngium foetidum* L. from India. International Journal of Current Science, 2 (2013) 1-5. https://www.semanticscholar.org/paper/Quantitative-determination-of-secondary-compounds-Bhavana-Saraswathi/9362e552ffe1536a9823bc5b39e3d3ad9453c8ac

[7] Rodrigues T.L.M., Castro G.L.S., Viana R.G., Gurgel E.S.C., Silva S.G., de Oliveira M.S., Andrade E.H.A., Physiological performance and chemical compositions of the *Eryngium foetidum* L. (Apiaceae) essential oil cultivated with different fertilizer sources. Natural Product Research, 35(23) (2021) 5544-5548. https://doi.org/ 10.1080/14786419.2020.1795653

[8] Silalahi M., Essential oils and uses of *Eryngium foetidum* L. GSC Biological and Pharmaceutical Sciences, 15 (2021) 289-294. https://doi.org/10.30574/gscbps.2021.15.3.0175

[9] Manabi P., Roktim G., Neelav S., Samarjeet S., Sanjoy K.C., Himangshu L., Mohan L., Amazonian *Eryngium foetidum* leaves exhibited very high contents of bioactive compounds and

high singlet oxygen quenching capacity. International Journal of Food Properties, 23.1 (2020) 1452-1464. https://doi.org/ 10.1080/10942912.2020.1811311

[10] Shavandi M.A., Haddadian Z., Ismail M.H.S., Eryngium foetidum L. Coriandrum sativum and Persicaria odorata L.: a review. Journal of Asian Scientific Research, 2(8) (2012) 410. https://archive.aessweb.com/index.php/5003/article/view/3365

[11] Aung H.T., Sein M.M., Aye M.M., Thu Z.M., A review of traditional medicinal plants from Kachin State, Northern Myanmar. Natural Product Communications, 11(3) (2016) 353-64. https://doi.org/10.1177/1934578X1601100310

[12] Matiusha K., Grytsyk A., Hrytsyk R., Raal A., Koshovyi O., Phytochemical screening and antimicrobial activity of leaf extracts of *Eryngium foetidum* L.(Apiaceae). Indo American Journal of Pharmaceutical Research, 6(2) (2016) 4339-4344. https://doi.org/10.3390/app15031433

[13] Singh S., Singh D.R., Banu S., Salim K.M., Determination of Bioactives and Antioxidant Activity in Eryngium foetidum L.: A Traditional Culinary and Medicinal Herb. Proceedings of the National Academy of Sciences, India Section B: Biological Sciences, 83(3) (2013) 453-460. https://doi.org/ 10.1007/s40011-012-0141-y

[14] Pham Van So and B.T.N. Thuan, Food testing. 1991: Hanoi University Of Science And Technology.

[15] Hà Duyên Tư, Kỹ thuật phân tích cảm quan thực phẩm. 2006, Hà Nội: Khoa học kỹ thuật.

[16] Daisy P., Santosh K., Rajathi M., Antihyperglycemic and antihyperlipidemic effects of *Clitoria ternatea* L. in alloxan-induced diabetic rats. African Journal of Microbiology Research, 3(5) (2009) 287-291. http://www.academicjournals.org/ajmr

[17] Lepcha T.T., Upadhyay S., Sharma L., Muddarsu V.R., Proximate and nutritional analysis of Culantro (*Eryngium foetidum*), Journal of Pharmacognosy and Phytochemistry, 7(1S) (2018): 3129-3134. https://www.semanticscholar.org/paper/Proximate-and-nutritional-analysis-of-Culantro-(-)-Lepcha-Upadhyay/0cbaab9c5a0ab0d8e9ac28676eca941ae327118d

[18] Saha, D. and A. Tamrakar, Xenobiotics, Oxidative Stress, Free Radicals Vs. Antioxidants: Dance Of Death to Heaven's Life. Asian J Res Pharm Sci, 1 (2011). https://ajpsonline.com/AbstractView.aspx?PID=2011-1-2-2

[19] TCVN 7975:2008, Chè thảo mộc túi lọc. https://isocus.vn/tieu-chuan-quoc-gia-tcvn-79752008-che-thao-moc-tui-loc

[20] Nguyễn Kim Phụng, Nguyễn Thị Hiền, and Phạm Bảo Nguyên, Khảo sát khả năng chống hóa nâu của Acid Ascorbic và ảnh hưởng nhiệt độ sấy đến chất lượng trái bần (*Sonneratia caseolaric* L.). Tạp chí Công thương, (2020) 342-348. https://tapchicongthuong.vn/khao-sat-kha-nang-chong-hoa-nau-cua-acid-ascorbic-va-anh-huong-nhiet-do-say-den-chat-luong-trai-ban--sonneratia-caseolaric-l-69395.htm

[21] Abhay S.M., Hii C.L., Law C.L., Suzannah S., Djaeni M., Effect of hot-air drying temperature on the polyphenol content and the sensory properties of cocoa beans. International Food Research Journal, 23 (2016). http://www.ifrj.upm.edu.my/23%20(04)%202016/(19).pdf

[22] Trương Quốc Tất, Phùng Thị Thúy Liễu, Nguyễn Thị Phương Trang, Nguyễn Duy Khánh, Ảnh hưởng của nhiệt độ sấy đến hàm lượng hợp chất polyphenol, sắc tố carotenoids, chlorophyll và hoạt tính chống oxy hóa của cây rau càng cua (*Peperomia pellucida* L.) thu ở tỉnh Tiền Giang. Ho Chi Minh City Open University, 16 (2021) 25-33. https://doi.org/10.46223/HCMCOUJS.tech.vi.16.1.1891.2021

[23] Nguyễn Thị Hiền, Nguyễn Kim Phụng, Nguyễn Thị Hồng Thảo, Nguyễn Hồng Phúc, Khảo sát ảnh hưởng tỷ lệ phối trộn các thành phần nguyên liệu đến chất lượng trà túi lọc bần (*Sonneratia caseolaris* L.). Tạp chí Công thương, 2 (2021) 299-303. https://scholar.dlu.edu.vn/thuvienso/bitstream/DLU123456789/150473/1/CVv146S22021299.pdf

[24] Rathi M.H., Abdulhay H.S., Total phenolic contents and antioxidant activity of extracts of tea (black, green and white), Asian Jr. of Microbiol. Biotech. Env. Sc, 20 (2018) 45-49. https://scholar.google.com.tw/scholar?q=total+phenolic+contents+and+antioxidant+activity+of+extracts+of+tea+(black,+green+and+white)&hl=vi&as_sdt=0&as_vis=1&oi=scholart

[25] Ruenwai, R., Analysis of phenolic content and free radical scavenging capacity of Chrysanthemum flower extracts. Health Sci Tech Rev, 13 (2020) 16. https://li01.tci-thaijo.org/index.php/journalup/article/view/221859

Green Materials, Processing & Characterization – ICoGMPaC 2025 | Materials Research Forum LLC
Materials Research Proceedings 68 (2026) 166-173 | https://doi.org/10.21741/9781644904237-22

Strength and Carbonation of Sand Cement Brick Containing Fly Ash as Partial Cement Replacement Under CO_2 Curing

Faisal Sheikh KHALID[1,a]*, Ming Quan CHONG[1,b], Syazwi Hakimi SAAIDIN[1,c], Connie Hui Jun KONG[1,d], Syafiqa AYOB[2,e], Zahir ZAKI[1,f] and Raden Mohd Farhan Helmy RADEN ISMAIL[1,g]

[1]Faculty of Civil Engineering and Built Environment, Universiti Tun Hussein Onn Malaysia, 86400, Batu Pahat, Malaysia

[2]Faculty of Engineering and Built Environment, Lincoln University College, Malaysia

[a]faisalsh@uthm.edu.my, [b]chong000804@gmail.com, [c]syzwihkmi@gmail.com, [d]conniekong2935@gmail.com, [e]syafiqa@lincoln.edu.my, [f]zahirzaki91bebe@gmail.com, [g]radin@uthm.edu.my

Keywords: CO_2 Curing, Sand-Cement Brick, Coal Fly Ash, Magnesium Oxide

Abstract. Human activities, such as burning fossil fuels (including gas, coal, and oil) and deforestation, have significantly raised atmospheric CO_2 levels since the Industrial Revolution. Sand cement bricks, which were commonly used in construction due to their durability and affordability, were mainly manufactured from cement and sand. Global process emissions from cement production in 2017 amounted to 1.48 ± 0.20 Gt CO_2, equivalent to about 4% of emissions from fossil fuels, prompting exploration of alternative materials and methods to enhance sustainability in construction. This study examined the strength and carbonation characteristics of sand-cement bricks incorporating fly ash (FA) and magnesium oxide (MgO) under accelerated carbonation conditions of 50% CO_2, 27°C, for 7 and 28 days. Sand-cement bricks with dimensions 215mm x 103mm x 65mm were formulated with FA replacing cement at 20%, 40%, and 60%, along with a fixed 2% MgO content. Compressive strength tests conducted before and after carbonation revealed that a 20% FA replacement with 2% MgO achieved the best balance between strength and carbonation resistance via carbonation depth. R20 sample showed compressive strength of 47.16MPa and 1.24mm of carbonation depth at 28 days, which was the highest compressive strength and lowest carbonation depth among all samples with FA and MgO replacement. Higher FA replacements (40% and 60%) showed reduced early strength but increased carbonation depth, indicating their limited use in load-bearing applications. This study demonstrated the potential of utilizing industrial by-products like FA in sustainable construction while minimizing environmental impacts such as waste accumulation and CO_2 emissions.

Introduction

Sand cement bricks were commonly used in construction due to their durability and affordability. They were generally more affordable than other types of bricks, such as clay or fly ash bricks. The materials used in sand cement bricks included sand, cement, and water, which were relatively inexpensive and readily available in many regions, such as Malaysia. Cement-sand bricks were used as an alternative for low-cost housing in Malaysia [1]. However, conventional cement production significantly contributed to carbon dioxide (CO_2) emissions. Studies showed that global process emissions from cement production in 2017 amounted to 1.48 ± 0.20 Gt CO_2, equivalent to about 4% of emissions from fossil fuels [2], prompting exploration of alternative materials and methods to enhance sustainability in construction. Carbonation was a natural chemical process in which atmospheric CO_2 reacted with calcium hydroxide, $Ca(OH)_2$ in cementitious materials to form calcium carbonate, $CaCO_3$. This process occurred over the lifespan

Materials Research Proceedings 68 (2026) 166-173 https://doi.org/10.21741/9781644904237-22

of concrete structures, influencing their mechanical properties and durability. Previous studies showed that magnesium-modified sodium carbonate-activated slag/fly ash concrete demonstrated acceptable mechanical strength, high resistance to sulfate attack, and greater resistance to acid attack, offering a low carbon footprint and an environmentally friendly alternative to conventional cements [3]. Understanding the carbonation behavior of sand cement bricks containing fly ash and MgO was essential for predicting their long-term performance and sustainability.

In recent years, one of the most critical global issues faced was climate change. Human activities, such as burning fossil fuels (including gas, coal, and oil) and deforestation, significantly raised atmospheric CO_2 levels since the industrial revolution. According to the United Nations' Intergovernmental Panel on Climate Change (IPCC), it was predicted that by 2050, CO_2 concentrations would exceed 550 parts per million (ppm). This increase was expected to raise the planet's average temperature by 1.4 to 5.7 degrees Celsius, leading to severe weather events, including droughts, wildfires, floods, sea-level rise, and food shortages [4]. The additional CO_2 in the atmosphere intensified the greenhouse effect, trapping heat and causing global temperatures to rise. Most climate scientists agreed that there was a direct link between rising CO_2 levels and increased temperatures [5]. One promising approach was the development of new brick blocks blended with special materials to help reduce atmospheric CO_2. Magnesium oxide and fly ash showed success in cement-like materials, forming a blend known as MgO-FA [6]. This concept involved using magnesium oxide (MgO) and coal fly ash in the production of brick blocks, which could absorb CO_2 during their formation process (carbonation). By increasing the production of these CO_2-absorbing brick blocks, the overall amount of CO_2 in the atmosphere could gradually be reduced. It was crucial to find the best combination of materials for these blocks to efficiently capture high levels of CO_2. The main goal of this research was to study how accelerated carbonation affected cement hardened with reactive magnesium oxide and coal fly ash. Firstly, to investigate the density and compressive strength of sand-cement bricks containing different proportions of fly ash (FA) and a fixed concentration of magnesium oxide (MgO) after exposure to accelerated carbonation. Next, to evaluate the carbonation effects on sand-cement bricks containing FA and MgO as partial cement replacement materials.

Methodology

The materials used for this study included Ordinary Portland Cement (OPC), magnesium oxide (MgO), coal fly ash (CFA), and river sand. OPC served as the primary binding agent, while MgO and CFA were utilized as partial replacements to enhance sustainability. CFA was sourced from a local power plant, dried, sieved, and ground to the required particle size. MgO was purchased commercially and prepared according to the study's specifications. The sand was dried, sieved, and prepared to ensure uniformity. Tap water, clean and free from impurities, was used for mixing. Sand-cement bricks were cast using standardized molds (215mm × 103mm × 65mm), following the specifications of MS 76 (1972) and BS 3921 (1985), with cement-sand ratio of 1:6, water-binder ratio of 0.60, varying CFA proportions (0%, 20%, 40%, and 60%) and a fixed MgO concentration (2%). 2% of MgO can increase compressive strength and lead to interaction in curing [7]. 2% MgO also has good water permeability and superior compressive strength [8]. Hence, studies had confirmed that 2% MgO encouraged the absorption of free lime and formation of C_3S, thus increasing strength development and shortening setting time. Table 1 showed the mix proportion of single brick with a sand-cement ratio of 1:6.

Table 1: Mix proportion of single brick with sand cement ratio of 1:6.

Mix Design	Sand [kg]	w/c	Water [kg]	MgO [%]	Fly Ash [%]	Fly Ash [kg]	Cement [kg]
Control	1.92	0.60	0.38	2.00	0.0	0.00	0.63
R20					20.0	0.13	0.50
R40					40.0	0.25	0.38
R60					60.0	0.38	0.25

Once molded, the samples were subjected to accelerated CO_2 curing with conditions of 50% CO_2, 27°C, for 7 and 28 days as shown in Fig. 1. After curing for specified periods, the bricks were tested for density, compressive strength, and carbonation depth to evaluate their performance. The research investigated the impact of substituting conventional cement and sand in bricks with CFA and MgO, aimed to determine the optimal mix ratio for these materials. Specimen preparation, accelerated CO_2 curing, and testing were carried out at the Heavy Structures Laboratory and Advanced Materials Laboratory in Universiti Tun Hussein Onn Malaysia. Table 2 showed the number of specimens with different percentages of fly ash.

Table 2: Number of specimens with different percentages of fly ash.

Mix Design	Percentage of Fly Ash [%]	Compression Test		Density Test		Carbonation Test
		7 Days	28 Days	7 Days	28 Days	28 Days
Control	0.0	3	3	3	3	3
R20	20.0	3	3	3	3	3
R40	40.0	3	3	3	3	3
R60	60.0	3	3	3	3	3
No. of Specimen		12	12	12	12	12
Total Specimen		60				

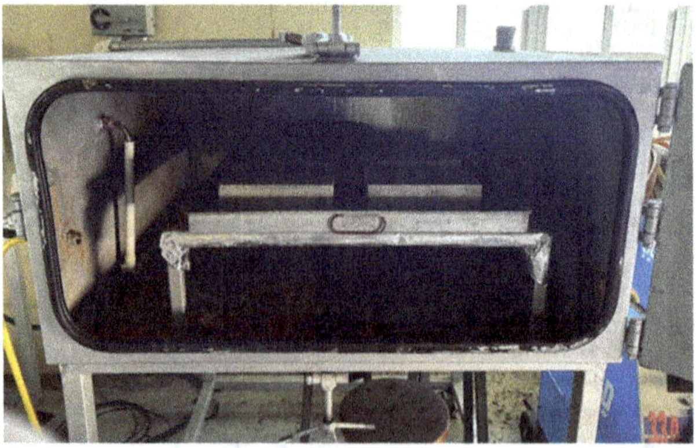

Fig. 1: Chamber of controlling CO_2 concentration, temperature, and relative humidity.

The density test was conducted to determine the mass-to-volume ratio of the sand-cement brick specimens, which reflect their structural compactness and quality. The test involved measuring the mass of the cured bricks using a calibrated digital balance and recording their dimensions to calculate the volume.

The compressive strength test was performed to evaluate the load-bearing capacity of the sand-cement brick specimens. This test involved placing the cured brick specimens in a compression testing machine, following the guidelines of BS EN 772-1:2011 [9]. Each brick was subjected to a gradually increasing compressive load until failure occurred, and the maximum load at failure was recorded. The carbonation test was conducted to evaluate the depth of carbonation of the sand-cement brick specimens. The bricks were split, and a phenolphthalein indicator solution was applied to the freshly broken surfaces. The non-carbonated regions turned pink due to the alkaline environment, while the carbonated zones remained colourless [10]. The carbonation depth was measured using a digital caliper, and the results were analyzed to assess the extent of carbonation in bricks with varying proportions of coal fly ash and MgO

Results and Discussion

Fig. 2 showed the density of concrete brick for 7 days and 28 days with Malaysian Standard. The density results of sand-cement bricks incorporating fly ash and MgO demonstrated compliance with Malaysian Standards (BS EN 772-13:2000), which specify a density range of 1800-2200 kg/m^3 for masonry units. At 7 days, the densities of the specimens ranged from 2035.5 kg/m^3 to 2107.3 kg/m^3, while at 28 days, they ranged from 2028.5 kg/m^3 to 2111.3 kg/m^3. These values confirmed that all bricks, regardless of the percentage of fly ash substitution, were suitable for structural applications. The slight increase in density over time was attributed to ongoing hydration and carbonation processes, which contributed to the densification of the material matrix and enhanced performance [12].

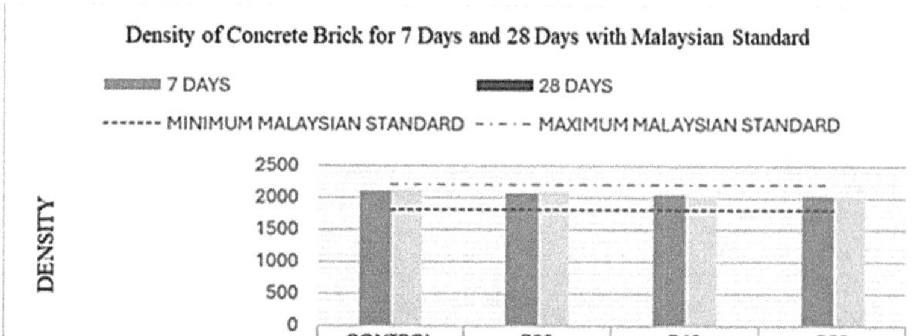

	CONTROL	R20	R40	R60
7 DAYS	2107.3	2067.9	2047.1	2035.5
28 DAYS	2111.2	2074.3	2036.9	2028.5
MINIMUM MALAYSIAN STANDARD	1800	1800	1800	1800
MAXIMUM MALAYSIAN STANDARD	2200	2200	2200	2200

Fig. 2: Density of concrete brick for 7 days and 28 days with Malaysian Standard.

Fig. 3 presented the comparison of the compressive strength of concrete brick at 7 and 28 days against BS 6073-2:2008 Standard. The compressive strength results of sand-cement bricks incorporating fly ash and MgO were evaluated against BS 6073-2:2008 standards, which set a minimum requirement of 3.5MPa for individual units and 5MPa for the average of tested bricks. At 7 days, all mixes exceeded these requirements, with the control mix achieving the highest strength of 28.26MPa, followed by the R20 mix at 26.30MPa. The R40 and R60 mixes, with strengths of 23.00MPa and 21.78MPa respectively, also demonstrated compliance with the standards, confirming their suitability for load-bearing applications. These results highlight the effectiveness of the materials and curing process in achieving strong early performance. At 28 days, the compressive strength of all bricks improved further due to continued hydration and carbonation processes [13-14]. The control mix reached the highest strength of 49.54MPa, with the R20 mix closely following at 47.16MPa. The R40 and R60 mixes, although lower at 32.33MPa and 31.38MPa respectively, remained well above the minimum requirement for load-bearing bricks. These findings validate the use of fly ash as a partial replacement for cement, demonstrating that even with higher substitution levels, the bricks maintained sufficient structural integrity for construction purposes.

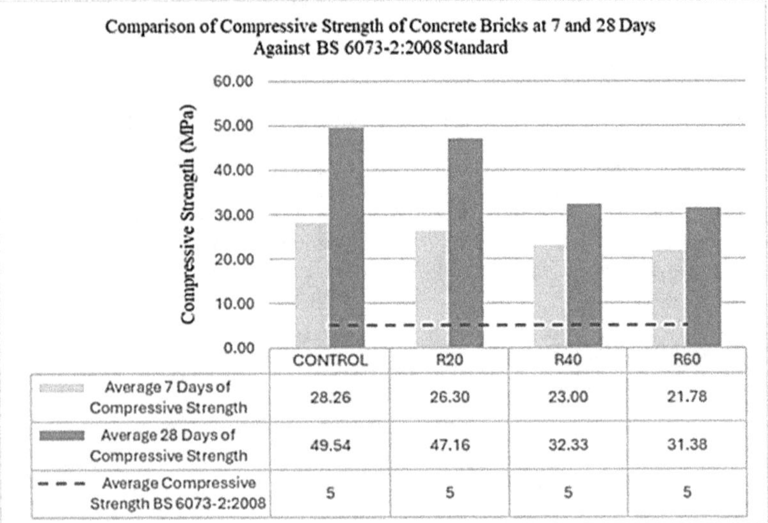

Fig. 3: Comparison of compressive strength of concrete brick at 7 and 28 days against BS 6073-2:2008 Standard.

Carbonation Depth. Carbonation depth tests were performed for 28 days to evaluate the material's resistance to carbon dioxide absorption. The 28-day carbonation results provide an accurate reflection of the material's long-term resistance to carbonation, as the curing process is complete at this stage. These results are essential for assessing the material's durability over time, especially in environments where carbonation can affect its structural integrity and long-term performance. Fig. 4 illustrated the effects of CFA replacement on carbonation depth across mix design. The carbonation results for sand-cement bricks incorporating varying proportions of fly ash (20%, 40%, and 60%) and MgO revealed a clear relationship between fly ash content and carbonation depth. As fly ash content increased, carbonation depth progressively rose from 1.21mm in the control mix to 1.32mm in the R60 mix. This trend reflects the reduced availability of calcium hydroxide in fly ash-based mixes, which leads to lower resistance to carbonation penetration [15]. The control mix, with its dense cement matrix and high calcium hydroxide content, demonstrated the lowest carbonation depth and thus the highest resistance to CO_2 diffusion. These results align with the objective of investigating the impact of fly ash proportions on carbonation resistance.

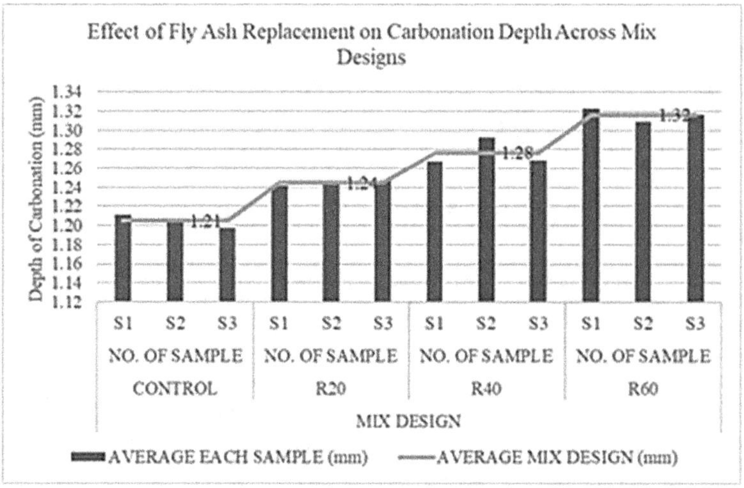

Fig. 4: Effect of CFA replacement on carbonation depth across mix design.

Conclusion

The study successfully evaluated the effects of incorporating fly ash and MgO into sand-cement bricks, focusing on their compressive strength, density, and carbonation behaviour under controlled curing conditions. The findings confirmed that bricks with up to 60% fly ash replacement and 2% MgO met the minimum requirements for structural applications, as specified by BS 6073-2:2008. The compressive strength results demonstrated that all mixes maintained sufficient load-bearing capacity, with the control and R20 mixes achieving the highest strengths, while R40 and R60 showed slightly reduced but acceptable performance. These results highlight the potential of fly ash as a sustainable partial replacement for cement, reducing cement dependency without compromising the mechanical performance of cement brick.

The study also revealed that carbonation depth increases with higher fly ash proportions, indicating a reduction in carbonation resistance due to increased porosity and lower calcium hydroxide availability. Despite this, carbonation contributed to surface densification through calcium carbonate formation, improving durability in the carbonated zones. The relationship between density and carbonation depth demonstrated that control and R20 mixes achieved a better balance between densification and carbonation resistance, making them ideal for applications requiring long-term durability. The findings emphasize the importance of optimizing fly ash content to balance structural performance, durability, and sustainability.

In conclusion, the use of fly ash and MgO in sand-cement bricks demonstrates significant potential for sustainable construction applications. By optimizing material proportions and curing conditions, it is possible to produce eco-friendly bricks that meet industry standards while contributing to reduced carbon emissions and industrial waste disposal management. This study provides a foundation for further research into the long-term performance of these materials and their application in diverse construction environments.

Acknowledgment

This research was supported by Universiti Tun Hussein Onn Malaysia (UTHM) Tier 1 (vot Q854). This research also was supported by Ministry of Higher Education (MOHE) through Fundamental

Research Grant Scheme (FRGS/1/2022/WAB02/UTHM/03/1). The assistance from the Research Management Centre (RMC), UTHM is greatly appreciated.

References

[1] Chow, M. F., & Rosidan, M. a. K. (2020). Study on the Effects of plastic as admixture on the mechanical Properties of Cement-Sand Bricks. IOP Conference Series Materials Science and Engineering, 713(1), 012016. https://doi.org/10.1088/1757-899x/713/1/012016

[2] Andrew, R. M. (2018). Global CO2 emissions from cement production, 1928–2017. Earth System Science Data, 10(4), 2213–2239. https://doi.org/10.5194/essd-10-2213-2018

[3] Abdalqader, A., Jin, F., & Al-Tabbaa, A. (2019). Performance of magnesia-modified sodium carbonate-activated slag/fly ash concrete. Cement and Concrete Composites, 103, 160–174. https://doi.org/10.1016/j.cemconcomp.2019.05.007

[4] IPCC. (2005). Carbon dioxide capture and storage. In B. Metz, O. Davidson, H. de Coninck, M. Loos, & L. Meyer (Eds.), Cambridge University Press. Retrieved from https://www.ipcc.ch/report/special-report-on-carbon-dioxide-capture-and-storage/

[5] Wang, D., Noguchi, T., & Nozaki, T. (2019). Increasing efficiency of carbon dioxide sequestration through high-temperature carbonation of cement-based materials. Journal of Cleaner Production, 238, 117930. https://doi.org/10.1016/j.jclepro.2019.117930

[6] Sequeira, L., Forero, J., Bravo, M., Evangelista, L., & De Brito, J. (2023). Durability of Concrete with Partial Replacement of Portland Cement by Incorporating Reactive Magnesium Oxide and Fly Ash. Materials, 16(7), 2670. https://doi.org/10.3390/ma16072670

[7] Supriya, J., & Raut, A. (2021). Performance Parameter Analysis of Magnesia based cement Products – a review. IOP Conference Series Materials Science and Engineering, 1197(1), 012078. https://doi.org/10.1088/1757-899x/1197/1/012078

[8] Yang, X. Q., Wang, Z., Song, J. G., & Zhong, L. (2022). Effect of sintering AIDS on performance of manganese tailings permeable bricks. Diffusion and Defect Data, Solid State Data. Part B, Solid State Phenomena/Solid State Phenomena, 331, 221–226. https://doi.org/10.4028/p-2u7ps0

[9] Salem, N., Brar, K. K., Asgarian, A., Kaur, K., Magdouli, S., & Perreault, N. N. (2025) Advancements in Carbon Capture, Utilization, and Storage (CCUS): A Comprehensive review of technologies and prospects. Clean Technologies, 7(4): 109. https://doi.org/10.3390/cleantechnol7040109

[10] Soomro, M. H., Abdullah, U. R., Ghani, W. a. W. a. K., & Kusin, F. M. (2025) Mineral carbonation pathways in accelerating sequestration capacity of industrial waste as feedstock for long-term carbon capture and storage. Journal of Environmental Chemical Engineering, 13(5): 118696. https://doi.org/10.1016/j.jece.2025.118696

[12] Zhang, X., Liu, S., Wu, K., & Yuan, Z. (2024). Experimental investigation into the self-carbonation mechanism of magnesium oxide carbon sequestration foamed concrete. Cement and Concrete Composites, 148, 105486. https://doi.org/10.1016/j.cemconcomp.2024.105486

[13] Oke, J. A., & Abuel-Naga, H. (2025). Compressive strength properties of green bricks cured with hydration and accelerated carbonation curing. Emergent Materials, 8(7), 5961–5977. https://doi.org/10.1007/s42247-025-01220-7

[14] Raihan, A. (2025) Carbon capture, utilization, and storage (CCUS) in the context of earth energy systems: A multidisciplinary review. Geoscience Frontiers, 16(6): 102177. https://doi.org/10.1016/j.gsf.2025.102177

[15] Khalid, F. S., Chong, M. Q., Kong, C. H. J., Saaidin, S. H., Ayob, S., Altowayti, W. A. H., Juki, M. I., & Shahidan, S. (2025) Impact of CO2 Curing on the Mechanical Properties of Cement Sand Bricks Incorporating Magnesium Oxide. Journal of Advanced Research in Micro and Nano Engineering, 41(1): 1–9. https://doi.org/10.37934/armne.41.1.19

Mechanical Performance of POFA-Modified Mortar Under Varying Curing Regimes

Nurul Hana MOHAMAD ZAINI[1,a], Nuril Izzeaty ISHAK[1,b*] and Mohamad Nidzam RAHMAT[1,c]

[1]Faculty of Built Environment, Universiti Teknologi MARA, 40450 Shah Alam, Selangor, Malaysia

[a]hana.zaini@yahoo.com, [b]Izzeaty@uitm.edu.my, [c]mnidzam676@uitm.edu.my

Keywords: Mechanical Performance; POFA, Steam Curing; Green Mortar

Abstract. The accelerated urbanization rate has intensified global demand, raising concerns over the environmental impact of CO_2 emissions from cement production. Palm oil fuel ash (POFA), a silica-rich agro-industrial by-product, offers potential as a sustainable partial cement replacement due to its pozzolanic properties. This study evaluates the effects of POFA particle sizes of 212 μm and 150 μm on mortar performance, incorporating manufactured sand (M-sand) retained on a 600 μm sieve. Mortar cubes with dimensions of 50 × 50 × 50 mm were prepared with POFA replacement levels of 5%, 10%, 15%, and 20% by weight of cement, alongside a control mixture. Specimens were subjected to normal and steam curing regimes and tested for compressive strength at 7 and 28 days. Results demonstrate that 10% POFA replacement yields an optimal compressive strength under steam curing regimes. The findings highlight POFA's efficacy in enhancing mortar performance while reducing cement dependency, supporting environmentally responsible construction practices in alignment with Sustainable Development Goal 12. Integrating POFA into cementitious systems represents a forward-thinking approach, fostering environmental responsibility and advancing innovation in construction practices.

Introduction

The rapid pace of global urbanization and infrastructure expansion has significantly increased the demand for concrete, a material predominantly dependent on Ordinary Portland Cement (OPC) as its primary binder [1,2]. However, the escalating consumption of OPC has raised substantial environmental concerns, primarily due to its high energy requirements and the considerable carbon dioxide (CO_2) emissions, with the cement industry accounting for approximately 5–8% of total global CO_2 emissions, largely resulting from the calcination of limestone and the combustion of fossil fuels during clinker production [3,6,5]. In response, the construction industry has shifted toward sustainable strategies, such as partially replacing OPC with Supplementary Cementitious Materials (SCMs), typically industrial by-products with pozzolanic or latent hydraulic properties. Materials such as fly ash (FA), ground-granulated blast furnace slag (GGBS), silica fume, and palm oil fuel ash (POFA) have shown considerable potential to reduce the carbon footprint of cementitious systems while enhancing certain aspects of concrete performance [6,7,8,9.10]. The integration of SCMs contributes to emissions reduction and offers technical advantages including improved workability, reduced heat of hydration, and a lower risk of deleterious reactions such as alkali-silica reaction [11]. Palm oil fuel ash (POFA), a byproduct of palm oil processing, has emerged as a promising and environmentally sustainable alternative to OPC in mortar and concrete applications. Due to its pozzolanic nature, finely processed POFA can react with calcium hydroxide (C-H) in cementitious systems to form additional calcium silicate hydrate (C-S-H) gel, enhancing concrete's long-term strength and durability [12]. Recent research has increasingly focused on the feasibility of using POFA as a partial replacement for cement, driven by both environmental and performance considerations [9,13,14]. Its utilization will reduce dependence on OPC, lowering associated CO_2 emissions, and offer an effective solution for managing agro-industrial waste within the palm oil industry [15,16]. This integration aligns with global sustainability goals, particularly in promoting low-carbon construction practices and advancing responsible waste management [17].

In response to the depletion of natural river sand depletion and environmental concerns, manufactured sand (M-sand) has become a sustainable alternative in mortar production [18,19]. In this context, M-sand offers consistent particle size and improves aggregate quality, enhancing workability, packing density, and mortar performance. It shows synergistic effects with pozzolanic materials like POFA, leading to more uniform particle distribution and denser matrices, enhancing the fresh and hardened properties. [20,21]. Curing regimes play a critical role in determining the performance of mortar, as it facilitate the complete cement hydration and contribute to the mortar's strength development. The hydration process of Portland cement is significantly influenced by curing temperature, with higher temperatures accelerating the reaction [22,23]. Previous studies show that OPC with SCM slows early hydration due to low pozzolanic reactivity with C-H, making it unsuitable for applications requiring high early-age strength. Therefore, steam curing accelerates early strength development, especially in precast applications [18,24]. The controlled steam curing temperatures help mitigate the moisture loss and maintain consistent curing conditions, reducing shrinkage and cracking of the concrete structure [19,25]. Enhanced quality of steam curing improves durability and lifespan. But, normal curing, based on ambient conditions, is more cost-effective, environmentally friendly, and common in construction for its simplicity [20].

Although POFA has gained attention as a pozzolanic material in cement replacement, limited studies have explored how different POFA particle sizes affect mortar's mechanical performance under various curing regimes. Most research focuses on curing temperature, time, or POFA inclusion alone, without examining their combined effects on mortar strength. This study addresses this gap by investigating the mechanical properties of mortar modified with two POFA particle sizes under both normal and steam curing regimes, aiming to determine the optimal formulation for enhanced strength performance.

Materials and Methods

Materials. Several materials were used in this study, including Ordinary Portland Cement (OPC), Palm Oil Fuel Ash (POFA), Manufactured Sand (M-sand), water, and superplasticizer. The OPC complied with the MS EN 197-1:2014 CEM I 52.5N standards, and the brand used was Panda Blue, manufactured by Hume Cement Sdn. Bhd. POFA is a by-product material obtained from United Oil Palm Industries Sdn. Bhd. located in Nibong Tebal, Pulau Pinang. The as-received POFA was sieved, and the ash retained on the 212 μm and 150 μm sieves was used in this study. The fine aggregate used was manufactured sand (M-sand) with a particle size of 600 μm, sourced from YTL Cement. The superplasticizer Master Rheo Build 1100 RM was included as a chemical admixture by 1% of the total weight of binder to improve the flowability of the mixture.

Mix Proportions and Curing Process. This study highlighted the influence of POFA as a partial replacement of cement on the strength development of high-strength green mortar, subjected to normal and steam curing regimes. The mix design for high-strength green mortar was based on a trial mix approach, with OPC being partially replaced by POFA at varying levels of 5%, 10%, 15%, and 20% by weight of binder. The water-to-binder ratio kept constant at 0.3 for all mixes, and the superplasticizer content was consistently maintained at 1%, as shown in Table 1. A total of 60 mortar cube specimens, with dimensions of 50 mm × 50 mm × 50 mm, were prepared for the compressive strength test. The mixing process used a heavy-duty stand mixer, in which water containing a superplasticizer was gradually added to the dry mix until a homogeneous mortar was produced. The mortar was then cast into cube moulds and left at the ambient temperature for 24 hours. The specimens were then demoulded and underwent two curing regimes: 1) the normal curing, in which the specimens were fully immersed in water at ambient temperature to maintain continuous hydration. 2) Under steam curing regimes, specimens were initially exposed to a controlled steam environment at 55 °C, followed by immersion in the normal water temperature at the designated testing ages of 7 and 28 days.

Table 1: Mix design for the production of the POFA-modified mortar.

Mixes	Binder				
	OPC (%)	POFA (%)	Sand(kg/m^3)	W/b ratio	SP (%)
1	100	0	840	0.3	1
2	95	5	840	0.3	1
3	90	10	840	0.3	1
4	85	15	840	0.3	1
5	80	20	840	0.3	1

Testing Procedures. The chemical composition of OPC and POFA was analyzed using X-ray fluorescence (XRF) spectroscopy with the S1TITAN Handheld device. The samples were oven-dried and sieved to a fine powder, then pressed into pellets for analysis. The XRF process involved exposing the sample to high-energy X-rays, which excited the atoms and caused the emission of secondary (fluorescent) X-rays. These emissions were detected and quantified to determine the elemental composition, mainly focusing on oxides such as SiO_2, Al_2O_3, Fe_2O_3, CaO, and other components to assess pozzolanic activity.

Compressive strength tests were performed on mortar cubes cured for 7 and 28 days to assess the mechanical performance of mixes with OPC and POFA. The tests were conducted using a calibrated compression testing machine at the Concrete Laboratory, Faculty of Built Environment, Universiti Teknologi MARA (UiTM), Shah Alam, which followed ASTM C109 with a loading rate of 0.9 kN/s until failure. Mortar specimens were surface-dried for 3 hours before testing to reduce surface moisture effects. The maximum load at failure and the related compressive stress were recorded for each specimen.

Result and Discussion

X-ray fluorescence (XRF) Analysis. X-ray fluorescence (XRF) analysis was performed on OPC and POFA to determine their oxide compositions, as summarized in Table 1. The XRF data indicate that POFA has a distinct chemical profile compared with OPC, with elevated contents of MgO, Al_2O_3, Fe_2O_3, K_2O, and SO_3, which could enhance its pozzolanic behavior and contribute to strength development in blended systems if properly processed. The OPC sample was mainly composed of CaO (67.31%), SiO_2 (20.19%), and Al_2O_3 (3.04%), indicating strength performance through the formation of C-S-H. However, POFA has significantly less CaO (42.63%), showing lower inherent cementitious properties, but it contributes to pozzolanic reactivity, especially at long-term ages. The SiO_2 content of POFA is below the LOD (limit of detection), possibly due to testing limitations or sample inconsistency in grinding fineness. The combination of high Al_2O_3 (7.18%) and Fe_2O_3 (11.1%) still qualifies POFA as a reactive pozzolan under ASTM C618 Class F. P_2O_5 and K_2O contents are typical of agro-industrial waste ashes and may influence hydration reactions at long-term ages.

Table 1: XRF data analysis of OPC and POFA.

Chemical composition	Materials (%)	
	Ordinary Portland Cement (OPC)	Palm Oil Fuel Ash (POFA)
MgO	0.8406	8.4451
Al_2O_3	3.0423	7.1781
SiO_2	20.1854	<LOD
P_2O_5	<LOD	7.7026
SO_3	3.2977	7.8473
K_2O	0.5456	13.6243
CaO	67.3142	42.6281
TiO_2	0.2098	0.9166
MnO	0.0768	0.5197
Fe_2O_3	4.4877	11.1382

Compressive Strength Results. Fig. 1 and 2 show the results of the compressive strength tests on mortar cubes with various replacement levels and particle sizes of POFA. The results show that these variables affected the development of compressive strength under two different curing processes of normal and steam curing regimes, at testing ages of 7 and 28 days.

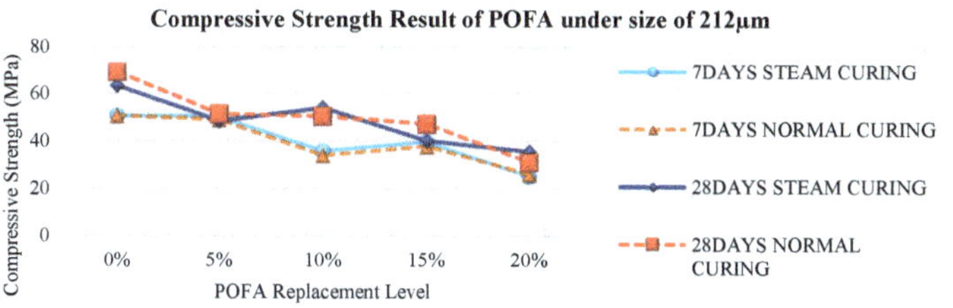

Fig. 1: Compressive Strength Result of POFA with particle size of 212μm under normal and steam curing regimes.

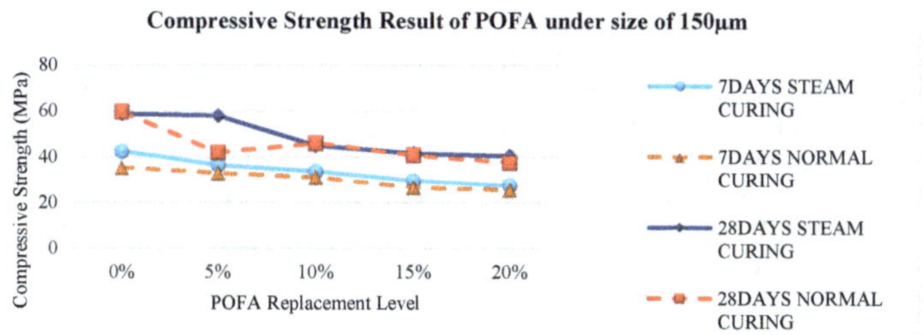

Fig. 2: Compressive Strength Result of POFA with particle size of 150μm under normal and steam curing regimes.

Based on the results presented in both Figures, the compressive strength of mortar was significantly influenced by the particle size of POFA (212 μm and 150 μm) and by the curing regime. In most cases, the control mix (0% POFA) consistently exhibited the highest compressive strength across both curing ages, corroborating findings by Ohwofasa et al. [26], who reported that control mixes typically give better performances than those incorporating SCMs at an early stage of hydration. Under normal curing regimes, 212 µm POFA showed a marked reduction in early-age strength, particularly at higher replacement levels. For example, at 7 days of curing, the control mix achieved 50.6 MPa, whereas the mixture containing 10% POFA replacement recorded a lower strength of 33.9 MPa. At 28 days of curing, the 10% POFA replacement mixture demonstrated a significant strength increment to 45.9 MPa, surpassing the 5% POFA replacement mixture with 41.9 MPa. This strength gain over time highlights the delayed pozzolanic activity of POFA, in which amorphous silica (SiO_2) reacts with C-H, a hydration byproduct of cement, to form additional C–S–H gel, which is enhanced at long-term age [27].

As the POFA replacement increases, the quantity of cement is reduced, which limits the amount of C-H to react with the silica content. Since C-H is an essential component for initiating the pozzolanic reactions, its reduction limits the formation of secondary C–S–H gel, which could compromise the compressive strength as the POFA in research indicates 5-10% POFA is optimal, providing enough C-H for ongoing pozzolanic activity and strength [28]. In addition, material fineness also played a crucial role in compressive strength performance. Mortars incorporating finer POFA of 150 µm consistently exhibited superior strength performance compared to those with coarser POFA of 212 µm. For example, at 28 days of curing, the difference in strength between the two particle sizes was about 8 MPa, attributed to the increase in specific surface area of POFA particles, which improves particle packing, accelerates hydration reactions, and enhances further pozzolanic reactivity [26]. Yet, under normal curing conditions, a 10% replacement of POFA with 150 µm particle size achieved a compressive strength of 31.0 MPa at 7 days of curing, and the value is further increased to 45.9 MPa at 28 days of curing, surpassing the performance of the 5% POFA replacement mixture. The improved performance of finer POFA is linked to its reactivity, which promotes a more effective conversion of C-H into additional C–S–H gel, thereby improving the microstructural integrity of the mortar matrix, as highlighted by Liew et al. [27]. The combination use of M-sand and POFA also helps offset the dilution effect of SCMs by maintaining matrix density. These findings align with earlier research demonstrating that finer SCMs enhance particle packing, strengthen interfacial bonding, and improve microstructural cohesion.

Despite the steam curing giving an improvement on early-age strength for mixes with coarser POFA of 212 µm, its influence on the long-term strength development was rather limited. Contrary to the specimens under the normal curing regime, which is more effective in promoting sustained strength development in mortars containing finer POFA. This observation shows that the specimens with finer POFA consistently improved under normal curing regimes, compared to the coarser samples under steam curing regimes, due to the lower reactivity of the finer particles. This trend supports the previous studies indicating that the finer POFA, produced through advanced grinding techniques, can significantly enhance the pozzolanic activity and long-term mechanical properties of cementitious materials [28]. Thus, optimizing POFA particle size with the inclusion of M-sand is a critical factor in maximizing its performance as a sustainable binder, particularly when subjected to different curing conditions.

The specimens cured under steam curing regimes have accelerated early strength gain, particularly for coarser POFA mixes. For example, the specimens containing 5% replacement of 212 µm POFA achieved 50.2 MPa at 7 days, nearly to the control mixture with the strength value of 50.5 MPa, indicating that elevated temperature promotes rapid cement hydration and stimulates the pozzolanic reaction of composite material. However, the results indicate that the specimen under steam curing regimes was less effective at higher replacement levels of 10 to 20% due to the dilution effect, which gives limited strength gain. In contrast, finer POFA with 150 µm gives favourable performances to both curing methods. For example, the 5% replacement of POFA under steam curing regimes provides the strength value of 36.4 MPa and 57.9 MPa at 7 and 28 days, respectively, similar to the control mixture with 59.6 MPa. It was proved by Zhou et al. [21] that the steam curing regimes accelerate the C-H consumption and enhance C–S–H formation in finer reactive materials. However, the specimens with 10% replacement of POFA with 150 μm under normal curing regimes give an optimum value compared with the specimens under steam-curing regimes at 28 days, suggesting that continuous hydration is more pronounced in long-term pozzolanic activity.

The overall results indicate that the specimens containing POFA under both curing regimes show some reduction in strength performance compared to control mixture. However, the POFA specimens under steam-cured POFA shows significant early and late strength development. At 7 days, steam-cured samples have higher strength due to enhanced pozzolanic activity of POFA and accelerated hydration. By 28 days, the control achieves highest strength, indicating normal curing supports long-term strength. Notably, 10% POFA with steam curing reaches optimum strength similar to control, suggesting optimal POFA levels could yield high-performance mortar under accelerated curing conditions.

Conclusion

This study confirms that incorporating POFA as a partial replacement for cement in mortar mixtures significantly influences compressive strength development, with performance outcomes closely linked to both the replacement percentage and the applied curing regime. This study also demonstrates that partial replacement of OPC with POFA and M-sand significantly affects mortar compressive strength, influenced by replacement level, curing method, and POFA fineness. XRF analysis revealed that POFA contains high amounts of MgO, Al_2O_3, K_2O, SO_3, and Fe_2O_3, with lower CaO and undetectable SiO_2, indicating limited initial cementing ability but strong pozzolanic potential through secondary reactions, especially at later ages. Compressive strength decreased at early ages with increasing POFA content, particularly at levels beyond 10%, due to slower pozzolanic activity and dilution. However, at 28 days, OPC with up to 10% POFA showed strength recovery due to enhanced C–S–H formation. Finer POFA consistently show improved performances compared to coarser particles through enhancing the microstructure and strength development, as the surface area and reactivity of the SCM material increase. Overall, 10% POFA with fine particles under steam curing offers a good balance of performance and sustainability. While finer POFA particles may delay early-age strength due to a slower initial hydration process, they contribute more effectively to long-term strength through increased pozzolanic activity and microstructural densification via C–S–H gel formation. These results reinforce the potential of POFA as a sustainable supplementary cementitious material, contributing to Sustainable Development Goal 12 (SDG 12) by promoting efficient material usage, reducing dependence on OPC, and mitigating CO_2 emissions in the construction sector. Furthermore, a comprehensive understanding of curing impacts enables more strategic decision-making in construction practices, advancing both performance and environmental stewardship in infrastructure development.

Acknowledgement

This author would like to acknowledge the Faculty of Built Environment, Universiti Teknologi Mara, Shah Alam, Selangor, for their technical support and the use of laboratory facilities.

References

[1] Pranavan, S. and G. Srinivasan, Investigation on behaviour of M-sand and sea sand based concrete. Materials Today: Proceedings, 2021. 45: p. 7079-7085. https://doi.org/10.1016/j.matpr.2021.01.930

[2] Lehne, J. and F. Preston, Making concrete change. Innovation in Low-carbon Cement and Concrete, 2018.

[3] Worrell, E., et al., Carbon dioxide emissions from the global cement industry. Annual review of energy and the environment, 2001. 26(1): p. 303-329. https://doi.org/10.1146/annurev.energy.26.1.303

[4] Sheheryar, M., R. Rehan, and M. Nehdi, Estimating CO2 Emission Savings from Ultrahigh Performance Concrete: A System Dynamics Approach. Materials, 2021. 14(4): p. 995. https://doi.org/10.3390/ma14040995

[5] Khaiyum, M.Z., S. Sarker, and G. Kabir, Evaluation of carbon emission factors in the cement industry: An emerging economy context. Sustainability, 2023. 15(21): p. 15407. https://doi.org/10.3390/su152115407

[6] Hamada, H.M., et al. Applications of Nano palm oil fuel ash and Nano fly ash in concrete. in International Conference on Innovative Technology, Engineering and Sciences (iCITES). 2018. Univ Malaysia Pahang, Pekan Campus Lib, Gambang, MALAYSIA: Iop Publishing Ltd. https://doi.org/10.1088/1757-899X/342/1/012068

[7] Abd Elghani, K.A., F.A. Ali, and O.S. Hussien, Effect Of Agricultural Wastes On Mechanical Properties Of Green Concrete. Egyptian Journal of Chemistry, 2023. 66(13): p. 1193-1199. https://doi.org/10.21608/ejchem.2023.228035.8397

[8] Sathiparan, N., D.H.H.P. Dassanayake, and D.N. Subramaniam, Utilization of supplementary cementitious materials in pervious concrete: a review. International Journal of Environmental Science and Technology, 2024. https://doi.org/10.1007/s13762-023-05440-4

[9] Sidek, M.N.M., et al. Workability and Performance of High-Performance Concrete by Using Palm Oil Fuel Ash as a Cement Replacement Material. in World Conference on Byproducts of Palms and their Applications. 2020. Springer.

[10] Muhedin, D.A. and R.K. Ibrahim, Effect of waste glass powder as partial replacement of cement & sand in concrete. Case Studies in Construction Materials, 2023. 19: p. e02512. https://doi.org/10.1016/j.cscm.2023.e02512

[11] Raghav, M., et al., Review of the Effects of Supplementary Cementitious Materials and Chemical Additives on the Physical, Mechanical and Durability Properties of Hydraulic Concrete. Materials, 2021. 14(23): p. 7270. https://doi.org/10.3390/ma14237270

[12] Ahmadi, R., et al., MECHANICAL PERFORMANCE OF POLYPROPYLENE FIBRE REINFORCED MORTAR INCORPORATING PALM OIL FUEL ASH. JOURNAL OF SUSTAINABILITY SCIENCE AND MANAGEMENT, 2022. 17(6): p. 92-103. https://doi.org/10.46754/jssm.2022.06.008

[13] Hamada, H., et al. Effect of incorporation POFA in cement mortar and desired benefits: a review. in IOP Conference Series: Earth and Environmental Science. 2019. IOP Publishing. https://doi.org/10.1088/1755-1315/365/1/012060

[14] Hasan, N.M.S., et al., Eco-friendly concrete incorporating palm oil fuel ash: Fresh and mechanical properties with machine learning prediction, and sustainability assessment. Heliyon, 2023. 9(11). https://doi.org/10.1016/j.heliyon.2023.e22296

[15] Hamada, H.M., et al., Sustainable use of palm oil fuel ash as a supplementary cementitious material: A comprehensive review. Journal of Building Engineering, 2021. 40: p. 102286. https://doi.org/10.1016/j.jobe.2021.102286

[16] Hamada, H.M., et al., The present state of the use of palm oil fuel ash (POFA) in concrete. Construction and Building Materials, 2018. 175: p. 26-40. https://doi.org/10.1016/j.conbuildmat.2018.03.227

[17] Ismail, M.H., et al., Performance of High Strength Concrete Containing Palm Oil Fuel Ash and Metakaolin as Cement Replacement Material. Advances in Civil Engineering, 2022. 2022: p. 1-11. https://doi.org/10.1155/2022/6454789

[18] Shi, J., et al., Properties evolution of high-early-strength cement paste and interfacial transition zone during steam curing process. Construction and Building Materials, 2020. 252: p. 119095. https://doi.org/10.1016/j.conbuildmat.2020.119095

[19] Li, C., et al., Effect of Maintenance and Water-Cement Ratio on Foamed Concrete Shrinkage Cracking. Polymers, 2022. 14(13): p. 2703. https://doi.org/10.3390/polym14132703

[20] Naas, A., et al., Combined effect of powdered dune sand and steam-curing using solar energy on concrete characteristics. Construction and Building Materials, 2022. 322: p. 126474. https://doi.org/10.1016/j.conbuildmat.2022.126474

[21] Zhou, Y., et al., A review of the effects of raw material compositions and steam curing regimes on the performance and microstructure of precast concrete. Materials, 2022. 15(8): p. 2859. https://doi.org/10.3390/ma15082859

[22] Martínez-Ramírez, S. and M. Frías, The effect of curing temperature on white cement hydration. Construction and Building Materials, 2009. 23(3): p. 1344-1348. https://doi.org/10.1016/j.conbuildmat.2008.07.012

[23] Pang, X., et al., Influence of curing temperature on the hydration and strength development of Class G Portland cement. Cement and Concrete Research, 2022. 156: p. 106776. https://doi.org/10.1016/j.cemconres.2022.106776

[24] Yu, Y., et al., Evolution of temperature stress and tensile properties of concrete during steam-curing process. Construction and Building Materials, 2021. 305: p. 124691. https://doi.org/10.1016/j.conbuildmat.2021.124691

[25] Yao, B., et al., Influence of self-ignition coal gangue on properties of foam concrete with steam curing. Case Studies in Construction Materials, 2022. 17: p. e01316. https://doi.org/10.1016/j.cscm.2022.e01316

[26] Ohwofasa, J., C. Ikumapayi, and C. Arum, Effects of Fly Ash on Compressive Strength and Durability Properties of Lean Concrete. Journal of Applied Sciences and Environmental Management, 2023. 27(11): p. 2597-2610. https://doi.org/10.4314/jasem.v27i11.32

[27] Chinnu, S., et al., Influence of palm oil fuel ash in concrete and a systematic comparison with widely accepted fly ash and slag: A step towards sustainable reuse of agro-waste ashes. Cleaner Materials, 2022. 5: p. 100122. https://doi.org/10.1016/j.clema.2022.100122

[28] Hamada, H., et al., Effects of nano-palm oil fuel ash and nano-eggshell powder on concrete. Construction and Building Materials, 2020. 261: p. 119790. https://doi.org/10.1016/j.conbuildmat.2020.119790

Green Materials, Processing & Characterization – ICoGMPaC 2025 Materials Research Forum LLC
Materials Research Proceedings 68 (2026) 182-188 https://doi.org/10.21741/9781644904237-24

The Synthesis of Graphitic Carbon from Coconut Shell Waste by Catalytic Graphitization

Yap Yee WEN[1,2,a*], Norsuria MAHMED[1,2,b], Midhat Nabil AHMAD SALIMI[1,2], Azlin Fazlina OSMAN[1,2], Kamrosni ABDUL RAZAK[1,2], Shayfull Zamree ABD RAHIM[2] and Mohd Yusry MOHAMAD YUNUS[3]

[1]Faculty of Chemical Engineering & Technology, Universiti Malaysia Perlis), Kompleks Pusat Pengajian Jejawi 2, Taman Muhibbah, 02600 Jejawi, Perlis, Malaysia

[2]Center of Excellence Geopolymer and Green Technology, Kompleks Pusat Pengajian Jejawi 2, Taman Muhibbah, 02600 Jejawi, Perlis, Malaysia

[3]Terkemuka Solution Sdn Bhd, Seksyen U10 Alam Budiman, 40170 Shah Alam, Selangor Malaysia

[a]yapyeewen1129@gmail.com, [b]norsuria@unimap.edu.my

Keywords: Graphitic Carbon, Catalytic Graphitization, Coconut Shell Waste

Abstract. The graphitic carbon has been synthesized by catalytic graphitization process. There were three types of catalyst that have been used during the study, namely potassium hydroxide (KOH), ferric nitrate ($Fe(NO_3)_3$ and nickel chloride ($NiCl_2 \cdot 6H_2O$) solutions. During the experiment, the coconut shell carbon (CSC) powder was impregnated in the catalyst solutions and stirred at 60°C until all the solution evaporated. The samples then were transferred into the crucible and graphitized at 1000°C in a muffle furnace for 3 hours of holding time. From the CHNS analysis, it was found that the KOH catalyst improved the carbon content of CSC from 84.8% to 86.5%. From the X-ray diffraction (XRD) and the Raman spectroscopy analysis, it was proved that by using 1M of KOH catalyst, the amorphous phase of CSC was transformed into a graphitic carbon. The field-emission scanning electron microscopy (FE-SEM) images of the CSC further support the findings as the KOH-CSC was approaching a layering structure. Therefore, it is suggested that KOH is feasible to transform the amorphous CSC into graphitic carbon/graphite structure.

Introduction

Over the last few years, the production of synthetic graphite is drawing a tremendous of attractions. Graphite is the crystalline form of carbon and thus, the production of graphite need to begin from its basic element, carbon. There are a lot of research had used bio-waste materials as the potential precursors for carbon in order to synthesize graphite [1,2]. Agriculture activities are one of the most popular areas where they are producing a large number of bio-waste materials over the years. An agricultural bio-waste material or an agricultural biomass material is an unwanted waste product that is discarded from various agricultural activities [3]. The utilization of agricultural biowaste is important as it can contribute to the agricultural waste management, which is urgently in need recently. The application of bio-waste materials as the precursor for graphite production is popular due to the reasons that are omnipresent, environmentally benign [4], renewable and cheap.

Biomass waste with a long carbon chain and high lignocellulosic structure is preferable to be used as the carbon precursor when it comes to the conversion of waste materials studies [5]. Although there are a lot of study suggestions on the methods in producing graphitic carbon, they will remain as the challenging issues since the synthesis of graphite or even graphene from different waste materials involved various parameters. Factors such as the carbonization and graphitization temperatures and atmospheres, and the concentrations and types of catalyst used, had different reactions depending on the types of biomass waste used. Each type of biomass waste will have its feasibility with a specific reaction temperature, chemical and even the pressure needed. Also, in order to avoid further harmful to the environment, most researchers suggest producing graphitic carbon at lower temperature [6].

This study demonstrated the synthesis of graphitic carbon by catalytic graphitization. The coconut shell (CS) waste was used as the raw lignocellulosic biomass carbon precursor since it naturally has about 50% carbon content. Different types of catalyst were used, under a fixed graphitization temperature in order to investigate their effects on the final products. Thus, the main discussion of this paper is the effect of different catalyst used in synthesizing the graphitic carbon from the coconut shell waste. The phase, structure and morphology of the synthesized samples will be investigated and compared with the commercial graphite.

Materials and Methods

Materials. Coconut shell waste (CSW) was collected from Pasar Besar Kangar, Perlis Malaysia. The catalyst solution used were iron (III) nitrate solution, $Fe(NO_3)_3$ and potassium hydroxide solution, KOH from Sigma Aldrich, Germany while nickel (II) chloride hexahydrate solution, ($NiCl_2 \cdot 6H_2O$) was from Merck, USA. The hydrochloric acid, HCl was supplied by HmbG Chemical, Germany.

Carbonization of Coconut Shell Carbon (CSC). The coconut shell (CS) was first cleaned from the husk and fibers. Then, it was washed and rinsed with distilled water and heated in the oven at 110°C for 12 hours for moisture removal from the coconut shell. The dried coconut shell was crushed and sieved into smaller pieces ranging in size from 10 to 20mm. The dry weight of the metal crucible and coconut shell pieces were measured and recorded. The carbonization temperature was set at 1000°C, with a heating rate of 5°C/min, a holding period of 3 hours, and a cooling rate of 10°C/min. To provide an oxygen-limiting atmosphere for the carbonization process, the metal crucible was tightly covered. The dry weight of the CSC produced was also measured and recorded.

Catalyst Graphitization. In order to graphitize the CSC powder at lower temperature, the iron (III) nitrate (Fe(NO3)3), potassium hydroxide (KOH) and nickel (II) chloride hexahydrate (NiC12 · 6H2O) solutions were used as catalyst. During the experiment, the CSC powder was immersed in 1M of Fe(NO3)3 solution. The mixture was stirred at 60°C until the solvent evaporated. Then, the impregnated CSC powder was placed into a crucible and graphitized at 1000°C under an oxygen-limited condition with 5°C/min of heating rate, 3 hours holding time, and a cooling rate of 10°C/min. After the graphitization process, the sample was washed with 2M of hydrochloric acid (HCl) solution under continuous stirring overnight to remove the residual iron metal and impurities in the CSC powder. Then, the powder was rinsed with deionized water until pH 7 and dried in an oven at 60°C. The experiment was repeated by replacing the iron catalyst with 1M KOH and 5mmol NiC12 · 6H2O solutions, respectively. The prepared samples were denoted as Fe-CSC, KOH-CSC and Ni-CSC respectively. The commercial graphite powder was used as reference sample.

Characterization. The proportion of (C, H, N, S) contained in the samples were determined by performing an ultimate analysis 6. The CSC sample was placed in a silver capsule and tested by an elemental analyzer, Vario Micro Cube SN 15135057 with the procedure outlined in BS EN ISO 16948:2015 (solid biofuels). X-ray diffraction (XRD) patterns of CSC samples ranging from (2Θ) 10° to 90° were produced by using a multifunctional diffractometer (Bruker D2 Phaser) with Cu-Kα (λ =1.5406 Å) radiation at an accelerating voltage of 40 kV and an applied current of 30 mA. The scan rate used was 0.1 s/step. The Raman spectroscopy was performed using Renishaw inVia

Raman spectrometer. The excitation laser beam (633 nm) with ~20 mW was focused on the CSC sample to capture the backscattered Raman data. Each Raman spectrum was obtained between 100 cm-1 and 4000 cm-1. Leo Supra 50vp field-emission scanning electron microscope (FESEM) was used to obtain the morphology of the CSC samples. The working voltage used was 10 kV.

Results and discussion

Table 1: The CHNS analysis of raw CS and CS carbon and catalyzed CS carbon.

Sample	Nitrogen (%)	Carbon (%)	Hydrogen (%)	Sulfur (%)
Raw Coconut	0.0	47.3	6.1	0.6
CS carbon	0.2	84.8	1.3	1.0
Ni-CSC	0.2	81.4	1.6	1.0
KOH-CSC	0.2	86.5	1.1	0.9
Fe-CSC	0.2	82.1	1.5	1.0

Table 1 gives a summary of the C, H, N and S content for raw coconut shell and the coconut shell carbon samples. A significant increase from 47.3% to 84.8% is observed in the CSC after carbonization at 1000°C. This situation showed that the carbonization process is important to increase the carbon content in the sample. This is due to the great extent of decomposition of organic matters occurred during the process [7]. After graphitization, it was found that the KOH-CSC sample marked the highest carbon content, 86.5% followed by Fe-CSC, 82.1% and Ni-CSC, 81.4%, respectively. The carbon contents of Ni-CSC and Fe-CSC were also decreased after the graphitization process. Compared to KOH-CSC sample, the carbon content increased about 2%. Therefore, the KOH catalyst is the most feasible catalyst for coconut shells in the production of high carbon content since it can enhance the degree of carbonization and more volatile materials are released [8].

Graphitic carbon has a distinct structure from amorphous carbon. The XRD pattern is displayed in Fig. 1 and Fig. 2 to better comprehend the phase of the CSC samples. Fig. 1 depicted the commercial graphite sample with a sharp and high-intensity carbon (002) peak located at 26.4805°. A highly crystalline structure sample will exhibit an intense and sharp XRD peak [9]. All three CSC samples displayed a broad peak at $2\theta = 20°\text{-}30°$ in Fig. 2, except for KOH-CSC sample, indicates that the CSC samples are all in amorphous carbon structure as no prominent graphitization characteristic was found. Nevertheless, based on Fig. 2, the peak of KOH-CSC is located at 26.0141° which is approximately to the carbon (002) graphite peak. Left-shifted peak indicating the presence of disordered or amorphous stage carbon is still present in KOH-CSC [10]. KOH-CSC also exhibited the sharpest peak among all four CSC samples indicating that most of the amorphous CSC structures are approaching being converted into crystalline graphitic carbon structures. KOH catalyst is the optimum catalyst among the catalyst used as it improved the graphitization degree of the CSC. It can also be noticed that the Ni-based catalyst produced the CSC carbon with a more intense peak as compared to $Fe(NO_3)_3$ catalyst. Ni catalyst is considered to be more feasible than Fe catalyst for converting amorphous CSC into graphitic carbon as it approaches to form a sharper and more intense XRD peak. The phenomenon that the XRD peak of Ni-CSC remains as an amorphous broad peak might be due to the insufficient carbonization temperature. Temperature played an important role in transforming amorphous to graphitic nanostructure. As reported in [11], a more intense graphitic peak was found when heating Ni catalyzed carbon sample under 1200°C. Therefore, it can be explained that the broad peak formed by the Ni-CSC sample is due to insufficient heating temperature.

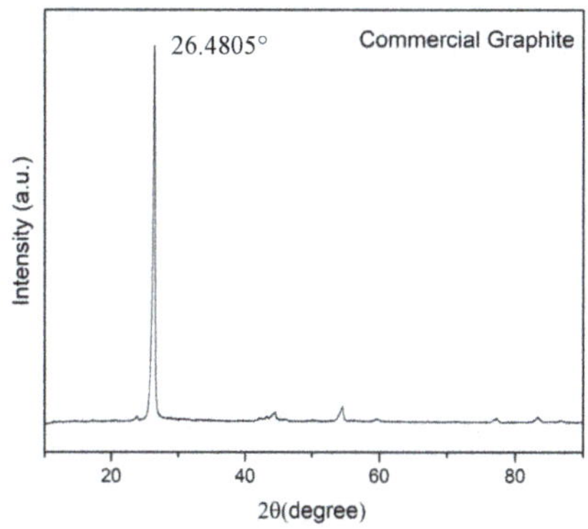

Fig. 1: The XRD pattern for commercial graphite.

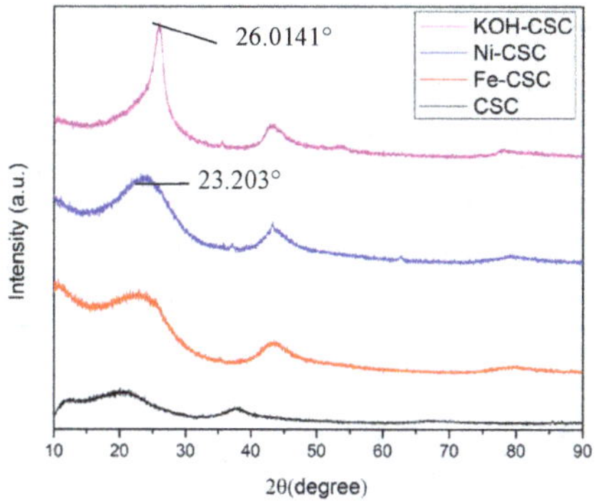

Fig. 2: The XRD pattern for CSC and catalyzed CSC samples.

The Raman spectra of CSC and KOH-CSC as shown in Fig. 3 were displayed for monitoring the graphitic structure formation in the CSC sample. As reported in [1], the band located between 1320 cm^{-1} to 1365 cm^{-1} in the Raman spectroscopy is defined as the D peak while the band located at 1520 cm^{-1} to 1600 cm^{-1} was considered the G peak. The D peak can be used in indicating the disordered state in the carbon structure while the G peak represents the sp^2 bond in graphite carbon [12]. Both the Raman spectra of CSC and KOH-CSC fall in between the mentioned range. However, as compared to the commercial graphite Raman spectra [10], both CSC and KOH-CSC are in an amorphous structure as the formation of the sharp and intense peak is the sign of a highly crystalline graphitic structure. Nevertheless, The I_D/I_G ratio calculated from the Raman spectra can be used to measure the degree of defects and disordered of the carbon sample [13]. The higher the ratio, indicating the higher extent of defects or disorders in the carbon sample. As compared, the I_D/I_G ratio of KOH-CSC, 0.74 is lower than that of the CSC, 1.12. This indicates that the use of a KOH catalyst had improved the disorder issue of a CSC. The presence of a 2D peak which is located around 2690 cm^{-1} [10] is a significant characteristic of graphite. There is no 2D peak that will be marked by an amorphous carbon [2]. Although having a broad D and G peak, the presence of the 2D peak further confirms that the KOH-CSC was on the route to transform into a crystalline graphitic carbon.

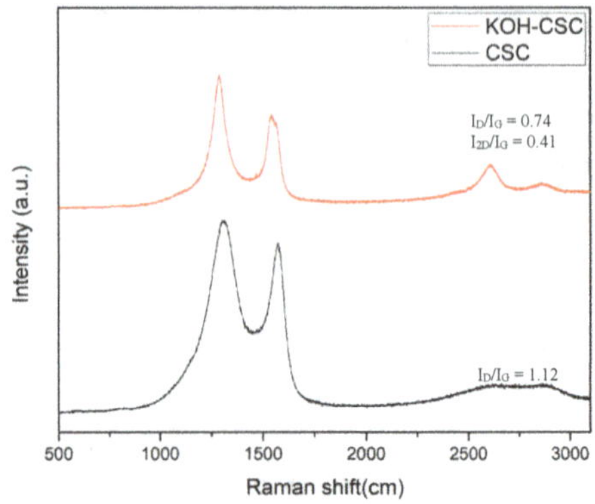

Fig. 3: The Raman spectra of CSC and KOH-CSC samples.

Graphite with its naturally occurring stable form displayed a layering structure [14]. As shown from the FE-SEM image from Fig. 4(a), it can easily observe that the layering structure is found in commercial graphite. While as from the FE-SEM image from Fig. 4 (c), under the same magnification at 20 Kx as in Fig. 4 (a), no obvious layering structure was found from the KOH-CSC. The layer structure of graphite plays an important role in terms of its functionality. Under high temperatures with a high graphitization degree, the synthetic graphite should also display a layering structure which is alike to naturally occurring graphite [15].

Based on the FE-SEM image in Fig. 4 (b), the KOH-CSC sample is having a layer-like structure that approaches the commercial graphite. This situation can be explained together with the XRD peak and Raman spectra by KOH-CSC as it is approaching a crystalline sharp intense peak with the position (2θ=26.0141°) nearby to the peak position of the commercial graphite (2θ=26.4805°). The slightly left shift in the XRD peak and the structure that approaches the layered graphite structure suggested that the KOH-CSC is soon to meet the characteristic of a crystalline graphite. Therefore, further improvement in the parameters of the KOH catalytic graphitization process will be needed to achieve the synthesis of a CS graphitic carbon that has the same physical and chemical properties as commercial graphite.

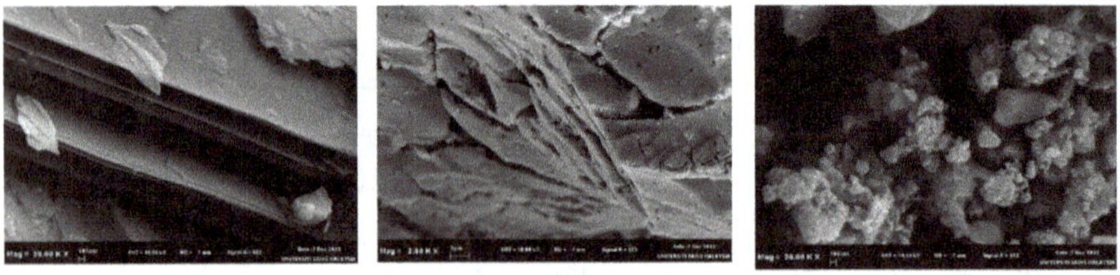

Fig. 4: The FE-SEM image of (a) Commercial graphite, (b) KOH-CSC with 2.0 Kx magnification and (c) KOH-CSC with 20.0 Kx magnification.

Conclusion

In summary, this study was done to synthesis the graphitic carbon from coconut shell waste by catalytic graphitization process. The carbon content of the CSC was further improved from 84.8% to 86.5% under the usage of the KOH catalyst. As compared to the commercial graphite XRD peak, all CSC samples exhibited a broad peak, indicating the CSC sample was still in the amorphous structure. However, with the use of KOH catalyst, it produced the XRD peak which is relatively sharp and intense, located at 2θ = 26.0141° which is nearby the C (002) peak of the commercial graphite at 2θ=26.4805°. From the Raman spectroscopy, KOH-CSC marked the peak that falls in the range of the graphite peak. The disordered issue of the CSC was successfully reduced by the KOH catalyst as the I_D/I_G ratio of KOH-CSC, 0.74 is lower than that of the CSC, 1.12. Through the FE-SEM image, no layering structure like from the commercial graphite was found in the CSC sample. However, KOH-CSC depicted a layering-like structure, which shows that the KOH-CSC is semi-crystalline graphitic carbon.

Acknowledgement
This work was financially supported by the Terkemuka Solution Sdn Bhd (UniPRIMA Industry Grant 9002-00127) and Universiti Malaysia Perlis (UniPRIMA Grant 9001-00668).

References

[1] L. Gai, J. Li, Q. Wang, R. Tian, K. Li, Evolution of biomass to porous graphite carbon by catalytic graphitization, J. Environ. Chem. Eng. 9 (2021) 106678. https://doi.org/10.1016/j.jece.2021.106678

[2] Z. Sun , D. Yao , C. Cao , Z. Zhang , L. Zhang , H. Zhu , Q. Yuan , B. Yi, Preparation and formati on mechanism of biomass-based graphite carbon catalyzed by iron nitrate under a low-temper ature condition, J. Environ. Manage 318 (2022) 1-28. https://doi.org/10.1016/j.jenvman.2022.115555

[3] N. Karić, A.S. Maia, A. Teodorović, N. Atanasova, G. Langergraber, G. Crini, A.R.L. Ribeiro, and M. Đoli, Bio-waste valorisation: Agricultural wastes as biosorbents for removal of (in)organic pollutants in wastewater treatment, Chem. Eng. J. Adv. 9 (2022) 100239. https://doi.org/10.1016/j.ceja.2021.100239

[4] J. Hoekstra, A.M. Beale, F. Soulimani, M. Versluijs-Helder, J.W. Geus, and L.W. Jennesken, Base Metal Catalyzed Graphitization of Cellulose: A Combined Raman Spectroscopy, Temperature-Dependent X-ray Diffraction and High-Resolution Transmission Electron Microscopy Study. J. Phys. Chem. C 119(19) (2015) 10653-10661. https://doi.org/10.1021/acs.jpcc.5b00477

[5] M. T. Safian, U. S. Haron, and M. N. Mohamad Ibrahim, A review on bio-based graphene derived from biomass wastes, BioRes. 15(4) (2020) 9756-9785. https://doi.org/10.15376/biores.15.4.Safian

[6] O. Adeoye, R. O. Quadri, and O. S. Lawal, Assessment of biofuel potential of tenera palm kernel shell via fixed bed pyrolysis and thermal characterization. Results Surf. Interfaces 9 (2022) 100091. https://doi.org/10.1016/j.rsurfi.2022.100091

[7] Tomczyk, Z. Sokołowska, and P. Boguta, Biochar Physicochemical Properties: Pyrolysis Temperature and Feedstock Kind Effects, Rev. Environ. Sci. Biotechnol. 19 (2020) 191-215. https://doi.org/10.1007/s11157-020-09523-3

[8] K. Macdermid-Watts, E. Adewakun, O. Norouzi, T.D. Abhi, R. Pradhan, and A. Dutta, Effects of FeCl3 Catalytic Hydrothermal Carbonization on Chemical Activation of Corn Wet Distillers' Fiber, ACS Omega 6(23) (2021) 14875-14886. https://doi.org/10.1021/acsomega.1c00557

[9] S. Supriya, G. Sriram, Z. Ngaini, C. Kavitha, M. Kurkuri, I.P. De Padova, and G. Hegde, The Role of Temperature on Physical Chemical Properties of Green Synthesized Porous Carbon

Nanoparticle Waste Biomass Valor. 11 (2020) 3821-3831. https://doi.org/10.1007/s12649-019-00675-0

[10] F. Destyorini, Y. Irmawati, A. Hardiansyah, H. Widodo, I. N. D. Yahya, N. Indayaningsih, R. Yudianti, Y. I. Hsu, H. Uyama, Formation of nanostructured graphitic carbon from coconut waste via low temperature catalytic graphitisation, Eng. Sci. Technol. Int. J. 24 (2021) 514 -523. https://doi.org/10.1016/j.jestch.2020.06.011

[11] F. Destyorini, R. Yudianti, Y. Irmawati, A. Hardiansyah, Y. I. Hsu, H. Uyama, Temperature driven structural transition in the nickel-based catalytic graphitization of coconut coir Diam Relat, Mater. 117 (2021) 108443. https://doi.org/10.1016/j.diamond.2021.108443

[12] J. Dennison, M. Holtz, and G. Swain, Raman Spectroscopy for carbon materials, Spectroscopy 11 (1996) 38-45.

[13] P.H. Guo, H.M. Cheng, J.S. Chen, Y.S. Wu, C.C. Yang, F.Y. Zeng, R.J. Chung, and P.T. Chen, IEEE Access (2021).

[14] H.L. Ip S O N, Proc. R. Soc. London. Ser. A. Math. Phys. Sci. 181, 101 (1942).

[15] T. Qiu, J.G. Yang,X. J. Bai and Y. L. Wang, The preparation of synthetic graphite materials with hierarchical pores from lignite by one-step impregnation and their characterization as dye absorbents. RSC Advances, 9(22) (2019) 12737-12746. https://doi.org/10.1039/C9RA00343F

Biogas Potential from the Investigation of Paddy Straw Co-Digestion with Cow Dung

T.Y.T. ROSLINA[1,a*], M. MARIAM[1], A.R. NOR' AINI[2], M.S.M. FAIZ[3] and C.M. HASFALINA[4]

[1]Faculty Technology of Mechanical Engineering, Universiti Malaysia Perlis (UniMAP), Sungai Chuchuh, Padang Besar, 02100 Perlis Indera Kayangan, Malaysia

[2]Department of Bioprocess, Faculty of Biotechnology and Biomolecular Sciences, Bioprocessing and Biomanufacturing Research Centre Faculty of Biotechnology and Sains Biomolekul, Universiti Putra Malaysia, 43400 UPM Serdang, Selangor, Malaysia

[3]Institute of Systems Biology, Universiti Kebangsaan Malaysia, 43600 Bangi, Selangor, Malaysia

[4]Department of Biological and Agricultural Engineering, Faculty of Engineering, Universiti Putra Malaysia, 43400 UPM Serdang, Selangor, Malaysia

[a]tengkuroslina@unimap.edu.my

Keywords: Paddy Straw Waste, Anaerobic Digestion, Biogas Production

Abstract. Paddy straw waste (PS) is an organic waste that is disposed in open land after preparation of rice harvest that is generated in equal or greater quantities than the rice itself. Generally, it is disposed in an open land which increases anthropogenic gases. Converting it into useful energy or value added product may reduce disposal problem and anthropogenic activity. In this study, PS was co-digested with cow dung (CD) for obtaining biogas by anaerobic digestion. For this purpose, PS was mixed with CD at different proportions, namely 50:50, 40:60, 30:70, 20:80, and 0:100 percentages on a mass basis, the samples were used in five different anaerobic digesters. The samples were kept in different anaerobic digesters for the study. The effect of important input parameters like pH, Carbon to Nitrogen (C/N) ratio on the biogas production were studied. A maximum biogas production obtained from the co digestion of substrate containing 30% ps and 70% Cd, for a digestion time of 20 days and d3 shows a max pH value of 7.16, Further, the collected biogas from the digesters were characterised to ensure the suitability for use as a renewable fuel. Furthermore, the digested slurry was also analysed for its use in agriculture sector. The results are presented in this paper.

Introduction

The increasing concentrations of greenhouse emission effect from the global warming and ozone depletion. The global warming affected from the continuously increasing energy crises, depletion of fossil fuel resources and severe environmental pollution (Raj Singh et al., 2022). Biogas is use as an alternative due to the more friendly environmental energy sources, clean energy production and alternate to resolve environmental pollution issues. The waste that can produce biogas such as animal manure, agriculture waste and other agriculture residues through anaerobic digestion (Shubhra Singh et al., 2021) [1]. Biogas has been looked upon as a totally controlled and promising source that can be produced from agricultural and animal wastes.

Paddy straw is the preferred substrate for bioenergy production. Recently, the demands for reduction and utilization of paddy waste have rapidly increased in Malaysia. Annually about 3.66 million tonne of paddy residue is left in the fields. Towards year 2020 this value is forecasted to increase to 7 million tonne per year due to emerging technology development in agriculture industries. The increasing of paddy residue production in Malaysia, will cause the abundant of availability of this resources as well as create the problem of waste management if this residue cannot be manage properly (Shafie, 2015) [3]. The burning of paddy residue in the field will cause the pollution (Singh, 2015) [4]. By used the paddy residue in electricity generation can be one solution for waste management and also the pollution problem. Paddy straw is biodegradable in nature and effective for biomethane or biogas production through AD process in action of microbial consortium activity (Abraham et al., 2016) [5]. AD process produces biomethane which come from biomass which is is a clean, renewable energy source that currently represents approximately 15% of the global energy consumption (Sapp, 2017). Fewer air pollutants and less CO_2 per unit of energy are released when compared with non-renewable fossil fuels (Nguyen et al., 2016).

A number of research studies have suggested that paddy straw is one of the most suitable feedstock for AD process owing to several advantages, such as their renewable nature, huge availability, low-cost and high bio-degradability (Goodman, 2020; Selvarajoo et al., 2022; Srivastava, 2019) [7,8,9]. Further, co-digestion of paddy straw feedstock with the combination of other wastes may increase the biogas production yield due to the improvement in nutrient balance (e.g., C/N ratio), adjustment of additional buffering capacity, and dilution of toxic chemicals (Kainthola et al., 2019) [10]. Paddy straw (PS) is considered as one of the most prominent feedstocks for the AD process, and the paddy crop secures its position as the third-largest crop in the global market. The main aim of this study was to make an attempt to use the paddy straw (PS) as a co-substrate in combination with cow dung for production of biogas. For this purpose, PS was mixed with CD at different proportions in five different anaerobic digesters like AD1, AD2, AD3, AD4, and AD5. Further, the study was aimed to evaluate the various affecting parameters on the biogas production. Finally, the samples of biogas obtained from the digesters were characterised to ensure the quality for using it as an alternative gaseous fuel. Further, the digested slurry was also analysed to utilize as a fertilizer for growth of crops.

The main aim of this study was to make an attempt to use the paddy straw (PS) as a co-substrate in combination with cow dung for production of biogas. For this purpose, PS was mixed with CD at different proportions in five different anaerobic digesters like AD1, AD2, AD3, AD4, and AD5. Further, the study was aimed to evaluate the various affecting parameters on the biogas production. Finally, the samples of biogas obtained from the digesters were characterised to ensure the quality for using it as an alternative gaseous fuel. Further, the digested slurry was also analysed to utilize as a fertilizer for growth of crops.

Materials and Methods

Feed Stock. The fresh cow dung (Cd) was collected from a local farm house at Perlis. The visible straws present in the CM were removed manually. The collected CM was mixed with water in the ratio 1:1, stirred for 10 min at 2000 rpm. On the other hand, PS was collected from paddy field around Perlis. The collected PS was dried for 48 h at 80 OC and stored at room temperature in a dry place for further experiments.

Experimental Setup. In this research work, experiments were conducted in laboratory scale anaerobic digesters D1, D2, D3, D4 and D5 containing 50:50, 60:40, 70:30, 80:20, and 100:0 on a mass basis of CM: PS respectively. The prepared samples of CM:PS mixture were diluted with water at a ratio of 1:1 and 1:3 respectively. The characteristics of the feed stocks at different proportions of CM and PS are given in Table 1. The prepared samples were in a kept in 150 mL serum bottle with 100 mL of working volume. The bottle were tightly cap with rubber septa and aluminum cap. The anaerobic digesters used in this study are shown in Fig. 1. The experiments were conducted at a mesophilic temperature 35 oC. All the reactors were flushed with nitrogen for 5 min before sealing of the digesters. Since, there is no mechanical stirrer available in the laboratory for stirring the mixture. Therefore, for obtaining a better reaction mixture, manual shaking was done. Each digester was shaken manually for 1 min twice a day prior to the measurement of biogas volume (Zhai et al., 2015) [11].

Fig. 1: Anaerobic digester used in this study.

Biogas Measurement and its Composition Analysis. To measure the composition of gas produced, a sample of gas was collected daily from the headspace of each digester using gas tight syringe (25 mL Perkin Elmer). Biogas composition was measured daily, in terms of percentage volume.

Analytical Methods and Calculations. Firstly, feedstock characterization was carried out to ensure the suitability of the feed stock for anaerobic digestion, which included the proximate and ultimate analyses. The proximate analysis of all the samples was calculated, as recommended by APHA [29] and are given in Table 1. The ultimate analysis of the samples was done by a C-H-N-S elemental analyser shows in Table 2. The C/N ratio was determined by dividing the total carbon content to the total nitrogen content. The pH value of the feed stock was measured with a pH meter.

Table 1: Proximate analysis of CM and PS.

Substrate	Weight % dry basis			
	Moisture content	Volatile matter	Ash	Fixed carbon
Cow dung	73.4	14.3	4.1	8.2
Paddy straw	8.4	39.3	9.2	43.1

Table 2: Ultimate analysis of CM and PS.

Substrate	Weight % dry basis						
	C	H	N	S	O	P	K
Cow dung	38.24	4.63	1.64	0.05	51.34	0.06	0.05
Paddy straw	65.3	7.45	0.81	0.43	16.81	0.7	1.50

Results and Discussions

Effect of pH. pH value is most of the parameter effect on biogas production of the slurry. pH value changes depending on the different stages of anaerobic digestion process. Pradeshwaran et al., reported that, the pH value of the digester should kept 6.8-7.2 which is suitable for the anaerobic bacteria performs in a good yield. At initial stages of anaerobic digestion, the pH value of the substrate rapidly drops because of digestible in hydrolysed mode and converted into fatty acids. Due to the formation of higher amount of fatty acids, sometimes, pH value may fall below 5 in the digester. Fig. 2 portrays the variation of pH with the digestion time by considering the average value of pH in every five consecutive days.

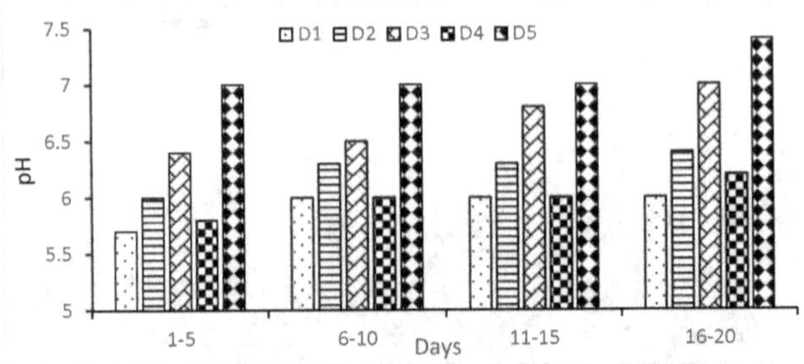

Fig. 2: Variation of pH value at different time.

Based on the Fig. 2, among five anaerobic digesters, D5 gives a maximum pH. During 1-15 days of the digestion time, a maximum ph value of 7 and 6 and 6.5 are obtain for the D2 and D3 respectively. which is about 7.19% and 14.97% less than the digester containing cow manure alone (D5) respectively. Because of hydrolytic and acetogenic bacterial are observed to be more during those days. During the anaerobic digestion process, the digester D1, D2, D3, D4, and D5 show its peak value on 16th day, 17th day, 18nd day, 19st day, 20th day respectively.

Effect of C/N. The carbon to nitrogen (C/N) ratio is the most influencing parameter for the biogas production. It was reported that, the C/N ratio should lie in the range of 20-20:1 (Athanasoulia et al., 2012) [13], for the optimum biogas production. If the C/N ratio is much higher than the range, then the biogas production will be drop. This is because, nitrogen will be consumed rapidly by methanogenic bacteria for meeting their protein requirements, and will no longer react on the left carbon remaining in the sample. If the C/N ratio is very low, it will accumulate in the form of ammonia (NH_3). As a fact, NH_3 will intern increase the pH value of the slurry. A significant amount of release of NH_3 may cause toxic to methanogenic bacteria in the slurry, thus biogas production will be decrease (Dioha et al., 2013) [14]. Fig. 3 depicts the cumulative biogas production at different C/N ratios for a 20 days of digestion time.

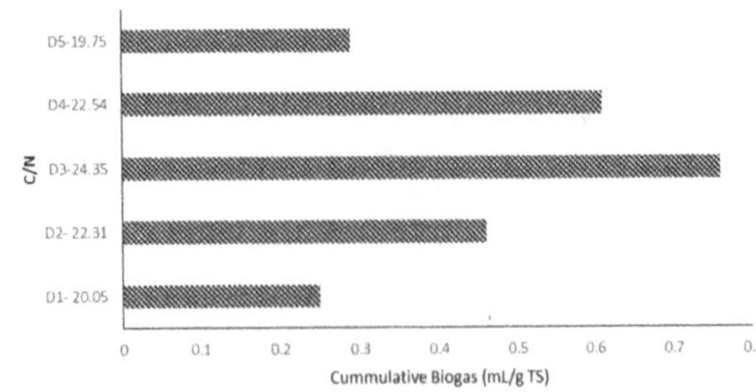

Fig. 3: Cumulative biogas production at different C/N ratio.

It can be observed from the Fig. 3, that the anaerobic digester D3 gives a maximum biogas production of 0.76 ml/g TS than D4 gives 0.61 ml/g TS than D2 gives 0.46 ml/g TS, than D5 gives 0.29 ml/g TS. It can also be noticed that, the lowest yield for the digester D1 is 0.25 ml/g TS. The C/N ratio is important parameter in the sense in the feedstock with the higher value makes more carbon available for biogas production. However, if the nitrogen level is too low, it can effect on microbial activity due to the microorganisms need nitrogen to maintain the growth and metabolic activity. On the other hand, low C/N ratios present high nitrogen levels which can result in ammonia inhibition. The co-digestion of PS content and CD can help to balance and maintain optimum C/N levels and consequently, adequate nitrogen and alkalinity levels (Karki et al., 2021) [15]. The lowest biogas production for D1 might be due to formation of NH3, which may lead to toxic to bacteria. It was also reported by the Khalid et al., 2011, the C/N ratio of 22-25 is the best condition for carrying out anaerobic digestion. The digester D3 has a C/N ratio of 24.35, which is well under the optimum range. But, on the addition of PS more than 30%, the C/N ratio slightly shifts from the optimum range and tends to decrease. Because of which, excessive nitrogen will be liberated that could inhibit the methanogenic bacteria; thus the biogas production decreases (Weerayutsil et al., 2016) [16].

Conclusion

To excess the possibility of using paddy straw for energy use and biogas production, PS was co-digested with CD, and the performance is assessed by considering the input parameters like pH and C/N ratio. A maximum biogas production can be obtained from the co digestion of substrate containing 30% ps and 70% Cd, for a digestion time of 20 days and d3 shows a max pH value of 7.16, which indicates good for anaerobic digestion in biogas production.

Acknowledgment

The authors acknowledge School of Mechanical Technology Engineering, Universiti Malaysia Perlis for the lab facilities. Special thanks to those who contributed to this project directly or indirectly.

References

[1] S. Singh, N. Hariteja, S. Sharma, N. J. Raju, T. J. R. Prasad, Production of biogas from human faeces mixed with the co-substrate poultry litter and cow dung, Environ. Technol. Innov. 23(2021) 101551. https://doi.org/10.1016/j.eti.2021.101551

[2] S. Oranusi, and S. O. Dahunsi, Co-digestion of food waste and human excreta for biogas production, Br. Biotechnol. J. 3(4) (2013) 485-499. https://doi.org/10.9734/BBJ/2013/4476

[3] S. M. Shafie, Paddy residue based on power generation in Malaysia: environmental assessment using LCA approach, J. Eng. Appl. Sci. 10(15) (2015) 1819-6608.

[4] J. Singh, Overview of electric power potential of surplus agricultural biomass from economic, social, environmental and technical perspective : A case study of Punjab, Renew. Sustain. Energy Rev. 42 (2015) 286-297. https://doi.org/10.1016/j.rser.2014.10.015

[5] A. Abraham, A. K. Mathew, R. Sindhu, A Pandey, P. Binod, Potential of rice straw for bio-refining: An overview. Bioresour. Technol. 215 (2016) 29-36. https://doi.org/10.1016/j.biortech.2016.04.011

[6] H. V. Nguyen, C. D. Nguyen, T. V. Tran, H. D. Hau, N. T. Nguyen, M. Gummert, Energy efficiency, greenhouse gas emissions, and cost of rice straw collection in the Mekong River Delta of Vietnam. Field Crop. Res. 198 (2016) 16-22. https://doi.org/10.1016/j.fcr.2016.08.024

[7] B. A. Goodman, Utilization of waste straw and husks from rice production: A review, J. Bioresour. Bioprod. 5(3) (2020) 143-162. https://doi.org/10.1016/j.jobab.2020.07.001

[8] A. Selvarajoo, Y. L. Wong, K. S. Khoo, W. H. Chen, P. L. Show, Biochar production via pyrolysis of citrus peel fruit waste as a potential usage as solid biofuel. Chemosphere, 294 (2022) 133671. https://doi.org/10.1016/j.chemosphere.2022.133671

[9] R. K. Srivastava, Bio-energy production by contribution of effective and suitable microbial system. Mater. Sci. Energy Technol. 2 (2019) 308-318. https://doi.org/10.1016/j.mset.2018.12.007

[10] J. Kainthola, A. S. Kalamdhad, V. V. Goud, Enhanced methane production from anaerobic co-digestion of rice straw and hydrilla verticillata and its kinetic analysis. Biomass Bioenergy 125 (2019) 8- 16. https://doi.org/10.1016/j.biombioe.2019.04.011

[11] N. Zhai, T. Zhang, D. Yin, G. Yang, X. Wang, G. Ren, Y. Feng, Effect of initial pH on anaerobic co- digestion of kitchen waste and cow manure. Waste Manag. 38 (2015) 126-131. https://doi.org/10.1016/j.wasman.2014.12.027

[12] P. Vijayakumar, S. Ayyadurai, K. D. Arunachalam, G. Mishra, W. H. Chen, J. C. Juan, S. R. Naqvi, Current technologies of biochemical conversion of food waste into biogas production: A review, Fuel, 323 (2022) 124321. https://doi.org/10.1016/j.fuel.2022.124321

[13] E. Athanasoulia, P. Melidis, and A. Aivasidis, Optimization of biogas production from waste activated sludge through serial digestion, Renew. Energy 47 (2012) 147-151. https://doi.org/10.1016/j.renene.2012.04.038

[14] I.J. Dioha, C.H. Ikeme, T Nafi'u, N. I. Soba, M. B. S. Yusuf, Effect of Carbon to Nitrogen Ratio on Biogas Production, Int. Res. J. Nat. Sci. 1 (2013) 30-39.

[15] R. Karki, W. Chuenchart, K. C. Surendra, S. Shrestha, L. Raskin, S. Sung, A. Hashimoto, S. Kumar Khanal, Anaerobic co-digestion: current status and perspectives. Bioresour. Technol. 330 (2021) 12500. https://doi.org/10.1016/j.biortech.2021.125001

[16] P. Weerayutsil, U. Khoyun, K. Khuanmar, Optimum ratio of chicken manure and napier grass in single stage anaerobic co-digestion. Energy Proc. 100 (2016) 22-25. https://doi.org/10.1016/j.egypro.2016.10.141

Performance of Thermosiphon Effect Using Different Flat Plate Collector of Solar Water Heater

M.M. NORUL AHSANAH AULIA[1,a*], A. MARIAH[1,b], H. NURAMIDAH[1,c], D.S. ZALEHA[2], M.Y. FATIMAH[1], A.R.M. ASHRAF[1], M.M. KAMARUZAMAN[1] and A. FARIDAHANIM[3]

[1]Faculty of Engineering Technology, Universiti Tun Hussein Onn Malaysia, Kampus Pagoh, 84600 Muar, Johor, Malaysia

[2]Faculty of Built Environment & Surveying, Universiti Teknologi Malaysia, 81310 Johor Bahru, Johor, Malaysia

[3]Faculty of Civil Engineering, Universiti Teknologi Malaysia, 81310 Johor Bahru, Johor, Malaysia

[a]auliamahani98@gmail.com, [b]mariah@uthm.edu.my, [c]nuramidah@uthm.edu.my

Abstract. Solar thermal energy has become one of the most sustainable and low-carbon technologies for empowering our future water heater systems. Nowadays, water heating is the second biggest portion of residential energy use after air conditioning systems. Previous researcher study a wide range of solar water heater designs have been suggested and produced within two methods whereas forced circulation using a mechanical pump and natural circulation without the use of a pump also known as thermosiphon solar water heater. This research was to analyse most efficient Flat Plate Collector (FPC) in terms of heat absorption on Thermosiphon Effect due to energy collection by different colour, material and glazing. This experiment uses Solar Energy Trainer to compare the impact of several forms FPC on heat absorption of thermosiphon effect in solar water heater. The most efficient FPC in terms of colour is black collector with higher efficiency of 9.3% compared to white while most efficient FPC in terms of material and glazing are copper and double glazing collector with 10.86% and 11.15% respectively. The experiment and analysis show that various types of FPC with good heat absorber have higher efficiency.

Introduction

Solar thermal energy is one of the energy circumscribe sustainable and low-carbon technologies which will play a critical role in these years ahead. The thermal energy can be used directly as heat for domestic or industrial uses, depending on the range of temperature and production scale [1]. The increased energy use for heating is having a significant impact on the national system, causing an oversupply of electricity and negatively impacting electricity prices [2]. The solar water heater can efficiently convert solar energy into concentrated heat, with the features of established technology, low global climate change danger, and reduced life cycle cost [3]. Solar water heaters (SWHs) are the most practical solar energy harvesting devices for both household and industrial use [4]. Solar collectors and heat storage tanks are compulsory among the solar system part. When it is available, the storage tank performs an important function in the solar power system, delivering heat when it is needed [5]. The water (working fluid) is cycled between the collector and the tank using one of two methods: forced circulation using a mechanical pump or natural circulation without the use of a pump. Water heating systems in most properties still largely depend on electricity from the power grid for heating water due to low efficiency of solar thermal collectors and inefficient functioning of the heating system [6].

Flat plate collectors are commonly implemented for water heating because of their simple construction. The flat plate collector's lower efficiency is caused by the fact that there are comparably fewer regions that receive direct sunlight [7]. Therefore, first problem statement is low efficiency of flat plate collector due to small area of receiving solar radiation. The objectives to be achieve is analyse most efficient Flat Plate Collector in terms of heat absorption on Thermosiphon Effect due to energy collection by different colour, material and glazing. Moreover, solar water heater with natural circulation (thermosiphon effect) is frequently favoured due to its simplicity for rural places where access to professional workers and electricity is limited [8]. Solar water heater performance can be improve by increasing the efficiency of solar flat plate collector. This study conduct experiment using various types of FPCs to measure which are the most efficient to be used. Its features are well-known, and it is the simplest and least expensive to build, operate, and maintain when compared to other collector types. FPC can process adequate high temperature of heat for commercial use such as swimming pools, residential and domestic hot water [9]. A solar water heater is a device that captures and uses solar heat energy to provide hot water for part or all of a household's needs [10]. Water heating accounts for around 20% to 30% of total energy usage in a housing complex. As a result, utilizing solar water heaters on a yearly basis can supply 70% of the energy required for water heating [11].

Moreover, solar water heater can effectively transform solar energy to concentrated heat, with the benefits of proven technology, distinct advantage of producing clean and sustainable energy, and economic feasibility [3, 12]. The three core elements that build up the device are solar collectors, a storage tank, and a heat exchanger device that transfers the heat generated from the solar collectors to the water tank [12]. Several research of different types for solar water heater systems were conducted in various regions [14,15,16]. Basically, thermosiphon SWHs are commonly applied which solar energy delivers energy to the system in the collector absorber. Temperature differences cause a density difference, and water circulates naturally (thermosiphon effect), with hot water rising and cold water going down [13]. Unbalanced gravitational forces come from the temperature of the water in the collector rising and the density falling, which causes fluid migration into the storage tank. All components of the loop flow because water is nearly incompressible and continuity is guaranteed [14]. The appropriate operation of a thermosiphon effect solar collector is driven by a range of interrelated parameters, including solar radiation, ambient temperature, the water flow rate through collectors, the container structure (vertical or horizontal), component materials and certain other aspects [5]. In need for the thermosiphon phenomenon to work in a thermosiphon system, the water tank must be higher than top of the solar collectors. It is necessary to initiate and sustain temperature stratification in the tank to improve thermal efficiency [15]. This system also come with flat plate collectors which is an instrument that capture solar thermal energy. Flat plate collectors are the most frequent and also the oldest type of collector [16]. It was previously reported that the collector's design has a significant impact on its thermal performance [17]. The elements impacted include flow tube, absorber plate and glazing cover [18]. The sun's energy absorbed was converted into heat and then transferred to the heat transfer fluid where most commonly is water before being carried to storage tank or used [19].

Methodology

This chapter describe method for experimental setup of flat plate collector (FPC) in investigating the influence of heat absorber colour, material, and glazing on the collector efficiency of thermosiphon effect on Solar Water Heater. This experiment uses Solar Energy Trainer which may indeed easily move to a location with direct thesis sunlight.

Experimental Setup. Experimental will be conduct in three terms of thermosiphon effect on solar water heater using different collector plates, which are colour absorption capacity level, material absorption capacity level, glazing effect on heat absorption. It is importance to understand the working principle of thermosiphon system. It is also important to understand how heat is transferred by thermal conduction. The influence of colour, material and glazing on heat absorption experiment was to see if there were any significant temperature variations between the collector panels. The data will be collected on five distinct days for each collector, with the average values being used in the analysis. The entire field experiment will start at 9:00 a.m. and end at 4:00 p.m. During the experiment, the Solar Energy Trainer will automatically tabulate the water tank (T1), water temperature at inlet (T2), and water temperature at output (T3) every 15 minutes onto the computer. The data for ambient temperature (T) will be taken every hour using a 4 in 1 meter. After all of the data has been collected, the efficiency of FPC will be calculated. The current study examines the performance of a flat plate solar collector in response to a variety of features as well as energy efficiency reported by [20]. The energy that calculates percentage of the incoming radiation delivered as usable energy to the working fluids can be used to measure the flat plate collector's efficiency [21]. The incident solar energy is transformed to heat and transported to a medium like water in the flat plate collector [22]. Besides, [23] stated that on the top surface of the solar collector, a continuous heat flux (solar irradiance) is supplied. In this experiment, the solar radiation intensity, I = 722 W/m2 was compared to a study by [24]. This is due to the fact that [24] used the same methods as this experiment to be conducted in Malaysia. The thermosyphonic mass flow rate is quite difficult to quantify using conventional equipment. The current work proposes a new approach for measuring mass flow rate based on collector intake and outlet temperatures, water dynamic viscosity, and riser diameter. Aside monitoring the temperature of the collector's input and output, the equation may then be used to estimate the thermosyphonic mass flow rate. The suggested equation looks like this:

$$\dot{m} = 189.498 * D * \mu * \left(\frac{T_{out}}{T_{in}}\right)^{1.19} \quad (1)$$

Where, ṁ = mass flow rate, kg/s, D = riser diameter, m, μ = dynamic viscosity, Pa.s, Tout = Outlet water temperature, °C, Tin = Inlet water temperature, °C. The instantaneous solar collector's efficiency [23] in terms of inlet and outlet water temperature is calculated by the expression:

$$\eta = \frac{\dot{m}C_p\,(T_o - T_i)}{IA_c} \quad (2)$$

Where, η = Thermal efficiency, ṁ = mass flow rate, kg/s, Cp = Specific heat of water, kJ/kg.°C, To = Outlet water temperature, °C, Ti = Inlet water temperature, °C, I = Solar insolation, W.m-2
Ac = Effective collector area, m^2

Results and discussion

Colour Absorption Capacity Level. As for colour absorption capacity level experiment, figure below show the average temperature of the collectors with time on five different days started from 14th to 18th August 2022 (black) and 11th to 15th September 2022 (white).

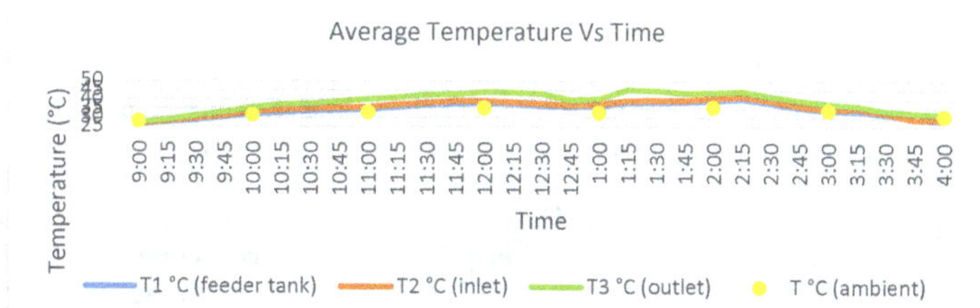

Fig 1: Average temperature of water with time for black plate collector.

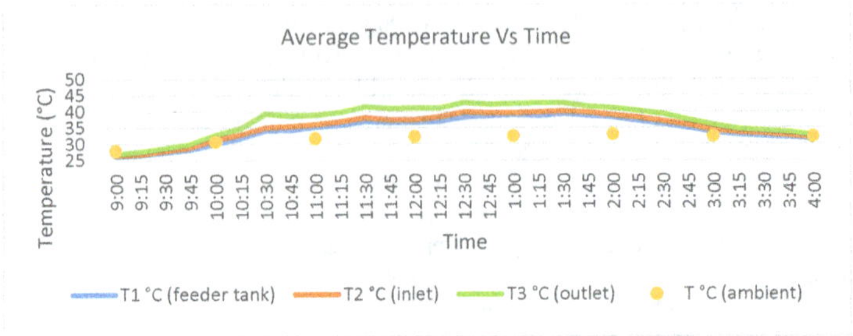

Fig. 2: Average temperature of water with time for white plate collector.

Fig. 1 and Fig. 2 show average temperature of water with time for colour absorption capacity level black versus white collector experiment. In the morning and evening the ambient temperature found to be low and humid with air and wind movement for both collector. The temperature in both figure started to increase after 9:00am due to the increases of solar radiation supplied until 2:00pm. Both plate collectors had average temperature decrease after 2:00pm. Temperature decreases happened due to the weather as it plays an important role during the experiments. The weather's high solar radiation and beginning water temperature are also comprehensible given that the effective supply volumes are linked to the same months but different forms of heating [25]. There also less exposure of sunlight in evening due to location of Solar Energy Trainer that been blocked by tall campus building. Solar Energy Trainer was placed and tested in an open space but due to limit in cable length which need to be connect to an electrical source, it in some way need to be placed nearby building. The impacts of ambient temperature, humidity, and wind speed were fully considered [26]. This research simplified that the black collector absorbs more solar heat energy. It is the best energy absorber for both light and heat, making it an excellent option for solar water systems. All light waves are absorbed by black collector. White collector reflect all visible light wavelengths, absorbing the least heat.

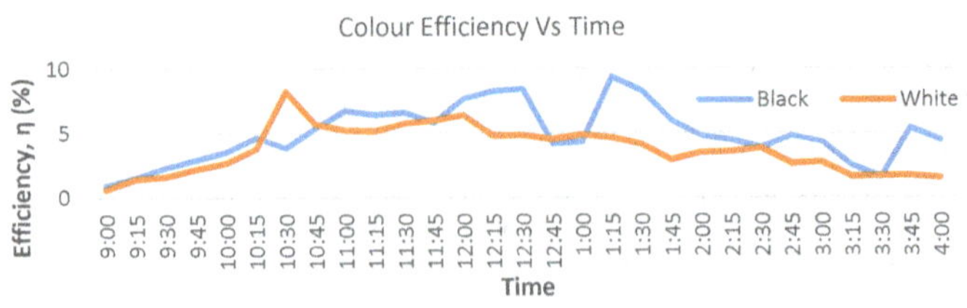

Fig. 3: Efficiency with time for black and white flat plate collector.

When the water flow rate in the plate collector tube is higher, the collector is more effective. For example, the efficiency of black plate collector rises progressively until it reaches its peak at 1.15 pm with 9.3% and start to fall afterward. In between 12.30 pm to 1.15 pm, there was an overcast weather that lead to drop or relatively low efficiency with 4.12% (12.45 pm) and 4.27% (1.00 pm) for black collector plate. The efficiency for white plate collector was high especially at 10.30 am with 8.2% and drop afterward until 4:00 pm. The decreases of efficiency in white collector was happened because of weather as the month in rainy season and poor absorber of heat. The decreases in efficiency for both black and white collector after 2:00 pm was occurred as a result of the Solar Energy Trainer's location. Therefore, the efficiency of FPC grows along with an increase in solar radiation.

Material Absorption Capacity Level. As for material absorption capacity level, the graphs display average temperature of collectors over five days, which were measured on 25 to 29 September 2022 (copper) and 18 to 22 September 2022 (polypropylene).

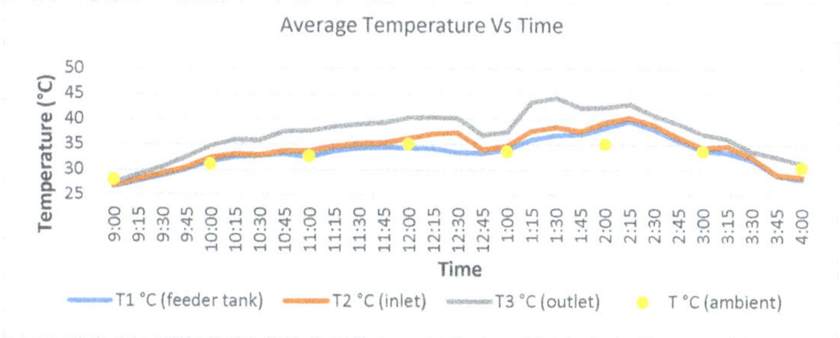

Fig. 4: Average temperature of water with time for copper plate collector.

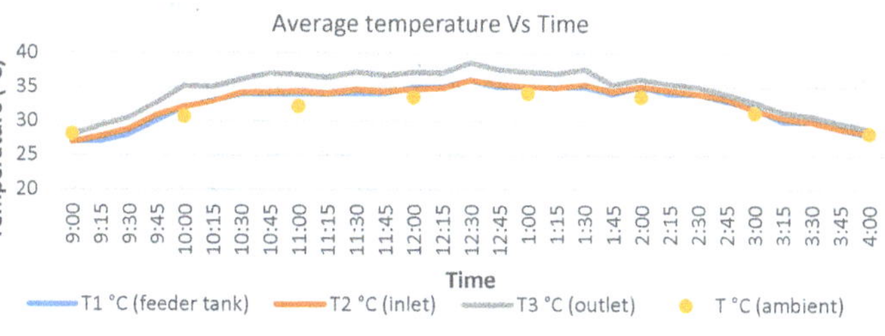

Fig. 5. Average temperature of water with time for polypropylene plate collector.

For material absorption capacity level experiment, the graphs in Fig.4 and 5 show variety average temperature of feeder tank (T1), inlet (T2), outlet (T3), and ambient temperature (T) versus time of collectors. From Fig.4, it shown that average temperature at outlet (T3) had significantly higher value compared to the other average temperature. This data from collector capacity level performance was correlated to temperature rise in ambient temperature (T) since (T) can reduce operational efficiency leading to heat loss. The average temperature values in Fig.5 were lower when compared to the average temperature value from Fig.4. From both Fig.4 and Fig.5, all temperature involved were relatively low during the morning in respect of low solar radiation. Then, as a result of the increased sun radiation, temperature of the water began to rise between 9:00 am and 2:00 pm. Maximum peak for temperature especially (T3) occurred differently for both plate collector due to the material condition. This experiment can be summarised as copper collector absorbs more solar heat energy. This occurs because copper has high heat-conducting properties than polypropylene with a weak ability to conduct heat.

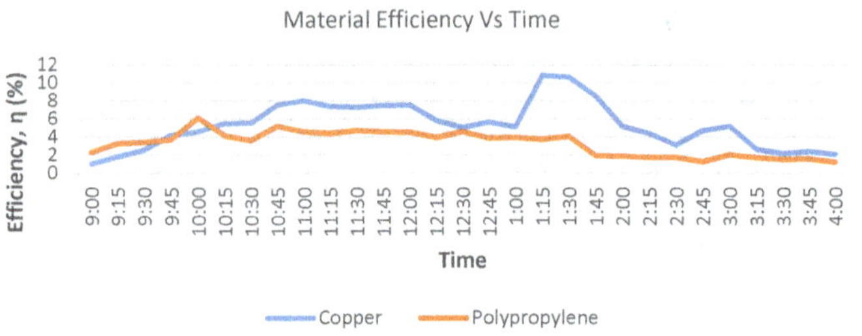

Fig. 6. Efficiency with time for copper and polypropylene flat plate collector.

The FPC is more effective when the water flow rate, or velocity at which the water moves through the tube, is higher. The efficiency for both plate collector will increases gradually with time until reach the maximum efficiency. But, happened instead was both copper and polypropylene plate collector had inconsistency graph of efficiency with time for average of five days. The inconsistent result was happened due to the unstable and overcast weather during day of the experiments take placed. Copper flat plate collector had gradually increased in efficiency with time, but had a few falls before reach the maximum efficiency at 1.15 pm with 10.86%. Meanwhile, polypropylene plate collector efficiency with time started higher than copper and continue increasing until 10.00 am with 6.1% before decreases moderately afterward. The fundamental reason for the investigation is variation in solar irradiation values.

Glazing Effect on Heat Absorption. Experiment for glazing effect on heat absorption, the graphs depict the average of collectors over five different days started from 21th August 2022 for single glazing cover and double glazing cover started from 27th August 2022.

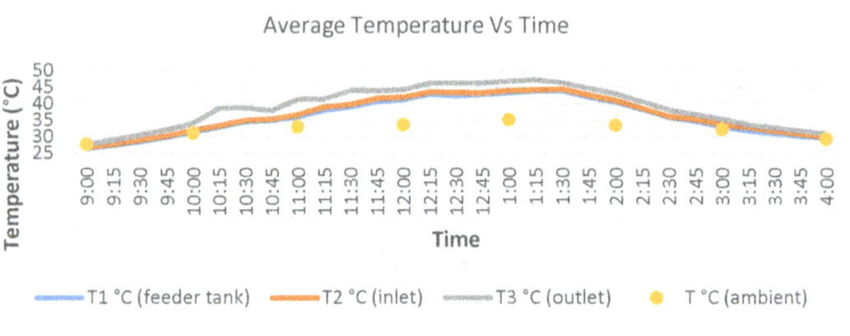

Fig. 7. Average temperature of water with time for single glazing plate collector.

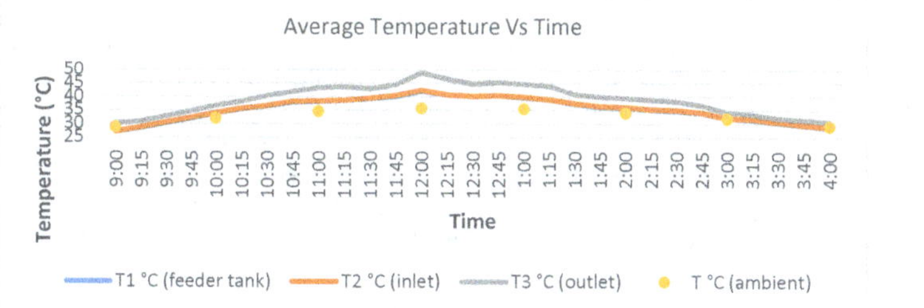

Fig. 8 : Average temperature of water with time for double glazing plate collector.

The graphs in Fig. 7 and Fig. 8 depict the variation of water temperature of glazing effect on heat absorption experiment at feeder tank (T1), inlet (T2), outlet (T3), and ambient temperature (T) versus time on five different days. Both Fig. 7 and Fig. 8 show that there was low solar radiance as the water temperature started low during 9:00am. The temperature started to increase afterward due to increases of solar radiation supplied. Single glazing cover increases progressively until reach the highest temperature at 1.15 pm with 47.2°C and decreases gradually later. Both single and double glazing cover had average temperature decrease after both reach the maximum average temperature in a day. The average temperature at outlet (T3) for double glazing cover had higher temperature value. The temperature started to increase subsequently until 12:00 pm with 48.8°C due to solar radiation. Average temperature in double glazing cover reach it highest temperature in a day early in the afternoon. It can be summed up as double glazing cover collector captures higher heat energy from sunlight. The amount of heat loss from the collector will decrease as the number of covers increases, resulting in the flat plate collector operating at its highest efficiency.

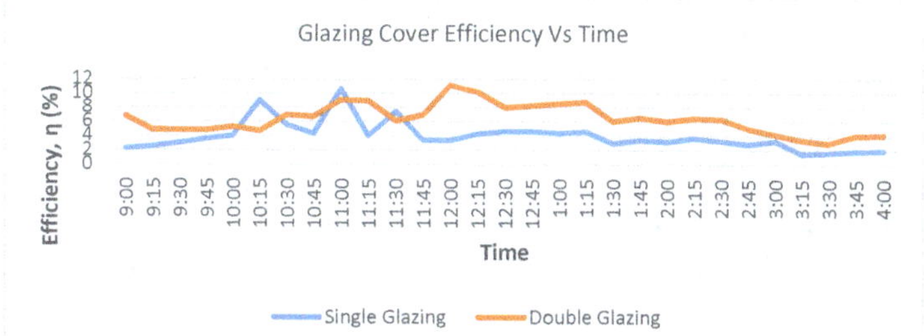

Fig. 9. Efficiency with time for single glazing and double glazing plate collector.

The FPC is more effective when the water flow rate, or velocity at which water moves through the tube is higher. Single glazing cover had an acutely increasing efficiency at 10:15am and a fluctuation graph eventually. Double glazing cover collector started with high efficiency at 9:00am but decreases a bit later due to the dim weather. The efficiency increases constantly to maximum efficiency before decreases later. In previous study, according [27] for single glazing cover collector highest efficiency was 12% while in the current study, the single glazing cover collector maximum efficiency is 10.6%. While the maximum efficiency for double glazing cover collector in the previous report [27] was 20.8%, the maximum efficiency for double glazing cover collector in this research was 11.15%. It can be concluded that water temperature and heat absorption increases when the efficiency of the flat plate collector increases.

Conclusion

In conclusion, as for the colour collector type, black collector absorbed more heat compared to the white collector. This resulted in high efficiency of black than white collector. Copper collector absorbed more heat than polypropylene collector. Make the copper collector efficiency highest than polypropylene collector. Lastly, double glazing collector absorbed more heat and have less heat loss and make double glazing collector the most efficient compared to single glazing collector. Therefore, a thorough experiment was conducted to simulate the FPC temperature and forecast its efficiency. The experiment and comparison show that various types of FPC have varying degrees of efficiency.

Acknowledgment

This research was supported by Ministry of Education (MOE) through Fundamental Research Grant Scheme (FRGS/1/2020/TK0/UTHM/03/35).

References

[1] S.Fereres, Solar Thermal Energy Systems. (2018). https://doi.org/10.2174/97816810871911l8030003

[2] S. B. Amin and N. Nuarey Mithila, J. Electron. 1-47 (2017).

[3] T. S. Ge, R. Z. Wang, Z. Y. Xu, Q. W. Pan, S. Du, X. M. Chen, T. Ma, X. N. Wu, X. L. Sun and J. F Chen, Renew. Energy 126, 1126-1140 (2018). https://doi.org/10.1016/j.renene.2017.06.081

[4] B. Freegah and A. A. Al-Tabbakh, UPB Sci. Bull. Ser. D Mech. Eng. 81, 117-132 (2019).

[5] Y. Amirgaliyev, M. Kunelbayev, W. Wójcik, B. Amirgaliyev, A. Kalizhanova, O. Auelbekov, N. Kataev and A. Kozbakova, J. Ecol. Eng. 19, 176-181 (2018). https://doi.org/10.12911/22998993/91882

[6] W. T. Li, W. Tushar, C.Yuen, B. K. K.Ng, S. Tai and K. T. Chew, Energy Build, 224 (2020).

[7] T. Rajashekaraiah, M. J. Mura and G. S. Sharma, "Optimizing the Efficiency of Solar Flat Plate Collector with Trapezoidal Reflector" (AIP Publishing, 2019), p. 030008 https://doi.org/10.1063/1.5092911

[8] M. Feilizadeh, M. R. K. Estahbanati, M. Khorram and M. R. Rahimpour, Renew. Energy 143, 328-334 (2019). https://doi.org/10.1016/j.renene.2019.05.013

[9] B. K. Gond, S. Mittal, P. Prajapati and R. Khare, Analysis of Solar Flat Plate Collector 105-110 (2016).

[10] O. Touaba, M. S. Ait Cheikh, M. E. A. Slimani, A. Bouraiou, A. Ziane, A. Necaibia and A. Harmim, Sol. Energy 199, 630-644 (2020). https://doi.org/10.1016/j.solener.2020.02.064

[11] R. Alayi, N. Khalilpoor, S. Heshmati, A. Najafi and A. Issakhov, Int. J. Photoenergy (2021). https://doi.org/10.1155/2021/6838138

[12] A. Jamar, Z. A. A.Majid, W. H. Azmi, M. Norhafana and A. A. Razak, Int. Commun. Heat Mass Transf. 76,178-187 (2016). https://doi.org/10.1016/j.icheatmasstransfer.2016.05.028

[13] Y. M Liu, K. M. Chung, K. C. Chang and T. S. Lee, Energies 5, 3266-3278 (2012). https://doi.org/10.3390/en5093266

[14] A. Brew-Hammond and F. O. Akuffo, "A Review of Experimental Studies on Traditional Thermosiphon Solar Water Heater.Pdf," (1989), p. 12.

[15] I. Michaelides, P. Eleftheriou, G. Roditis, P. Kyriacou and G. A. Siamas, "On the Night Heat Losses in Thermosyphon Solar Water Heaters," (2016), pp. 1-7. https://doi.org/10.18086/eurosun.2010.16.22

[16] K. Murari and R. Chaurasiya, Renew. Sustain. Energy Rev., 67, 641-650 (2017). https://doi.org/10.1016/j.rser.2016.09.078

[17] R. Shukla, K. Sumathy, P. Erickson and J. Gong, Renew. Sustain. Energy Rev., 19, 173-190 (2013). https://doi.org/10.1016/j.rser.2012.10.048

[18] N. Ikmal, S. Azha, H. Hussin and M. S. Nasif, "Thermal Performance Enhancement in Flat Plate Solar Collector Solar Water Heater : A Review," (2020), p. 14.

[19] S. Chowdhury and B. Salam, "Flat Plate Solar Collector Using Nanofluid" (2019).

[20] S. K. Verma, A. K. Tiwari and D. S. Chauhan, Energy Convers. Manag., 134, 103-115 (2017). https://doi.org/10.1016/j.enconman.2016.12.037

[21] B. A. Sup, "EFFECT OF ABSORBER PLATE MATERIAL ON FLAT PLATE COLLECTOR EFFICIENCY," (2010).

[22] H. Dagdougui, A. Ouammi, M. Robba and R. Sacile, Renew. Sustain. Energy Rev., 15, 630-638 (2011). https://doi.org/10.1016/j.rser.2010.09.010

[23] D. G. Gunjo, P. Mahanta and P. S. Robi, Renew. Energy 106, 24-36 (2017). https://doi.org/10.1016/j.renene.2016.12.041

[24] R. M. B. Nurhanis, "THE COMPARISON OF PERFORMANCES BY USING COLOUR , MATERIAL AND GLAZING OF SOLAR FLAT PLATE COLLECTOR," (2019), p. 140.

[25] M. M. Dhongadi, S. S. Sajjan, Turkish J. 12, 211-220 (2021).

[26] T. Zhang, Z. W. Yan, L. Y. Wang, W. J. Zheng and Y. H. Su, Sol. Energy 197, 433-442 (2020). https://doi.org/10.1016/j.solener.2020.01.019

[27] H. Vettrivel and P. Mathiazhagan, Int. J. Renew. Energy Res., 7, 267-274 (2017).

Transformation of Cellulose Allomorph of *Musa acuminata balbisiana* Peel Using Ethylenediamine

A.H. JASNI[1,a]*, A.S. AZMI[1,b], N.I. MOHAMAD PUAD[1,c], F. ALI[1,d] and Y. AHMAD NOR[1,e]

[1]Department of Chemical Engineering and Sustainability Kulliyyah of Engineering International Islamic University Malaysia, Jalan Gombak 53100 Kuala Lumpur, Malaysia

[a]ahawa.jasni@live.iium.edu.my, [b]azlinsu76@iium.edu.my, [c]illi@iium.edu.my, [d]fathilah@iium.edu.my, [e]yusilawati_ahmadnor@iium.edu.my

Keywords: Agricultural Waste, Ethylenediamine, Structural Analysis, Cellulose Allomorph

Abstract. Despite the high cellulose, hemicellulose, and lignin compounds in agricultural wastes of Banana Peel (BP), the effects of ethylenediamine (EDA) on the cellulose transformation were unclear. But to be able to use these lignocellulosic substrates efficiently for fermentation by *Rhizopus sp.* as substrates, a pretreatment process is required. This study will be the first time to use EDA in treating waste of BP to be used in lactic acid (LA) production by *Rhizopus sp.* later. Furthermore, the pretreatment method using EDA is needed as a simple process to achieve high enzymatic digestibility and fermentable sugar accessibility of the utilized agricultural wastes. The main objective of this study is to determine the optimum pretreatment conditions of lignocellulosic substrates for maximum cellulose allomorph conversion and lignin removal in BP pretreated waste. The transformation of the cellulose was characterized in this paper via scanning electron microscope (SEM), X-ray diffractometer (XRD) and Fourier transform infrared spectroscopy (FTIR). It was found that BP contain high cellulose III of 19.86 % after EDA pretreatment thus, potential ready accessible fermentable sugars.

Introduction

Among other pretreatment techniques, alkaline pretreatment is one of the most powerful methods that can improve the enzymatic hydrolysis of fibers by disrupting lignin and breaking the bond of lignin carbohydrates [1]. This method also showed the advantages of effectively reducing the crystallization rate by swelling, excluding the acetyl group and uronic acid replacements from hemicelluloses, and breaking down the ester bonds. The main hurdle of inorganic alkaline pretreatment is the release of large amounts of salt from black liquor. To overcome this problem, biomass was pretreated by the organic amine ethylenediamine (EDA) [2]. Additionally, EDA has been generally used in converting cellulose allomorph as well as ammonia. Cellulose Iβ and cellulose Iα are the official allomorphic forms of cellulose found in plants and microbes, respectively [3].

EDA can enter the natural crystalline cellulose, it cracks the hydrogen bonds between adjacent cellulose chains, make new hydrogen bonds between cellulose and EDA, and transform cellulose I to EDA–cellulose I complex. After eliminating EDA by drying under vacuum or washing with polar/non-aqueous solutions (e.g., ethanol), the hydrogen bonds rebuild between cellulose unit [3]. When compared to other pretreatment techniques like dilute acid, EDA has a number of advantages: it can be carried out at ambient pressure, which lowers process costs; it can be done without the addition of water (a dry-to-dry process), which avoids solid-liquid separation after pretreatment and reduces energy consumption during pretreatment; it protects hemicellulose; and it does not release furans, which are a fermentation inhibitor [4].

Methods

Preparation and Pretreatment of Substrate by EDA Aqueous Solution. The BP (*Musa acuminata balbisiana*) was collected from the local grocer in Pasar Borong Selayang, Selangor, Malaysia. The substrate was dried in an oven at 60°C for three days, then milled (< 0.25 mm). The substrates were mixed separately with dilute EDA 1, 2.25, 5.5, 7.75 and 10 % solutions in glass beakers. The solid-to-liquid ratio was kept at 1:10. Aluminum Foil was covered on the beakers. Then the mixture was heated by an oven under ambient pressure and at different temperatures (30, 60 and 90°C). After 1 h of pretreatment, the solid was recovered by filter paper without any washing steps. Then the substrates were dried at 60°C overnight. The substrates were stored in screw cap test tubes prior to analytical characterization.

X-Ray Diffraction Analysis (XRD). The crystalline index (CI) and cellulose allomorph (CA) had been estimated using a D2 Phaser Bruker Benchtop X-ray Diffractometer. Samples were milled and scanned at a speed of 0.250 time/step, range from 2θ = 10-40°. The samples were put on a quartz sample holder, and CuKα radiation operated at 30 kV, 10 mA. PeakFit (SeaSolve Software Inc.) software was to deconvolve the resulted diffractogram. Curve fitting analysis was carried out using Gaussian functions. The content of amorphous cellulose, cellulose I, and cellulose III will be calculated by using [3] equations:

$$I\% = \frac{A_I}{A_{AM}+A_{III}+A_I} \times 100\% \, I\% = \frac{A_I}{A_{AM}+A_{III}+A_I} \times 100\% \quad (1)$$

$$III\% = \frac{A_{III}}{A_{AM}+A_{III}+A_I} \times 100\% \, III\% = \frac{A_{III}}{A_{AM}+A_{III}+A_I} \tilde{A} - 100\% \quad (2)$$

$$AM\% = \frac{A_{AM}}{A_{AM}+A_{III}+A_I} \times 100\% \, AM\% = \frac{A_{AM}}{A_{AM}+A_{III}+A_I} \times 100\% \quad (3)$$

where AM %, III %, and I % are the contents of amorphous cellulose, cellulose III, and cellulose I in samples, respectively;. AAM, AIII, and AI are the peak areas of amorphous cellulose, cellulose III, and cellulose I, respectively. The Bragg angles of peak (004), (020), (1$\bar{1}$0), and (110) belonging to cellulose I are 34.5°, 22.3°, 16.3°, and 14.8°, respectively. The Bragg angles of peak (020) and (110) belonging to cellulose III are 20.0° and 11.3°, respectively. The Bragg angle of the amorphous peak is around 20.5°. The crystallinity index CI is calculated by equation (4) [5]:

$$CI = \frac{I_{002}-I_{AM}}{I_{002}} \times 100\% \, CI = \frac{I_{002}-I_{AM}}{I_{002}} \times 100\% \quad (4)$$

where IAM is the peak intensity at 18.5° and I002 is the peak intensity at 22.5°.

Fourier Transform Infrared Spectroscopy (FTIR). The cellulose, hemicellulose, and lignin content of untreated and pretreated substrate had been determined by FT-IR Spectrometer INVENIO - Bruker Spectrometer equipped with an Attenuated Total Reflectance (ATR) sampling accessory. Samples (20 mg) were pressed uniformly against the crystal surface via a spring anvil. Each sample's spectrum was recorded separately by averaging 32 scans collected from wavenumbers of 4,000 to 400 cm^{-1} at a resolution of 4 cm^{-1}. OPUS BRUKER software was used to record and analyze the FTIR-ATR data.

Scanning Electron Microscope (SEM). A scanning electron microscope (SEM) brand Jeol JSM IT100 in Department of Materials Engineering in Kuliyyah of Engineering had been used to evaluate the microstructures of the untreated and pretreated substrates.

Results and Discussion

The CI/cellulose ratio increased from 0.47 in untreated BP to maximum value of 1.69 after pretreatment in Table 1. It illustrated that while the crystallinity of the pretreated BP increased due to the removal of amorphous components, the crystallinity of cellulose itself decreased as the cellulose content dramatically increased. The deconvolution of an X-ray diffractogram was used to determine the change of allomorphic forms between different EDA pretreatment settings. The calculated curves were remarkably like the observed data (R^2 was greater than 0.95 for each sample). Untreated substrate cellulose was found to be cellulose I, with typical peaks (004), (020), (10) and (110) with Bragg angles of roughly 34.5°, 22.3°, 16.3°, and 14.8°, respectively (Fig. 1) maximum intensity of around 21°, the amorphous cellulose peak was substantially broader. Untreated BP contained 51.69 % cellulose I, and 48.31 % amorphous cellulose, as determined by peak areas with CI values. Then the CI increased from 24.37 % to 38.06 %. The increase of CI could be due to the removal of amorphous components (hemicellulose and lignin) by EDA pretreatment [5,6]. The highest cellulose III content in pretreated BP was at 19.86% was achieved at 60°C, 5.5%, EDA.

Table 1: Crystalline cellulose, amorphous cellulose, and crystallinity index contents of untreated and pretreated BP at different pretreatment conditions.

No.	EDA (%)	Temp. (°C)	BP				
			I (%)	III (%)	AM (%)	CI (%)	CI/ Cellulose
1	0	0	51.69	0	48.31	24.37	0.47
2	1	30	35.45	7.35	57.2	20.32	0.47
3	1	60	31.62	5.72	62.66	23.79	0.64
4	1	90	58.13	0.24	41.63	27.01	0.46
5	2.25	30	24.35	7.32	68.33	38.6	1.58
6	2.25	60	32.24	4.90	62.86	38.56	1.19
7	2.25	90	34.61	3.45	61.94	39.42	1.14
8	5.5	30	32.24	6.08	61.68	12.33	0.38
9	5.5	60	32.41	**19.86**	47.73	16.56	0.51
10	5.5	90	47.01	4.02	48.97	35.87	0.76
11	7.75	30	45.22	7.06	47.72	38.06	0.84
12	7.75	60	28.59	8.18	63.23	48.6	1.69
13	7.75	90	50.89	9.45	39.66	49	0.96
14	10	30	45.08	4.77	50.15	23.2	0.47
15	10	60	38.45	9.14	52.41	26	0.55
16	10	90	27.16	0.34	72.5	32.48	0.89

Materials Research Proceedings 68 (2026) 204-209 https://doi.org/10.21741/9781644904237-27

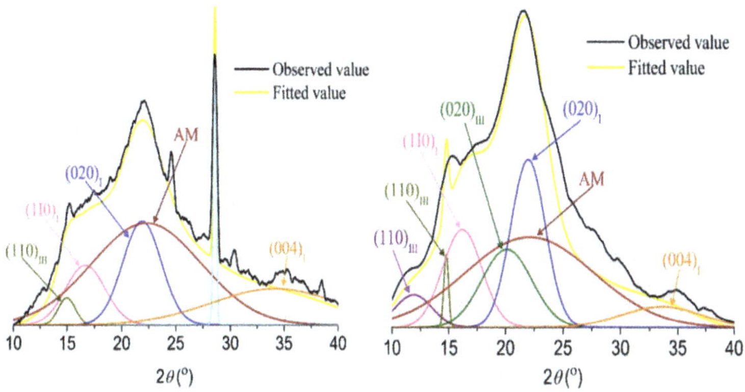

Fig. 1: X-ray diffraction profiles. A) Untreated BP and pre-treated BP, respectively.

All the substrates showed almost similar peak of FTIR after pretreatment with elevated peak under different EDA concentrations and obviously a tremendous difference in the untreated sample based on Fig. 2 and the functional groups can be referred in Table 2.

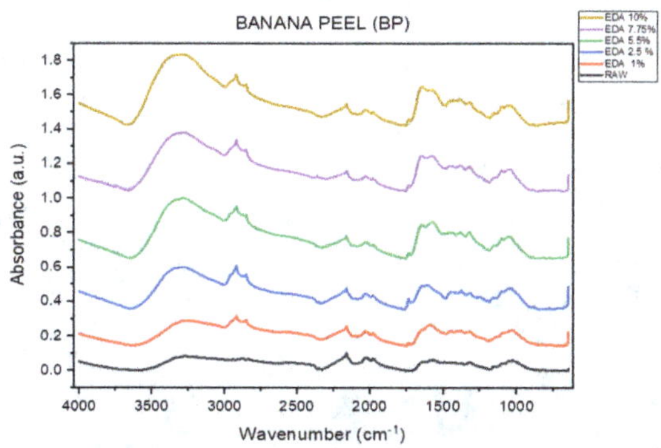

Fig. 2: FTIR spectra of BP in at different EDA concentrations.

Table 2: FTIR frequency range and functional groups present in BP samples.

NO.	SAMPLE BP	FUNCTIONAL GROUP	REMARKS AFTER EDA PRETREATMENT
1	3400-3458.96	Hydroxyl (O-H)	Indicating the explosion of cellulose and hemicelluloses after EDA treatment [7].
2	2849.70-2962.21	Aliphatic (C-H)	Referring to increased cellulose content due to the removal of lignin and hemicelluloses after EDA treatment.
3	1643.41-1643.43	Alkene (C≡C) & Carboxylic (C=O)	Corresponding to the acetyl groups (C=O) of hemicellulose and ester connecting with lignin) decreased after pretreatment, indicating that a part of lignin has been removed after EDA treatment.
4	922.46-1147.79	Alkane or alkyl (C–H)	Referring to the increased of cellulose content due to the removal of lignin and hemicelluloses and the destruction of glycosidic bonds and the bonding of EDA molecules with cellulose and hemicellulose after EDA treatment.

The SEM images (Fig. 3) of untreated substrate showed highly ordered smooth and intact surfaces. In contrast, the surfaces of EDA pretreated substrates were rougher, which might be due to high lignin and hemicellulose elimination, which increased the cellulose-exposed surface area and roughness [5]. The pretreated substrates were also covered with volcano-like holes. The gasification of EDA probably formed these holes in the biomass during the drying process [3]. These holes increased the structural porosity, which might improve enzymatic accessibility.

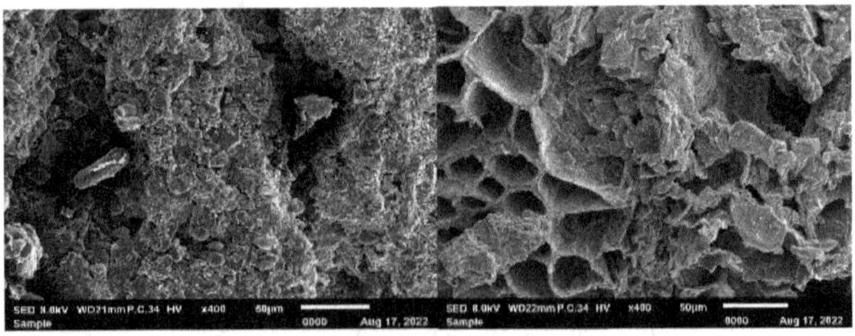

Fig. 3: SEM images. A] untreated BP and B] pretreated coconut husk at 5.5% EDA under 60, respectively.

Conclusion

This study is helpful in enhancing the quality of fermentation substrates produced where the objective is to repurpose discarded agricultural waste of BP through upcycling laboratory research. The results also show that by using EDA, the significant cellulose changes can be obtained to calculate CI value with reasonable accuracy. From the XRD and FTIR results, temperature and EDA concentration is the most significant factor influencing the cellulose I to cellulose II transformation. The optimal processing parameter obtained was at the temperature of 60 °C and 5.5% of EDA which it happened to transform the highest amount of cellulose III in BP.

Acknowledgment

This study was financially supported by the Ministry of Higher Education Malaysia (MOHE) (FRGS/1/2021/TK0/UIAM/02/19).

References

[1] L. Qin, X. Li, J. Q. Zhu, W. C. Li, H. Xu, Q. M Guan, M. T. Zhang and B. Z. Li, Y.J. Yuan, Ind. Crops Products, 102, 51-57.2 (2017). https://doi.org/10.1016/j.indcrop.2017.03.026

[2] H. Chen, W. Huo, B. Wang, Y. Wang, H. Wen, D. Cai, Ch. Zhang, Y. Wu and P. Qin, Ind. Crops Products, 141,111749 (2019). https://doi.org/10.1016/j.indcrop.2019.111749

[3] L. Qin, X. Zhao, W. C. Li, J.Q. Zhu, L. Liu, B. Z. Li and Ying-Jin Yuan, Biotechnol. Biofuels 11, 1-10 (2015).

[4] L. Qin, X. Zhao, W. C. Li, J. Q. Zhu, L. Liu, B. Z. Li and Y. J. Yuan, Biotechnol. Biofuels 11, 118 (2018). https://doi.org/10.1186/s13068-018-1118-8

[5] J. Q. Zhu, X. L Wu, W. C Li, L. Qin, S. Chen, T. Xu, H. Liu, X. Zhou, Xia Li, C. Zhong, B. Z. Li and Y. J. Yuan, Bioresource Technology 267, 227-234 (2018). https://doi.org/10.1016/j.biortech.2018.07.030

[6] L. Xu., W. C. Li, J. Xin, S. J. J. Zhang, B. Z. Li and Y. J. Yuan, ACS Sustain. Chem. Eng., 8, 13007-13018 (2020). https://doi.org/10.1021/acssuschemeng.0c03884

[7] H. Chen, W. Huo, B. Wang, Y. Wang, H. Wen, D. Cai, C. Zhang, Y. Wu and Peiyong Qin, Ind. Crops Products, 141, 111749 (2019). https://doi.org/10.1016/j.indcrop.2019.111749

Experimental Study on the Viscosity Properties of Modified Vegetable Oil with Hybrid hBN/TiO_2 Nanoparticles as Metalworking Fluid

A. M. SABRI[1,b], N. TALIB[1,a], H. ABDULLAH[2,c], A. S. A. SANI[2,d] and S. KUNAR[3,e]

[1]Faculty of Mechanical and Manufacturing Engineering, Universiti Tun Hussein Onn Malaysia, 86400 Batu Pahat, Johor, Malaysia

[2]Centre for Diploma Studies, Department of Mechanical Engineering, Universiti Tun Hussein Onn Malaysia, Hab Pendidikan Tinggi Pagoh, KM 1, Jalan Panchor, 84600 Panchor, Johor, Malaysia

[3]Faculty of Manufacturing and Mechatronic Engineering Technology, Universiti Malaysia Pahang, 26600 Pekan, Pahang, Malaysia

[4]Department of Mechanical Engineering, Aditya Engineering College, Surampalem-533437, India

[a]fazillah@uthm.edu.my, [b]ainaamardhiahsabri@gmail.com, [c]haslinaa@uthm.edu.my, [d]amirils@ump.edu.my, [e]sandipkunar@aec.edu.in

Keywords: Metalworking Fluid, Modified Vegetable Oil, Viscosity Properties

Abstract. Todays, the vegetables-based oil for metalworking fluid has gain attention from researchers to be used as alternative of commercial metalworking fluid (MWFs). This is due to the increase of minerals oil price and concerns of using mineral oil-based metalworking fluid. In this research, crude RBD palm oil (RBDPO) and crude Jatropha oil (CJO) are chosen as the based oil because of its suitability for metalworking fluid standards. The oils are chemically modified and combined with hybrid nanoparticles to improve its thermal-oxidative stability. The product of this process is modified RBD palm oil with 0.025 wt% hBN + TiO_2 nanoparticles (MRPOht) and modified Jatropha oil with 0.025 wt% hBN+TiO_2 nanoparticles (MJOht). The physichochemical properties of the oils is evaluated and compared with synthetic ester as the benchmark oil. The dynamic viscosity of the oils is evaluated by using Viscolite device and the viscosity index of the oils is calculated by using interpolation method. The findings show that MRPOht and MJOht exhibit higher viscosity index compared to synthetic ester and crude oils. The addition of hybrid nanoparticles has significantly improved the physichochemical properties of MRPOht and MJOht as both oils has possessed highest kinematic viscosity at 100°C stating the fact that both oils can withstand the high temperature and provide lubricity at high temperature.

Introduction

Metalworking fluids (MWFs) have been used in various metal-cutting procedures over the past 200 years. By forming a layer of lubricant, MWFs lower friction, serve as a cooling medium to lower heat generation, and limit the elution of metal chips, preventing metal pick-up resulting in improvement of workpiece's surface quality [1]. The metalworking fluid's lubricating and cooling properties boost manufacturing output while maintaining or improving product quality. However, following disposal, they have several harmful consequences on the environment [2]. The world's reserves of mineral oil are running out, and environmental worry about mineral oil's adverse effects is rising. In the lubricant sector, the hunt for environmentally benign alternatives to mineral oils as base oils has emerged as a frontier of study [3]. Due to several intrinsic technical characteristics and their capacity for biodegradation, vegetable oils are seen as substitutes for mineral oils for lubricant base oils [4].

Vegetable oils have taken over as the preferred lubricant for many purposes, intending to create an environmentally benign material with qualities like those of conventional lubricants. Vegetable oils are less expensive than mineral oils, have good tribological attributes, and are biodegradable [5]. Vegetable oils, such as Calophyllum inophyllum [6], Jatropha [7], sunflower, soybean [8], sesame [9], neem [10], peanut, maize, rapeseed, palm, and castor [11], have been reported to be used in metal-cutting operations. Because they are non-toxic, renewable, biodegradable, and highly viscous, vegetable oils are appropriate for machining applications. They are composed of triglycerides, characterized by the ester linkages connecting three long-chain fatty acids to the hydroxyl groups [12]. The triglyceride form contributes desirable characteristics to the lubricant, such as increased lubricity, viscosity index, shear stability, decreased volatility, and load-bearing capacity [13]. The "oiliness" or "lubricity" of vegetable oils and fats is a quality that is attributed to their ability to adsorb to metallic surfaces and create a tenacious monolayer, with the polar head sticking to the metals interfaces and the hydrocarbon links orienting in nearly normal orientations to the surface [4]. However, the oxidizing tendency of vegetable oils is a detriment to their use [14]. The oil oxidizes and gets sticky, which causes the chips to stick to equipment, components, and accessories [5]. Oxidation will affect the physical and chemical qualities, which could radically affect how well lubrication functions. Depending on the fatty acid composition, the vegetable oils will degrade both during storage and during use [15]. Vegetable oil was found to promote wear at high temperatures and under sliding circumstances, depending on the lubricated metals' unique fatty acids, the load, and the rate of sliding [16,17]. Vegetable oil needs to be modified to overcome its drawbacks regarding oxidation stability, high friction, high viscosity, thermal stability, and corrosion resistance. The use of non-edible oils, additives, chemical alterations, and thermal modifications are among ongoing efforts to get around these restrictions.

The production of eco-friendly bio-lubricants using palm and jatropha oil was researched by Heikal et al. [18]. The two-stage transesterification process was used to create the bio-lubricants and viscosity pour point, and heat stability was assessed using ASTM standard procedures. The jatropha oil-based bio lubricant was discovered to have a high viscosity index (140), low pour point (-3°C), and moderate thermal oxidation stability, which allowed it to satisfy the standards for commercial, industrial oil ISO VG46 grade. However, the bio lubricant made of palm oil has a high pour point of 5°C, which is undesired and needs to be improved. Nevertheless, the bio-lubricants viscosity, viscosity index, and flash point are equivalent to standard industrial oils ISO VG32 and VG46. In addition, Karanja oil that underwent chemical modification had a higher viscosity index of 194cp, lower pour point at -36°C and higher flash point at 212°C compared to Karanja oil before modification. Singh and Chibber [19] found that modified Karanja oil was suitable as a metalworking fluid. In addition, modified palm oil was examined as a metalworking fluid by Chang et al. [20]. They discovered that the modified palm oil performs similarly to the brand names of trimethylolpropane trioleate. It has an excellent viscosity index of 219 and a high viscosity of 44.3 mm^2/s at 40°C.

Moreover, adding additives to chemically modified vegetable oil improves the oil properties and performance as a metalworking fluid. The oxidative stability can also be increased by using the right additives. Nagarajan et al. [21] mentioned that the addition of nanoparticles improved the oxidation of the base oil y 61.15% as well as thermal conductivity. Talib et al. [22] developed MJO5a, a vegetable oil-based metal cutting fluid produced by transesterification process of Jatropha methyl oil (JME) with trimethylolpropane (TMP) combined with 0.05% hexagonal boron nitride. MJO5a was successfully utilized to cut AISI 1045. It was discovered to produce superior machining performances by lowering the cutting force, cutting temperature, chip thickness, tool chip contact length, and specific energy. Talib et al. [23] also study on the physicochemical properties of modified jatropha oil (MJO) with the addition of 0.05 wt.% hBN and molybdenum disulphide (MoS_2) nanoparticles respectively as metalworking fluid. The results of this

Materials Research Proceedings 68 (2026) 210-217 https://doi.org/10.21741/9781644904237-28

investigation demonstrate that MJO+0.05 weight percent hBN (MJOh) demonstrated outstanding physicochemical properties since it performed better in terms of kinematic viscosity. They concluded that MJOh has the capability to be an environmentally friendly metalworking fluid for machining.

MWFs carrying a particular type of nanoparticle in machining has been the subject of much research. To the author's knowledge, however, there have been very few investigations of hybrid nanofluids (i.e., colloidal fluid that contains two distinct kinds of nanoparticles). Regarding hybrid nanofluids, no standard model can accurately predict the viscosity values given a variety of factors, including temperature, particle concentration, particle size, and shape.

Wanatasanapan et al. [24] study on the thermal conductivity, rheological properties and dynamic viscosity of TiO_2-Al_2O_3 -water based hybrid nanofluids. They found that hybrid nanofluids have the higher thermal conductivity and higher dynanamic viscosity. Besides that, Yu et al. [25] mentioned that water with hybrid TiO_2-MWCNT nanoparticles depicts decreased in viscosity as temperature increased. Kumar et al. [26] found that the thermal conductivity of the copper-titanium dioxide hybrid nanofluids increased together with the temperature and nanofluid viscosity. Sirin et al. [27] undergoes study on the viscosity of hybrid hBN/Graphene nanoplatelets (GNP) nanofluids and found that hBN-GNP hybrid nanofluid had a viscosity value that was 7.66% and 5.59% higher than that of the hBN and the GNP mono nanofluids, respectively. This is due to the combination of hBN and GNP nanoparticles enhanced the viscosity value of the nanofluid because of their different geometric forms.

This study aimed to create a novel type of nanofluids by combining modified Jatropha Oil and modified RBD palm oil (MRPO) with hybrid nanoparticles, respectively. To develop sustainable MWFs, MJO and MRPO were mixed with hybrid nanoparticle additives of hexagonal boron nitride (hBN) nanoparticles and titanium dioxide (TiO_2) at a concentration of 0.025 wt% nanoparticles. Samples of hybrid nanofluids were analyzed to evaluate their viscosity index and kinematic viscosity. The outcomes were compared with pure MRPO, MJO, and SE. Novel hybrid formulation nanofluids may present new opportunities as MWFs in the lubrication industry.

Methodology

Sample Preparation. In this research, two oil which is crude jatropha oil (CJO) and RBD palm oil (RPO), were chemically altered respectively and combined with hybrid additives in nano-sized particles to increase lubricant performance in terms of thermal-oxidative stability and viscosity properties. The first stage entailed an esterification procedure that attempted to decrease the concentration of fatty acids (FFA) to less than 1% to generate a high yield of fatty acid methyl ester (FAME). First, CJO undergoes a two-step acid-based transesterification process to obtain Jatropha Methyl ester (JME). CJO undergo an acid esterification process in which CJO react with methanol (CH_4O) with the presence of 0.5 wt.% of sulphuric acid (H_2SO_4) as a catalyst and produces Esterified Jatropha Oil (EJO). After that, EJO and RBD PO undergo a base transesterification process to produce JME and Palm Oil Methyl ester (POME) (FAME). EJO and RBDPO react respectively with CH_4O at a molar ratio of 6:1 with one wt.% of sodium hydroxide (NaOH). Next, modified Jatropha Oil (MJO) and Modified RBD Palm oil (MRPO) created a chemical reaction of generated FAMEs with TMP in the presence of 1% (wt/wt) sodium methoxide ($NaOCH_3$). The FAME to TMP molar ratio (FAME: TMP) was 3.5:1. The reaction was conducted in an oil bath at 120°C for 24 hours under a vacuum with a constant pressure of 20 kPa.

Then, as indicated in Table 1, MJO and MRPO, known as modified vegetable oils (MVOs), were blended with 0.025% by weight of titanium dioxide (TiO_2) and hexagonal boron nitride (hBN) hybrid nanoparticles. Fig. 1 shows the samples for this research, and the properties of the nanoparticles are summarized in Table 2. The MVOs were mixed with the 0.025 wt.% of hybrid nanoparticles for 30 minutes at 700 rpm and heated at 60°C using a magnetic stirrer. The mixture was then homogenized for 30 minutes using an ultrasonic homogenizer of the Bandelin HD3200 type (at 20 kHz frequency and 200 W). The next step was to visually inspect the combinations for any precipitation or layer separation. This process made sure that hybrid nanoparticles were dispersed uniformly throughout MVOs. The sedimentation technique was used to investigate the sample over time to evaluate the stability of the dispersion. Hybrid nanoparticles tend to settle over time because their density is higher than MVOs. However, because there is no discernible sedimentation over a short observation time, it is believed that the dispersion of hybrid nanoparticles in MVOs has excellent dispersion stability.

Table 1: Detail of samples.

Samples	Description of samples
SE	Synthetic Ester
MJO	Modified Jatropha Oil
MJOht	MJO + 0.025 wt.% (hBN + TiO_2)
MRPO	Modified RBD Palm Oil
MRPOht	MRPO+0.025 wt.% (hBN + TiO_2)

(a) (b) (c) (d)

Fig. 2 : Samples (a) MJO, (b) MJOht, (c) MRPO, and (d) MRPOht.

Table 2 : Properties of nanoparticles.

Properties	hBN	TiO_2
Appearance	Colourless crystal powder	White solid powder
Density (g/cm^3)	2.3	4.23
Thermal expansion coefficient (10^{-6}/K)	1	8.4
Thermal conductivity (W/m K)	8.4	4.8

Physicochemical Testing. Kinematic viscosity and viscosity index measurements were used to determine the physicochemical characteristics of hybrid nanofluid samples (VI). Viscosity is one of the critical factors for the function of a lubricant. A viscometer device (Viscolite 700) was used to determine the kinematic viscosity at temperatures ranging from 30°C to 100°C according to ASTM D445. The kinematic viscosity of the lubricant measured at 40°C and 100°C was used to calculate the viscosity index (VI) using the calculation method described by the interpolation data from the ASTM D2270 standard. The viscosity index (VI) also revealed a connection between viscosity and temperature. The kinematic viscosity and viscosity index of the samples were compared to benchmark oil, Unicut Jinen MQL, synthetic ester, SE.

Result and Discussion

Kinematic Viscosity and Viscosity Index. The testing results for the kinematic viscosity of the lubricating sample in the range of 30°C to 100°C for all samples are shown in Fig. 3. The kinematic viscosity of the samples appears to decrease significantly as the temperature rises. This is due to as the temperature rose, the rate of intermolecular change also rose. Additionally, there are additional strong attractive and cohesive forces between a liquid's molecules (which are much closer together than those of a gas). As the temperature rose, the cohesive force decreased and the liquid's viscosity decreased [28]. Sharma et al. [29] and Sukkar et al. [30] mentioned that the kinetic energy of the molecules increases and contact time to their nearest neighbor particles reduces as the temperature of the nanofluids rises. As a result, as the temperature rises, the average intermolecular forces weaken and the viscosity decreases. As can be seen from Figure 3, compared to MRPOs and MJOs, SE had the highest kinematic viscosity 29.5 mm^2/s at 30°C. The kinematic of SE, however, was at its lowest with 5 mm^2/s at 100°C. This result demonstrates that SE's kinematic viscosity significantly decreases from the highest temperature to the lowest temperature, correlating with a smaller viscosity index (VI). It is also can be seen that at higher temperatures between 80°C and 100°C compared to a lower temperature, the kinematic viscosity changes for all sample are smaller. This is because at high temperatures, molecular forces weaken, and molecular bonds are more easily broken [31]. Talib et al. [22] claim that the unreacted FAME was the reason why MRPOs and MJOs had lower kinematic viscosity than SE. Due to the viscosity improver, corrosion inhibitor, and anti-wear additives that have already been added to the lubricant, SE has a higher viscosity [32]. Moreover, compared to MRPO and MJO without the addition of nanoparticles, MRPO and MJO with hybrid nanoparticles show higher kinematic viscosity. This phenomenon happens as a result of the addition of nanoparticles, which create larger nanoclusters that obstruct the flow of fluid layers on top of one another. As a result, the layers migrate closer together and the intermolecular forces increase, increasing the viscosity [31]. Interestingly, compared to MRPO and MJO as well as SE, MRPOht and MJOht, respectively, showed greater kinematic viscosities. The addition of TiO_2 increases the viscosity of the nanofluid. Similar findings were made by Nik Roselina et al. [33], who noted that adding well-dispersed TiO_2 nanoparticles increased flow resistance and kinematic viscosity.

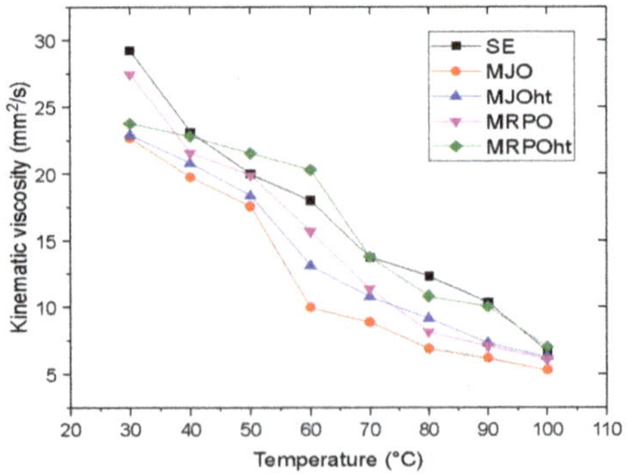

Fig. 3 : Kinematic viscosity at a range between 30 to 100°C of all samples.

The viscosity index (VI) was calculated for each sample based on kinematic viscosity at 40°C and 100°C and is displayed in Fig. 4. Temperature-dependent viscosity stability is improved with a higher viscosity index (VI) (independent to temperature) [34]. The viscosity index (VI) of a lubricating fluid describes how much its viscosity changes with temperature. A high VI denotes that temperature variations have little impact on the fluid's viscosity, whereas a low VI denotes a significant shift in viscosity. The graph shows that MRPO and MJO exhibited higher VI values than SE (261 and 227, respectively). Compared to SE, the viscosity index increased by 74% and 52% for MRPO and MJO, respectively. Higher VI is caused by the mass of the TMP ester and the formation of extended molecular chains [35]. MRPOht had the highest viscosity index of 303, followed by MJOht (288), MRPO (260), MJO (227) and lastly, SE (149). Compared to the MRPO and MJO base vegetable oils, adding hBN and TiO_2 nanoparticles dramatically improves the VI by 16% and 27%, respectively. It was significantly influenced by hBN's lower thermal expansion coefficient (1X 10^{-6}/C), which impacted the oil's thermal stability properties and was related to a higher level of contact that sustained a more extensive thermal network [36]. Additionally, the excellent thermal stability of the hBN nanoparticles was attributed to their hexagonal crystalline structure, where boron and nitrogen atoms were closely packed and firmly bonded together [37]. The viscosity index is also improved by the inclusion of hybrid TiO_2/hBN nanoparticles. TiO_2 also had a lower thermal expansion coefficient, and when working at higher temperatures and with greater friction, materials with a lower thermal expansion coefficient are preferable [38]. The presence of two different additives will improve the qualities of the vegetable-based oil, as demonstrated by the combination of hybrid hBN and TiO_2. This is confirmed by Moghaddam & Motahari [39], who claimed that based lubricant with a variety of two different additives had a higher viscosity than based oil alone. MRPOht has smaller changes in kinematic viscosity at 40°C and 100°C that contribute to higher Viscosity Index value. Smaller changes in kinematic viscosity as temperature increased resulting in better oxidative stability of nanofluids [40].

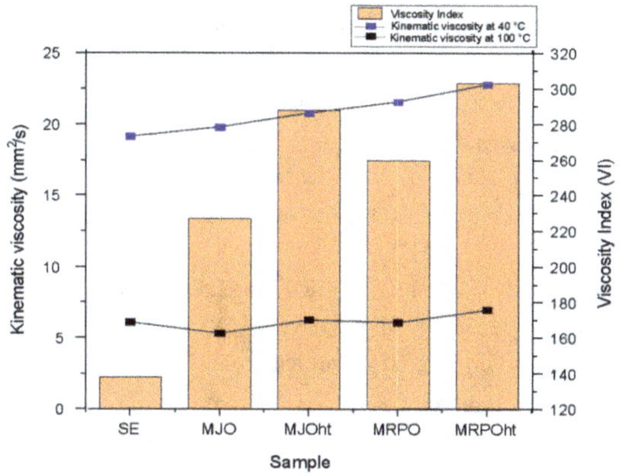

Fig. 4 : The calculated viscosity index of all samples based on kinematic viscosity at 40 and 100°C.

Conclusion

In general, the present study evaluated physicochemical parameters to compare the results with commercial SE to determine the relative performance of each sample of lubricant as a sustainable lubricant. The physicochemical of MJO and MRPO were greatly enhanced by chemical modification and hybrid nanoparticle addition. The kinematic viscosity of MJOs and MRPOs decreased with increasing temperature. Furthermore, MJOht and MRPOht show the highest VI of 303 and 288, respectively. By adding hybrid hBN and TiO_2 nanoparticles with MJO and MRPO enhanced the physicochemical properties of based oil as nano lubricants. The oxidative stability of samples depends on kinematic viscosity change as temperature increased. Among MJOs and MRPOs, MRPOht demonstrates remarkable viscosity and viscosity index, thus it is appropriate to replace the SE in terms of viscosity properties.

Acknowledgement
This research was supported by Ministry of Higher Education (MOHE) through Fundamental Research Grant Scheme (FRGS) (FRGS/1/2025/TK10/UTHM/02/1) (K592). The author gratefully acknowledges the Faculty of Mechanical and Manufacture Engineering at UTHM Johor for the lab facilities.

References

[1] A. S. A. Sani, S. Baharom, N. A. Mamat, A. S. M. Rozlan, and N. Talib, Mater. Today Proc. 48(6) , 1783-1788 (2022) https://doi.org/10.1016/j.matpr.2021.08.232

[2] S. Gugulothu and V. K. Pasam, Aust. J. Mech. Eng. 20(3), 814-824 (2022). https://doi.org/10.1080/14484846.2020.1756067

[3] S. A. Lawal, I. A. Choudhury, I. O. Sadiq, and A. Oyewole, Manuf. Rev. 1 (2014). https://doi.org/10.1051/mfreview/2014021

[4] T. Y. Woma, S. A. Lawal, A. S. Abdulrahman, M. A. Olutoye, and M. M. Ojapah, Tribol. Online 14(2), 60-70 (2019). https://doi.org/10.2474/trol.14.60

[5] G. Gaurav, G. S. Dangayach, M. L. Meena, and A. Sharma, Lubricants 10(6), (2022). https://doi.org/10.3390/lubricants10060126

[6] A. S. A. Sani, N. A. Jamaluddin, A. S. Jamaludin, and N. Talib, Calophyllum-Inophyllum from Pahang Malaysia as Biolubricant Feedstock for Industrial Application, Springer Singapore, 2021.

[7] N. Talib, A. M. Sabri, A. A. Zolkefli, K. S. Tan, and S. Kunar, J. Tribol. 33, 113-124 (2022).

[8] M. F. Trajano, E. I. F. Moura, K. S. B. Ribeiro, and S. M. Alves, Mater. Res. 17(5), 1124-1128 (2014). https://doi.org/10.1590/1516-1439.228213

[9] W. K. Shafi and M. S. Charoo, Ind. Lubr. Tribol. 73(2), 297-307 (2021). https://doi.org/10.1108/ILT-06-2020-0217

[10] E. U. Jabba, E. Aminu, and B. Usman, IOSR J. Mech. Civ. Eng. 15(2), 16-20 (2018).

[11] S. Guo, C. Li, Y. Zhang, Y. Wang, B. Li, M. Yang, X. Zhang and G.Liu, J. Clean. Prod. 140, 1060-1076 (2017). https://doi.org/10.1016/j.jclepro.2016.10.073

[12] A. M. Sabri, N. Talib, A. Sahab, A. Sani, and K. Kamdani, Jurnal Tribologin37, 84-97 (2023)

[13] N. A. Jamaluddin, N. Talib, and A. S. Abdul Sani, Int. J. Eng. Adv. Technol. 9(3) 3144-3149 (2020).

[14] A. Fadda, D. Sanna. E.H. Sakar, S. Gharby, S. Mulas, S. Medda, N.S. Yesilcubuk, A.C. Karaca, C.K. Gozukirmizi, M. Lucarini, G. Lombardi-Boccia, Z. Diaconeasa, A. Durazzo, Sustain. 14(2) 1-29 (2022). https://doi.org/10.3390/su14020849

[15] M. Maszewska, A. Florowska, E. Dłuzewska, M. Wroniak, K. Marciniak-Lukasiak, and A. Zbikowska, Molecules 23(7), 15-17 (2018). https://doi.org/10.3390/molecules23071746

[16] U. S. Choi, B. G. Ahn, O. K. Kwon, and Y. J. Chun, Tribol. Int. 30(9) 677-683 (1997). https://doi.org/10.1016/S0301-679X(97)00039-X

[17] F. P. Bowden and D. Tabor, The friction and lubrication of solids. Clarendon Press, 2001. https://doi.org/10.1093/oso/9780198507772.001.0001

[18] E. K. Heikal, M. S. Elmelawy, S. A. Khalil, and N. M. Elbasuny, Egypt. J. Pet. 26(1), 53-59 (2017) https://doi.org/10.1016/j.ejpe.2016.03.003

[19] C. P. Singh and D. V. K. Chibber, J. Deliv. Ther. 3(3), 117-122 (2013)

[20] T. S. Chang, R. Yunus, U. Rashid, T. S. Y. Choong, D. R. A. Biak, and A. M. Syam, J. Oleo Sci. 64(2), 143-151 (2015). https://doi.org/10.5650/jos.ess14162

[21] T. Nagarajan, M. Khalid, N. Sridewi, P. Jagadish, S. Shahabuddin, K. Muthoosamy and R. Walvekar, Sci. Rep. 12(1), 1-12 (2022) https://doi.org/10.1038/s41598-022-16026-4

[22] N. Talib, H. Sasahara, and E. A. Rahim, Int. J. Adv. Manuf. Technol. 92, 371-391 (2017) https://doi.org/10.1007/s00170-017-0148-x

Materials Research Proceedings 68 (2026) 218-224 https://doi.org/10.21741/9781644904237-29

pH Dependent Growth of ZnO Seed Template in Alkaline Solution

M. MARINA[1,a*], M.Z.M. ZAMZURI[1,b], N.H. ABDUL HALIM[2,c], R.N. AHMAD[1,d], Z. NOORAIZEDFIZA[1,e], S.S. AHMAD[1,f], N. NISHA RAZALLI[1] and Z. IBERAHIM[1]

[1]Faculty of Mechanical Engineering Technology, Universiti Malaysia Perlis, Pauh Putra, 02600 Arau, Perlis, Malaysia

[2]Institute of nanoelectronics Engineering, Universiti Malaysia Perlis, 01000 Kangar, Perlis, Malaysia

[a]marina@unimap.edu.my, [b]mzamzuri@unimap.edu.my, [c]nhamidah@ unimap.edu.my, [d]rozie@ unimap.edu.my, [e]nooraizedfiza@ unimap.edu.my, [f]syarahsyazwaniabdullah@gmail.com

Keywords: ZnO Nanostructure, Sol Gel Alkalinity, Elemental Analysis

Abstract. To comprehend the relaying factors behind the growth optimization of ZnO nanostructures, the needs to understand the cause of preferences in the enhancement of desired properties is essential. Sol gel alkalinity level is one of the critical processing parameters to be focused on as the ionic equilibrium of the Zn^{2+}, OH^- and their surrounding conditions would determine the ZnO seed film properties. In the present work, the systematic changes on the properties of ZnO seed thin film upon pH increase of the sol gel are studied. From the morphology observation, sol gel alkalinity up to 9 maintained the particle dispersity. The alkalinity increase resulted in lattice parameters expansion up to pH 9. At pH 10, parameters shrinking is observed due to dissolution of newly formed ZnO induced by the excessive OH- ions. Correspondingly, the light transmittivity and energy band gap of the seed thin film is seen highest at pH 9 of ~85% and 3.28 eV compared to pH7 film (~83% and 3.27 eV).

Introduction

Sensors composed of oxide-based nanostructures such as ZnO nanowires (NWs) or nanorods (NRs) with improved sensing properties owing to their high effective surface-to-volume ratio [1,2,3,4,5]. These nanostructures have been reported grown by highly complex and costly methods such as sputtering methods[6] and vapor phase method such as MOCVD [7]. Because of that, a low-cost and simplified fabrication method for depositing ZnO thin films by sol–gel method is approached. Sol-gel method provides several advantages in terms of reliability, repeatability, cost, low temperature, and ease of composition control. On the other hand, successfully grown ZnO nanowires/nanorods depends on their seeding template that allows good epitaxial growth[8]. Seed layer is crucial in growing single-oriented or vertically aligned ZnO NRs as the seed geometry would affect the growth alignment[9]. Previous researches have covered on the effect of seed layer thickness [10,11] and types of catalyst incorporated[12], however few studies have reported on the effect of sol-gel synthesis parameters. Furthermore. in past reported works, researchers have applied their own unique set of sol-gel parameters and variables to synthesize oxide-based nanostructures hence no systematic explanation could be supplied when comparing the attributes of ZnO nanostructure synthesized by different procedures in sol-gel process. Particularly, solution's pH is one of the critical parameters to be focused on as the chemical interaction is the main part of the ZnO formation. The pH value influences the amount of H^+ or OH^- ions that defines the acidity and alkalinity of the sol; and give major impact on the hydrolysis and condensation mechanism during sol-gel formation, and thus to the morphology, grain size and amount of formed ZnO[13]. Previous literatures on ZnO nanostructure as electrode materials have focused on the porous ZnO nanoparticles of about 4 to 50 nm by sol-gel in sol's condition of pH 6-11 using

$Zn(CH_3COO)_2.2H_2O$ precursor in CH_3OH solvent and NaOH as hydrolysis catalyst [14,15,16]. Accordingly, in this study, spin coating method was used to deposit the sol-gel prepared at different solution's pH. The aim of this current study is to investigate the effect of the sol-gel pH on the properties of ZnO seed film.

Methodology

The seed solution was prepared with $Zn.CH_3(COO)_2.2H_2O$ (98%, Sigma-Aldrich) as the precursor, C_2H_7NO (Sigma Aldrich), NaOH and C_3H_8O (Merck) were used as stabilizer, catalyst and solvent; respectively. 0.1 M Zn^{2+} solution was prepared by dissolving the precursor in 50 ml of solvent with vigorous stirring for 30 minutes. The alkalinity of the solutions is varied between neutral (pH 7), 8.9 and 10 by titrating 1M NaOH in separate solutions. The solutions were allowed to rest for 48 hours to allow complete condensation and sol-gel formation. Afterwards the solutions were centrifugated at 3000 rpm for 30 minutes and the obtained precipitates were dried in oven. The dried precipitates were then washed with deionized water, dried and ground. For seed thin film deposition, the ground precipitate was mixed with CH_3COOH and Triton X-100 to form a thick suspension. A drop of the suspension is deposited onto cleaned glass substrate and spin coated at 3000 rpm for 30 s. This procedure was repeated 3 times. The deposited films were annealed in air at temperatures of 400°C, 500°C and 600°C. The annealed films were structurally analysed by XRD (Bruker, Germany) with CuKα radiation at deflection angle (2θ) from 20° to 80° and 2°/ min rate. The XRD spectra of blank ITO substrate is included in the analysis for reference. The formed nanostructure was observed by FESEM (ZEISS GeminiSEM 360) and the optical properties were obtained using Uv-Vis spectroscopy (Perkin Elmer Lambda 35, USA) between 200- 800 nm wavelength.

Result and Discussion

Morphology. Fig.1 illustrates the SEM morphology of ZnO nanostructures synthesized at pH 7,8,9 and 10. The nanoparticles were seen mostly spherical in shape. Overall, at pH 7, 8, and 9 the synthesized ZnO nanoparticles are dispersed and at pH 10 the particles appeared smaller and aggregated. The alkalinity level affected the hydrolysis and condensation behaviour of the solution during gel formation and therefore influences the ZnO attributes. Beside the growth of nanoparticles induced by OR mechanism, the aggregation presumably to be induced by oriented attachment (OA) mechanism which was introduced by Penn et. al [17]. OA mechanism has been widely associated with the specific assembly of quantum systems into aggregated nanocrystals, including ZnO nanoparticles [18]. The coalescence of the nanoparticles was induced by the surface free energy created on the ultra-small particles, which explain the radical aggregation in pH 10's nanoparticles. During the sol-gel activity, ZnO particles nucleation usually occurs by precipitation. This process includes the reaction between the soluble $Zn(CH_3COO)_2.2H_2O$ with the hydroxide ions supplied by NaOH. The nucleation reaction between the Zn salt and NaOH is described as follows [19]:

$$ZnCH_3COO)_2.2H_2O + 2NaOH \rightarrow ZnO + 2Na^+ + 2\,CH_3COO^- + H_2O \quad (1)$$

The growth process resumed in the supersaturated solution until solid saturation concentration is attained. Following the nucleation and growth, the average particle size is influenced by aging. The particle size is seen uniformly grown in pH 7, 8 and 9. A slight estimate of ZnO nanoparticle size expansion at pH 9 compared to a measure of size in pH 7 (Fig. 1a). The increase particle size reflected that sufficient OH- supply in the hydrolysis and condensation stage of the sol. At pH 10 it is seen that smaller ZnO particle agglomeration into coarse particles of about 150 -200 nm. ZnO nanoparticles synthesized at pH 10 presumably was partly dissolved due to the excessive OH- ions. The dissolution yields smaller ZnO particles (Fig. 1d) that initiated the OA mechanism which resulted in the particle agglomeration. The aggregation and coarsening induced by aging affect the final ZnO particle size. On the other hand, stable dispersion is preferred for maximizing the effective surface area of individual particles. The dispersity of a colloid system is an important aspect for implementation into device applications as it ensures higher response and efficiency [20].

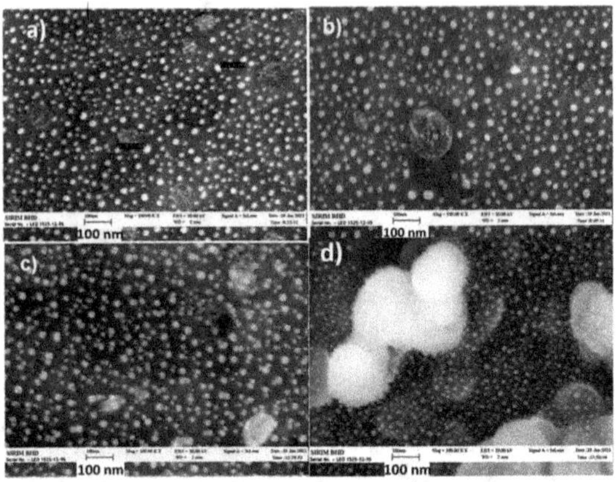

Fig. 1: Morphology of ZnO seed films synthesized at a) pH 7, b) pH 8, c) pH 9 and d) pH 10.

The growth mechanism of the ZnO nanoparticles by the sol-gel process is described as follows; first, during aging stage, condensation and gelation induce crystal growth. At the same time, OR starts with the initial small crystalline nuclei formation that evolved into a larger particle size to some extent in a supersaturated reaction solution. The separation of the viscous solvent facilitates the conservation of ZnO precursor within the gel. Next, due to the evaporation of the gel during drying at 80°C, the concentration of Zn^{2+} increases, which facilitates the hydrolysis of Zn^{2+} and thus lead to the formation of Zn (OH), and the concentration depend upon the solution concentration of Zn and the pH value. The obtained nanoparticle is deposited as thin film and annealed. During the temperature ramping in the furnace, the remaining sol within the slurry would change into gel and further spontaneously combust, releasing H_2O vapours. The annealing increases the localized temperature of the slurry which favours the crystallization of ZnO[21].

Structural. Fig. 2a exhibits the XRD spectra of ZnO synthesized at pH 7,8,9 and 10. The observed peaks on all specimens correspond to ZnO wurzite hexagonal structure with preferential growth orientations of (100), (002) and (101) planes. Due to the dissolution of ZnO nanoparticles into small ZnO crystallites, the peak intensities of pH 10 ZnO is seen to be markedly reduced (Fig. 2b). Similar to the constants of varied temperature synthesized ZnO, lattice spacing *d* on all specimens showing the highest in the (100) plane, indicating the preferential growth orientation of the nanostructures (Fig. 3a). The *d*- spacing, which is the distance between the planes of atoms that give rise to (100) diffraction peaks, compared to (002) and (101) growth planes. The hexagonal wurzite constants, *a* and *c* are shown in Fig. 3b. Expansion of *a* is seen on the (100) growth plane of pH 8 and 9 synthesized ZnO while expansion in c is noted on the pH 8 synthesized ZnO. The expansion agrees with the particle size expansion seen in the micrographs of pH 8 and 9 synthesized ZnO. The expansion leads to the growing size of individual crystallites. Subsequently, crystallite size, *D* of the specimens is shown in Fig. 4.13 c). By comparing with the size estimation in the micrographs of the particles, the calculated *D* seems agreeable. The crystallite size corresponds to the size of an individual crystals inside the particles; hence the particle size is equivalent to the total crystallites inside the particle [19]. Therefore, the calculated *D* should be equal or smaller than the estimated size from the micrograph.

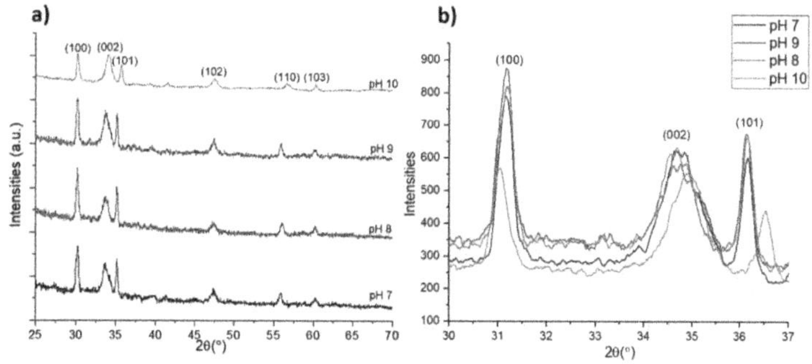

Fig. 2: Zoom in of preferential growth peaks of ZnO synthesized in pH 7. 8, 9 and 10 solutions.

On the other hand, further increase of OH^- concentration in pH 10 synthesized ZnO reduce the *D* and estimated particle size due to the partial dissolution of larger ZnO molecules into smaller ZnO crystallites (Fig. 3d). The size reduction denotes that the amount of OH^- played a crucial part in determining the crystallinity of ZnO nanoparticles. After reaching a sufficient amount of OH^- ions as seen in pH 8 and 9, the particle size decreased due to the dissolution. This is supported by the observations of ZnO dissolution reactions with excessive amount of OH^- ions as reported by [22]. Furthermore, the smaller particles due to dissolution facilitated the particle agglomeration as seen in Fig. 4.11d). It is also suggested that beyond pH 9, the solution reached supersaturation and Zn^{2+} ions started to react with the OH^- ions to form monomeric hydroxyl species such as $Zn(OH)^+$ (in aqueous form) or $Zn(OH)_2$ (in aqueous or solid) [23].

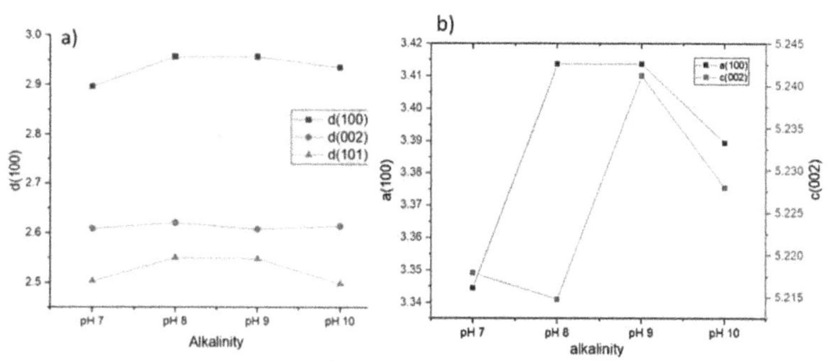

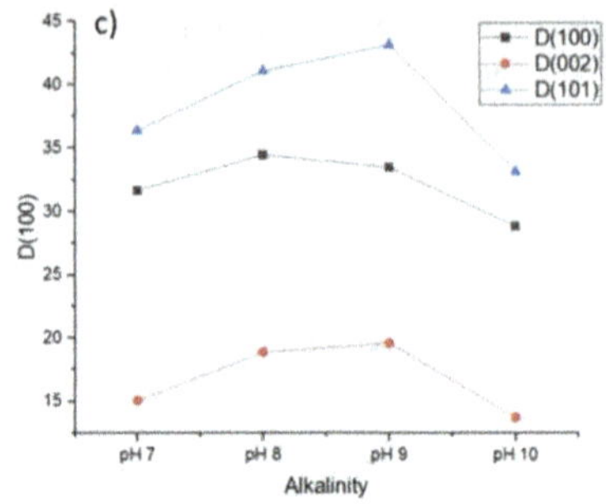

Fig. 3: Estimated lattice parameters of a) lattice spacing, d; b) lattice parameters a and c; and c) crystallite size, D of ZnO synthesized at different solution pH.

Elemental. Fig. 4 illustrates the atomic percentage of the synthesized ZnO at different solution's pH. The dominance of Zn and O elements confirmed the achieved synthesis of ZnO. No traces of Na from NaOH were detected implies that element was probably remained in the aqueous part of the sol gel that is separated from the solid precipitates during centrifugation. The centrifugation step effectively isolates the precipitates from undesired impurities such as $Zn(OH)_2$ that help to enhance the Zn-O interaction with solvent [19]. It is seen that Zn and O elements in pH 7 prepared sol are comparatively lower than pH 8 and 9. This finding could be explained by insufficient amount of OH^- during the synthesis. On the other hand, the excessive OH- in pH 10 sol reacted with newly formed ZnO that caused dissolution. This reversed reaction produced $Zn(OH)_2$ which explains the imminent change in the main element ratio.

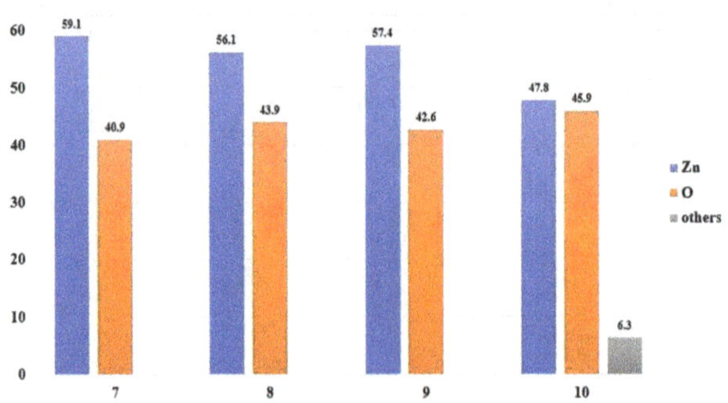

Fig. 4: EDX atomic percentages of the ZnO.

Optical. The light transmittivity of the ZnO is shown in Fig.5a. All films exhibit high transparency across the visible spectrum above 80%. Light transmittance of the seed films is dependent to the nanoparticle size, distribution and agglomeration. Since steady increase of size and dispersity of the nanoparticles were noted from pH 7 to 9, transmittance increased from ~83% (pH 7) to ~85% (pH 9) although a drop is noted in pH 10 seed film due to particle agglomeration. The bandgap, E_g of the ZnO films with respect to pH of the sol is shown in Fig. 5b. The energy bandgap of the films was of about 3.27 - 3.28 eV, which are in the range of optimum values of nanostructured ZnO [19]. At pH 10 the was slightly shifted to the lower energy than the standard range. The blue-shifting tendency is believed to be associated with the quantum confinement effect of the smaller pH 10 ZnO [24]. In optoelectronics application, the most relevant parameter of ZnO is the characteristics direct E_g, which can reach up to 3.45 eV at room temperature [25].

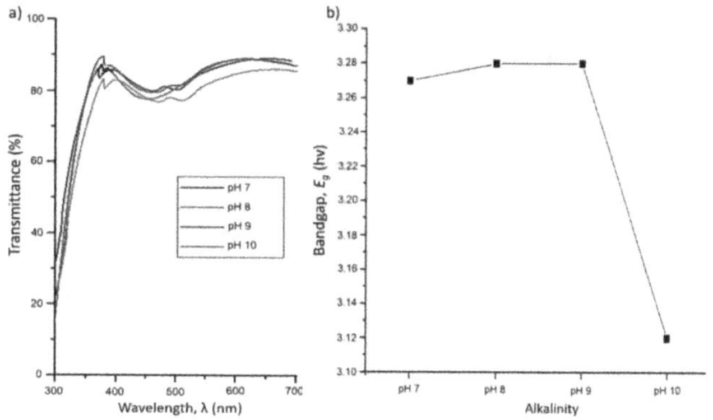

Fig. 5: a) Transmittance spectra, and b) energy bandgap of ZnO thin films synthesized at varied solution pH.

Conclusion

ZnO nanoparticles have been effectively synthesized by varying the pH of the sol gel from 7-10. ZnO synthesized at pH 9 was considered as having the best characteristics owing to the largest achieved crystallites with dispersity. Elemental analysis revealed that the dominance of Zn and O atoms were preserved up to pH 9 indicating the complete synthesis of ZnO. Optical analysis showed highest light transmittance with direct band gap enhancement to 3.28 eV compared to pH 7 ZnO.

Acknowledgment

The authors acknowledge Faculty of Mechanical Engineering Technology (FTKM), Universiti Malaysia Perlis for the lab facilities. Special thanks to Advanced Materials Processing & Design group, FTKM for the contribution to this project.

References

[1] E. S. Jang et al., Appl Phys Lett, 88(2), 1-3(2006). https://doi.org/10.1063/1.2213200

[2] A. Manoharan, Synthesis of Zinc Oxide Nanorods and use in Biosensor Applications, (2011).

[3] Y. H. Navale, S. T. Navale, F. J. Stadler, N. S. Ramgir, and V. B. Patil, Ceram Int 45(2), 1513-1522 (2019). https://doi.org/10.1016/j.ceramint.2018.10.022

[4] S. Park et al., Ceram Int, 40(7), 11051-11056 (2014). https://doi.org/10.1016/j.ceramint.2014.03.120

[5] S. Zhao, Y. Shen, A. Li, Y. Chen, S. Gao, and W. Liu, Ceram Int, 47(17), 24218-24226 (2021). https://doi.org/10.1016/j.ceramint.2021.05.133

[6] R. Nandi and S. S. Major, "ppl Surf Sci. 399, 305-312 (2017). https://doi.org/10.1016/j.apsusc.2016.12.097

[7] M. F. Melendrez, P. Flores, A. F. Jaramillo, A. Fundora, and P. Tijerina, Ceram Int. (2015).

[8] Y. Kim et al., Nature, 544(7650), 340-343 (2017). https://doi.org/10.1038/nature22053

[9] A. Bramantyo, K. Murakami, M. Okuya, A. Udhiarto, and N. R. Poespawati, Journal of Engineering (United Kingdom), 2019, (2019). https://doi.org/10.1155/2019/2793853

[10] H. Ghayour, H. R. Rezaie, S. Mirdamadi, and A. A. Nourbakhsh, Vaccum, 86(1), 101-105 (2011) https://doi.org/10.1016/j.vacuum.2011.04.025

[11] T. A. N. Peiris and H. Alessa, J Nanopart Res. 15, 2115 (2013).

[12] M. Kashif et al., J Nano, 2012 (2012).

[13] S. S. Alias, A. B. Ismail, and A. A. Mohamad, J Alloys Compd, 499(2), 231-237 (2010). https://doi.org/10.1016/j.jallcom.2010.03.174

[14] S. Arya et al., ECS Journal of Solid State Science and Technology, 10(2), 23002 (2021). https://doi.org/10.1149/2162-8777/abe095

[15] Y. Rilda, D. Damara, Syukri, Y. E. Putri, Refinel, and A. Agustien, (IOP Conference Series: Earth Environmental Science, 2019), 217(1). https://doi.org/10.1088/1755-1315/217/1/012015

[16] G. L. Kabongo, G. H. Mhlongo, T. Malwela, B. M. Mothudi, K. T. Hillie, and M. S. Dhlamini, J Alloys Compd., 591, 156-163 (2014). https://doi.org/10.1016/j.jallcom.2013.12.075

[17] L. Penn and J. L. Banfield, American Mineralogist, 83, 1077-1082 (1998). https://doi.org/10.2138/am-1998-9-1016

[18] N. Akmal, C. Lah, A. Kamaruzaman, and S. Trigueros, Nanomaterials, 11(34), (2021). https://doi.org/10.3390/nano11010034

[19] S. S. Alias and A. A. Mohamad, SpringerBriefs in Materials (2014).

[20] T. E. H. S. Jeff, "The Dispersity Of Zinc Oxide Colloids Prepared By Sol-Gel Spin," Multimedia University Malaysia, (2016).

[21] H. Zhang, X. Ma, J. Xu, J. Niu, and D. Yang, Nanotechnology, 14, (2003). https://doi.org/10.1088/0957-4484/14/4/303

[22] N. Daneshvar, S. Aber, M. S. Seyed Dorraji, A. R. Khataee, and M. H. Rasoulifard, Sep Purif Technol, 58(1), 91-98 (2007). https://doi.org/10.1016/j.seppur.2007.07.016

[23] N. S. Ridhuan, K. Abdul Razak and Z. Lockman, Sci Rep, 8(1), 1-13 (2018). https://doi.org/10.1038/s41598-018-32127-5

[24] W. L. Liu and Y. F. Zhang, Integrated Ferroelectrics, 188(1), 112-120 (2018). https://doi.org/10.1080/10584587.2018.1454222

[25] T. Yildirim and R. Lubart, Physica E, 27 (2005).

Neck Formation Properties of SS316L Particle Powder Against Time in Conventional Sintering Process

M.R. MAZLAN[1,b], M.A.M. YUSOFF[1,c] and N.H. JAMADON[1,a *]

[1]Department of Mechanical and Manufacturing Engineering, Faculty of Engineering and Built Environment, Universiti Kebangsaan Malaysia, Bangi 43600, Selangor, Malaysia

[a]rhafiq95@gmail.com, [b]muazaiman17@gmail.com, [c]nashrahhani@ukm.edu.my

Keywords: Powder Metallurgy, Conventional Sintering, Microstructure Analysis

Abstract. More than 70% of metal powder production is aimed at the automotive industry because its use saves energy, can improve material performance, can produce flexible designs, and save costs. Small and round shaped powders provide advantages in terms of high flowability and material density. However, the cost for producing small and round shape of SS316L powder is very expensive compared to large and uneven shape of powder. Therefore, bimodal particles were introduced to overcome these problems. For this paper, the study emphasizes on the effect of bimodal mixture size and sintering time on the necking size between the particles. SS316L spherical shape steel powder with size of 5µm, 30 µm and 50 µm was mixed with uneven shape 100 µm powder size with a ratio of 75:25 and the binder used was stearic acid and Polyvinyl Alcohol (PVA) with a ratio of 5% of the total feedstock. Then, the mixture was compressed before sintered under various sintering holding time of 1 hour, 2 hours and 3 hours. Studies have found that sintering time plays a very important role in producing interparticle neck formation. This is because the 3-hour sintering time for 100-5 µm mixed powders produces the largest particle neck size compared to 100-50 µm. However, the smallest necking formed under 100-50 µm bimodal mixture due to wedging effect. The necking size for 100-50 µm for 3 hours was found out to be smaller than 100-5 µm for 1 hours. Thus, particle distribution plays a bigger role in the necking formation.

Introduction

The automotive industry is one of the sectors main economy due to its value chain on global and is a key sector in the use of Powder Metallurgy (PM). However, car manufacturer faces a lot of difficulties in producing comfortable and safe vehicles yet low in cost which are lightweight and durable [1]. Hence, PM technology has been introduced because it allows the use of materials optimal further reducing the cost of production light products and components in the automotive industry. PM is a technique metal production by compressing metal powder to the desired shape and then perform the sintering process at a temperature that appropriate [2]. In PM process, the process started by applying pressure onto the powder in order to produce green body. Then, the sintering process, which is a heat treatment process will be applied in order to join the particles together. This conventional sintering process is recommended for high production value compared to the alternatives one such as laser sintering process. However, the quality of the final product is quite poor due to the present porosity in the parts. However, this porosity problems can be reduced by using proper parameter selection [3].

Nowadays, metal powder technology is in high demand by the industry world engineering because there is almost no waste in the use of materials to produce a product in the manufacturing process [4]. Metal powder milling is available through some of these processes are gas atomization, water atomization and plasma atomization [5]. Each of the following processes produces different morphologies, sizes, and shapes of particles [6]. However, the use of sphere particles is more popular than irregular particles even though irregular particles are proven to produce denser materials. This is because spherical particles have the greatest critical solids loading and the least shrinkage potential. Until today, several studies have been done on bimodal particles with the same particle shape to increase the density of the sintered material. It was found that the mixing of small particles into larger particles at a certain amount succeeded in increasing the density of the sintered material. However, studies on the use of bimodal particles of different shapes are still limited. The use of irregular particles is recommended because of the simpler and cheaper manufacturing costs compared to round particles. Furthermore, these irregular particles have also been confirmed to be able to produce nanoparticles under high heating temperatures [7].

Therefore, bimodal particles of different shapes with a high size difference will be used to produce high particles. This is due to the fact that study of neck growth on bimodal materials of different shapes using the laser sintering process has not yet been studied. Research needs to be done to ensure whether these uneven particles will be able to produce nanoparticles and further increase neck growth through the absorption process [8] or increase the pore size due to agglomeration or evaporation due to high laser temperature.

Some previous studies have found that mixing materials of different sizes results in a low green material density [9]. This is because the presence of smaller nanoparticles will increase the space or distance between the larger particles. However, in Oh et al in 2017 study, the mixing of 75% micro particles with 25% nano particles proved to increase the density of green matter [8]. Previous studies agree that large particle size differences result in higher density [9]. From another point of view, the density of green materials for the use of particles of different shapes has not yet been studied. Hence, the best ratio of bimodal materials of different sizes will be used and then the comparison between the density and microstructure of the material will be made.

Following this compaction process, the green material will go through the manufacturing process. In a previous study, the use of bimodal particles of different sizes proved to produce materials with high density values [10]. However, material selection, power and laser scanning speed play an important role in producing dense materials. From the study, the use of high laser energy density above 200 W (81.2 J/mm3) was found to reduce the density of bimodal particle material. This is likely due to the smaller particles in the composite vaporizing due to the high energy input, thus leaving larger voids between the larger particles [10]. Therefore, the selection of appropriate laser parameters needs to be identified. Thus, in this study, the ability of laser power to supply energy for irregular particles to produce necking or joining between particles will also be investigated.

Methodology

Powder Selection, Characterization and Preparation. The powders used in this study were 316L stainless steel. 5 μm, 15 μm, 15-53 μm, and 53-100 μm powders which were provided by Changsha Tijo Metal Material Co. Ltd. In this study, 3 different mixtures of particle size and shape will be used in this study in order to determine the effect of bimodal particle on the necking formation of the interparticle. Regularly, compared to irregular shaped particles, gas atomized powder or sphere particles are highly recommended for compaction process. This is because spherical powder have excellent packing density and requires minimal binder to form necking between particles [11,12].

In this study, bimodal mixtures were produced using mechanical stirrer by combining fine powder of (5 μm, 15 μm, and 30 μm) with coarse powder (100 μm). The powders were mixed by using mechanical stirrer for 1 hour with speed of 240 RPM. Meanwhile, the mixture percentage ratio that have been use was 75:25 [13]. Table 1 shows the composition for the mixture with ratio of the material. For the characterization of each powder, the powder's size was analyzed by using FESEM. Particle size distribution was also analyzed by particle size analyzer (HORIBA, LA-960). The binding material used is Polyvinyl powder Alcohol (PVA) and stearic acid by volume respectively are 4% and 1%.

Table 1: Simulation analysis results for each run

Sample	Size (shape)	
	Irregular	Spherical
1	100 µm	30 µm
2		15µm
3		5 µm

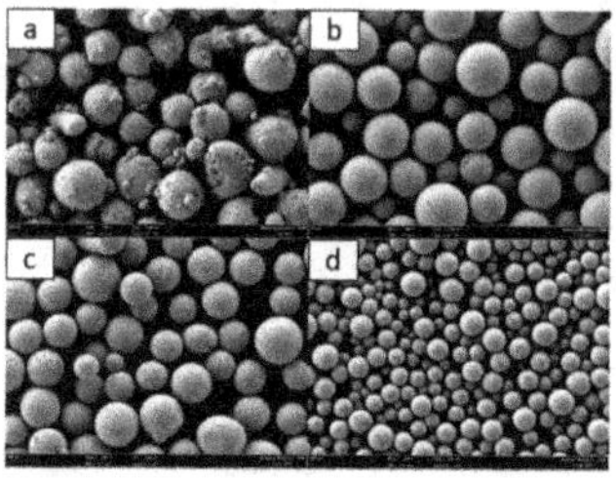

Fig. 1: Scanning electron microscopy (SEM) micrographs of the powders used in the bimodal mixture (a) 100 µm (b) 50 µm (c) 30 µm (d) 5µm.

Compaction and Sintering Process. The compression process will be done using a Kennedy model hydraulic press. The pressure force that will be used in this study is 180 bars with the surrounding temperature being room temperature for 5-10 seconds. Thus, a green body of a cylinder with a diameter of 15mm and a thickness of 5mm were formed. Then, for the sintering process, the sintering process was done by using Hot Temperature Tube machine that can heat up to 1500°C as shown in Fig. 2. Sintering temperature for all samples are maintained at 1350°C but the holding time is varied from 1 to 3 hours as shown in Table 2. In this process, Argon gases was added and act as a shielding gas. Argon gases was used in order to avoid formation of oxide layers that could cause defect in final products [14,15,16].

Samples that have been completed through the grinding and polishing process will be analyzed. The effect of neck formation in SS316L for different size particle under different sintering holding time of 1 to 3 hours was observed by using Scanning Electron Microscope (SEM). The interparticle joining that was observed under SEM will be evaluated by using Image J application.

Table 2: Sintering time for different samples.

Particles	**A**	**B**	**C**
	100 – 5	**100-15**	**100-50**
1 hour	✓	✓	×
2 hours	✓	✓	×
3 hours	✓	✓	✓

Results and Discussion

Results. In order to observe the necking behaviour in different sintering time, the samples that have been sintered was observed under SEM. Fig. *2* shows the mixing of two powders 100-5 size SS316L particles sintered at 3 hours which are observed under SEM. Meanwhile, Fig. *3* shows the necking formation of different bimodal mixture under different sintering time. The holding time in sintering process play an important role in necking formation. It has been found out that higher necking size can be found under longer holding time of 3 hours. Fig. 3(a) shows that smaller particle with size range between 5 μm exist between larger particles. It has been noted that the smaller particle accumulated between the larger particles did not undergo proper sintering process. Small necking size in one hour indicates that the particles absorb insufficient energy for proper joining process. As the sintering time increase, the smaller particles start to diffuse with larger particle which help in producing larger necking size. This due to the fact that longer holding time allows particle to absorb enough heat for proper joining process. Larger necking size is required in sintered part as it could leads to more dense materials [17]. shows schematic diagram on how necking mechanism for bimodal part form under progressing sintering time. As shown in Fig. 4, when 100-5 μm bimodal particles were used, it can be noticed that size of necking for 3 hours is larger than 1 hours. The necking size for the 100-5 particle that was sintered under 1 hour is 51.4 μm while the necking size for the same particle mixture which was sintered for 3 hours is 74.6 μm.

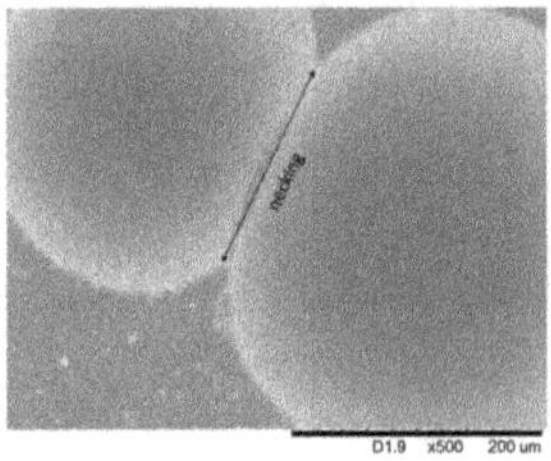

Fig. 2: Necking formation between particles of 100 μm -5 μm under 3 hours.

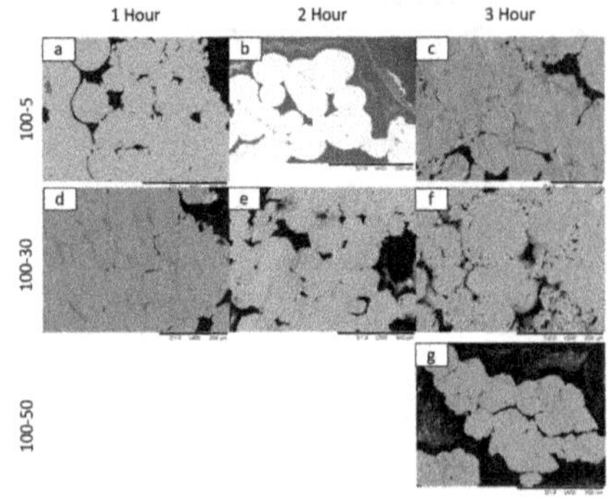

Fig. 3: Necking formation observed under SEM (a) 100-5 μm for 1 hours (b) 100-5 μm for 2 hours(c) 100-5 μm for 3 hours (d) 100-30 μm for 1 hours (e) 100-30 μm for 2 hours (f) 100-30 μm for 3 hours (g) 100-50 μm for 3 hours.

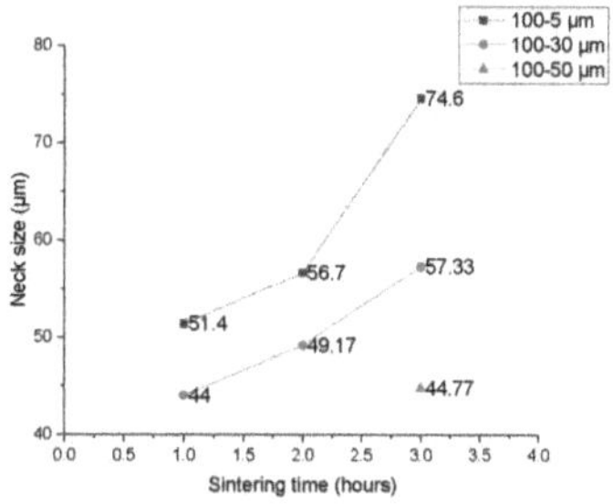

Fig. 4: Necking size for different bimodal mixture size under different sintering time.

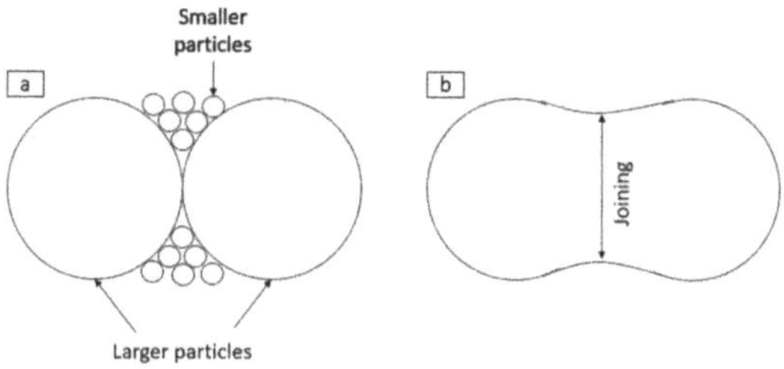

Fig. 5: Schematic diagram of diffusion between bimodal particles.

In term of bimodal ratio, large gap ratio also plays a crucial role in necking formation. As shown in Fig. 2, necking size for bimodal mixture of 100-5 µm under 3 hours is 72 µm. Meanwhile, necking size for 100-15 for 3 hours is 44.77 µm. It can be concluded that higher necking growth can be observed when smaller ratio of powder was used in bimodal mixture. This theory has been proven where bimodal mixture ratio of 100-50 shows a smaller necking size of 57.33 µm. It was found out that poor bimodal size selection cause in few defects such as wedging effect, wall effect, and loosening effect.

Figure 6 shows a schematic diagram on the particle distribution in bimodal particle. As shown in Fig. 6 (a), smaller powder particles in proper particle distribution will eventually help in filling the gaps between larger particles. Hence, it will help in increasing the necking size and density of the final products. Meanwhile, Wong & Kwan in 2014 and Ye et al in 2018 find out that, excessive smaller particle in bimodal will cause in wall effect amd loosening effect [18,19]. However, poor particle distribution as shown in Fig. 6, this effect eventually will cause gaps exist between the particles and hence causing poor necking formation. Mzorik in 2019 stated that insufficient smaller particles will eventually causing wedging effect [20].

Besides, small gap size between particles will also lead to increase in porosity point. Fig. 6 resulted in more porosity point compared to unimodal particles. Gap exist between the particle eventually will cause gases to trap in them which could cause restriction in necking growth during the sintering process. This can be seen in bimodal mixture of 100-50 µm and 100-30 µm where the necking size for it is much smaller than 100-5 µm mixture. Fig. 3 shows that the size difference between 100-30 µm lead to more porosity point which lead to poor necking process. Hence, it could be agreed that larger bimodal particle mixture causing in poor particle distribution which could led to poor necking size formation.

In summary, larger necking growth can be found when 100-5 μm was sintered for 3 hours. This study found out that proper necking was contributed to proper particle distribution and longer sintering time. In previous study, bimodal particles have been proven to be providing better necking growth than unimodal particles. Therefore, it has been found out in the study that particle distribution has significant effect compared to sintering time. This bimodal mixture particles are required not only due to providing better necking process and density, but also due to less cost required in production as smaller particles are more complicated to produce hence costing more than larger particles. Meanwhile, necking size can be improved significantly by using proper compacting process with higher pressure. Besides, smaller bimodal particles in nanoscales particles also can be introduced for better necking process.

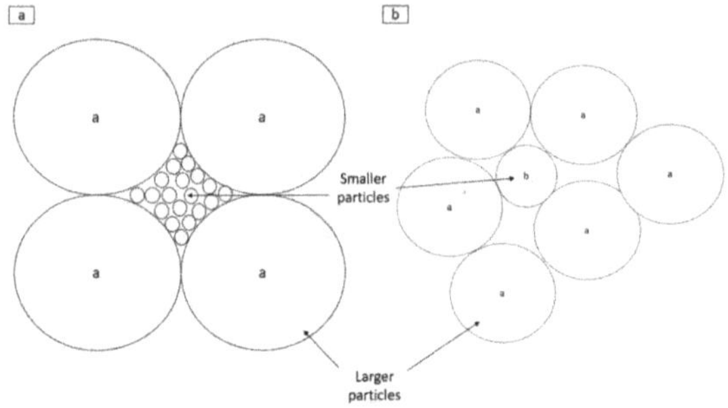

Fig. 6: Schematic diagram of (a) proper particle distribution and (b) poor particle distribution.

Conclusion

The effect of bimodal particle size and sintering time on the necking size was investigated. In this investigation, We found that the longer holding time in sintering process can improve the necking size of the particle. The necking size increase perpendicularly as the time increase. Longer sintering time allow the particles to absorb more heat for better diffusion process. Meanwhile, proper bimodal size also one of the important criteria in necking formation. The use of bimodal particle in sintering process not only could reduce the preparation cost, but also improve the necking formation and led to higher density product. Smaller particles in bimodal mixtures will fill up the gap between the particles. When sufficient heat was absorbed by the smaller particles, the smaller particles will sintered and engulf by the bridge formed between the particles. Hence, larger necking size was produced. However, from the necking analysis that have been done, particle distribution play more crucial role in necking formation. This can be seen in the graph where bimodal mixture of 100-50 μm for 3 hours have smaller necking size than necking for 100-5 μm for 1 hours. Hence, powder particle distribution has higher impact in necking formation compared to sintering time. This defects was contributed by wedging effect that produce higher porosity points between the particles.

Acknowledgment

The authors wish to gratefully thank and acknowledge the Ministry of Higher Education Malaysia (MOHE) for their financial support to complete this study under grant number FRGS/1/2020/TK0/UKM/03/5.

References

[1] D. Ramakrishnan, "Automotive applications of powder metallurgy", (Woodhead Publication Limited, 2013). https://doi.org/10.1533/9780857098900.4.493

[2] G. B. Jang, M. D. Hur, and S. S. Kang, J. Mater. Process. Technol. 100(1), 110-115 (2000) https://doi.org/10.1016/S0924-0136(00)00426-X

[3] J. M. Torralba, "Improvement of Mechanical and Physical Properties in Powder Metallurgy", 3, (Elsevier, 2014). https://doi.org/10.1016/B978-0-08-096532-1.00316-2

[4] M. Attaran, Bus. Horiz. 60(5), 677-688 (2017). https://doi.org/10.1016/j.bushor.2017.05.011

[5] A. S. Baskoro, S. Supriadi, and Dharmanto, "Review on Plasma Atomizer Technology for Metal Powder," (MATEC Web Conference, 2019), pp. 5004. https://doi.org/10.1051/matecconf/201926905004

[6] K. Kassym and A. Perveen, "Atomization processes of metal powders for 3D printing" (Material Today Proceeding, 2019), pp. 1727-1733. https://doi.org/10.1016/j.matpr.2020.02.364

[7] C. Chen et al., J. Mater. Sci. 54(20), 13344-13357 (2019). https://doi.org/10.1007/s10853-019-03813-0

[8] J. W. Oh, S. K. Ryu, W. S. Lee, and S. J. Park, Powder Technol. 322, 1-8 (2017). https://doi.org/10.1016/j.powtec.2017.08.055

[9] Y. Bai, G. Wagner, and C. B. Williams, "Effect of Bimodal Powder Mixture on Powder Packing Density and Sintered Density in Binder Jetting of Metals," (Proceeding - 26th Annual International Solid Freeform Fabrication Symposium - An Additive Manuacturing Conference SFF 2015, 2020), pp. 758-771.

[10] H. G. Coe and S. Pasebani, J. Manuf. Mater. Process. 4(1), 1-16 (2020). https://doi.org/10.3390/jmmp4010008

[11] H. K. Hassun, B. H. Hussein, and B. K.h.al-Maiyaly, "Modeling the Effect of Particle Packing Density for Sintering" (Journal of Physic Conference Series, 2021) vol. 1879 no. 3. https://doi.org/10.1088/1742-6596/1879/3/032080

[12] M. J. Cima, a. Lauder, S. Khanuja, and E. Sachs, "Microstructural Elements of Components Derived From 3D Printing" (Solid Freeform Fabrication Symposium, 1992) pp. 220-227.

[13] K. V. V. Nagaraju, S. Kumaran, and T. Srinivasa Rao, "Microwave sintering of 316L stainless steel: Influence of sintering temperature and time" (Materials Today Proceedings, 2019), vol. 27, pp. 2066-2071 https://doi.org/10.1016/j.matpr.2019.09.062

[14] M. I. and S. R. A. I. K.I. Yaakob, Angew. Chemie Int. Ed. 6(11), 951-952., 11(3), 2895-2905 (2017).

[15] M. Naqiuddin, M. Salleh, M. Ishak, and K. Yamasaki, Advanced Research in Fluid Mechanics and Thermal Sciences 1(1), 91-100 (2021).

[16] M. Ishak, K. Maekawa, and K. Yamasaki, Mater. Sci. Eng. A. 536, 143-151 (2012). https://doi.org/10.1016/j.msea.2011.12.092

[17] M. A. Almomani, A. M. Shatnawi, and M. K. Alrashdan, Adv. Mater. Res. 1064, 32-37 (2014). https://doi.org/10.4028/www.scientific.net/AMR.1064.32

[18] V. Wong and A. K. H. Kwan, Powder Technol. 268, 357-367 (2014). https://doi.org/10.1016/j.powtec.2014.08.036

[19] X. Ye, Y. Li, Y. Ai, and Y. Nie, Adv. Powder Technol. 29(9), 2280-2287 (2018). https://doi.org/10.1016/j.apt.2018.06.012

[20] L. Mzorik, "Shaping The Structure of Low Water-Binder Ratio Cement Pastes", (Engineering Mechanics, 2017), pp. 1-5

Effect of Air Abrasion Surface Treatment on the Bonding Behavior at Zirconia-Resin Interface

N.S. AMINUDDIN[1,a], M.F.A. RAHIM[1,b] and N.H. JAMADON[1,c*]

[1]Department of Mechanical and Manufacturing Engineering, Faculty of Engineering & Built Environment, Universiti Kebangsaan Malaysia, 43600 UKM Bangi, Selangor Darul Ehsan, Malaysia

[a]shayhiera.work@gmal.com, [b]rrfaredzuan@gmail.com, [c]nashrahhani@ukm.edu.my

Keywords: Air Abrasion, Bonding Behaviour, Zirconia-Resin

Abstract. Zirconia 3YSZ is a superior material for dental restoration because it has the best mechanical strength, aesthetics and biocompatibility. Creating a reliable and durable bond to the surface of zirconia is difficult due to its inert chemical nature and has limited its use. However, the introduction of functional monomers such as methacryloyloxydecyl dihydrogen phosphate (MDP) in cementation materials has increased the bond strength to zirconia. The main objective of this study is to investigate the effects of surface treatment on the bond strength of novel zirconia and commercially available adhesive cements. The prepared zirconia samples were divided into two groups: without surface treatment and with surface treatment (air-particle abrasion, 50 μm). Surface roughness was tested with an electronic profilometer (Ra). Adhesive cement (RelyX Ultimate Clicker, 3M) is applied to the zirconia surface. The microstructure of the zirconia is also characterised. After surface treatment, the zirconia exhibited higher surface roughness (Ra = 17.19 ± 4.13) than without surface treatment (Ra = 3.72 ± 0.51). The zirconia phase is mostly in a tetragonal phase with a few monoclinic phases. The change in surface roughness further promotes cell attachment. Sintering samples at 1500 °C gave good microstructural density and does not cause the grains to pull out; this is the most suitable temperature for producing zirconia with good mechanical properties. The bonding effect of the cement is related to its ability to penetrate the surface irregularities of the substrate, the mechanical properties being influenced by the total filler content. The presence of 43 wt% total filler can improve mechanical properties and reduce solubility.

Introduction

Advances in science and technology have continued to introduce new materials in industry to fulfill it purpose. Advances in the dental restorative industry are all about better aesthetic value. This have led to the increasing in metal-free dental restoratives demand. It can be seen by the gradual replacement of amalgam teeth with composite teeth. This was followed by full metal (all-metal) and porcelain fused to metal before the advent of ceramic dentistry i.e. zirconia [1, 2]. Due to its superior mechanical qualities in comparison to other dental materials and almost resembles the color of natural teeth, zirconia stabilized with 3 mol% yttria (3 mol% yttria stabilized zirconia, 3YSZ) is a remarkable ceramic material that is frequently used as prosthetic materials in dental restorations [3,4].

Few studies show that monoclinic phase increases when the air abrasion used on the sintered zirconia. In contra with the result of sintering after applying surface treatment may decreases the percentage of the monoclinic phase as well as flaws and cracks in ceramic during sintering can be effective in increasing the bond strength [5]. Thus, this could be as a result that sandblasting the zirconia bonding surface increases surface roughness and undercuts. While, the addition of resin may result in a rough surface for micromechanical interlocking and an increase in surface area for chemical bonding with reactive substances [6].

The development and regular use of adhesives has begun to revolutionize many aspects of restorative dentistry. Since the introduction of adhesives, there have been significant changes in cavity preparation techniques, as labor-intensive and difficult stages are no longer required. Furthermore, it has eliminated negative user effects such as micro leakage, which is a major issue in the dental field. Dental anatomy, on the other hand, must be taken into account because there are two major tissue structures, enamel and dentin, that must be studied in order to understand how these tissues affect the adhesive bond. Previous research has discovered that several factors, including surface morphology and zirconia ceramic composition, contribute to the strength of the bond [7]. Furthermore, the ceramic surface treatment outperformed the untreated ceramic surface in terms of adhesion [8].

The adhesiveness of zirconia is also affected by the nature of the monomers found in the adhesive. Monomers are a type of small molecule that can bond together when exposed to light, heat, or pressure to form more complex molecules known as polymers. [9]. Therefore, the effect of adhesion using commercial cement on the novel zirconia was investigated to determine the reliability of adhesion. The purpose of this research is to study the effect of surface treatment of novel zirconia on commercial adhesive cementand analyze the cross-sectional morphology on the interfacial bond of zirconia and adhesive cement.

Materials and Methodology

Sample Preparation. Zirconia nano powder stabilized with 3 mol% yttria (3 mol% Yttria-Stabilized Zirconia, 3YSZ) (Inframat Advanced Materials, USA) is the main material used in this study. Polyethyleneimine (PEI, Sigma-Aldrich) with an average molecular size of 50,000 was used as a dispersing agent during the suspension preparation process. Dental adhesives namely Single Bond Universal (SBU) and RelyX Ultimate Clicker (RelyX) (3M) are applied to zirconia. The zirconia suspension is prepared using 12% powder loading. The colloidal processing was carried out in 2 stages, namely the addition of 0.5 bt. % PEI and pH2 adjustment using nitric acid. The suspension was stirred for 45 minutes and then placed in an ultrasonic machine for 10 minutes at a frequency of 50 Hz to break up soft agglomerates. Then, it was poured into a teflon mold (diameter 15 mm & height 21 mm) and left to dry for 24 hours to 48 hours at room temperature. The zirconia sample was milled to get the flat surface before undergoes Cold Isostatic Pressing (CIP) process at pressure of 250 MPA for 2 minutes. The temperature of 1500°C with a heating rate of 3°C/min was sintered on the samples.

Surface Treatment. The surface treatment was applied using an air particle abrasion machine (Renfert, USA) with alumina particles of 50 μm size at a distance of 10 mm and a pressure of 3 bar for 10 seconds. This process is done to increase the roughness of the zirconia surface. In this study there are two groups of zirconia which is with surface treatment and without surface treatment.Next, the SBU adhesive was first applied to the zirconia with the adhesive sweep process carried out for 20 seconds. Then, the process of air drying, and light curing for 10 seconds was carried out to dry the SBU. RelyX cement is then applied over the SBU layer and undergoes a light curing process for 10 seconds.

Testing analysis. Surface roughness refers to the small spacing and unevenness of peaks and valleys on the surface of a material. The distance between the peaks or valleys is too small i.e. <10 μm which cannot be seen by the naked eye. This test was conducted using a Mitutoyo Formtracer SV-C3100 machine.For the morphological analysis of the zirconia surface, the sample has gone through a thermal etch process at 900°C with a heating rate of 20°C/min. While for cross-sectional morphological analysis of zirconia and adhesive cement, a non-conductive process is performed because the sample has been placed in the resin. All zirconia samples were analyzed using a SEM machine (Zeiss Merlin, Germany).

Materials Research Proceedings 68 (2026) 232-238 https://doi.org/10.21741/9781644904237-31

Result and Discussion

Characterization of Zirconia Initial Powder Size. The zirconia powder has a homogeneous spherical shape with the average size of the zirconia powder measured is 30 nm based on the TEM image in Fig. 1. The presence of agglomerates on the zirconia powder can also be seen based on this characterization. The agglomerate presence factor was analyzed through Brunauer-Emmet-Teller (BET) analysis by measuring the specific surface area of zirconia powder through nitrogen adsorption. This analysis shows that the surface area of zirconia powder is large which is approximately 32 m^2/g [10].

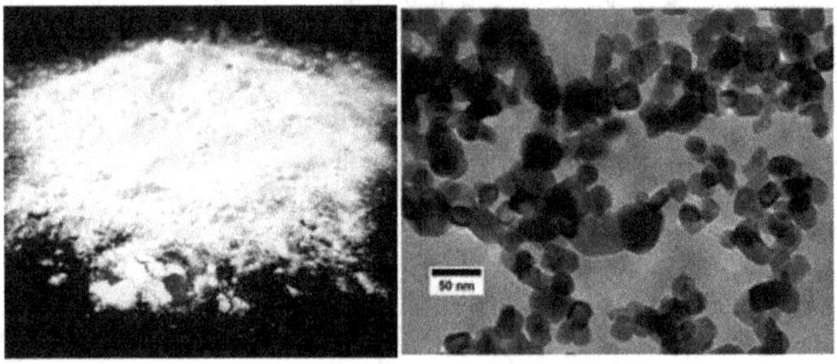

Fig. 1: TEM image of the initial 3YSZ zirconia powder as received.

Zirconia powder that has a large surface area becomes a factor for agglomerate to occur. The larger the surface area of the zirconia powder, the higher the surface energy of the powder. This surface energy can be reduced and more stable by clumping together to form larger particle sizes [11]. As a result of the agglomeration, the surface energy decreases due to the increase in the contact surface area.

Surface Morphology of Zirconia via Scanning Electron Microscope (SEM). Morphological analysis of zirconia surface was carried out via Scanning Electron Microscope (SEM) for zirconia subjected to CIP pressure at 250 MPa. The results of the SEM micrograph can be seen in Fig. 2. Through Image J software, this micrograph image is analysed to obtain a measurement of the pore size. The presence of a high pore size in zirconia samples can affect the final properties of the material. Through this CIP pressure, the porosity content found in the zirconia sample is reduced and produces a minimal pore size.Through the study of Amat et al. 2018, the same results can be observed on samples subjected to CIP pressure at 200 MPa and above 250 MPa have a decreasing pore size (15 nm) compared to samples that did not go through any CIP process [12]. For the CIP process at 300 MPa, the pore size did not show significant changes with the sample subjected to CIP pressure at 250 MPa. Based on the SEM results, the CIP process at a pressure of 250 MPa is used because it is able to provide sufficient pressure and the density of zirconia blocks can be increased.Through the slip casting technique at the beginning of the zirconia production process and CIP pressure at a sufficient level, the zirconia sample becomes denser, and this process is proven to reduce the pore size.

Materials Research Proceedings 68 (2026) 232-238 https://doi.org/10.21741/9781644904237-31

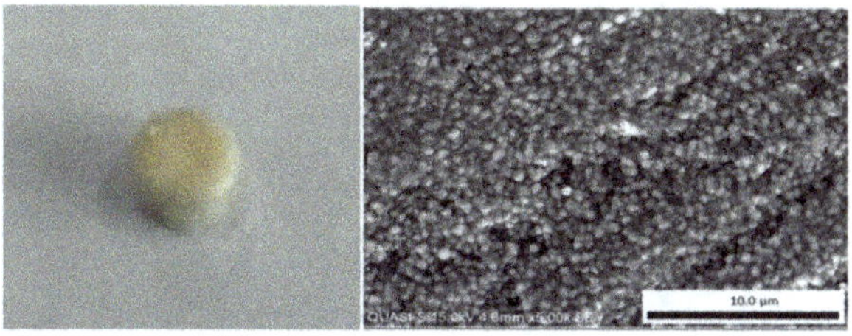

Fig. 2: SEM micrograph of zirconia at a CIP pressure of 250 MPa.

Chemical Composition via Energy Dispersive X-Ray (EDX). The chemical composition of zirconia after colloidal technique and cold static pressing was analyzed using EDX. In this study, the amount of dispersing agent (polyethyleneimine, PEI) and pH were optimized to achieve high colloidal stability during the suspension preparation process. A dispersing agent is an additive that acts to disperse the ceramic powder in the solvent and helps stabilize the slurry against flocculation or settling. These additives are organic which are then removed through the sintering process.The elements found in PEI are composed of carbon C, hydrogen (H), and nitrogen (N). While pH regulation through nitric acid has elements H, N and oxygen (O). The main material used, which is zirconia powder, has the elements zirconium (Zr) and O. Each element has a unique atomic structure, producing a unique set of peaks in the X-ray spectrum that correspond to the elemental composition of the sample. Through the sintering process at a temperature of 1500°C, the additive will be removed and will affect the zirconia grain size. Table 1 shows the average value for the EDX analysis of zirconia ceramics shows the chemical composition of zirconia 3YSZ has the main element which is Zr with the presence of O and C elements.

Table 1: Mean value ± standard deviation of zirconia surface roughness for without surface treatment and with surface treatment.

Element	Weight (%)	Atom (%)
O	23.71	52.43
Zr	69.19	26.82
C	7.10	20.75
Total	100	100

Phase Change via X-Ray Diffraction (XRD). The XRD peak intensity of the monoclinic and tetragonal phases for the sample sintered at 1500°C is shown in Fig. 3. The condition of zirconia sample used in this study is before aging process. A low monoclinic peak intensity can be seen on the -111 lattice plane. The monoclinic phase occurring at this sintering temperature can be attributed to the increase in grain size. As the grain size increases, the t-m transformation at the initial stage occurs due to the thermal stress of sintering. This situation results in volume expansion which in turn produces compressive stress on the affected zone.

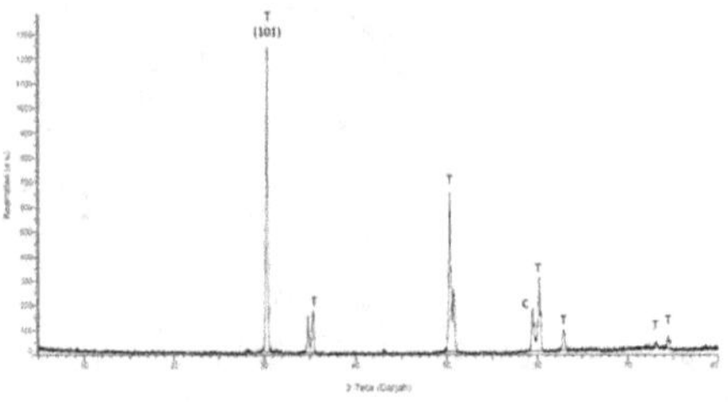

Fig. 3: Phase change of zirconia at 1500°C sintering temperature.

The highest peak can be seen in the tetragonal phase (101) with better mechanical properties. It is reported that the mechanical properties of zirconia in terms of toughness and fracture strength as well as wear resistance increase on the stability of the tetragonal phase. Through the study of Amat et al. 2018, at a sintering temperature of 1600°C shows that the monoclinic phase is seen relatively high on the (111) peak before aging. This is because the t-m transformation has occurred at this stage followed by the pull out of grain on the zirconia surface.

Based on these results, it can be concluded that the sintering temperature has an important role in producing zirconia that has good mechanical properties because in the monoclinic phase, the mechanical properties of zirconia are weak and contribute to the decrease in the mechanical properties of the material.

Zirconia Surface Roughness. Surface roughness value (R_a) for zirconia samples without surface treatment and with surface treatment are shown in Table 2.

Table 2: Mean value ± standard deviation of zirconia surface roughness for without surface treatment and with surface treatment.

Group	Surface Roughness (μm), R_a
Without surface treatment	3.72 ± 0.51
With surface treatment	17.19 ± 4.13

Based on the table above, there is a significant difference between the two groups. The group without surface treatment showed very low surface roughness values. While the group with surface treatment showed a significant improvement because the original surface of the zirconia was changed after the air particle abrasion method. The rough surface of zirconia can help the resin adhesive to provide contact areas and micropores, thereby increasing bond strength.

Cross Sectional Morphology of Zirconia Blocks with Adhesive Cement by SEM. The bond interface is the area with the highest stress and the bond strength measurement between the adhesive cement and the substrate is more accurate if taken in this area. There are claims that zirconia that has an inherent roughness is sufficient for bonding with adhesive cement through micromechanical interlocking and any further surface treatment process is unnecessary [13].

The presence of methacrylate monomers containing phosphoric acid in the chemical composition of the adhesive cement allows bonding to zirconia and dentin. Fig. 4 shows SEM micrographs of adhesive cement, zirconia, and zirconia bond with adhesive cement. Based on the micrograph, the gap found at the interface is caused by the rheological properties contained in the chemical composition of the adhesive. The viscosity of RelyX Ultimate Clicker is low because new rheological modifiers have been incorporated into the adhesive cement.

According to the adhesive cement manufacturer, the filler content in RelyX Ultimate Clicker is 43 bt%. These fillers act to improve mechanical properties and reduce solubility. Therefore, this adhesive cement has good mechanical strength because it has a high filler content. Through the study of Youm et al. 2015, the bond capacity of cement is related to the ability to infiltrate into uneven surface substrates with mechanical properties being influenced by filler content [14].

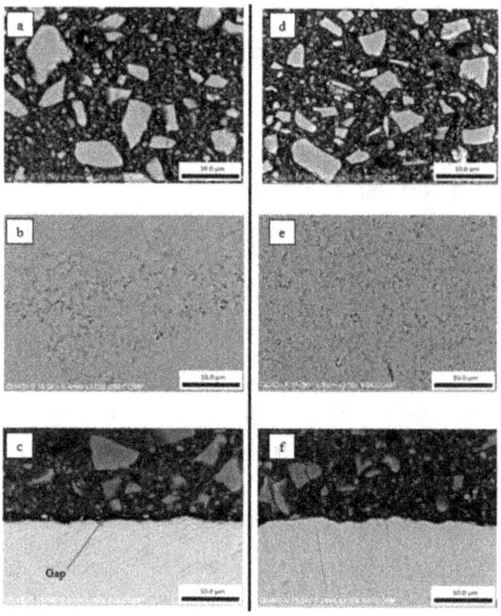

Fig. 4: SEM micrograph; (a) and (d): adhesive cement, (b) dan (e): zirconia, (c): adhesive cement bond with zirconia surface without treatment, (f): adhesive cement bond with zirconia surface with treatment.

Conclusion

This study has shown that the zirconia surface has been altered by the surface treatment method, the roughness of the surface can be increased, which is much more conducive to cell attachment. The application of an air abrasion method with aluminum oxide particles of 50 m at a pressure of 3 bar and a distance of 10 mm for 10 seconds resulted in changes in the zirconia surface structure without cracking. Furthermore, the SEM micrographs of zirconia and adhesive cement cross-sections show that gaps are more pronounced in zirconia without surface treatment than in zirconia with surface treatment. This could be due to poor dispersion of the adhesive cement on untreated zirconia. The presence of 43 wt% fillers in the chemical composition of the adhesive cement improves the mechanical properties while reducing solubility. In conclusion, the zirconia performs well even when used with commercial adhesive cement, as demonstrated in this study, and has the potential to be used in real dental applications.

Acknowlegment
The authors would like to thank the Faculty of Engineering and Built Environment, The National University of Malaysia for providing the laboratory facilities and funding for the research project (GUP GUP -2021-020).

References

[1] S. C. Bayne, J. L. Ferracane, G. W. Marshall, S. J. Marshall, and R. van Noort, J. Dent. Res. 98(3), 257-265 (2019). https://doi.org/10.1177/0022034518822808

[2] L. Zhao, Y. T. Jian, X. D. Wang, and K. Zhao, J. Prosthet. Dent. 116(5), 790-796 (2016). https://doi.org/10.1016/j.prosdent.2016.03.020

[3] I. Denry and J. R. Kelly, Dent. Mater. 24(3), 299-307 (2008). https://doi.org/10.1016/j.dental.2007.05.007

[4] N. F. Amat, A. Muchtar, N. Yahaya, and M. J. Ghazali, Aust. J. Basic Appl. Sci. 6(12), 9-13 (2012).

[5] A. Ghasemi et al., J. Adhes. Dent. 16(4), 377-382 (2014).

[6] R. M. Dawood and A. F. Ibraheem, J. Baghdad Coll. Dent. 27(1), 77-85 (2015). https://doi.org/10.12816/0015268

[7] S. C. Lin, W. C. Lin, T. C. Hu, M. Yan, and C. M. Tang, Coatings 11(9), (2021). https://doi.org/10.3390/coatings11091017

[8] R. M. Melo, R. O. A. Souza, E. Dursun, E. B. C. Monteiro, L. F. Valandro, and M. A. Bottino, Oper. Dent. 40(6), 636-643 (2015). https://doi.org/10.2341/14-144-L

[9] D. M. D. Moura et al., J. Mech. Behav. Biomed. Mater. 88, 69-77 (2018).

[10] N. Faeizah Amat Andanastuti Muchtar, M. Jameelah Ghazali, and N. Yahaya, J. Mech. Eng. 4(3), 10-19 (2017).

[11] M. Seipenbusch, S. Rothenbacher, M. Kirchhoff, H. J. Schmid, G. Kasper, and A. P. Weber, J. Nanoparticle Res. 12(6), 2037-2044 (2010). https://doi.org/10.1007/s11051-009-9760-5

[12] N. F. Amat, A. Muchtar, M. S. Amril, M. J. Ghazali, and N. Yahaya, Ceram. Int. 44(6), 6409-6416 (2018). https://doi.org/10.1016/j.ceramint.2018.01.035

[13] R. Gamal, Y. F. Gomaa, and M. A. Abdellatif, Futur. Dent. J. 3(2), 55-60 (2017). https://doi.org/10.1016/j.fdj.2017.07.002

[14] S. H. Youm, K. H. Jung, S. A. Son, Y. H. Kwon, and J. K. Park, J. Adv. Prosthodont. 7(4), 317-322 (2015). https://doi.org/10.4047/jap.2015.7.4.317

Oxidation Behavior of Aluminium Granules and Aluminium Alloy Chips During In-Situ Casting

Ain Zubaidah Binti SAMIAN[1,a*] and Aslinda Binti SALEH[1,b]

[1]Department of Mechanical Engineering, Faculty of Mechanical and Manufacturing Engineering, Universiti Tun Hussein Onn Malaysia, 86400 Parit Raja, Johor, Malaysia

[a]hd200057@siswa.uthm.edu.my, [b]aslinda@uthm.edu.my

Keywords: Oxidation Behaviour, Aluminium Casting, Physical And Elemental Analysis

Abstract. A revolutionary casting technique, in-situ melting, eliminates the requirement for pouring and reduces the production of porosities in aluminium (Al) castings, which impacts their melting processes. This study intends to investigate the reaction of aluminium granules and chips made of the alloy (AlSi7Mg) to oxygen during in-situ melting. The granules and chips were heated for 30 minutes in a laboratory furnace at temperatures ranging from 600°C to 850°C using the in-situ casting technique. The physical properties of granules and chips changed from metallic-like silver to greyish-coloured with a matte appearance. The heated samples were analysed for surface morphology and elemental analyses using a Scanning Electron Microscope (SEM) equipped with Energy Dispersive Spectroscopy (EDS). The presence of oxygen on the chip's surface was disclosed by EDS analysis, and XRD analysis indicated that the chips had been oxidised, resulting in the creation of aluminium oxide (Al_2O_3).

Introduction

Aluminium and aluminium alloys have excellent qualities such as high electrical and thermal conductivity, high reflectivity, adaptability, workability, recyclability, high ductility, moderately high strength, and comparatively low cost. Due to their very low density and corrosion resistance, they are highly appealing for various purposes, from soft foils to the most demanding technical applications in the industrial sectors [1]. Machining, extrusion, forging, welding, and casting are some of the manufacturing methods that may be used to manufacture aluminium alloys. Nevertheless, the casting technique is excellent for generating precisely designed items that other forms do not [2]. The casting performance is hampered by the oxidizability, which introduces many oxide flaws and ruins the continuity and integrity of the casting matrix. Therefore, it is critical to comprehend how oxide defects originate and change to enhance casting quality, traditionally ascribed to oxidation slag inclusions [3]. Investment casting is the most effective method for producing complicated small-scale objects with excellent dimensional and geometrical correctness. On the other hand, traditional tilt pouring and turbulent flow enhance gas entrapment, oxide development, and mould wall erosion, which may encourage the creation of porosities and inclusions in castings [4]. As a consequence of oxidation at the surface of the aluminium melts throughout the melting, transferring, and pouring operations, and due to the nature of the metal, which is drawn to oxygen and absorbs hydrogen in its liquid form, the melt now has a thin oxide layer on it. The bifilm is a later folding of the aluminium oxide (Al_2O_3) surface oxide film caused by turbulence. An inventive substitution for in-situ melting or casting is a pouring-free casting technique. The procedure is based on the fundamental notion that when metal is heated in a furnace in a more diminutive form packed in a mould, liquid metal fills the mould cavity as the metal melts [5]. With less than 2% porosity, the in-situ microwave casting technique produced aluminium castings. However, the setup and the experiment design are complex [6]. Magnesium components may be made using the investment casting method by using the cutting-edge in-situ melting technique, which uses small magnesium alloys like chips, turnings, and swarf [7]. This research purposed to investigate the oxidation behaviour of aluminium granules and aluminium alloy chips during in-situ casting.

Methodology

The preparation of the ceramic mould. The typical method for producing investment casting moulds was used to construct a cylindrical mould with a diameter of 34mm and a height of 65mm. Combining 75-micron Zircon ($ZrO_2.SiO_2$) flour with colloidal silica concentration (SiO_2) yields the ceramic slurry. After that, the design is immersed in slurry and left to dry for an hour. After being dipped into the slurry again to form the first backup coat, the structure is immediately stuccoed with a fine grain (0.3-0.7mm) alumino silicate ($Al_2O_3.SiO_2$) sand. After curing for an hour, the design is dipped into the slurry and stuccoed with coarser (0.7-1.0mm) alumino silicate ($Al_2O_3.SiO_2$) sand with six additional coatings. The pattern is dipped into the slurry and left to dry for 24 hours to produce the final coat. The dewaxing process is finished in a furnace heated to 200°C for one hour. The ceramic mould's wall thickness ranged between 6 and 7mm.

The Heating process of pure aluminium granules and aluminium alloy (Al-Si7Mg) chips. Aluminium (Al) granules (99.4% pure) with dimensions 4mm diameter and 5mm length and aluminium alloy (Al-Si7Mg) chips with dimensions 1mm width and various lengths due to the machining process were used. The investment casting ceramic mould containing the aluminium granules and aluminium alloy chips was put into the laboratory furnace KSL 1800X in preparation for the in-situ melting procedure. As indicated in Table 1, the aluminium granules were heated for 30 minutes at temperatures ranging from 700°C to 850°C. While the aluminium alloy chips were heated for 30 minutes at temperatures ranging from 600°C to 750°C. After that, the samples were cooled to room temperature. The samples of after-heated aluminium alloy chips and aluminium granules were then ready for metallography testing.

Table 1: Heating time and temperature for each type of material.

Material	Time (minutes)	Temperature (°C)			
Aluminium granules	30	700	750	800	850
Aluminium alloy chips	30	600	650	700	750

Physical evaluation and oxidation analysis. The heated samples were examined using visual for physical observation, scanning electron microscope (SEM) with Energy Dispersive Spectroscopy (EDS) model Hitachi SU1510 to analyse the oxide structure and morphology, as well as the elemental analysis and X-ray diffraction (XRD) model Bruker D8, to identify the compounds and phases present in the oxide layer of pure aluminium granules and aluminium alloy chips during in-situ melting. The British Standard Color Guides were used to determine the colours using the colour names (BS 5252 Colour Chart).

Result and Discussion

Physical evaluation. Fig. 1(a-d) show images of aluminium granules, while Fig. 1(e-h) shows images of aluminium chips, respectively, after being heated for 30 minutes at four different temperatures. Visual inspection revealed that even at 850°C and 750°C, the granules and chips could not melt completely into a pool of molten metal. As a result of the interaction between aluminium and oxygen, it was discovered that aluminium granules and aluminium alloy chips experience oxidation during heating in the furnace at the respective temperatures. Despite being heated over the melting point of pure aluminium, 660°C [8], the granule preserved its structure in granule form; however, the colour was changed. While the aluminium alloy chips that heated beyond the melting point of 610°C [9], also changed their colour, the chips' form remained the same. This might be due to the active surface oxidation of granules and chips during the melting process, which creates a thin oxide layer enclosed inside the melt, preventing the melting of the tiny aluminium [5]. The oxidation of the material is hastened by the high surface-to-volume ratio of the granules and machining chips, making them stand out among other types of trash for remelting [10].

Fig. 1: The heated aluminium granules were for 30 minutes at temperatures of (a) 700°C (b) 750°C (c) 800°C (d) 850°C and aluminium alloy chips were for 30 minutes at temperatures of (e) 600°C (f) 650°C (g) 700°C (h) 750°C.

After being heated for 30 minutes at four different temperatures, which are 700, 750, 800, and 850°C, respectively, as shown in Fig. 1(a-d), the colour of the aluminium granules changed from a metallic silver before heating to a dusty grey after heating. According to the chart in the British Standard Color Code (BS 5252), it was discovered that the heated aluminium granules took on the appearance of squirrel grey, blue mink, pearl grey, and corvette blue at temperatures of 700, 750, 800, and 850°C, respectively. The granules' entire transformation into a greyish colour indicates that their surface has been oxidised to its fullest extent. While the aluminium alloy chips were silvery grey before heating and finally changed to dark grey with a matte appearance. Although the milling process has made the surface of the chip slightly rough, the original physical nature of this aluminium alloy is still plainly visible. The colour of the chip samples has changed as a result of the heating process in the furnace. Fig. 1(e) and 1(f) show that the colour of heated aluminium alloy chips changed from shiny silvery grey to dull olive after 30 minutes of heating at the temperature of 600°C and changed to lapwing grey-green colour at the temperature of 650°C. Despite having turned a dark grey tone, certain areas of the chip remain shiny. Fig. 1(g) and 1(h) show that after 30 minutes of heating at 700°C and 750°C, the heated aluminium alloy chips changed from shiny silvery grey to moss grey and graphite grey-green, respectively. The heated chip samples had the same appearance with a matte, dusty texture. The aluminium granules and chip's surface developed a thin oxide layer due to oxidation during heating in the furnace, turning the glossy finish dusty. Depending on its phases, the crystalline oxide condition caused the colour of the aluminium alloy to shift [11].

Oxide structure, surface morphology and elemental analysis. Fig. 2 shows SEM pictures of the surface morphology of the oxides on aluminium granules that had been heated for 30 min at temperatures of 700, 750, 800, and 850°C. On the surface of the granule, an oxide nodule was seen to develop and grow throughout the observation. On the surface of the granules were scattered in a random pattern, both solitary oxide nodules and agglomerated oxide nodules that made up the oxides. When heated to 700°C, the oxide began to develop as a solitary nodule. At 800°C, it expanded close to the first nodules and agglomerated on the surface of the granules. It was noticed that the morphology of the oxides on the surface of the heated granules increased at 850°C, but it is comparable to the morphology at 800°C.

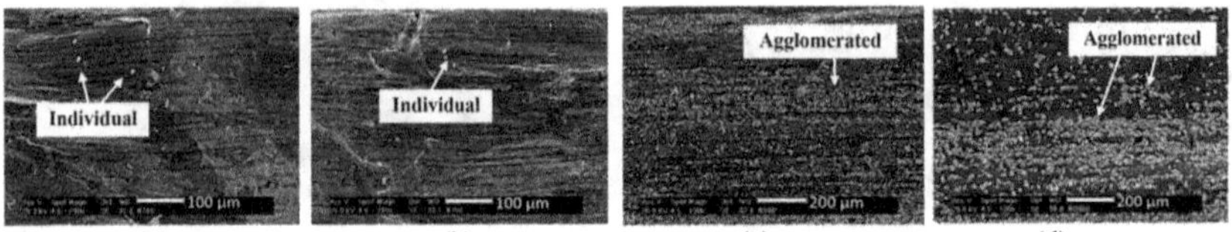

(a) (b) (c) (d)

Fig. 2: Images of oxides morphology on the surface of heated aluminium granules for 30 minutes at temperatures of (a) 700°C (b) 750°C (c) 800°C (d) 850°C.

Fig. 3 shows SEM images of the oxide morphology of the surface of the aluminium alloy chips after being heated for 30 minutes at temperatures ranging from 600ºC to 750ºC, which revealed a wrinkled surface and oxide nodular. On the surface of the chips, an oxide nodule was observed to develop and grow. The oxides were made up of individual and agglomerated oxide nodules randomly distributed on the surface of the chips. These figures clearly show that the wrinkles on the surface of chips increased when the heating temperatures increased.

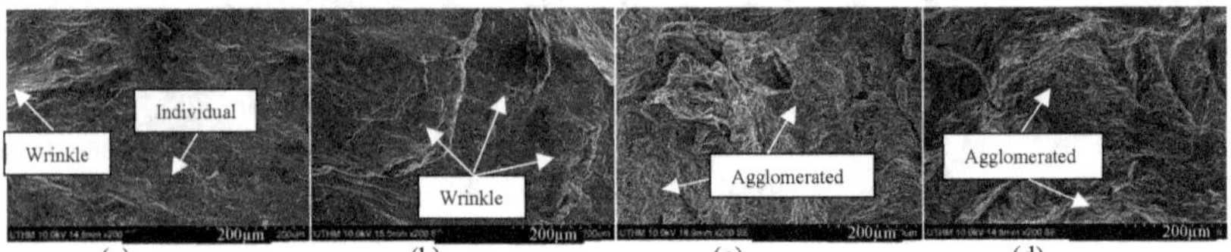

(a) (b) (c) (d)

Fig. 3: SEM images of the oxide morphology of the surface aluminium alloy chips after being heated for 30 minutes at temperatures of (a) 600°C (b) 650°C (c) 700°C (d) 750°C.

Fig. 3(a) shows the oxide generated as an individual nodule, and wrinkles appear after 30 minutes of heating at 600ºC. At 650ºC, the morphology of the oxides on the surface of the heated chips is similar to 600ºC, but its wrinkles were seen as more significant, as shown in Fig. 3(b). Fig. 3(c) shows the oxide nodules that agglomerated, and wrinkles occur in almost all areas of the chip surface after heating at 700ºC. The morphology of the oxides on the surface of the heated chips is comparable to 700ºC for the same duration at 750ºC, but the oxides nodules and wrinkles increase, as shown in Fig. 3(d). This happens when the chips are heated to a high temperature, which causes the oxide form to grow enormously.

EDS analysis was used to detect the presence of the element on the surface of the aluminium granules and aluminium alloy chips after 30 minutes of heating at temperatures 850ºC and 750ºC, respectively. As shown in Fig. 4(a) and 4(b), the results revealed the existence of Al and O peaks, indicating that the heated chips were oxidised. The sample's oxygen concentration of 31.21wt.% showed that the aluminium granules had oxidised during heating. Fig. 4(b) displays the EDS spectrum of an aluminium alloy surface heated for 30 minutes at 750°C. The spectrum shows maxima for aluminium (Al) and oxygen (O) at 32.92 weight percent and 28.92 weight percent, respectively. When heated in a furnace for in-situ melting, aluminium alloy chips oxidised, as evidenced by the oxygen on the oxide morphology of the heated sample. The chips' surface oxidation, which stopped the melted entity from bursting out and creating a molten pool of alloy, impeded total in-situ melting. This may be due to the active surface oxidation of chips and granules during melting, which creates a thin oxide layer enclosed inside the melt [5].

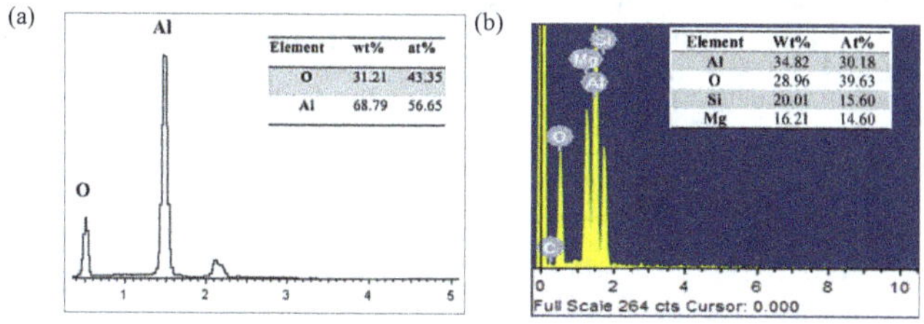

Fig. 4: EDS analysis of the surface of (a) aluminium granules after 30 minutes of heating at 850°C (b) aluminium alloy chips after 30 minutes of heating at 750°C.

Oxide Compounds and phases analysis. Fig. 5 and Fig. 6 depict the XRD analysis of the aluminium granule and aluminium alloy chips before and after 30 minutes of heating and at temperatures 750°C and 850°C. Crystalline aluminium peaks were found in the XRD spectrum for the aluminium granule sample before heating at 2θ angles of 38.49°, 44.68°, 65.04°, 78.19°, and 82.44°, as shown in Fig. 5(a). The oxidation product is thermodynamically stable α-Al_2O_3. The intensity of the aluminium crystalline peaks found before heating (Fig. 5(a)) was decreased, suggesting that the aluminium had oxidised during heating, as seen in Fig. 5(b) and (c).

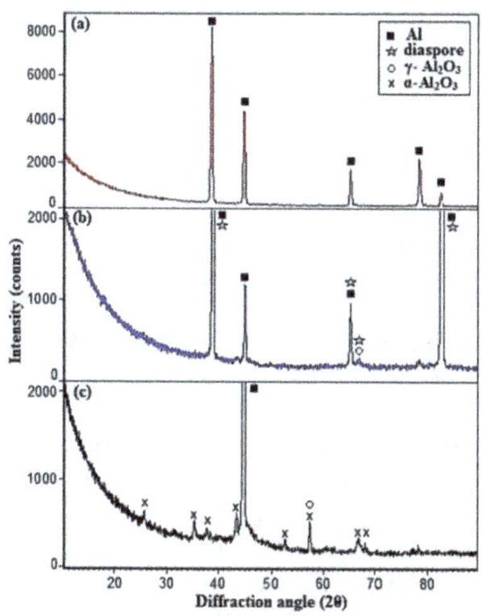

Fig. 5: XRD patterns of aluminium granules (a) before heating and (b) after heating for 30 minutes at 750°C and (c) at 850°C.

Fig. 6 shows the XRD analysis for aluminium alloy chips before and after heating with durations of 30 minutes and at temperatures 650°C and 750°C. The trend of the analysis graph is almost the same as aluminium granules. Crystalline aluminium peaks were found in the XRD spectrum for the aluminium alloy chips sample before heating at 2θ angles of 38.47°, 44.74°,

65.14°, 78.23°, and 82.44°, as shown in Fig. 6(a). Diaspore [α-AlO(OH)] was found in the XRD pattern, which coincided with the crystalline aluminium peaks for the samples heated. The hydration of aluminium during heating caused the production of diaspore. It can also be shown that [γ-Al_2O_3] was identified in both XRD patterns of the heated samples. The transition of boehmite [γ-AlO(OH)] that was produced when aluminium was hydrated resulted in the synthesis of metastable [γ-Al_2O_3]. The intensity of the aluminium crystalline peaks found before heating (Fig. 6(a)) was decreased, suggesting that the aluminium had oxidised during heating, as seen in Fig. 6(b) and (c).

(a)
(b)
(c)

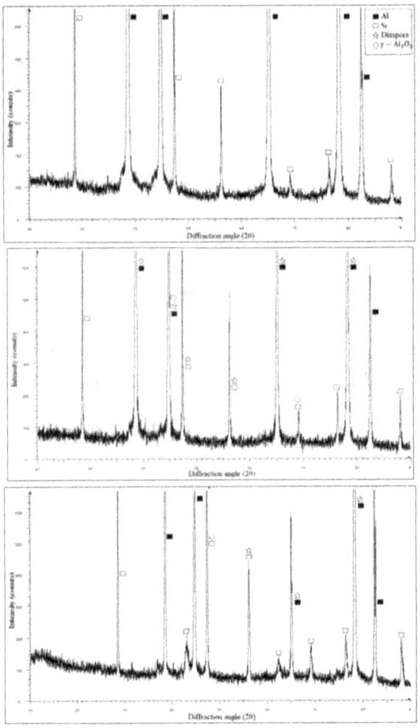

Fig. 6: XRD patterns of aluminium alloy chips (a) before heating and (b) after heating for 30 minutes at 650°C and (c) at 750°C.

Conclusion

When the temperature rises, the heated aluminium granules change colour in the following order: squirrel grey, blue mink, pearl grey, and corvette blue. The heated aluminium alloy chips also changed from shining silvery grey after 30 minutes of heating to dull olive-lapwing grey green-moss grey-graphite grey-green in that sequence. Both aluminium in granules and chips formed were oxidised when heated during in-situ casting experiments forming Al_2O_3 oxide's encapsulating the granules and chips kept them from melting into a pool of molten aluminium. The active oxidation of the granules and chips' surface creates the Al_2O_3 oxide crust that keeps the molten aluminium within. In contrast, concurrent oxidation of the molten aluminium protrusion prevents the granule and chips from completely merging in the ceramic investment casting mould.

Acknowledgement
This research was supported by the Ministry of Higher Education (MOHE) through Fundamental Research Grant Scheme (FRGS) (FRGS/1/2020/TK0/UTHM/03/34) Grant No. K322. Special thanks to those who contributed to this project directly or indirectly.

References

[1] D. de la Fuente, Metals and Alloys, 160-169, 2022. https://doi.org/10.1016/B978-0-12-819726-4.00047-8

[2] S. Kalpakjian, S. R. Schmid and K. S. V. Sekar, "Manufacturing Engineering and Technology, 7th Edition" (Pearson, 2014).

[3] H. Song, L. Zhang, F. Cao, X. Gu and J. Sun, Mat. Lett., 285 (2021). https://doi.org/10.1016/j.matlet.2020.129089

[4] J. Barbosa and H. Puga, J. Mater. Process. Technol. 244, 150-156 (2017). https://doi.org/10.1016/j.jmatprotec.2017.01.031

[5] A. Saleh, International Journal of Advanced Trends in Computer Science and Engineering 9(1.4), 224-228 (2020). https://doi.org/10.30534/ijatcse/2020/3391.42020

[6] R. R. Mishra and A. K. Sharma, Mater. Sci. Eng. A 688, 532-544 (2017). https://doi.org/10.1016/j.msea.2017.02.021

[7] H. Jafari, M. H. Idris and A. Ourdjini, Appl. Mech. Mater. 465-466, 1087-1092 (2014). https://doi.org/10.4028/www.scientific.net/AMM.465-466.1087

[8] Information on https://pubchem.ncbi.nlm.nih.gov/compound/Aluminum.

[9] K. Logesh, P. Hariharasakthisudhan, A. Arul Marcel Moshi, B. S. Rajan and K. Sathickbasha, Materials Research Express 7(1), 015004 (2020). https://doi.org/10.1088/2053-1591/ab587d

[10] M. I. A. Kadir, M. S. Mustapa, N. A. Latif and A. S. Mahdi, Procedia Eng. 184, 687-694 (2017). https://doi.org/10.1016/j.proeng.2017.04.141

[11] A. Saleh and M. H. Idris, Adv. Mater. Res. 1125, 28-32 (2015). https://doi.org/10.4028/www.scientific.net/AMR.1125.28

The Potential of Autonomous Self-Healing Microcapsules for Concrete Cracking

N.M. ZAHARI[1,2], Z. ITAM[1,2 a *], A. SYAMSIR[1], M.H. ZAWAWI[1,2], S.M.M.S.A. FADZIL[2], M.M. ZAINOODIN[2], Z.A.A. HAMID[3], and N.A.M. AZMAN[2]

[1]Institute of Energy Infrastructure (IEI), Universiti Tenaga Nasional, Jalan IKRAM-UNITEN, 43000 Kajang, Selangor, Malaysia

[2]Civil Engineering Department, College of Engineering,, Universiti Tenaga Nasional, Jalan IKRAM-UNITEN, 43000 Kajang, Selangor, Malaysia

[3]School of Materials & Mineral Resources Engineering, Engineering Campus, Universiti Sains Malaysia, Nibong Tebal 14300, Malaysia

[a]iZarina@uniten.edu.my

Keywords: Self Healing Concrete, Microencapsulation, Autonomous Healing

Abstract. Concrete damage and cracking can result from a variety of factors, including poor material selection during construction, design flaws, and environmental factors such freezing and thawing, carbonation, and alkali-silica interactions. Even worse is that damage repairs are more difficult when cracks are not evident or easily accessible. However, self-healing concrete can be utilized for filling the fractures after hardening, which is generated in buildings using the concrete microsphere reaction. Autonomous self-healing concrete is a phrase used for cemented materials, which are separated after the material or structure has been destroyed by a specific kind of degradation process. The development of calcium or calcium hydroxide, special sedimentation, persistent hydration, and swelling are probable factors for the self-healing process. This research was conducted with two sizes of microcapsules imbibed with either sodium or calcium nitrate added into concrete slabs measuring 50mm x 305mm x 305mm. The capsules were added at the volumetric fractions of 0 to 4% in unit intervals to cement. The water- to-cement ratio was taken as 0.4. From this research, it was found that the inclusion of microencapsulated sodium and nitrate silicate has showed that cracked specimens throughout a 28-day healing interval shows improvement in terms of healing. Based on the experimental results, the sodium silicate and calcium nitrate encapsulated in large capsules contribute to greater healing properties than the small capsules.

Introduction

Concrete is a highly important and valuable material for building work and key material utilized in the construction sector. It provides structural flexibility since it can be moulded into a variety of forms. Concrete is very resistant to compression but susceptible to tensile stresses. Therefore, reinforcements must be integrated into the concrete in order to absorb tension. Concrete cracking is one of the main problems with concrete. Concrete damage and cracking can result from a variety of factors, including poor material selection during construction, design flaws, and environmental factors such freezing and thawing, carbonation, and alkali-silica interactions [1]. These little breaks may orientate from a localized region and can propagate throughout the structure and morph into enormous breaks. Aggressive elements like chloride, carbonate, or sulphate can infiltrate concrete that has cracked for any reason. Both the concrete's degeneration and the reinforcements' corrosion may be results of these attacks. The steel reinforcements will consequently be exposed to air and water, and start to corrode, which results in the decrease of structural stiffness, service life, and dependability. Cracks are causing considerable damage to large projects or high-rise structures.

These catastrophic failures could have a big influence on new and existing structures' upkeep, rehabilitation, and repair. Evidently, the deterioration of goods or infrastructure is a persistent and alarming issue. If the harm is not contained at an early stage, the knock-on consequences could have far-reaching effects. Consequently, the economic, social, and environmental effects of this catastrophic failure may be significantly impacted. To stop major harm from occurring, preliminary action should be taken at the microlevel.

Self-healing in concrete cracking. Self-healing techniques, which relate to a system's capacity to recover following initial injury, have the capability to handle these failures. Self-healing technology can improve repairability, or extend product lives by reducing molecular level functioning. The idea of self-healing was influenced by biological or natural processes, such as blood clotting or the repair of broken bones. Later, a larger range of engineering applications adopted this approach. In some circumstances, hand repair work can be challenging, particularly if the cracks or damages are concealed or difficult to access [2]. For a wide range of applications, microencapsulation of certain elements, whether solid or liquid, has attracted the interest of numerous researchers in recent years. In the realm of civil engineering, microencapsulated self-healing cementitious elements have recently grown in popularity. They were inspired by the idea of biological regeneration of living tissue. Concrete cracking and damage problems may have an alternate solution thanks to the success of self-healing materials included in concrete constructions.

According to Lucianna Restuccia et al. (2017), concrete is the world's most extensively utilised construction material. In addition, damage repairs are more difficult when cracks are not evident or easily accessible. In the construction sector, for example, a substance known as liquid glass is often employed as a concrete waterproofing ingredient. It has previously been used in several studies on self-healing cementing materials, which produced rather excellent results [3]. Self-healing concrete is a compound that generates calcareous material on the surface of concrete buildings to mend breaks. The concrete materials are combined with the calcium nutrients known as calcium lactate, nitrogen, and phosphorus to precisely chosen strains of the Bacillus species. These self-healing chemicals will last in concrete for up to 200 years [4]. The microencapsulated capsules must first be manufactured before the self-healing microcapsules may be used in concrete. During the mixing process for concrete, these mechanically activated self-healing microcapsules and their catalysts are added to the cementitious mixture. The self-healing microcapsules are incorporated into the concrete structure after curing. Fig. 1 depicts the self-healing microcapsules in concrete and their mechanism. The imbedded self-healing microcapsules usually break when exposed to concrete cracking, releasing the healing ingredient that is enclosed inside. The healing agent polymerizes when it comes into contact with the implanted catalyst, sealing and repairing the cracks [5].

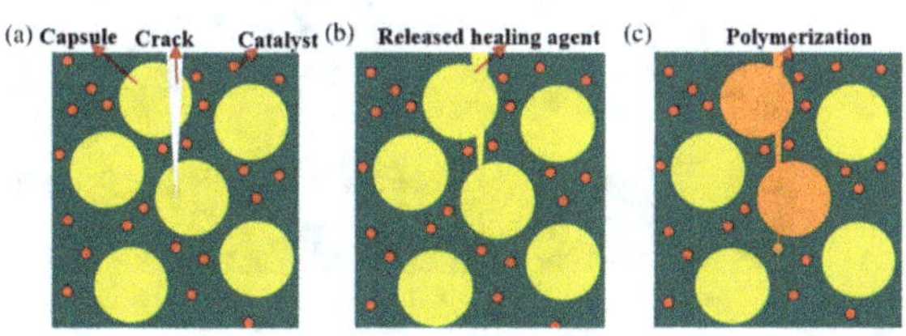

Fig. 1: Self-healing schematic concept for cementitious materials that are (a) crack forming, (b) healing agent releasing, and in (c) polymerization with catalysts [5].

Jose Milla et al. (2016) examined the usage of micro-encapsulated calcium nitrate in self-healing concrete. The compressive strength of microcapsulated concrete (by a percentage of the cement weight) was compared with that of control specimens without microcapsules of the same mix design. Using surface resistivity measurements the surface permeability of concrete specimens with and without microcapsules was assessed. The elasticity test module (ASTM C469) was employed for self-healing capacity, with measurements recorded after 14 days before and after injury Table 1 shows the concrete's compressive strength was directly influenced by the size and concentration of the microcapsules introduced. The strength of the concrete reduced as the microcapsule content rose. The strength of the concrete was also impacted by the microcapsule size. The cement matrix's air content rose as the microcapsule size increased [6].

Table 1 : Compressive Strength Results for Samples Before Autonomous Healing [6]

Agitation Rate (rpm)	Microcapsule Concentration (%)	Sample Number	Type Break	Compressive Strength (MPa)	Average (MPa)	SD
800		1	1	27.70		
	1.20	2	2	27.78	27.55	0.33
		3	1	27.18		
800		1	3	10.40		
	2.00	2	2	10.97	10.72	0.29
		3	3	10.79		
Control		1	6	42.90		
	na	2	6	40.69	41.06	1.69
		3	1	39.58		

NOTE: na = not applicable.

P. Giannaros et al. (2016) studied the sealing of cracks in cement using microencapsulated sodium silicate in order to measure the influence on rheological and mechanical qualities of two distinct sodium silicate microcapsules (T130 and L500). During a 28-day healing period, the inclusion of micro-encapsulated silicate of 4% (by cement volume) has shown a decrease in the cracking that occurred in concrete specimens as can be seen in Fig. 2 [7].

(a)

(b)

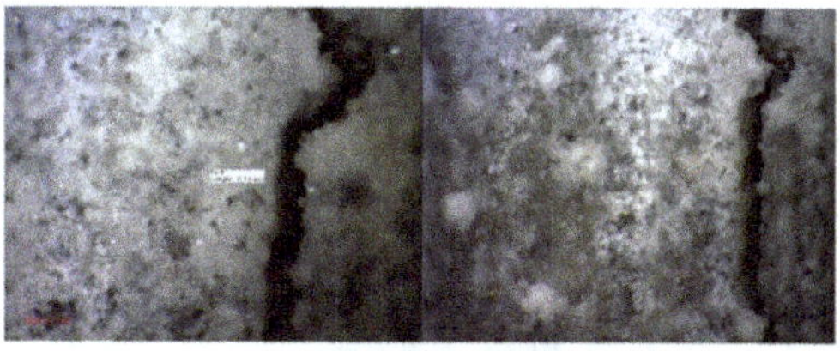

(c)

Fig. 2 : Crack formation in (a) cement specimens, (b) cement specimens of 4% L500 microcapsules, and (c) cement specimens of 4% T130 microcapsules. Images on the left show samples after 7 days of healing, while the ones on the right indicate a 28-day healing period [7].

Research Methodology

This study intends to examine the use of self-healing concrete microspheres. For observing the autonomous self-healing in the concrete, the drop weight impact test, and the Scanning Electron Microscopy (SEM) test will be conducted. The drop weight impact test is carried to initiate cracking in concrete containing self-healing microcapsules. SEM is a test procedure that scans an electron beam sample to provide a magnified picture for examination. The approach is also known as SEM analysis and SEM microscopy and is utilized in micro-analysis and failure analysis of solid inorganic materials extremely successfully.

For the preparation of the capsules, a heated and unheated capsule made from polylactic acid (PLA) will be employed. The capsules are cooked in an oven at 70°C for 20 to 40 minutes. A 20L planetary mixer is used to perform a mixing test and the method is depicted in Fig. 3. The surviving capsules will be studied using optical microscopy.

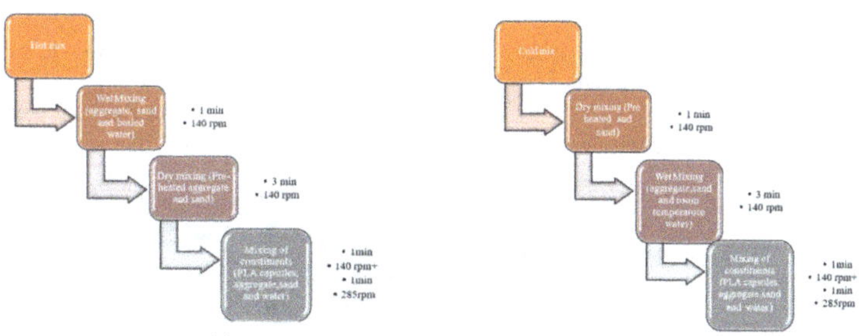

(a) (b)

Fig. 3: Mixing Processes of the polymeric capsules.

The cylindrical capsules were fabricated with dimensions 50mm x 7mm, and 50mm x 2mm for the self-healing agent to heal slab concrete from the crack on the surface as shown in Fig. 4.

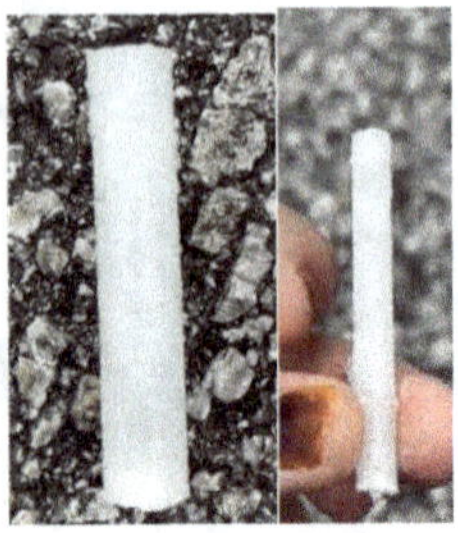

Fig. 4 : 50mm x 7mm and 50mm x 2mm PLA capsules.

The capsules were to encapsulate two solutions, which are calcium nitrate and sodium nitrate to assess its influence on rheological and mechanical qualities as seen in Fig. 5.

Calcium Nitrate (b) Sodium Nitrate

Fig. 5 : Capsule solutions.

The capsules were added at the volumetric fractions of 0 to 4% in unit intervals to cement, with the water-to-cement ratio of 0.4. Slab dimensions were set as 50mm x 305mm x 305mm as reflected in Fig. 6. Slab ingredients are volumetrically measured with the ratio for cement:coarse aggregate:fine aggregate taken as 1:1:9:3.

Fig. 6 : Slab sizing 50mm x 305mm x 305mm.

Samples sent for testing contain 4% of capsules for four slab mixtures, which individually contains large sodium silicate capsules, small sodium silicate capsules, large calcium nitrate capsules, and small calcium nitrate capsules. The drop impact testing was carried out after 28 days of curing to evaluate the impact of the microcapsule on the intrinsic characteristics of the concrete, after which, SEM test was to be conducted to obtain results before the healing process begins. The cracked samples will then be left to heal by immersion in water for 28 days. Then, the samples were examined in the regions containing both concrete cracking as well as the released capsule mixture.

Results and Discussion

The images shown in Fig. 7 and 8 show that the embedded microcapsules were big enough to be visible with the naked eye. Both the sodium silicate and calcium nitrate admixtures could also be seen to have escaped the capsules due to breakage, and has solidified into the surrounding locations, proving that the capsules have accomplished the expected process of self-healing on the 28 days of concrete lifespan.

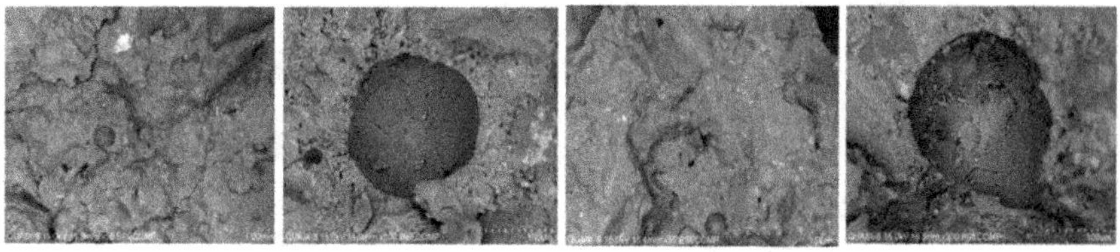

a) b) c) d)

Fig. 7 : Magnified Image Capsule of Calcium Nitrate, a) Large (x50), b) Large (x300), c) Small (x50), d) Small (x300).

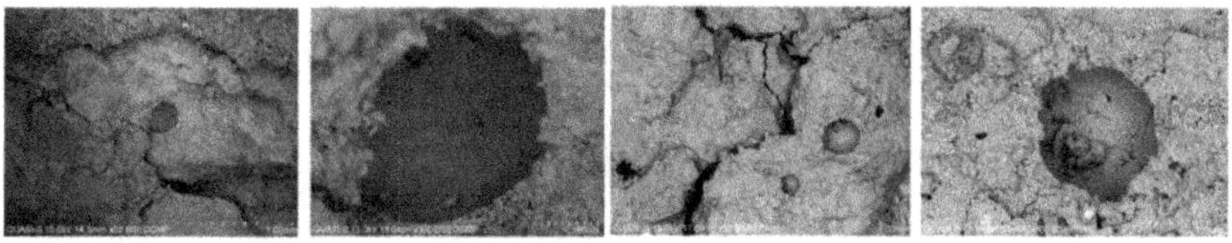

a) b) c) d)

Fig. 8 : Magnified Image Capsule of Sodium Silicate, a) Large (x50), b) Large (x300), c) Small (x50), d) Small (x300).

In terms of autonomous self-healing, it can be seen from Fig. 9 that the measurement of the crack on the specimen containing large capsules of calcium nitrate is 0.3 mm during initial cracking, which was reduced to 0.14 mm after the autonomous self-healing took place. The measurement of the crack on the specimen containing small capsules of calcium nitrate was 0.18 mm during initial cracking, and also showed reduction to 0.15 mm after the autonomous self-healing. As for the large capsules containing sodium nitrate, the crack width reduced from 0.21 mm to 0.1 mm after autonomous self-healing, while the crack width reduced from 0.2 mm to 0.18 mm for the small capsules, as seen in Fig. 10. This could be attributed to the fact that larger capsules tend to contain a greater amount of sodium and calcium nitrate when compared to the smaller sized capsules, causing more reaction and healing of concrete cracking to take place. The experiment also reveals that both sodium and calcium nitrate may be utilized to treat concrete cracking autonomously.

Materials Research Proceedings 68 (2026) 246-253 https://doi.org/10.21741/9781644904237-33

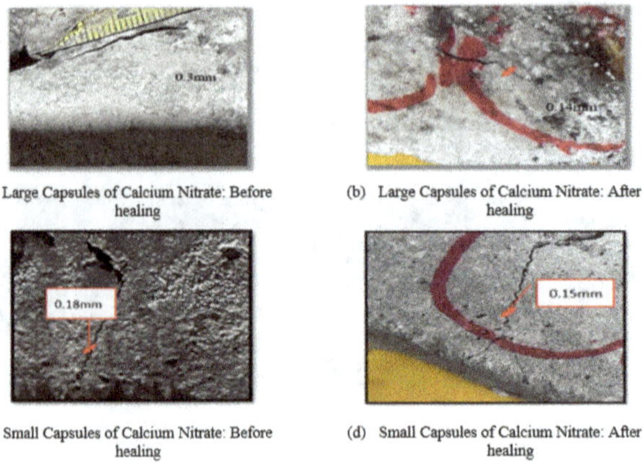

(a) Large Capsules of Calcium Nitrate: Before healing
(b) Large Capsules of Calcium Nitrate: After healing
(c) Small Capsules of Calcium Nitrate: Before healing
(d) Small Capsules of Calcium Nitrate: After healing

Fig. 9 : Before and After Concrete Cracking Autonomous Healing: Calcium Nitrate Microcapsules.

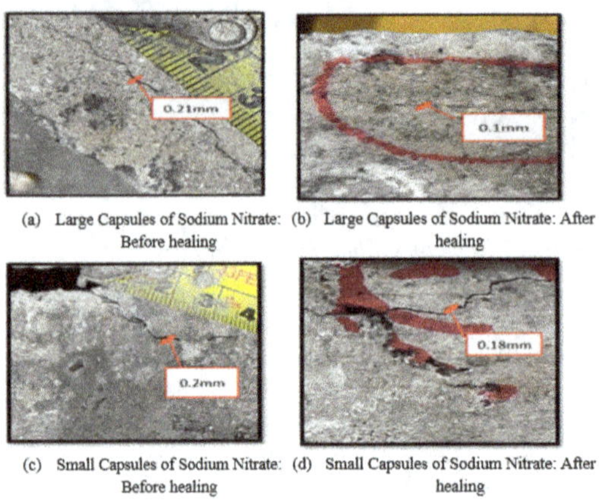

(a) Large Capsules of Sodium Nitrate: Before healing
(b) Large Capsules of Sodium Nitrate: After healing
(c) Small Capsules of Sodium Nitrate: Before healing
(d) Small Capsules of Sodium Nitrate: After healing

Fig. 10 : Before and After Concrete Cracking Autonomous Healing: Sodium Nitrate Microcapsules.

Conclusion

In conclusion, the microspheres are capable of healing concrete cracking. The inclusion of 4% microencapsulated sodium silicate (in terms of cement volume) has showed that cracked specimens throughout a 28-day healing interval shows improvement in terms of healing. Based on the experimental results, the sodium silicate and calcium nitrate encapsulated in large capsules contribute to greater healing properties than the small capsules since the amount of sodium silicate or calcium nitrate is higher. Future study could benefit in focusing on the optimization and amount of the volumetric fraction of microcapsules essential for a given healing degree.

Acknowledgement

This research was funded by Universiti Tenaga National (UNITEN) in collaboration with Universiti Sains Malaysia, grant number BOLD 2022 J510050002/2022024 awarded by UNITEN.

References

[1] Z. Itam, S. Beddu, D. Mohammad, N.L.M. Kamal, N.A. Razak, and Z.A.A. Hamid, Mater Today: Proc. 17(3), 717-726, (2019). https://doi.org/10.1016/j.matpr.2019.06.355

[2] N. N. F. N. M. N. Kahar, A. F. Osman, E. Alosime, N. Arsat, N. A. M. Azman, A. Syamsir, Z. Itam, and Z. A. A. Hamid, Polymers (Basel) 13(8), 1194, (2021). https://doi.org/10.3390/polym13081194

[3] L. Restuccia, A. Reggio, G. A. Ferro, and J. Tulliani, Procedia Struct. Integr. 3, 253-260, (2017). https://doi.org/10.1016/j.prostr.2017.04.016

[4] Information on https://www.academia.edu/19624898/Self_Healing_Concrete.

[5] Xue, Caihong, W. Li, J. Li, V. W Y Tam, and G. Ye, Struct. Concr. 20(1) 198-212, (2019). https://doi.org/10.1002/suco.201800177

[6] J. Milla, M. Hassan, and T. Rupnow, J. Mater. Civ. Eng. 29(12), (2017). https://doi.org/10.1061/(ASCE)MT.1943-5533.0002072

[7] P. Giannaros, A. Kanellopoulos, and A. Al-Tabbaa, "Damage recovery in self-healing concrete", (HealCON Conference, 2016).

Influence of Fusion Welding Processes on Microstructure and Mechanical Properties of Dissimilar Metal (AISI 304 SS-Copper) Weldment

Bikash Ranjan MOHARANA[1,3,b], Aezeden MOHAMED[2,a*], Sushant Kumar SAHU[3,c], Susanta Kumar SAHOO[3,d], Dillip Kumar BISWAL[4,e], Tapas Kumar MOHARANA[5,f] and Kamalakanta MUDULI[2,g]

[1]Department of Mechanical Engineering, C V Raman Global University, Odisha, India-752054

[2]Mechanical Engineering Department, PNG University of Technology, Lae 411, Papua New Guinea

[3]Department of Mechanical Engineering, National Institute of Technology, Rourkela, India-769008

[4]KMBB College of Engineering and Technology, Odisha, India-752056

[5]National Council of Science Museums, Kolkata, West Bengal, India-700091

[a]aezeden.mohamed@pnguot.ac.pg, [b]bikashrnjn@gmail.com, [c]sushanta.sahu@outlook.com, [d]sks@nitrkl.ac.in, [e]dillipkumarbiswal@gmail.com, [f]mechengineer.tapas@gmail.com, [g]kamalakantam@gmail.com

Keywords: Dissimilar Metal Wedding, Microstructure Analysis, Mechanical Analysis

Abstract. Dissimilar metal welding of AISI 304 stainless steel (304 SS) and commercially pure copper (CP copper) has been taken up to understand the influence of the different fusion welding processes on microstructure and mechanical properties of the weldment. The fusion welding processes specifically involve continuous wave CO_2 laser, pulsed wave Nd: YAG laser and pulsed wave tungsten inert gas (P-TIG) welding. The primary objective of the present work is to develop a successful weld between two dissimilar metals without using any filler material by different fusion welding processes. An overall comparison of the results obtained from different sets of analyses is represented. The optical image of the top surface of the weld revealed that the no spatter formation during CO_2 laser and P-TIG welding process, whereas minimal spatter for Nd: YAG laser. In comparison to laser processes, the weld face formed by P-TIG welding was much wider. However, 100% penetration was achieved with all three types of welding. The average widths of heat affected zone (HAZ) found to be 10 μm and 23 μm for CO_2 and Nd:YAG laser respectively. The way the microstructure changes at the fusion zone was the same in all cases. For example, copper channels form, dendrites grow epitaxially, etc. SEM pictures indicate the Cu tubular channels observed more often for P-TIG welding. The microhardness distribution in the weld zone was found to be larger than that of copper and close to that of 304 SS. The acceptable ultimate tensile stress (UTS) for dissimilar metal welding of SS-Cu couples was found 275 MPa, 278 MPa and 195 MPa for CO_2 laser, P-TIG and Nd:YAG laser respectively.

Introduction

Welding is a widely used manufacturing technology for fabrication and repair work all over the globe. A variety of welding techniques are available to meet the needs of various situations and materials [1]. The scientific community has paid varied amounts of attention to each approach, which has led to investigation and process improvement. Based upon the market research report, the global market for welding equipment, accessories, and consumables is expected to USD 32.5 billion in 2026, at a compound annual growth rate (CAGR) of 4.6% during the forecast period [2].

"Grand View Research" reported the global market size for welding products is expected to reach USD 23.51 billion by 2027 [3]. The market is expected to rise as a result of quick industrialization and rising investments in infrastructure projects in key economies including the countries like United States, the United Kingdom, China, India, and Brazil. The power generating, construction, military, food processing, and aviation industries are reportedly the biggest users of welding.

Dissimilar welding is the one in which two materials with different core properties are welded together. Because the two materials may not naturally fit together, this is often a difficult operation. Despite all of these challenges, dissimilar welding is rapidly being used for a wide variety of applications. Stainless steel (SS) and copper dissimilar couple is used in critical applications in energy, military, aviation, food processing sectors, etc. In all these applications, copper acts as an intermediary in heat transfer, while SS offers the required structural stiffness and corrosion resistance. Depending upon the specificities of the application, different grades of SS and copper are joined together. For example: SS 316 offers resistance to chlorides and sulphates. SS 309 is used for handling of similar acids at relatively higher temperatures. Type 321 and 347 are used where the fluid has high concentrations of chloride. Oxygen-free copper canisters are used for underground disposal of radioactive waste. The expected design life of such waste is assumed to be 100000 years [4, 5]. Use in such demanding applications necessitate that the welds be of extremely high quality and be durable.

Although, so many different types of welding techniques and their standards are prescribed, still many types of investigations are going on for method improvement, application to new materials, new heat source, etc. Each welding technique has its own advantages and limitations related to their applications. Choosing a suitable welding technique depends upon various factors viz. the materials to be welded, process parameters, cost, deposition rate, joint strength, tolerance towards distortion due to residual stress, etc.

Numerous researchers are investigated the weldability of particular austenitic stainless steel and copper combination by different fusion welding processes. Rowcliffe et al. [6] explained the behavior of the austenitic steel (316L) and copper alloys (GLIDCop Al25 and Cu-Cr-Zr) bi-layered couple in neutron environment. The microstructural damage, fracture behavior, and deformation for all cases due to radiation were summarized. As a first wall heat sink material, GLIDCop Al25 affects more mechanically due to its high fracture toughness and poor resistance to the crack growth parallel to the bi-layer interface. Yilmaz and Aksoy [7] investigated the weld possibilities between 304 SS and electrolytic Cu through diffusion welding method. Here authors investigated the microcrack occurrence condition by intrinsic and chemical diffusion coefficients, which have obtained by Heumann and Darken methods. Sabetghadam et al. [8] explained the welding between stainless steel (grade 410) and copper through a diffusion bonding process using nickel interlayer at a temperature range of 800–950 °C. The phase constitutions and their related microstructure at the Cu/Ni and Ni/SS diffusion bonding interfaces have been studied. Microvoids also found between Ni layer and Cu during microstructure analysis. The tensile strength increased with temperature up to 900^{0}C and measured to be 145 MPa. Higher microhardness peaks found in the Ni interface region compared to the base materials. Kundu et al. [9] also investigated the weld characteristics between CP Ti and AISI 304 SS with 300 μm thick copper layer by solid-state diffusion bonding process. Magnabosco et al. [10] explained the joining of copper plates with three different austenitic stainless steel plates by electron beam welding. Cracks and porosity have found in the weld pool from the morphological and microstructural study. They also suggested that a precise parametric optimization is required to achieve perfect welding. They explained about the presence of steel and copper globules inside the weld pool, but not in details. Chen et al. [11] investigated the effectiveness of processing parameters such as welding speed, laser power, offset and incline angle of the laser beam, etc., on mechanical as well as metallurgical properties. They explained that the tensile strength was weakly dependent on melting of the copper; however, the melting of the copper induced a decrease in the toughness of the joint. Excess of copper melting may cause the microcrack formation inside the fusion zone. The author found spherical copper globules formed inside the fusion zone, confirmed from the EDS analysis [12,13,14].

A very limited approach reported on joining of dissimilar metal combination via. TIG welding process without addition of filler material or interlayer. The most common and effective practice to add filler or interlayer like Ni, Fe, bronze, nickel based super alloys, etc. for joining ferrous and nonferrous metal combination [15,16,17,18]. P-TIG welding technique also an effective process of joining similar and dissimilar metal combination as reported by several researchers. The authors suggested that the P-TIG welding demonstrated superior weldability performance in every mechanical and metallurgical aspect, and it is highly reliable for joining dissimilar material combinations [19,20,21,22,23,24,25,26,27].

This article focuses on the comparative evaluation of the effect of welding processes, namely, continuous wave CO_2 laser, pulsed wave Nd: YAG laser and pulsed wave tungsten inert gas (P-TIG) welding on the microstructure and mechanical properties of dissimilar metal weldment of AISI 304 SS and CP Copper.

Experimental Procedure

For this present research work, three different fusion welding processes CO_2 laser, Nd:YAG laser and P-TIG considering two different metals i.e., AISI 304 SS and CP copper were performed. The chemical compositions of both the materials obtained through EDS analysis are listed in Table 1. The schematic of welded specimen has been shown in Fig. 1. The welding parameters taken for experimentations via CO_2 laser, Nd:YAG laser and P-TIG are given in Table 2, 3 and 4 respectively. All three fusion welding processes are of square-butt joint weld without addition of filler material. After completion of all experimental work, the weld specimens are gone through different mechanical and metallurgical examination for analysis.

Table 1: Chemical composition (weight fraction, %) of 304 SS and Cu.

Composition	C	Si	Mn	Ni	Cr	Cu	Fe
AISI 304 SS	0.04	0.29	0.72	7.13	17.71	---	Balance
CP Cu	---	---	---	---	---	99.63	---

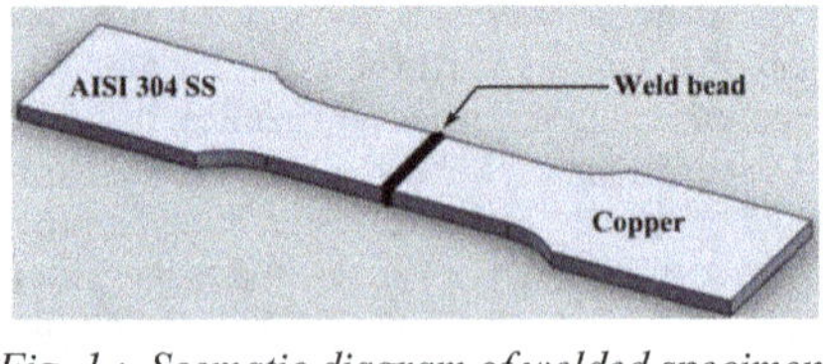

Fig. 1 : Scematic diagram of welded specimen.

Table 2 : CO_2 laser welding parameters for experimentations

Parameters	Values
Laser power	2 – 4 kW
Welding speed	1 – 4 m/min
Beam Diameter	0.18 mm
Gas Flow Rate	Argon, 20 lmp
Beam Profile	Gaussian Mode
Focal Position	At the surface
No. of pass	1

Table 3 : Nd:YAG laser welding parameters for experimentations

Parameters	Values
Laser power	3 – 4 kW
Pulse duration	2 – 6 ms
Welding speed	1 – 3 mm/sec
Beam Diameter	0.2 mm
Frequency	10 Hz
Gas Flow Rate	Argon, 10 lmp
Beam Profile	Gaussian Mode
Focal Position	At the surface
No. of pass	1

Table 4 : Pulsed TIG parameter for experimentations.

Parameter	Values
Peak current	80 – 110 Amp
Welding speed	90 – 150 mm/min
Frequency	500 Hz
No. of pass	1
Electrode size	1.6 mm
Polarity	DCEN (Straight polarity)
Gas flow rate	Argon, 10 lpm

Results and Discussion

An attempt has been made to compare and discuss the metallurgical properties such as macrostructure, microstructure, etc., of the weld pool geometry for all three types of fusion welding processes. Macrostructure refers to the top surface morphology and weld pool shape, whereas phase formation, compositional analysis, microhardness, etc., belongs to microstructure analysis as follows.

Weld Top Surface Morphology. Top surface of the weld pool in any types of welding process gives an approximate idea concerning the welding characteristics and suitability of the process. A smooth appearance not only gives a good surface finish but also indicates stable welding parameters. Optical images of top surface for all three types of welding processes are shown in Fig. 2. Surfaces look smooth in all cases but the presence of spatters are prominent in pulsed wave Nd:YAG laser welding process. As P-TIG and Nd:YAG laser welding processes are pulsed wave type, the pattern looks similar. The overall surface inequality of the weld pool compared to label of base metals are 80 – 140 μm and 120 – 230 μm in both the direction for laser and TIG welding process respectively, measured by surface profilometer. It is marked that both welding speed and input power enhance the weld crown width. Crown surface smoothness increases with frequency and decreases with weld speed, becomes smooth for CO2 laser. At the same time, with decrease in weld speed, weld area as well as the crown-to-root opening ratio decreases. The comparisons of results have shown that the input power and welding speed should be optimized in order to minimize the weld area and weld crown smoothness.

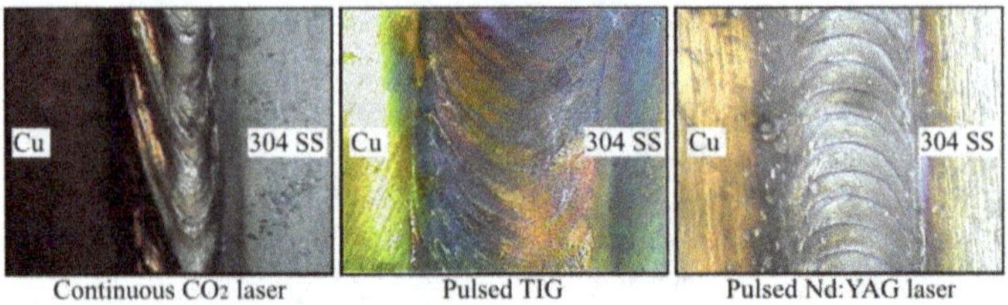

Fig. 2 : Weld pool top surface for different welding processes.

Weld Bead Geometry. The weld pool geometry for all three welding processes is different to each other depending on the nature of heat flux. The shape of the weld pool in CO2 laser welding process is of keyhole shape type, whereas in P-TIG and Nd:YAG process refers to conduction type as shown in Fig. 3.

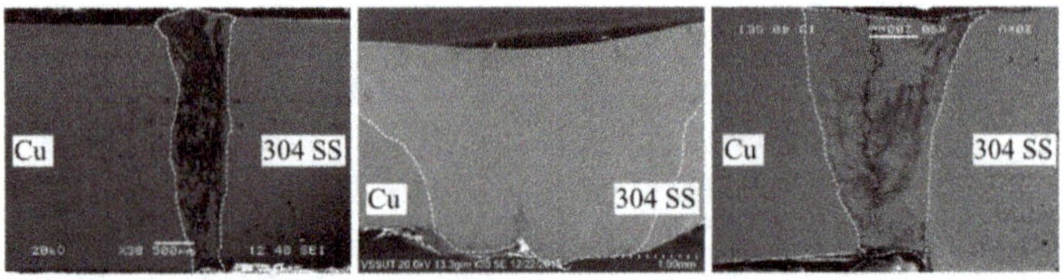

Fig. 3 : Weld pool geometry for different welding processes.

The width of the weld face and HAZ for TIG welding are quite larger compare to other two laser welding processes. The average face width for CO_2 and Nd:YAG laser are measured as 950 μm and 650 μm respectively. However, the HAZ is quite small in CO_2 laser welding process. The average widths of HAZ found to be 10 μm and 23 μm for CO_2 and Nd:YAG laser respectively. It is seen that differential material properties in the dissimilar welding processes prevent high-quality weld bead shape. Fundamentally, input power and welding speed selection is important to optimize the heat input to minimize the weld area, which would provide a satisfactory weld with reliable quality.

Microstructure Analysis. The microstructural behavior at the fusion zone is found to be similar in all the cases i.e. formation of copper channels, epitaxial growth of dendrites etc. The solidification growth of weld pool from the fusion boundary is more prominent in laser welding processes. The Cu tubular channels found more in case of TIG welding, confirms from SEM images. Fig. 4 shows the presence of Cu channels inside the fusion zone for different welding processes. These may be formed due to mutual immiscibility, and SS & copper in a molten state. Their characteristics are very much influenced by the Reynolds number of the flow active in that locality during melt surge.

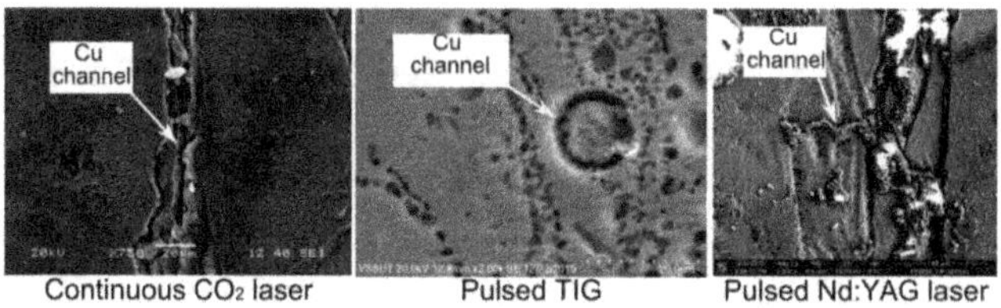

Fig. 4 : Cu tubular channels inside the fusion zone.

Lamellar dendrites are observed in CO_2 laser welding, as it indicates eutectic microstructure, which refers to higher yield strength. The average thickness of the lamellae is approximately 0.93 μm. Dendrites are formed mainly due to the diffusion of Cr, Ni, Fe and Cu resulting intermetallic formation. Rectangular shape nano-structured Cu grains are perceived inside the weld pool in pulsed Nd:YAG laser welding. The average size of those Cu grains are varies from 0.75 μm to 1.95 μm approximately. Unmixed zones are traced near the fusion boundary at the copper side in both pulsed TIG and Nd:YAG laser welding processes. The copper and steel arranged in layer by layer instead of mixing, due to a lack of cooling time. We know that one can get finer dendritic structure at higher cooling rate, but at the same time we require full penetration and less unmixed zone. On the other hand, copper has good thermal and reflectivity characteristics, which prevent it. Hence larger input power with increased welding speed is required to get better result.

Microhardness Analysis. Microhardness number also an indirect indicator of the tensile property of the material and identify the effect of microstructural heterogeneities. Microhardness values are measured individually for all types of welding processes. The average microhardness number for AISI 304 SS and CP copper are 50-80 HV and 190-215 HV respectively.

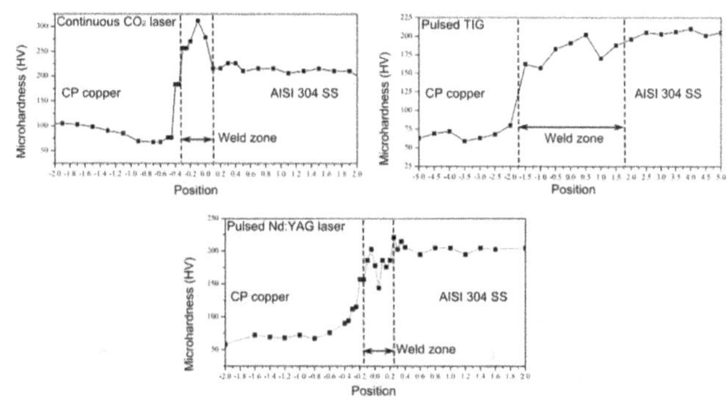

Fig. 5 : Microhardness distribution for different welding processes.

The variation in number observed in the fusion zone and HAZ for different welding processes. Highest microhardness values are obtained in CO2 laser welding process i.e. 290-310 HV, which is much higher than the parent materials. In the case of TIG and Nd:YAG laser welds, the microhardness distribution in weld zone found to be greater than copper and near to the 304 SS values. A central depression also observed at the middle of the weld pool. The comparisons between microhardness distributions for all welding processes are shown in Fig. 5. It is observed that there was a gradual increase in the hardness from copper plate to weld zone. It reached to the maximum at the middle of the joint and then remains steady to the SS side for TIG welding, decreases for CO_2 laser welding and some undulation for Nd-YAG laser welding. The mode of welding (TIG/continuous laser/pulse laser) and degree of copper dilution gives rise to different amount of cooling rate variation in the fusion zone from copper side to SS side. This control the solidification microstructure size and formation of IMCs to exhibit variation of hardness in the weld pool for different processes.

Tensile Strength Analysis. The mechanical property i.e., tensile strength is ultimate output to evaluate the acceptance of any welding. The tensile stress of the weld zone should greater than the base material. Because of two base materials, i.e., AISI 304 SS and CP copper, the stress value should be greater than copper tensile stress. The fracture should occur outside the weld zone.

The measured ultimate tensile stresses for dissimilar metal welding of SS-Cu couples by continuous wave CO_2 laser has been found 275 MPa, which is much higher than the stress value of copper. The fractures are occurred outside the welding zone. In pulsed TIG welding process, lots of trial experimentations have been carried out. However, the highest tensile stress value obtained 278 MPa. Such samples are also elongated more compared to others. Very poor results have been recorded for the dissimilar metal welding via pulsed wave Nd:YAG laser welding process. The maximum tensile stress result obtained 195 MPa.

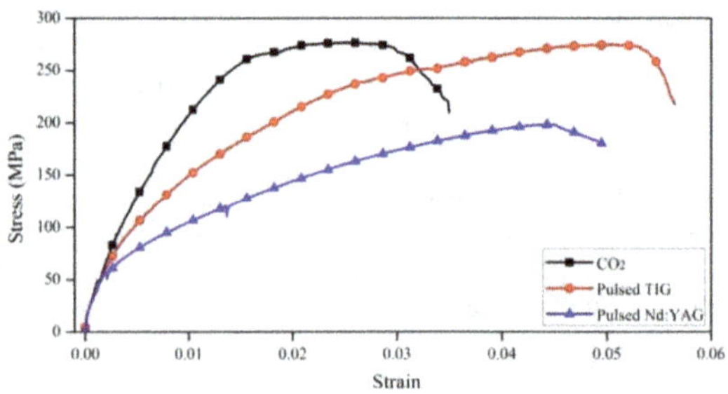

Fig. 6 : Tensile stress vs. strain graph for different welding processes.

The utmost tensile stress values obtained from each welding processes are shown in Fig. 6. The pulsed TIG welding process is showing premier strain rate followed by pulsed Nd:YAG and CO_2 welding process. It is observed that more heat input (at optimized input power and weld speed settings) with full penetration can give higher tensile strength.

Fracture Surface Morphology. Fracture surface morphology investigation has been carried out after completion of tensile test as shown in Fig. 7. The fracture surfaces are showing very similar result in all three types of welding processes. Both ductile and brittle fracture happens in the fusion zone. The cleavage fracture detected in TIG welding process. It is observed that two types of failure modes can take place in a tensile mode of failure. If the failures come about in the base metal, ductile type fractured surfaces with dimple and cone like features would arise. These features are a result of plastic deformation before failure. But brittle type fractured surface will be seen when failures happen in the area inside or closer to the weld pool interface.

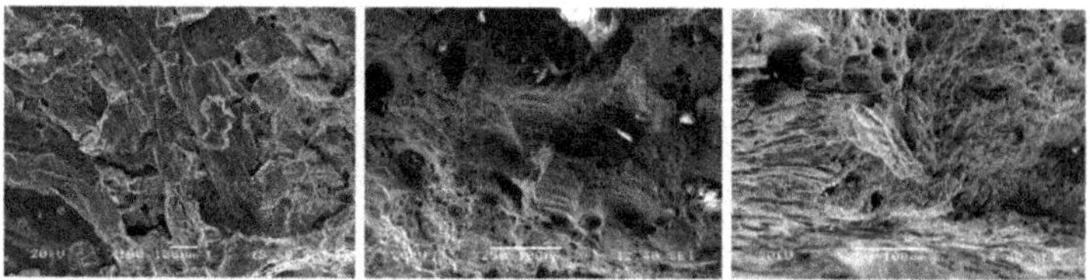

Continuous CO_2 laser Pulsed TIG Pulsed Nd:YAG laser

Fig. 7 : Fracture surface morphology of weld after tensile test for different welding processes.

Conclusion

The comparisons and similarities in mechanical as well as metallurgical aspects in welding of dissimilar metal couples "AISI SS 304 and CP copper" by three different types of fusion welding processes such as continuous wave CO_2 laser, pulsed wave TIG, and pulsed wave Nd:YAG laser have been explained. The following conclusions can be drawn from the observations, which is the top surface of the welded specimens for laser as well as TIG welding processes are found smooth enough as the bump or depression of the weld pool compare to the base metal. Surface profiler measurement confirms that the laser welded specimens are smoother than TIG.Full penetration weld pool is observed for all welding processes. However, the weld face is much wider in TIG welding compares to laser processes. As the welding materials are same for all processes, many similarities are found in microstructural aspects. The highest microhardness number has been observed for CO_2 laser welding process followed by pulsed TIG and pulsed Nd:YAG. The tensile stress results for CO_2 and TIG welding processes are found to be much higher than Nd:YAG. However, the strain rate in TIG welding is more than other two.Similar pattern fracture has been obtained for all three types of welding processes.

References

[1] L. Jeffus, "Welding" (Cengage Learning, 2011).

[2] "Welding Equipment, Accessories & Consumables Market by Equipment, Accessory, Consumable (Electrodes & Filler Materials, Fluxes & Wires Gases), Technology (Arc Welding, Oxy-Fuel Welding), End-Use Industry, and Region - Global Forecast to 2026", (Market Research Report, MarketsandMarkets Research Private Ltd., MM 4523, April 2022).

[3] "Welding Products Market Size, Share & Trends Analysis Report by Technology (Arc, Resistance), by Product (Stick Electrodes, Solid Wires), by Application, by Region, and Segment Forecasts, 2020 - 2027", (Grand View Research, ID: 4118822, March 2020).

[4] H. Aalto, R. Hannu and L. Lenni, "Production methods and costs of oxygen free copper canisters for nuclear waste disposal" (Finland: N. p., 1996. Web).

[5] A. Sanderson, C. S. Punshon and J. D. Russell, Fusion engineering and design 49, 77-87 (2000). https://doi.org/10.1016/S0920-3796(00)00407-5

[6] A. F. Rowcliffe, S. J. Zinkle, J. F.Stubbins, D. J. Edwards and D. J. Alexander, J. Nucl. Mater. 258, 183-192 (1998). https://doi.org/10.1016/S0022-3115(98)00333-X

[7] O. Yilmaz and M. Aksoy, J. Mater. Process. Technol. 121, 136-142 (2002). https://doi.org/10.1016/S0924-0136(01)01224-9

[8] H. Sabetghadam, A. Z. Hanzaki and A. Araee, Materials characterization 61, 626-634 (2010). https://doi.org/10.1016/j.matchar.2010.03.006

[9] S. Kundu, M. Ghosh, A. Laik, K. Bhanumurthy, G. B. Kale and S. Chatterjee, Mater. Sci. Eng. A 407, 154-160 (2005). https://doi.org/10.1016/j.msea.2005.07.010

[10] I. Magnabosco, P. Ferro, F. Bonollo and L. Arnberg, Mater. Sci. Eng. A 424, 163-173 (2006). https://doi.org/10.1016/j.msea.2006.03.096

[11] S. Chen, J. Huang, J. Xia, H. Zhang and X. Zhao, Metallurgical and Materials Transactions A 44, 3690-3696 (2013). https://doi.org/10.1007/s11661-013-1693-z

[12] B. R. Moharana, S. K. Sahu, S. K. Sahoo and R. Bathe, Engineering Science and Technology, an International Journal 19(2), 684-690 (2016). https://doi.org/10.1016/j.jestch.2015.10.004

[13] S. K. Sahu, B. R. Moharana, V. B. Shaibu and S. K. Sahoo, Int. J. Mechatronics and Manufacturing Systems 9(4), 310-326 (2016). https://doi.org/10.1504/IJMMS.2016.082862

[14] B. R. Moharana, S. K. Sahu, A. Maiti, S. K. Sahoo and T K Moharana. "An experimental study on joining of AISI 304 SS to Cu by Nd-YAG laser welding process" Materials Today: Proceedings 33, 5262-5268 (2020). https://doi.org/10.1016/j.matpr.2020.02.953

[15] M. Velu and S. Bhat, Materials & Design 47, 793-809 (2013). https://doi.org/10.1016/j.matdes.2012.12.073

[16] M. Velu and S. Bhat, Materials & Design 67, 244-260 (2015). https://doi.org/10.1016/j.matdes.2014.11.026

[17] L. M. Liu, S. X. Wang and M. L. Zhu, Science and Technology of Welding and Joining 11(5), 523-525 (2013). https://doi.org/10.1179/174329306X122794

[18] K. Suresh Kumar, G. Phanikumar, P. Dutta and K. Chattopadhyay, J. Mater. Sci. 37, 2345-2349 (2002). https://doi.org/10.1023/A:1015306408611

[19] F. Madadi, F. Ashrafizadeh and M. Shamanian, Journal of Alloys and Compounds 510(1), 71-77 (2012). https://doi.org/10.1016/j.jallcom.2011.08.073

[20] H. Yu, Y. Xu, N. Lv, H. Chen and S. Chen, J Mater Process Technol 213, 707-716 (2013). https://doi.org/10.1016/j.jmatprotec.2012.11.016

[21] V. Rajkumar and N. Arivazhagan N. Materials & Design, 63, 69-82 (2014). https://doi.org/10.1016/j.matdes.2014.05.055

[22] B. R. Moharana and S. K. Sahoo, Universal Journal of Mechanical Engineering 2(5), 169-173 (2014). https://doi.org/10.13189/ujme.2014.020504

[23] G. Zhao, V Du, Z. Wei, R. Geng and S. Xu, Journal of the Brazilian Society of Mechanical Sciences and Engineering, 41(1), 1-9 (2019). https://doi.org/10.1007/s40430-018-1563-0

[24] H. He, C. Wu, S. Lin and C. Yang, Journal of Materials Engineering and Performance, 28(2), 1180-1189 (2019). https://doi.org/10.1007/s11665-018-3848-y

[25] B. R. Moharana, S. K. Sahoo and K. Muduli, International Journal of Materials Engineering Innovation 13(4), 291-305 (2022). https://doi.org/10.1504/IJMATEI.2022.126516

[26] S. Pichumani, R. Srinivasan and V. Ramamoorthi, Journal of Mechanical Science and Technology 32(4), 1713-1723 2018. https://doi.org/10.1007/s12206-018-0130-4

[27] J. Sun, H. Wang and L. Liu, International Journal of Precision Engineering and Manufacturing 21(8), 1469-1477 (2020). https://doi.org/10.1007/s12541-019-00032-0

Flow Analysis of Eco-Friendly Geopolymer with Recycled Copper for Rapid Tooling Applications

Sharifah SALSABILA[1,a*], Z. SHAYFULL[1,2,b], S.M. NASIR[1,2,c], Leong Kean WEI[1,d], M. FATHULLAH[1,2,e]

[1]Universiti Malaysia Perlis, Faculty of Mechanical Engineering & Technology, Arau 02600, Malaysia

[2]Universiti Malaysia Perlis (UniMAP), Center of Excellence Geopolymer & Green Technology (CEGeoTech), Arau 02600, Malaysia

[a]shsalsabila@studentmail.unimap.edu.my, [b]shayfull@unimap.edu.my, [c]nasirsaad@unimap.edu.my, [d]kwleong@unimap.edu.my, [e]fathullah@unimap.edu.my

Keywords: Geopolymer, Recycled Copper, Rapid Tooling, Injection Moulding, Warpage, Autodesk Moldflow Insight, Sustainable Manufacturing

Abstract. Injection moulding is one of the most widely used manufacturing processes for producing plastic components with high precision and efficiency. However, warpage remains a critical defect, often caused by uneven cooling and differential shrinkage. The choice of mould material strongly influences cooling efficiency, cycle time, and dimensional stability. Traditionally, P20 tool steel has been the industry standard due to its durability, but its limited thermal conductivity and high machining cost have motivated the search for alternative solutions. This study aims to evaluate recycled copper powder composite (GGMC) as a sustainable alternative to P20 tool steel for mould fabrication. Using Autodesk Moldflow Insight (AMI), simulations of acrylonitrile butadiene styrene (ABS) parts were performed for fill, pack, cooling, and warpage analyses. Results showed that copper-based moulds reduced cooling time by up to 30% compared to P20, owing to superior thermal conductivity, while P20 maintained advantages in tool life for high-cycle production. Warpage analysis confirmed that packing pressure was the dominant factor influencing deformation in the X and Y axes, whereas melt temperature was most critical along the Z axis, with packing time having the least effect. Overall, the findings suggest that recycled copper powder composites provide an effective and sustainable option for rapid tooling and prototype moulds, while P20 remains more suitable for long-term, high-volume manufacturing.

Introduction

Injection moulding is among the most efficient manufacturing processes for producing thin-walled plastic components with high precision and repeatability. The quality of moulded products is strongly influenced by several factors, including processing parameters, material characteristics, and mould design [12,17,18]. Common defects such as shrinkage, warpage, weld lines, sink marks, and residual stresses are often attributed to variations in packing pressure, melt temperature, and mould temperature [8,10]. Numerous optimisation studies employing methods such as the Taguchi design, Response Surface Methodology (RSM), Genetic Algorithms (GA), and hybrid approaches have demonstrated that careful control of these parameters can substantially reduce dimensional errors and improve product stability [4]. For example, Chen et al. reported warpage reductions of up to 72.8% with shrinkage as low as 0.02%, while Nasir et al. identified packing pressure as the most influential variable affecting ABS parts [8]. Other researchers have confirmed that packing pressure and melt temperature remain the dominant contributors to dimensional deviation, with Autodesk Moldflow Insight (AMI) widely used to simulate and optimise filling, cooling, and warpage behaviour in injection moulding [2,6,11].

Rapid Tooling (RT) has further expanded the capability of injection moulding by enabling faster fabrication of mould inserts compared to conventional tooling, thereby reducing time-to-market in industries such as automotive, electronics, and medical devices [7,9,16]. P20 tool steel remains the most widely adopted material for mould inserts because of its machinability and durability [3]. However, its relatively low thermal conductivity results in prolonged cooling cycles and higher energy consumption. In contrast, copper and its alloys provide superior heat transfer but suffer from poor wear resistance, susceptibility to corrosion, and higher material cost, creating a trade-off between thermal efficiency and tool longevity. Geopolymers derived from industrial by-products such as fly ash have recently emerged as sustainable alternatives, offering high compressive strength, chemical stability, and a reduced carbon footprint [1,13]. When reinforced with recycled copper, these composites can achieve improved thermal conductivity while contributing to circular economy practices [15].

Despite these potential advantages, the use of geopolymer copper composites in injection mould tooling remains limited, especially when benchmarked against established materials such as P20 tool steel. Previous studies have focused primarily on polymer metal composites, with little exploration of geopolymer metal systems. Moreover, few investigations have applied Computer-Aided Engineering (CAE) tools to assess the performance of these composites in actual injection moulding applications.

Therefore, this study seeks to address this gap by comparing recycled copper geopolymer composites with P20 tool steel in terms of density, thermal conductivity, and compressive strength. Autodesk Moldflow Insight (AMI) is employed to simulate the filling, packing, cooling, and warpage stages, enabling a comprehensive evaluation of flow behaviour, cooling efficiency, and dimensional stability. This work provides new insights into the feasibility of geopolymer copper composites as eco-friendly alternatives for mould inserts.

Problem Statement

The injection moulding industry is increasingly pressured to shorten cycle times, reduce production costs, and adopt more sustainable manufacturing practices. Although P20 tool steel has long been the standard material for mould bases and inserts due to its strength, durability, and machinability, its relatively low thermal conductivity significantly extends cooling times and reduces overall process efficiency, with cooling often accounting for more than 60% of the total cycle [5]. In addition, the fabrication of P20 moulds requires extensive machining, heat treatment, and polishing, which increases cost and energy consumption, making it less attractive for small-batch or rapid tooling applications [14].

The reliance on virgin tool steel also contributes to high resource consumption and environmental impact, conflicting with the global shift toward green manufacturing and circular economy practices. Recycled copper powder composites represent a promising alternative because they offer higher thermal conductivity, lower cost, and the sustainable reuse of industrial waste [18]. However, these composites generally exhibit lower hardness and compressive strength than P20 steel, raising concerns about tool life and durability under repeated injection cycles.

To date, only limited research has directly compared recycled copper powder composites with P20 steel in terms of thermal, mechanical, and economic performance. Moreover, the application of advanced simulation tools such as Autodesk Moldflow Insight (AMI) to evaluate cooling efficiency, flow behaviour, and warpage in both materials has not been thoroughly explored. This gap underscores the need for a systematic investigation to determine whether recycled copper powder composites can serve as a viable and sustainable alternative to P20 steel in rapid tooling applications.

Methodology

This study was conducted using AMI, a widely used CAE simulation software for injection moulding analysis. The software was employed to evaluate the effect of mould material on recycled copper powder and tool steel P20 which on the warpage behaviour of ABS plastic parts.

Material. The moulded polymer selected was ABS, Polylac PA-777B (Chi Mei Corporation) was used to mould the specimens and the properties of ABS are shown in Table 1, a commercial grade ABS often used in automotive and consumer electronics due to its toughness and dimensional stability. Its properties were obtained from the AMI material database. The simulation setup used the Nissei NEX 1000 injection moulding machine, as specified in the AMI database. Machine parameters were imported to ensure realistic injection conditions.

Table 1 : Properties of ABS material.

Property	Unit	Value
Thermal conductivity, k_m	W/m.C	0.178
Melt density, ρ_m	kg/m3	956.17
Specific heat, C_p	J/kg.C	1464.7
Elastic modulus, E	MPa	2.24 × 103

Mould Design. Two mould materials were considered in this study which Tool Steel P20, and a GGMC, modelled as a high-conductivity insert material with a much higher thermal conductivity of about 200 W/m·K. The comparison between these materials highlights both the established traditional approach and a sustainable alternative, with the latter offering enhanced cooling performance due to its superior thermal properties. Fig. 1 presents the drawing of the plastic mould, and illustrates the 3D model of the plastic component.

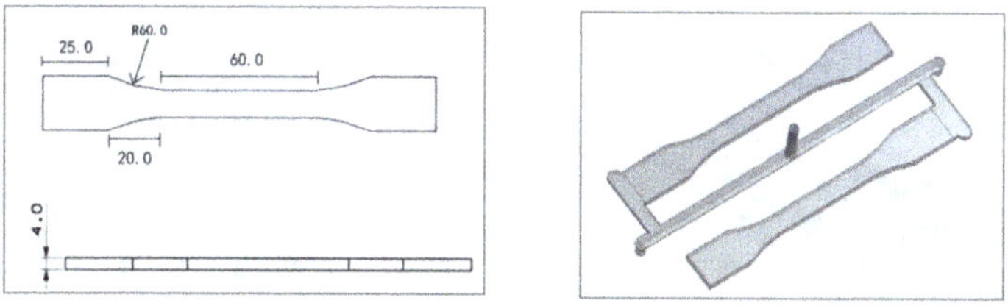

Fig. 1 : (a) Drawing of Plastic Mould, (b) 3D Plastic Mould

The simulation was based on the Nissei NEX 1000 injection moulding machine, with parameters imported from the AMI database to replicate realistic processing conditions. The melt temperature was set at 200 °C, while the mould surface temperature was maintained at 60 °C.

Eq. 1 and Eq. 2 describe the basic relationships used to determine the ram speed and screw stroke in the injection molding process. In Eq. 1 the ram speed is calculated as the ratio between the flow rate of the molten polymer and the effective cross-sectional area of the screw. This relationship shows that the ram speed increases proportionally with the flow rate when the screw area remains constant, allowing better control over the filling rate and cavity pressure during injection. Eq. 2 defines the screw stroke (SM), which represents the total displacement of the screw required for one molding cycle. It accounts for the combined volume of the molded part and the gating system, divided by the screw area, with an additional 5 mm allowance to compensate for melt compression and packing effects. Together, these equations provide a practical approach for optimizing machine settings, ensuring stable flow behavior, and achieving uniform part quality throughout the molding process.

$$Ram\ Speed = \frac{Flow\ rate}{A_{screw}} \tag{1}$$

$$SM = \left(\frac{V_{part}+V_{gs}}{A_{screw}}\right) + \left(\frac{V_{part}+V_{gs}}{A_{screw}}\right) + 5 \tag{2}$$

Table 2 show the velocity-to-pressure (V/P) switchover occurred at a ram position of 22.32 mm, followed by a packing phase at 80% of filling pressure for 10 s. Cooling was achieved using two circuits of 10 mm diameter channels, with a coolant inlet temperature of 50 °C and a Reynolds number of 10,000, ensuring turbulent flow and effective heat transfer.

Table 2 : Filling Control Profile.

Ram Position (mm)	Ram Speed (mm/s)
66.23	8.117
54.33	8.117
54.33	10.55
22.32	10.55

3D mesh is applied on the moulded parts with a thickness of 4 mm and 20507 of tetrahedral elements with beam elements applied for cooling circuit has been created for the single gating system as shown in Fig. 2.

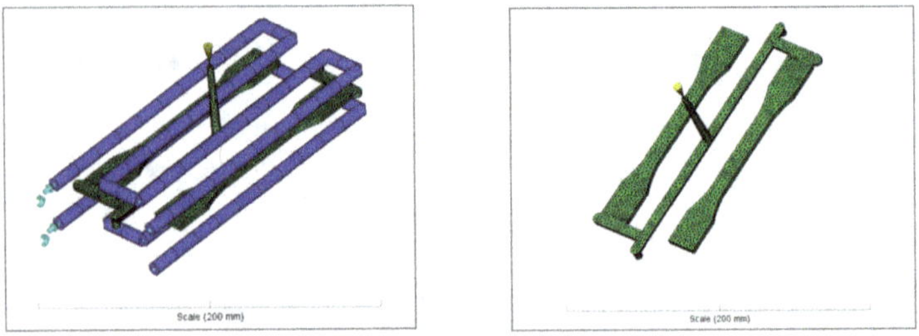

Fig. 2 : (a) Cooling Channel Design (b) Meshing

Finally, the performance of the recycled copper geopolymer composite mould was compared with that of P20 tool steel by analysing the results from simulation, alongside the physical and mechanical properties obtained experimentally. The comparison enabled a clear understanding of the trade-offs between thermal efficiency, mechanical strength, and sustainability for both materials, thus providing insights into the potential of recycled copper composites as a sustainable alternative in rapid tooling applications.

Result and Discussion

This section presents the comparative simulation results obtained using Autodesk Moldflow Insight (AMI) for the two tooling materials: conventional tool steel P20 (baseline) and geopolymer–copper composites (GGMC). Four simulation stages were considered: Fill, Fill + Pack, Cooling, and Fill + Pack + Warp.

Fill Analysis. The fill analysis showed that the cavity was fully filled without defects for both moulds. Fig. 3 shows GGMC mould filled faster (4.103 s) compared to P20 steel (4.515 s), due to the higher thermal conductivity of copper which delayed premature freezing of the melt.

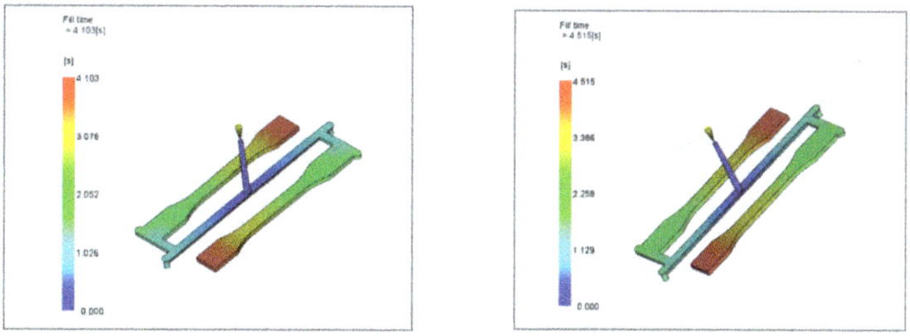

Fig. 3 : Fill time distribution, (a) GGMC Mould, (b) P20 Steel Mould.

At the velocity/pressure switchover, the injection pressure was about 129 MPa, with GGMC showing a slightly lower value, indicating reduced flow resistance. The maximum clamp force during filling was ~34 tonnes, within machine capacity. Both moulds reached ~99.8% volume filled at switchover, ensuring a smooth transition to the packing phase.

Predicted weld lines and air traps were observed in both moulds, but were less critical in the GGMC case due to more uniform filling.

In summary, the copper powder mould improved filling behaviour by reducing fill time and pressure, which is expected to lower differential shrinkage and help minimize warpage.

Table 3 summarizes the key findings from the fill analysis. Overall, the use of GGMC as a mould material improved flow behaviour by reducing filling time and injection pressure. These improvements are directly related to the superior thermal conductivity of the copper mould, which promotes uniform heat transfer during the filling stage. The improved filling behaviour is expected to reduce shrinkage differentials, thereby minimizing warpage in the final product.

Table 3 : Key results from Fill Analysis for P20 and GGMC moulds.

Parameter	GGMC	P20	Observation
Fill time (s)	4.103	4.515	Faster filling with GGMC
Injection pressure at switchover (MPa)	~125–127 (slightly lower)	129.3	Reduced flow resistance
Maximum clamp force (tonnes)	33.99	33.99	Within machine limit
Volume filled at V/P switchover (%)	99.85	99.85	Proper profile setup
Weld lines / Air traps	Present but less severe	Prese nt	Better melt temperature control with GGMC

Fill + Pack Analysis. The Fill + Pack analysis confirmed that the cavity was completely filled and packed under both mould conditions. The maximum clamp force reached was about 44 tonnes, within the machine's capacity, while the injection pressure during packing stabilized at around 103 MPa, with the GGMC mould requiring slightly less than P20 due to lower flow resistance. An average volumetric shrinkage of ~1.3% was recorded, with a maximum of 2.4% shown in Table 4, and shrinkage was more uniform in the GGMC mould, indicating a lower tendency for warpage. At the end of filling, the frozen layer fraction was ~51%, reflecting the effect of thermal conductivity on solidification. No significant sink marks were predicted, and the melt temperature distribution remained uniform at 200–206 °C, with the GGMC mould promoting faster heat dissipation. Overall, the copper powder mould improved filling efficiency, reduced injection pressure, and produced more uniform shrinkage, all of which are expected to reduce warpage compared to P20 steel.

Table 4 : Volumetric shrinkage distribution.

Volumetric Shrinkage	%
Max	2.4098
Min	-0.0992
Average	1.3097
Mean	0.5381

Cooling Analysis. The cooling analysis shown in Fig. 4 revealed that at 15.28 s, the part temperature varied between 50.16 °C and 106.8 °C depending on the location. The lowest values were measured at the thinner sections near the surface (≈ 50 °C), while higher temperatures remained in thicker regions (≈ 106 °C). This non-uniform cooling behavior highlights areas prone to differential shrinkage. However, compared to P20 steel, the GGMC mould exhibited a narrower temperature gradient, as the higher thermal conductivity of copper improved heat transfer and accelerated cooling. The more uniform cooling achieved with GGMC is expected to reduce internal stresses and minimize warpage in the final ABS product.

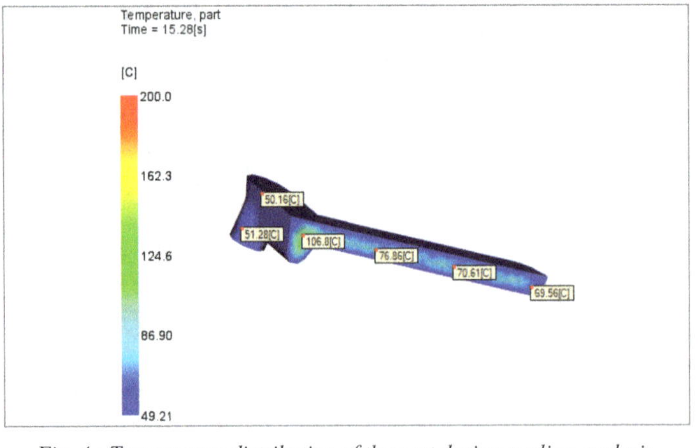

Fig. 4 : Temperature distribution of the part during cooling analysis.

Warpage Analysis. The warpage analysis showed that the ABS part experienced dimensional displacement in all three axes due to differential shrinkage after cooling. Fig. 5 shown the maximum deflections were recorded as 0.281 mm in the X-direction, 0.481 mm in the Y-direction, and 0.647 mm in the Z-direction, with the Z-axis contributing most significantly to the overall deformation. The total part weight stabilized at ~17.8 g, confirming complete cavity filling and consistent packing. Pressure profiles indicated stable behaviour during the packing phase, with no abnormal fluctuations. Comparison between mould materials revealed that the GGMC mould exhibited lower warpage magnitudes than P20 steel, as the higher thermal conductivity of copper provided more uniform cooling and reduced shrinkage mismatch. These results indicate that the use of recycled copper powder as a mould material can effectively minimize warpage, leading to improved dimensional stability of ABS parts.

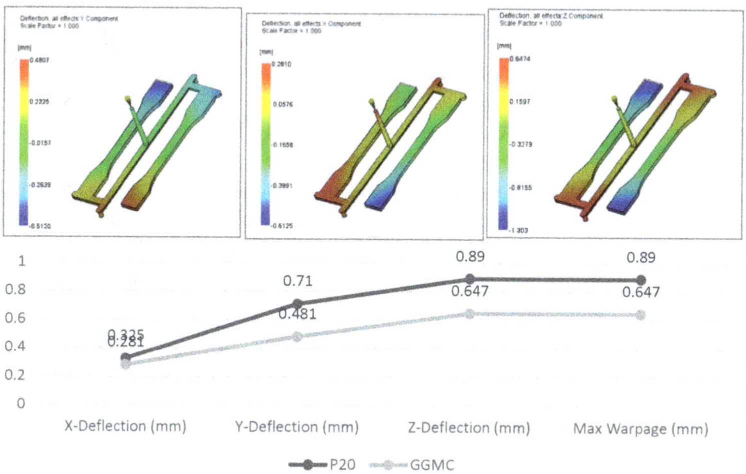

Fig. 5 : Deflection in X, Y and Z direction.

Conclusion

This study investigated the influence of mould material on the dimensional stability of ABS parts using Autodesk Moldflow Insight. Two mould materials were compare conventional P20 tool steel and recycled copper powder (GGMC). The Fill and Pack analyses showed that the GGMC mould reduced filling time and required slightly lower injection pressure, confirming improved flow behavior due to higher thermal conductivity. The Cooling analysis indicated that the GGMC mould provided faster and more uniform heat dissipation, reducing thermal gradients and shortening the cooling cycle time. Finally, the Warpage analysis demonstrated that maximum deflections occurred in the Z-direction, with overall warpage significantly lower in the GGMC mould compared to P20 steel. These improvements are attributed to the superior thermal conductivity of recycled copper powder, which enhances cooling uniformity and reduces shrinkage mismatch.

In conclusion, the use of recycled copper powder as a mould material not only offers a sustainable alternative to conventional tool steel but also improves part quality, dimensional accuracy, and productivity in injection moulding. Future work should focus on experimental validation of these findings and cost benefit analysis to assess industrial feasibility.

Acknowledgement

The author would like to acknowledge support from the Fundamental Research Grant Scheme (FRGS) under grant number FRGS/1/2023/TK10/UNIMAP/03/2 from the Ministry of Higher Education Malaysia. We would also like to thank the reviewers for their helpful advice and commitment. This study was supported by the Centre of Excellence Geopolymer and Green

Technology (CEGeoGTech), UniMAP, and the Faculty of Mechanical Engineering & Technology, UniMAP.

References

[1] Abdullah, M. M. A., Kamarudin, H., Mohammed, H., Khairul Nizar, I., Rafiza, A. R., & Zarina, Y. (2011). The relationship of NaOH molarity, Na2SiO3/NaOH ratio, fly ash/alkaline activator ratio, and curing temperature to the strength of fly ash-based geopolymer. Advanced Materials Research, 328-330. doi: 10.4028/www.scientific.net/AMR.328-330.1475 https://doi.org/10.4028/www.scientific.net/AMR.328-330.1475

[2] Ahmad, A. H., Leman, Z., Azmir, M. A., Muhamad, K. F., Harun, W. S. W., Juliawati, A., & Alias, A. B. S. (2009). Optimization of warpage defect in injection moulding process using ABS material. Proceedings - 2009 3rd Asia International Conference on Modelling and Simulation, AMS 2009. doi: 10.1109/AMS.2009.120 https://doi.org/10.1109/AMS.2009.120

[3] Buliński, Z., Pawlak, S., Krysiński, T., Adamczyk, W., & Białecki, R. (2019). Application of the ASTM D5470 standard test method for thermal conductivity measurements of high thermal conductive materials. Journal of Achievements in Materials and Manufacturing Engineering, 95(2). doi: 10.5604/01.3001.0013.7915 https://doi.org/10.5604/01.3001.0013.7915

[4] Chen, W. C., Nguyen, M. H., Chiu, W. H., Chen, T. N., & Tai, P. H. (2016). Optimization of the plastic injection molding process using the Taguchi method, RSM, and hybrid GA-PSO. International Journal of Advanced Manufacturing Technology, 83(9-12). doi: 10.1007/s00170-015-7683-0 https://doi.org/10.1007/s00170-015-7683-0

[5] Das, K., Sen, S., Joseph, A., Ghosh, A., Biswas, K., Mukherjee, S., & Ghosh, M. (2023). Investigation on the effects of pretreatment on the surface characteristics of duplex plasma-treated AISI P20 tool steel. Materialia, 27. doi: 10.1016/j.mtla.2023.101679 https://doi.org/10.1016/j.mtla.2023.101679

[6] Dimla, D. E., Camilotto, M., & Miani, F. (2005). Design and optimisation of conformal cooling channels in injection moulding tools. Journal of Materials Processing Technology, 164-165. doi: 10.1016/j.jmatprotec.2005.02.162 https://doi.org/10.1016/j.jmatprotec.2005.02.162

[7] Equbal, A., Sood, A. K., & Shamim, M. (2015). Rapid tooling: A major shift in tooling practice. Manufacturing and Industrial Engineering, 14(3-4). doi: 10.12776/mie.v14i3-4.325 https://doi.org/10.12776/mie.v14i3-4.325

[8] Kashyap, S., & Datta, D. (2015). Process parameter optimization of plastic injection molding: a review. International Journal of Plastics Technology, 19(1). doi: 10.1007/s12588-015-9115-2 https://doi.org/10.1007/s12588-015-9115-2

[9] King, D., & Tansey, T. (2002). Alternative materials for rapid tooling. Journal of Materials Processing Technology, 121(2-3). doi: 10.1016/S0924-0136(01)01145-1 https://doi.org/10.1016/S0924-0136(01)01145-1

[10] Kummara, H., & Gowd, G. H. (2014). Design and Fabrication of Plastic Injection Molding Tool for Pump Gaskets. International Journal of Advanced Engineering Research and Science (IJAERS), 1(3).

[11] Li, J., Bei, J., Liu, W., Xia, X., Zhou, B., Peng, X., & Jiang, S. (2021). Warpage prediction of rhcm crystalline parts based on multi-layers. Polymers, 13(11). doi: 10.3390/polym13111814 https://doi.org/10.3390/polym13111814

[12] Malloy, R. A. (2010). Plastic Part Design for Injection Molding. In Plastic Part Design for Injection Molding. doi: 10.3139/9783446433748.fm https://doi.org/10.3139/9783446433748.fm

[13] Mustafa Al Bakri, A. M., Kamarudin, H., Bnhussain, M., Rafiza, A. R., & Zarina, Y. (2012). Effect of Na2SiO3/NaOH ratios and NaOH molarities on compressive strength of fly-ash-based geopolymer. ACI Materials Journal, 109(5). doi: 10.14359/51684080 https://doi.org/10.14359/51684080

[14] Priyadarshini, M., Behera, A., Biswas, C. K., & Rajak, D. K. (2022). Experimental Analysis and Mechanical Characterization of AISI P20 Tool Steel Through Heat-Treatment Process. Journal of Bio- and Tribo-Corrosion, 8(1). doi: 10.1007/s40735-021-00607-3 https://doi.org/10.1007/s40735-021-00607-3

[15] Sandanayake, M., Gunasekara, C., Law, D., Zhang, G., Setunge, S., & Wanijuru, D. (2020). Sustainable criterion selection framework for green building materials - An optimisation based study of fly-ash Geopolymer concrete. Sustainable Materials and Technologies, 25. doi: 10.1016/j.susmat.2020.e00178 https://doi.org/10.1016/j.susmat.2020.e00178

[16] Shan, Z., Yan, Y., Zhang, R., Lu, Q., & Guan, L. (2003). Rapid manufacture of metal tooling by rapid prototyping. International Journal of Advanced Manufacturing Technology, 21(7). doi: 10.1007/s001700300055 https://doi.org/10.1007/s001700300055

[17] Shoemaker, J. (2006). Moldflow Design Guide : A Resource for Plastics Engineers. In Moldflow Design Guide. https://doi.org/10.3139/9783446418547.fm

[18] Tang, S. H., Kong, Y. M., Sapuan, S. M., Samin, R., & Sulaiman, S. (2006). Design and thermal analysis of plastic injection mould. Journal of Materials Processing Technology, 171(2). doi: 10.1016/j.jmatprotec.2005.06.075 https://doi.org/10.1016/j.jmatprotec.2005.06.075

The Tensile Properties and Microstructural Characteristics of Pulverised Silica-Coated Kenaf–Epoxy Composites

M.J. MEELAD[1,5], N.Z. NORIMAN[2,5a*], M.A. AZIZAN[4], M.F. OMAR[1,5], A.Z. ROMLI[3], R. HAMZAH[1], N. ISHAK[4], S.T. SAM[1,5]

[1]Faculty of Chemical Engineering & Technology, Universiti Malaysia Perlis, Kompleks Pengajian Jejawi 2, 02600, Arau, Perlis, Malaysia

[2]Faculty of Mechanical Engineering & Technology, Universiti Malaysia Perlis, Pauh Putra Campus, 02600, Arau, Perlis, Malaysia

[3]Centre of Chemical Synthesis and Polymer Composites Research & Technology, Institute of Science (IOS), Universiti Teknologi MARA, 40450 Shah Alam, Selangor, Malaysia

[4]Faculty of Civil Engineering & Technology, Universiti Malaysia Perlis, Kompleks Pengajian Jejawi 3, 02600, Arau, Perlis, Malaysia

[5]Center of Excellence Geopolymer and Green Technology (CEGeoGTech), Universiti Malaysia Perlis, Perlis 01000, Malaysia

[a]niknoriman@unimap.edu.my

Keywords: Kenaf Fibre, Epoxy Composite, Nanosilica Surface Modification, Tensile Properties, Optical Microscopy, Fiber-Matrix Interphase

Abstract: Natural-fibre reinforced polymer composites offer sustainable alternatives for lightweight structural applications, yet their mechanical performance is often limited by insufficient interfacial adhesion, poor resin wet-out, and microstructural defects. This study investigates the tensile behaviour and microstructural characteristics of pulverised silica-coated kenaf–epoxy composites with varying nanosilica content (1 %, 2 %, 3 %) and kenaf loadings (5–25 wt%). Kenaf fibres were surface-treated with an epoxy–acetone–nanosilica mixture to enhance interphase formation and subsequently pulverised to produce hybrid fiber-particulate reinforcements. Composites were fabricated via hand lay-up followed by compression molding, and tensile properties were evaluated according to ASTM D638. Fractured specimens were analyzed using optical microscopy to quantify void formation, debonding, fiber pull-out, and nanoparticle clustering. Results reveal a pronounced "optimum-then-penalty" behaviour: moderate nanosilica loadings (2 wt%) and intermediate kenaf content (15 wt%) yielded the highest tensile modulus, strength, and strain due to improved fiber–matrix bonding and efficient stress transfer. Excessive nanosilica or fiber content reduced mechanical performance through particle agglomeration, resin starvation, and increased porosity. Optical microscopy confirmed that low void fractions, short debond channels, and tortuous crack paths correlate with enhanced tensile response, whereas abundant voids and clustering at high filler fractions initiate premature failure. These findings demonstrate that careful control of both nanosilica and fiber content can optimize interphase quality, minimize microstructural defects, and maximize reinforcement efficiency. The combined mechanical and microstructural analysis provides a robust framework for designing high-performance, biomass-based epoxy composites with tailored tensile properties.

Introduction

Natural-fibre reinforced polymer composites are increasingly researched for sustainable lightweight structural applications due to their favourable specific properties and low ecological footprint [1]. However, their widespread structural use is still limited by insufficient stiffness, strength and durability compared with conventional glass or carbon fibre composites [2]. In particular, interfacial adhesion between plant fibre and thermoset resin, resin wet-out of the fibre network, fibre packing/porosity and filler dispersion remain key bottlenecks [3]. One strategy to mitigate these issues is incorporation of nanoscale silica particles into the resin or fibre–matrix interphase [4]. Nanosilica modification has been shown to stiffen the matrix, improve interphase bonding and retard crack propagation in fibre-reinforced epoxy systems [5]. For example, the inclusion of well-dispersed silica nanoparticles in epoxy resulted in significant improvements in tensile modulus, tensile strength and ductility in carbon- and glass-fibre composites, provided that agglomeration was avoided [2]. On the other hand, excessive nanoparticle loading or poorer dispersion can increase resin viscosity, promote cluster formation and microvoids, thus offsetting reinforcement benefits [4]. In parallel, increasing natural-fibre content enhances load-bearing capacity until a threshold where resin starvation, fibre–fibre contact, tighter packing and entrapped porosity degrade performance [6]. This optimum-then-penalty behaviour is routinely reported in the literature for both fibre loading and nanoparticle content [7]. In the present study we pursue a detailed tensile behaviour and optical microstructural analysis of pulverised silica-coated kenaf–epoxy composites with varying nanosilica (1 %, 2 %, 3 %) and kenaf loading (5, 10, 15, 20, 25 wt %). The goal is to clarify how the twin levers of fibre f raction and nanosilica content interact to define stiffness, strength and strain behaviour, and to link microstructural features (voids, debonding, pull-out and nanoparticle clustering) to the mechanical response. This combination of mechanical and optical microscopy evidence yields insight into interphase formation, optimum reinforcement windows and failure transitions for this novel biomass-based composite system.

Materials and Methods

Surface Treatment of Kenaf Fibers: Kenaf fibers (Hibiscus cannabinus) were used as the reinforcement phase. The raw fibers were manually combed without undergoing any chemical or aqueous washing, representing a novel pretreatment approach not widely reported in prior work. This combing process effectively aligned the fiber bundles, removed pith and debris, and allowed subsequent processing steps to proceed with consistent fiber dimensions and orientation. Before coating, the fibers were subjected to a surface treatment using an epoxy acetone mixture containing nanosilica particles. The use of acetone in this process served two critical functions: to dilute the epoxy resin, thereby reducing its viscosity for improved nanosilica dispersion and coating uniformity, and to promote straightening of the kenaf fibers upon drying, facilitating more homogeneous coating and alignment in later composite processing. The coating process was designed to create a robust interfacial layer between the fiber and the epoxy matrix. The presence of nanosilica within the diluted epoxy served a dual purpose: improving mechanical interlocking during curing and reducing stress concentrations by forming a semi-rigid interface around the irregular kenaf surface. This approach is expected to mitigate common interfacial weaknesses observed in untreated natural fiber composites and support better stress distribution under tensile and flexural loads. In particular, the coating strategy aimed to preserve silica presence during subsequent pulverization, enabling the nanoparticles to function as active reinforcement even in their fragmented state. Maintaining uniform silica coverage was therefore essential to ensuring consistency in later mechanical behavior across filler loadings. After drying, all coated kenaf fibers were pulverized using a high-speed rotary blade grinder. This process produced micro-particulate reinforcements ranging between 200-500 μm in length, providing a hybrid morphology with fiber-like aspect ratios and particle-like flowability. This form allowed for efficient dispersion in the epoxy matrix during composite fabrication, while retaining the reinforcing potential of the original kenaf material. Pulverized fibers were grouped based on nanosilica content (1 wt%, 2 wt%, and 3 wt%), and each group was further categorized by filler loading (5, 10, 15, 20, and 25 wt%), yielding a total of 15 composite formulations. These levels were selected to enable factorial analysis of both silica concentration and filler weight, consistent with hybrid bio-based composite design frameworks [8,9]. Each formulation was prepared by first mixing the epoxy resin and hardener in a 2:1 weight ratio to initiate uniform pre-polymerization. Once a homogenous mixture was achieved, the appropriate amount of pulverized fiber was gradually introduced into the epoxy-hardener blend under continuous mechanical stirring to ensure even distribution of the filler. This sequence minimized the likelihood of fiber clumping and promoted consistent dispersion throughout the matrix, aligning with recent strategies in silica/natural-fiber hybrid systems [10,11].

Tensile Test: Tensile testing followed ASTM D638 Type V specifications using an Instron 3365 universal testing machine. Samples were cut from composite panels, were each specimen's width, thickness, and gauge length were verified using a calibrated digital caliper [12]. Tests were conducted at a crosshead speed of 10 mm/min, with strain monitored via machine crosshead displacement. Average values of tensile strength, modulus of elasticity, and elongation at break were calculated based on five replicates per formulation [13]. Post-fracture surfaces were retained for later microstructural inspection. The optical microscopy (OM) was performed to examine filler distribution, fracture morphology, and interfacial quality. Fractured specimens from tensile and flexural tests were cross-sectioned and wet-polished using silicon carbide papers (P600 to P2000), rinsed with ethanol, and dried. A Nikon Eclipse MA200 microscope (20 × –100 × magnification) was used in reflective mode to observe surface features. Micrographs were analyzed to detect filler clustering, void formation, crack pathways, and matrix–fiber bonding. Key images were selected to represent each silica content level, focusing on distinctions in fracture mechanisms and microstructure [14]. OM was chosen due to its ability to visualize composite morphology without requiring conductive coatings or vacuum preparation, as is needed for SEM. Observations helped to correlate mechanical performance with microstructural integrity [15].

Results and Discussion

Tensile Modulus. The results in Fig. 1 showed a clear optimal point influenced by filler fraction and nanosilica content. The modulus increases as kenaf loading rises from low to moderate levels but decreases when the filler or silica content exceeds the matrix's dispersion and wetting capacity. Composites with moderate nanosilica demonstrate the highest stiffness, indicating better interphase formation and efficient load transfer. Excessive nanosilica encourages particle–particle interactions and micro-agglomeration, leading to microvoids and weakening the composite. Similarly, too much fiber content reduces available resin for proper wetting, increases packing density and porosity, and diminishes reinforcement effectiveness. These combined effects explain the observed optimal-then-penalty behavior and align with dispersion mechanical analyses reported in the literature [16].

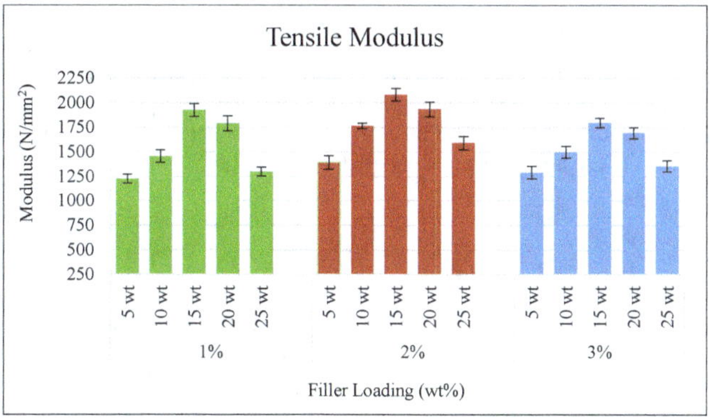

Fig. 1: Tensile Modulus of Kenaf Composites at Varying Fiber & Nano-Silica Loadings.

Tensile Stress. Tensile-stress behaviour follows a consistent, non-linear pattern driven by filler fraction and nanosilica content. Stress increases as kenaf content rises from low to intermediate levels, then declines once the filler fraction exceeds the matrix wetting and packing capacity. Composites containing a moderate nanosilica loading exhibit the highest tensile stress across the intermediate filler range, with lower silica giving the same trend at reduced magnitude and higher silica peaking similarly but falling away more sharply at high filler fractions. The poorer performance at excessive silica is attributed to increased particle interactions and micro-aggregate formation that act as stress concentrators and reduce the effective load-bearing cross section, a mechanism reported for epoxy silica systems [16]. At high fibre fractions the reduction in stress arises from resin starvation, tighter packing and higher porosity, which promote premature failure before the theoretical reinforcement limit is reached [17] Variability is greatest at the lowest and highest filler fractions, reflecting low reinforcement efficiency at the former and an increased risk of agglomeration or voiding at the latter.

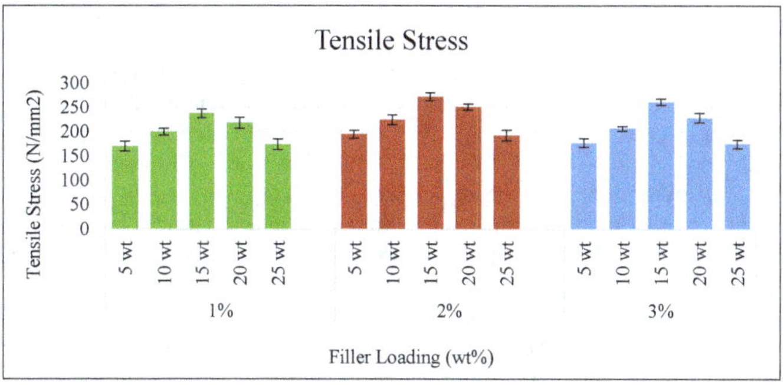

Fig. 2: Tensile Stress of Kenaf Composites at Varying Fiber and Nano-Silica Loadings.

Tensile Strain. Tensile-strain behaviour, as shown in Fig. 3, is governed by a balance between interphase reinforcement and particle dispersion. Strain capacity increases as fibre content rises from low to intermediate levels because more fibres engage in load bearing and a stronger interphase delays micro-debonding. Moderate nanosilica loadings strengthen the fibre matrix boundary and permit greater elastic-plastic deformation before macroscopic failure. However, excessive nanoparticle content reduces inter-particle spacing, promotes clustering and microvoid formation, and thereby trims composite ductility. Similarly, very high fibre fractions exceed the matrix wetting capacity, tighten packing, and generate entrapped porosity and premature debond sites that limit global extension. The combined "optimum-then-penalty" response therefore reflects two distinct failure drivers: agglomeration-driven damage initiation and wetting-limited porosity generation. These mechanisms are well documented for nanosilica-modified epoxies and for natural-fibre/epoxy composites.

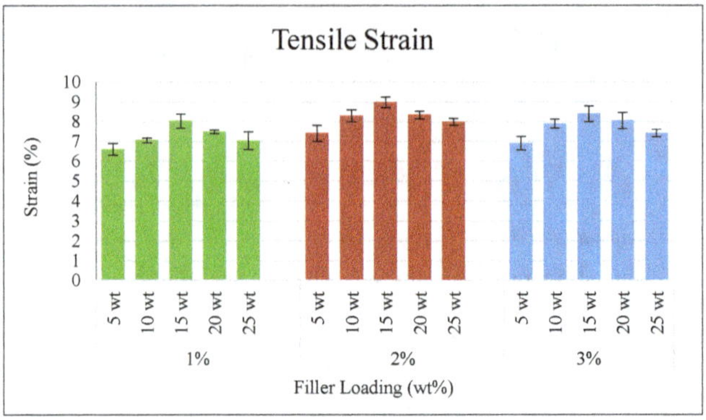

Fig. 3: Tensile Strain of Kenaf Composites at Varying Fiber and Nano-Silica Loadings.

Fig. 4 presents fractured-coupon optical micrographs arranged as a 3×3 grid (rows: SiO_2 = 1%, 2%, 3%; columns: kenaf = 5, 15, 25 wt%). Each panel captures the microstructural defects that control the corresponding mechanical datum. Three recurring signatures determine performance: (1) discrete circular voids within matrix ligaments, (2) crescent-shaped debond halos and pull-out channels at bundle contacts, and (3) local particle clusters or agglomerates that perturb crack paths. Panels that exhibit the best tensile response combine the fewest circular voids, the shortest pull-out channels, and the most tortuous crack paths. Panels showing poor response display abundant circular voids, long debond channels and flattened crack facets. These imaging observations explain the "optimum-then-penalty" tensile trends through two coupled failure mechanisms: void-seeded matrix cracking and nanoparticle-induced clustering that creates stress concentrators [18,19]. Across the SiO_2 rows, the change from 1% to 2% is characterised by void thinning and shorter debond halos at mid-kenaf loadings. At 2% silica, many mid-kenaf panels show fewer matrix bubbles and more branched, tortuous cracks. This morphology indicates improved wet-out and a stronger interphase that delays micro-debonding and promotes fibre bridging. The enhanced interphase at moderate nanosilica therefore maps directly to the tensile peaks observed in the mechanical plots. These effects reflect nanosilica's known role in stiffening the matrix and stabilising the interphase at modest loadings [20]. The 3% SiO_2 row re-introduces defects in localized zones. Small clusters and coalesced rings appear near visible particle aggregates. In those regions crack facets flatten, debond channels lengthen, and local failure initiates earlier despite the higher nominal particle fraction. This behaviour mirrors experimental and modelling studies that show mechanical declines when nanoparticle agglomeration outpaces dispersion [18,21]. In short, 3% reveals the particle-agglomeration penalty that reduces stiffness and strength.

Within each SiO_2 row, the kenaf progression (5, 15, and 25 wt%) follows a clear microstructural trajectory. Low filler (5 wt%) panels commonly show isolated matrix bubbles and long pull-out tracks reflecting weak fibre engagement. Mid filler (15 wt%) panels show the lowest void populations and the shortest debond channels; fibres share load and the interphase appears continuous. High filler (25 wt%) panels show resin-starved seams at bundle crossovers, increased circular void density and longer debond channels that bridge between fibres. These late-loading voids and seams explain the drop in modulus, strength, and strain at high kenaf fractions by producing early matrix cracking and reducing the effective load-bearing cross-section [22,23].

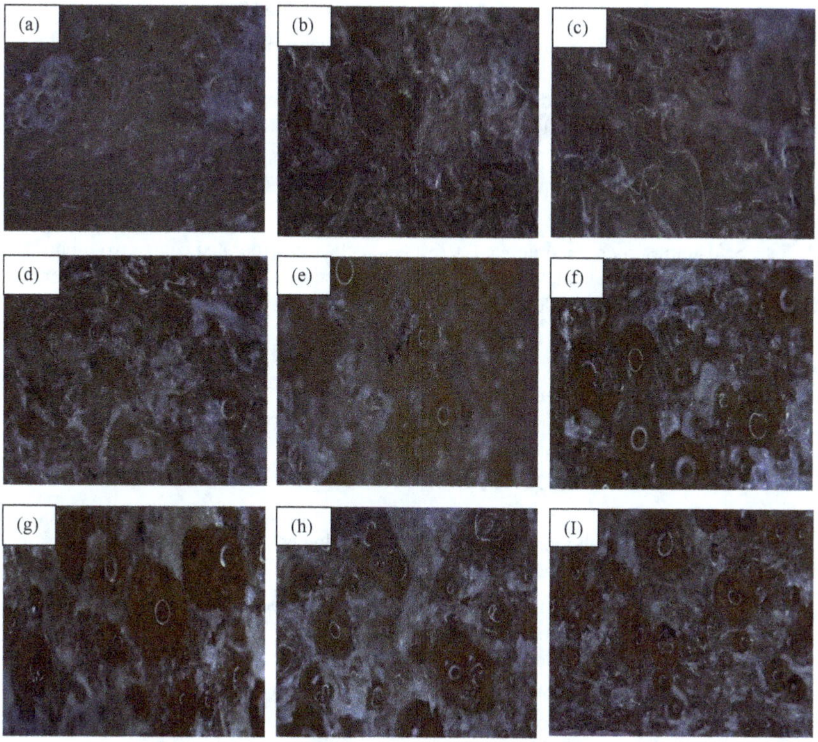

Fig. 4: Fractography (OM): 1%, 2%, 3% SiO_2 at 5 / 15 / 25 wt% kenaf.

Acknowledgement
The authors thank the Glide Marine Sdn Bhd for their fund to this project under the Industrial Research Grant 2025/2026.

References

[1] Munimathan, A., Muthu, K., Subramani, S., & Rajendran, S. (2024). Environmental behaviour of synthetic and natural fibre reinforced composites: A review. Advances in Mechanical Engineering. https://doi.org/10.1177/16878132241286020

[2] Sapiai, N., Salleh, M. S., Mustapha, F., & Yusof, N. M. (2020). Tensile and flexural properties of silica nanoparticles in fibre-reinforced polymer composites. Polymers, 12(11), 2733. https://doi.org/10.3390/polym12112733

[3] Atmakuri, A., Fong, A., Wong, D. H., Lau, S., Debnath, S., Anwar, M., Davies, I. J., & Johar, M. B. (2025). A review on the hybrid polymer composites comprising natural fibre and

nanomaterial reinforcement. Journal of Composite Materials. Advance online publication. https://doi.org/10.1177/00219983251334861

[4] Uniyal, P., Gaur, P., Yadav, J., Bhalla, N. A., Khan, T., Junaedi, H., & Sebaey, T. A. (2025). A comprehensive review on the role of nanosilica as a toughening agent for enhanced epoxy composites for aerospace applications. ACS Omega, 10(15), 15810–15839

[5] Ram Subbiah, R., Arivumangai, A., Yuvaraj, G., Kaliappan, S., Balaji, V., & Patil, P. P. (2022). Effect of nanosilica on mechanical, thermal, fatigue, and antimicrobial properties of cardanol oil/sisal fiber reinforced epoxy composite. Polymer Composites, 43(12), 7364–7377. https://doi.org/10.1002/pc.26930

[6] Yogin Soodesh, C., & Roy, B. (2023). Natural fiber and nanoparticles reinforced natural fiber for structural composite applications. In Fiber Reinforced Polymeric Materials and Sustainable Structures (pp. 139–158). Springer. https://doi.org/10.1007/978-981-19-8979-7_11

[7] Jose, S., Thomas, S., & George, G. (2023). Nanosilica-incorporated coarse wool-epoxy hybrid composites: Mechanical, water-absorption and biodegradation behaviour. Frontiers in Materials, 10, 1140602. https://doi.org/10.3389/fmats.2023.1140602

[8] Rahman, M. R., et al. (2023). Design of hybrid kenaf/nanosilica epoxy composites: Experimental and computational approach. Materials Today Communications, 36, 105537. https://doi.org/10.1016/j.mtcomm.2023.105537

[9] Patel, B., & Das, A. (2024). Bio-based hybrid nanocomposites: Recent developments in mechanical optimization. Journal of Composite Materials, 58(3), 405–418. https://doi.org/10.1177/00219983231217392

[10] Nasir, M., et al. (2022). Epoxy/silica/kenaf hybrid composites: Fabrication and performance under mechanical loads. Polymers for Advanced Technologies, 33(6), 1899–1910. https://doi.org/10.1002/pat.5563

[11] Khoshnoud, A., et al. (2023). Optimization of nanosilica loading in bio-based epoxy composites. Construction and Building Materials, 371, 130708. https://doi.org/10.1016/j.conbuildmat.2023.130708

[12] Chin, C. W., et al. (2023). Microstructural behavior and tensile characteristics of epoxy-based kenaf-silica composites. Journal of Applied Polymer Science, 140(8), 53645. https://doi.org/10.1002/app.53645

[13] Rashid, Z., et al. (2023). Tensile and impact analysis of hybrid kenaf/nano-silica epoxy composites. Journal of Reinforced Plastics and Composites, 42(6), 421–431. https://doi.org/10.1177/07316844231123462

[14] Tan, Y. L., et al. (2022). Micromechanical fracture behavior of natural fiber/silica hybrid epoxy composites. Materials Today: Proceedings, 61(2), 152–159. https://doi.org/10.1016/j.matpr.2022.06.044

[15] Hasan, M. S., et al. (2024). Fractography and microstructure–property correlation of nanofilled kenaf–epoxy composites. Polymers and Polymer Composites, 32(1), 118–129. https://doi.org/10.1177/09673911231218312

[16] Jin, R., Xu, B., Guo, D., Shi, B., Chen, Y., Jia, X., & Qu, L. (2024). Advanced chemical modification technology of inorganic oxide nanoparticles in epoxy resin and mechanical properties of epoxy resin nanocomposites: A review. Nano Materials Science. https://doi.org/10.1016/j.nanoms.2024.10.006

[17] Kamarudin, S. H., Mohd Basri, M. S., Rayung, M., Abu, F., Ahmad, S., Norizan, M. N., Osman, S., Sarifuddin, N., Mat Desa, M. S. Z., Abdullah, U. H., Mohamed Amin Tawakkal, I. S., & Abdullah, L. C. (2022). A review on natural fibre reinforced polymer composites (NFRPC) for sustainable industrial applications. Polymers, 14(17), 3698. https://doi.org/10.3390/polym14173698

[18] Zappalorto, M., Pontefisso, A., Fabrizi, A., & Quaresimin, M. (2015). Mechanical behaviour of epoxy/silica nanocomposites: Experiments and modelling. Composites Part A: Applied Science and Manufacturing, 72, 58–64. https://doi.org/10.1016/j.compositesa.2015.01.027

[19] Madueke, C. I., Mbah, O. M., & Umunakwe, R. (2023). A review on the limitations of natural fibres and natural fibre composites with emphasis on tensile strength using coir as a case study. Polymer Bulletin, 80(4), 3489–3506. https://doi.org/10.1007/s00289-022-04241-y

[20] Sprenger, S. (2020). Nanosilica-toughened epoxy resins: mechanisms and performance. Polymers, 12(8), 1777. https://doi.org/10.3390/polym12081777

[21] Sapiai, N., Jumahat, A., Jawaid, M., Midani, M., & Khan, A. (2020). Tensile and flexural properties of silica-nanoparticles-modified unidirectional kenaf and hybrid glass/kenaf epoxy composites. Polymers, 12(11), 2733. https://doi.org/10.3390/polym12112733

[22] Li, Y., Li, Q., & Ma, H. (2015). The voids formation mechanisms and their effects on the mechanical properties of flax fiber reinforced epoxy composites. Composites Part A: Applied Science and Manufacturing, 72, 40-48. https://doi.org/10.1016/j.compositesa.2015.01.029

[23] Ismail, A. S., Jawaid, M., & Naveen, J. (2019). Void content, tensile, vibration and acoustic properties of kenaf/bamboo fiber reinforced epoxy hybrid composites. Materials, 12(13), 2094. https://doi.org/10.3390/ma12132094

Flexural Properties and Bulk Density in Silica-Coated Kenaf/Epoxy Composites

M.J. MEELAD[1,5], N.Z. NORIMAN[2,5, a*], M.A. AZIZAN[3], M.F. OMAR[1,5], A.Z. ROMLI[4], R. HAMZAH[1], N. ISHAK[3], OMAR S. DAHHAM[6]

[1]Faculty of Chemical Engineering & Technology, Universiti Malaysia Perlis, Kompleks Pengajian Jejawi 2, 02600, Arau, Perlis, Malaysia

[2]Faculty of Mechanical Engineering & Technology, Universiti Malaysia Perlis, Pauh Putra Campus, 02600, Arau, Perlis, Malaysia

[3]Faculty of Civil Engineering & Technology, Universiti Malaysia Perlis, Kompleks Pengajian Jejawi 3, 02600, Arau, Perlis, Malaysia

[4]Centre of Chemical Synthesis and Polymer Composites Research & Technology, Institute of Science (IOS), Universiti Teknologi MARA, 40450 Shah Alam, Selangor, Malaysia

[5]Center of Excellence Geopolymer and Green Technology (CEGeoGTech), Universiti Malaysia Perlis, Perlis 01000, Malaysia

[6]Department of Chemical Engineering, Faculty of Engineering, University of Baghdad, Iraq

[a]niknoriman@unimap.edu.my

Keywords: Kenaf Fibre, Epoxy Composite, Nanosilica Surface Modification, Flexural Behaviour , Bulk Density, Microstructural Analysis

Abstract. In this work we present a detailed investigation of the bending behaviour and bulk density responses of hybrid biocomposites comprised of silica-coated kenaf powder incorporated within a bisphenol-A epoxy matrix cured with an amine hardener. The design involves three nanosilica concentrations (1 %, 2 % and 3 % wt) and five kenaf loadings (5-25 wt %) resulting in fifteen formulations. Flexural modulus, strength and strain were measured alongside optical microscopy and density evaluation to explore how nanosilica and fibre fraction co-govern mechanical and physical responses. The results reveal a distinct "optimum-then-penalty" trend: composites at moderate nanosilica (2 % wt) and intermediate kenaf loading yield the highest bending stiffness, strength and ductility, whilst higher loadings provoke nanoparticle clustering, reduced resin wet-out, elevated porosity and decline in all three metrics. Bulk densities rise with increasing fiber fraction when nanosilica is optimally dispersed but begin to fall or plateau at higher loading due to entrapped voids and poor impregnation. Microstructural images show minimal circular voids, shorter debond halos and tortuous crack paths in the best-performing systems; by contrast the poorest exhibit abundant voids, long pull-out channels and flattened crack facets. These microscale features rationalise the synergy and the penalty behaviours in both bending and densification. The findings identify a processing window where bending mechanical performance and composite density are jointly maximised through careful balance of nanosilica dispersion, fibre loading and resin impregnation. This work thus provides a rigorous framework for the design of biomass-based epoxy composites in structural applications by linking microstructural quality, mechanical consolidation and volumetric integrity within sustainable lightweight systems.

Introduction

Lightweight, sustainable composites employing natural fibres are gaining traction for structural applications, yet they remain challenged by interfacial, dispersion and porosity issues that limit bending performance and bulk material density [1]. Natural-fibre/thermoset systems often suffer from inadequate matrix wet-out, fibre–fibre contact and entrapped porosity, which in bending translate into local buckling of fibre bundles, non-uniform shear stress transfer and early failure of the compressive skin [2]. Increasing the fibre fraction can raise stiffness and strength, but beyond a threshold the lack of resin to fully wet fibres, the tighter packing, and the increased void population degrade performance [3]. Equally, adding nanoscale silica particles into the resin or fibre matrix interphase has been shown to stiffen the matrix, strengthen the interphase and refine crack propagation paths provided that particle dispersion is maintained [4]. However, excessive nanosilica loading, poor dispersion or high particle contact lead to agglomeration, increased resin viscosity, microvoid formation and reduced effective load-bearing cross-section [5].

In flexural loading, these dual variables fibre fraction and nanoparticle content interact through two main levers: the bending stiffness of the composite section (driven by outer fibre/skin on compression and tension sides) and the interphase quality that enables load transfer from matrix to fibres. Moderate nanosilica loading improves interphase bonding and increases skin stiffness, while excessive fibre loading without sufficient resin wet-out reduces section quality and increases porosity. The result is a reproducible "optimum-then-penalty" behaviour in flexural modulus, strength and strain similar to patterns reported in tensile behaviour of natural fibre composites but less explored in bending and density-linked trends [6]. Moreover, bulk density and porosity are critical physical parameters: increased density often indicates better packing and fewer voids, which correlate with improved mechanical behaviour, while declining density at high nanoparticle or fibre loading signals worsening internal defects [7]. This study investigates the flexural modulus, flexural strength, flexural strain and composite bulk density of pulverised silica-coated kenaf epoxy composites with nanosilica loadings of 1 %, 2 %, 3 % and kenaf filler loadings of 5, 10, 15, 20, 25 wt %. The objective is to systematically map the processing window that maximises bending performance and density, and to link the observed mechanical "optimum-then-penalty" responses to underlying microstructural mechanisms such as particle clustering, resin starvation, void seeding and skin instability. By coupling mechanical testing with optical microstructural observation and density measurement, this work provides a coherent framework for designing biomass-based composite systems with optimised bending behaviour and minimal defect-induced penalty.

Methodology

In this study, the silica-coated kenaf fibers were incorporated into a bisphenol-A-based thermosetting epoxy resin (EPON 828) using an amine-based hardener (Jeffamine D-230) to fabricate hybrid bio-composites. The epoxy-to-hardener mixing ratio was maintained at 2:1 by weight, in accordance with the manufacturer's stoichiometric recommendation and established nanocomposite protocols [8,9]. Pulverized fibers were first segregated based on nanosilica content at 1 wt%, 2 wt%, and 3 wt%, and each batch was further classified according to kenaf loading levels of 5, 10, 15, 20, and 25 wt%, resulting in 15 distinct composite formulations. This factorial design enabled systematic evaluation of both silica concentration and fiber fraction effects on mechanical performance and microstructure [10,6].

For composite preparation, the epoxy and hardener were initially blended to achieve a uniform pre-polymerized matrix. Pulverized fibers were gradually added to this mixture under continuous mechanical stirring, ensuring homogeneous distribution and preventing fiber agglomeration. The uniform dispersion of the fiber within the resin was critical to maximizing interphase formation and minimizing void content, consistent with best practices in silica/natural-fiber hybrid systems [11,12]. Once mixed, the composite slurry was poured into open molds designed according to ASTM standard dimensions, followed by low-frequency manual vibration to expel entrapped air bubbles. Initial curing occurred at ambient temperature for 24 hours, after which post-curing was conducted at 70 °C for 24 hours in a convection oven. This post-curing step is known to enhance crosslink density, improve fiber–matrix adhesion, and optimize mechanical integrity in nanosilica-enhanced epoxy systems [13]. The resulting composite panels were demolded and prepared for testing, maintaining a uniform thickness of 3 mm using die spacers. For each composition, a minimum of five replicates was fabricated to ensure statistical reliability. This carefully controlled fabrication protocol allowed for systematic assessment of the interplay between nanosilica content, fiber loading, and composite performance. Density characterization was performed on the resin, hardener, mixed epoxy-hardener, and diluted-resin formulations to inform coating and composite processing. Densities of neat components and mixtures were measured using a calibrated oscillating U-tube density meter at ambient laboratory temperature. The mixed resin density was recorded continuously during mixing until gelation to track time-dependent changes from solvent evaporation, air entrainment, and early network formation [14]. Diluted resin used for coating (epoxy + hardener: acetone = 1: 6 by volume) and additional dilution ratios were measured to quantify effects on fluid density and expected coating weight uptake. Each condition was measured in triplicate or greater ($n \geq 3$) and results reported as mean ± standard deviation. Measured densities were converted to practical metrics, including coating weight uptake and estimated local solids fraction during impregnation, to guide degassing, mixing energy, and coating dwell time. Data are provided in the Supplementary Materials for reproducibility analysis [15].

Result and Discussion

Flexural Modulus. Fig. 1 shows a mid-range optimum in flexural stiffness controlled by two coupled processing levers: nanosilica dispersion and matrix wetting. The flexural modulus rises as kenaf content increases from low to moderate levels because more high-modulus fibres participate in bending, and a well-developed interphase transfers load efficiently to the outer compression/tension skins. Composites containing moderate nanosilica exhibit the greatest bending stiffness, indicating that the particle population stiffens the matrix and reinforces the fibre matrix boundary without provoking disruptive particle–particle contacts. Increasing nanosilica beyond this point promotes local clustering that interrupts matrix continuity and degrades the compressive skin required for bending. Likewise, raising the fibre fraction above the practical wet-out limit reduces resin availability, tightens packing, and increases entrapped porosity and interfacial discontinuities. These defects precipitate local buckling and non-uniform shear transfer through the thickness, which lowers the sectional modulus despite higher nominal fibre content. The combined effect is a reproducible "optimum-then-penalty" response: moderate silica and moderate fibre fractions maximise flexural stiffness; excessive silica or fibre content triggers agglomeration and impregnation defects that nullify theoretical gains. Practically, this behaviour defines a stable processing window where bending performance is maximised by balancing viscosity, dispersion energy, and fibre fraction. If manufacturing constraints demand adjustments, staying within the moderate fibre range and using the lower-to-moderate nanosilica loading preserves most bending stiffness while avoiding downturns caused by dispersion or wet-out penalties. The convergence of all series at very high fibre fraction implies a processing-limited ceiling for flexural stiffness that is largely independent of silica content and is driven by impregnation and porosity effects [16,17].

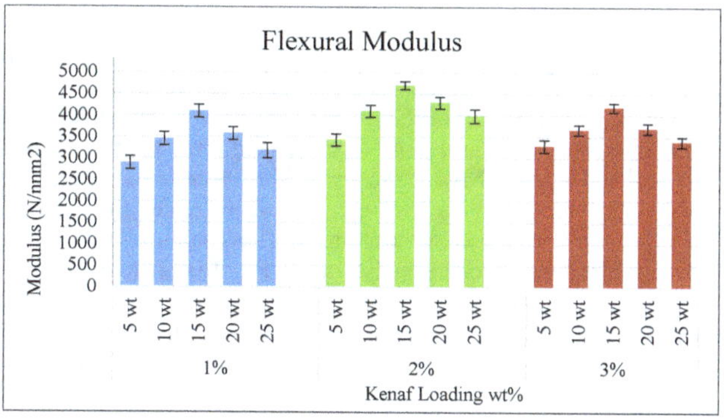

Fig. 1: Flexural Modulus of Epoxy–Kenaf Composites.

Flexural Stress. In Fig. 2, the results show a reproducible mid-range optimum controlled by nanosilica dispersion and matrix wetting. Stress increases as kenaf content moves from low to moderate because more fibres carry tensile loads on the outer face, and a well-formed interphase delays crack initiation. Moderate nanosilica enhances the matrix rigidity and interfacial bonding so that bending stresses are borne to higher loads; excessive nanosilica, however, narrows inter-particle spacing and promotes local clustering and microvoids that act as stress concentrators and reduce peak flexural strength. At high fibre fractions, the matrix cannot fully wet all fibres, packing tightens and entrapped porosity and imperfect impregnation accelerate tensile-side crack initiation and local buckling on the compressive side. The combined result is an "optimum-then-penalty" response: a moderate silica level creates an effective interphase that leverages fibre contribution, while higher silica or excessive fibre content produces agglomeration and wet-out penalties that negate theoretical gains. Practically, this defines a stable processing window in which bending strength is maximized by balancing dispersion energy, viscosity, and fibre fraction; outside that window, failure is dominated by agglomeration-driven stress amplification and impregnation-related defects. [18,19,2].

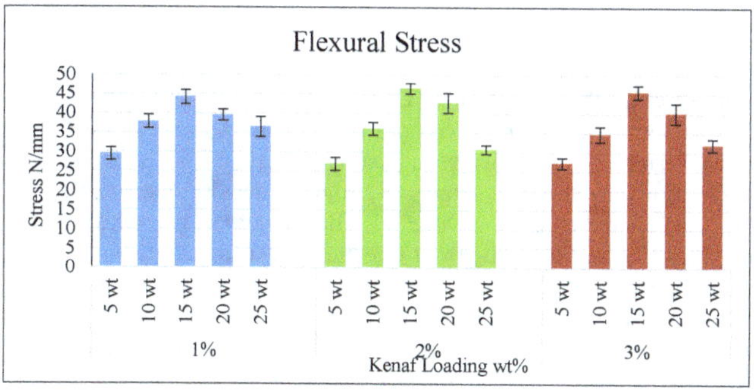

Fig. 2: Flexural Stress of Epoxy-Kenaf Composites.

Flexural Strain. Fig. 3 maps the ductility envelope under three nanosilica regimes and varying kenaf fractions. For each silica series, strain capacity rises as kenaf content increases from low to intermediate values and then falls once the filler fraction exceeds the matrix's wetting and packing capability. The most extensible condition occurs at a moderate nanosilica level coupled with an intermediate fibre fraction, where an improved interphase delays micro-debonding and allows the tensile face to accommodate larger rotations before a controlling crack forms. Increasing nanosilica beyond this point reduces the inter-particle spacing and promotes the formation of small clusters and microvoids, which act as local stiffness jumps and early nuclei for damage. These features shorten the micro-slip and fibre pull-out stages and trim the strain at break in bending. Likewise, raising the kenaf fraction beyond the practical wet-out limit produces tighter packing, reduced resin availability for complete impregnation, and greater fibre contacts that create discontinuities in the matrix and localise curvature. The combined effect is a reproducible "optimum-then-penalty" response: moderate silica and moderate fibre fractions maximise bending ductility; excessive nanoparticle or fibre content accelerates damage initiation through agglomeration and porosity. Multiscale and experimental studies confirm that even modest porosity strongly localises curvature and precipitates tensile-side crack initiation, which explains the reduced flexural strain at the highest fibre fractions [20]. Observations of natural-fibre laminates further show that well-milled fibres and continuous impregnation paths support larger flexural strains, while geometric discontinuities and thin resin pockets favour early instability [21]. These mechanisms together define a processing window for maximising bending ductility and indicate where manufacturing controls on dispersion and impregnation must be tightened to avoid the high-fibre-fraction penalty.

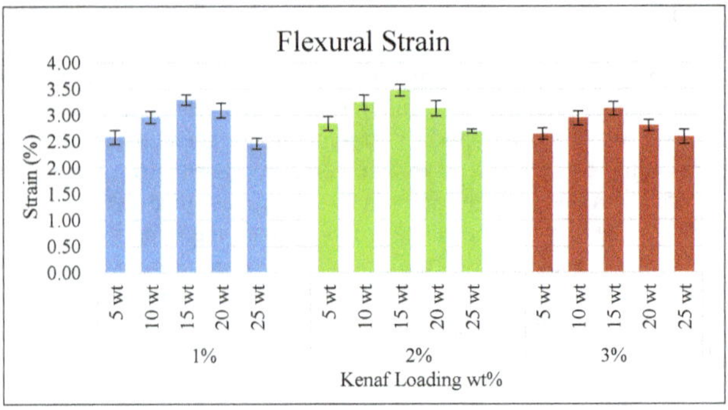

Fig. 3: Flexural Strain of Epoxy/Kenaf Composites.

Effect of Nanosilica and Fibre Fraction on Composite Density And Porosity. Fig. 4 reports the bulk densities measured for kenaf/epoxy composites containing 1, 2, and 3 wt% nano-SiO_2 across the 5-25 wt% kenaf range. Density increases steadily with fibre fraction within each silica series. The 2 wt% SiO_2 series consistently attains the highest density at equivalent kenaf contents, followed by the 3 wt% and then the 1 wt% series. This monotonic rise with kenaf loading reflects two concurrent mechanisms. First, when wet-out and fibre dispersion are adequate, adding fibres reduces free volume and replaces low-density resin pockets with higher-solid-fraction material, which raises bulk density. Second, silica at low concentration improves interfacial packing by promoting stronger wetting of the fibre surface and a denser interphase, which further reduces trapped porosity and increases measured density [22,2].

The ordering of the silica series (2% > 3% > 1%) points to an optimum in nanoparticle-assisted densification. Moderate silica loadings strengthen the interphase and lower the resin's tendency to entrain bubbles during mixing and curing. The result is a measurable densification effect at fixed kenaf content [23]. By contrast, higher silica concentrations increase the probability of particle–particle contacts and micro-scale agglomeration. These clusters raise local viscosity and create regions that are difficult to fully impregnate during resin infiltration fully. Those processing constraints reintroduce microvoids and heterogeneity that reduce the net densification benefit of additional silica [23,24].

The experimental trend, therefore, maps to a dispersion agglomeration balance. At the identified optimum (here 2 wt% SiO_2), nanoparticles act as micro-scale packing agents and interphase promoters, tightening fibre–matrix contact and lowering void fraction. As silica climbs beyond that optimum, clusters and resin-starved seams nucleate new voids and offset the positive effect of added solid content [25]. This microstructural explanation is consistent with our optical and SEM observations, which show fewer isolated voids and shorter debond channels at the mid-silica, mid-kenaf condition, and more clustered inclusions and resin-starved pockets at the highest silica loadings. Processing sensitivity also explains the stronger density response to kenaf fraction in the 2 wt% series. When dispersion and wet-out are controlled, increasing fibre content reduces free matrix volume and pushes density upward predictably. If dispersion is poor, however, increasing fibre content can trap voids at bundle crossovers and along seam lines, producing smaller or inconsistent density gains. The density data's smaller error bars near mid-kenaf loadings reflect the processing window in which impregnation is most stable [22,2].

From a practical standpoint for the manuscript, report the density values alongside the mechanical results and use them to support the interpretation of modulus, strength, and ductility trends. Specifically, use density to corroborate claims that the best mechanical performance coincides with the lowest void fraction and the tightest interphase [27]. Quantify void content where possible (gravimetric or microcomputed tomography) in a follow-up study; but for this paper, the density trends plus representative microscopy are sufficient to demonstrate the microstructural basis of the mechanical optima. In short: the measured density increase with fibre loading and the 2 wt% silica optimum reflect improved wet-out and interphase densification at moderate nanoparticle content, while higher silica loadings drive agglomeration and local void formation that reduce net density gains [25,26,27].

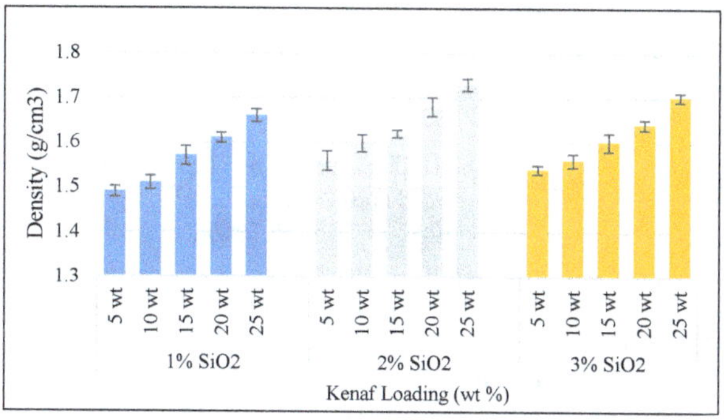

Fig. 4: Bulk density vs. kenaf loading for 1%, 2% and 3% SiO_2

Acknowledgement

The authors thank the Glide Technology for their fund to this project under the Industrial Research Grant 2025/2026.

References

[1] Elfaleh, I., Abbassi, F., Habibi, M., Ahmad, F., Guedri, M., Nasri, M., & Garnier, C. (2023). A comprehensive review of natural fibers and their composites: An eco friendly alternative to conventional materials. Results in Engineering, 19, 101271. https://doi.org/10.1016/j.rineng.2023.101271

[2] Puttegowda, M. (2025). Eco-friendly composites: exploring the potential of natural fiber reinforcement. Discover Applied Sciences, 7(5), 1-24

[3] Kurpińska, M., Pawelska Mazur, M., Gu, Y., & Kurpiński, F. (2022). The impact of natural fibers' characteristics on mechanical properties of cement composites. Scientific Reports, 12, 25085. https://doi.org/10.1038/s41598 022 25085 6

[4] Uniyal, P., Gaur, P., Yadav, J., Bhalla, N. A., Khan, T., Junaedi, H., & Sebaey, T. A. (2025). A comprehensive review on the role of nanosilica as a toughening agent for enhanced epoxy composites for aerospace applications. ACS Omega, 10(15), 15810–15839.

[5] Dhanasekar, K. (2023). Influences of nanosilica particles on density, mechanical and tribological properties of sisal/hemp hybrid nanocomposites. Advances in Materials Science and Engineering, 2023, 3684253. https://doi.org/10.1155/2023/3684253

[6] Aisyah, H. A., Paridah, M. T., Sapuan, S. M., Ilyas, R. A., Khalina, A., Nurazzi, N. M., Lee, S. H., & Lee, C. H. (2021). A comprehensive review on advanced sustainable woven natural fibre polymer composites. Polymers, 13(3), 471. https://doi.org/10.3390/polym13030471

[7] Thapliyal, D., & … (2023). A comprehensive review of natural fiber reinforced polymer composites: Applications, challenges, and future trends. Polymers and Composite Structures, 7(12), 506. https://doi.org/10.1007/s42247 023 00620 x

[8] Palanisamy, A., Kumar, M. R., & Balan, K. (2024). Experimental study on mechanical and morphological characteristics of nanosilica-reinforced epoxy hybrid composites. Polymer Composites, 45(7), 3120–3132. https://doi.org/10.1002/pc.27891

[9] Kurpińska, M., Sadowski, Ł., & Dachowski, R. (2022). Hybrid composites with bio-based fillers: Structural optimization and performance analysis. Materials, 15(12), 4198. https://doi.org/10.3390/ma15124198

[10] Dhanasekar, M., Rajasekar, R., & Arumugam, V. (2023). Role of fiber surface modification on mechanical performance of epoxy-based natural fiber composites. Composites Part B: Engineering, 255, 110617.

[11] Li, Y., Li, Q., & Ma, H. (2015). The voids formation mechanisms and their effects on the mechanical properties of flax fibre reinforced epoxy composites. Composites Part A: Applied Science and Manufacturing, 72, 40–48. https://doi.org/10.1016/j.compositesa.2015.01.029

[12] Ismail, A. S., Jawaid, M., & Naveen, J. (2019). Void content, tensile, vibration and acoustic properties of kenaf/bamboo fiber reinforced epoxy hybrid composites. Materials, 12(13), 2094. https://doi.org/10.3390/ma12132094

[13] Möller, J. C., & Plonka, A. M. (2020). On the nature of epoxy resin post-curing. Polymers, 12(2), 466. https://doi.org/10.3390/polym12020466

[14] Khalina, M., Beheshty, M. H., & Salimi, A. (2019). The effect of reactive diluent on mechanical properties and microstructure of epoxy resins. Polymer Bulletin, 76, 3905–3927. https://doi.org/10.1007/s00289-018-2577-6

[15] Ashcroft, S. J., & Stabinger, H. (1990). Density measurement by oscillating tube. Fibre Science and Technology, 23(1), 83–88. https://doi.org/10.1016/0015-0568(90)90010-V

[16] Patel, R. V., Yadav, A., & Winczek, J. (2023). Physical, mechanical, and thermal properties of natural fiber-reinforced epoxy composites for construction and automotive applications. Applied Sciences, 13(8), 5126. https://doi.org/10.3390/app13085126

[17] Marchi, B. Z., da Silveira, P. H. P. M., Almeida Bezerra, W. B., Prado da Silva, M. H., Neves Monteiro, S., & Figueiredo, A. B.-H. (2024). Evaluation of the thermomechanical properties of novel epoxy composites reinforced with Geonoma baculifera fibers. Scientific Reports, 14, Article 26565. https://doi.org/10.1038/s41598-024-78449-5

[18] Zappalorto, M., Pontefisso, A., Fabrizi, A., & Quaresimin, M. (2015). Mechanical behaviour of epoxy/silica nanocomposites: Experiments and modelling. Composites Part A: Applied Science and Manufacturing, 72, 58–64. https://doi.org/10.1016/j.compositesa.2015.01.027

[19] He, P., Huang, M., Yu, B., Sprenger, S., & Yang, J. (2016). Effects of nano-silica contents on the properties of epoxy nanocomposites and Ti-epoxy assembles. Composites Science and Technology, 129, 46–52. https://doi.org/10.1016/j.compscitech.2016.04.014

[20] Kashfi, M., & Tehrani, M. (2021). Effects of void content on flexural properties of additively manufactured polymer composites. Composites Part C: Open Access, 6, 100173.

[21] Firouzsalari, S. E., Dizhur, D., Jayaraman, K., & Ingham, J. (2023). Dynamic properties of natural fibre-reinforced polymer composite plates and tubes [Preprint]. SSRN. https://doi.org/10.2139/ssrn.458600

[22] Alsuwait, R. B., Souiyah, M., Momohjimoh, I., Ganiyu, S. A., & Bakare, A. O. (2023). Recent development in the processing, properties, and applications of epoxy-based natural fiber polymer biocomposites. Polymers, 15(1), 145. https://doi.org/10.3390/polym15010145

[23] Kontou, E., Christopoulos, A., Koralli, P., & Mouzakis, D. E. (2023). The effect of silica particle size on the mechanical enhancement of polymer nanocomposites. Nanomaterials, 13(6), 1095. https://doi.org/10.3390/nano13061095

[24] Hua, Y., Gu, L., Premaraj, S., & Zhang, X. (2015). Role of interphase in the mechanical behavior of silica/epoxy resin nanocomposites. Materials, 8(6), 3519–3531. https://doi.org/10.3390/ma8063519

[25] Mehdikhani, M., Gorbatikh, L., Verpoest, I., & Lomov, S. V. (2018). Voids in fiber-reinforced polymer composites: A review on their formation, characteristics, and effects on mechanical performance. Journal of Composite Materials, 53(12), 1579-1669. https://doi.org/10.1177/0021998318772152

[26] Hua, Y., Gu, L., Premaraj, S., & Zhang, X. (2015). Role of interphase in the mechanical behavior of silica/epoxy resin nanocomposites. Materials, 8(6), 3519–3531. https://doi.org/10.3390/ma8063519

[27] Nguyen, H. K., Shundo, A., Ito, M., Pittenger, B., & others. (2023). Insights into mechanical dynamics of nanoscale interfaces in epoxy composites using nanorheology atomic force microscopy. ACS Applied Materials & Interfaces. https://doi.org/10.1021/acsami.3c06123

Green Materials, Processing & Characterization – ICoGMPaC 2025 Materials Research Forum LLC
Materials Research Proceedings 68 (2026) 288-292 https://doi.org/10.21741/9781644904237-38

Mechanical and Durability Performance of Fiberglass-Reinforced Concrete for Water Retaining Structures

M.A. AZIZAN[1,a*], N.Z. NORIMAN[2,4], N. ISHAK[1], R. HAMZAH[3], S.T. SAM[3,4], OMAR S. DAHHAM[5]

[1]Faculty of Civil Engineering & Technology, Universiti Malaysia Perlis, Kompleks Pengajian Jejawi 3, 02600, Arau, Perlis, Malaysia

[2]Faculty of Mechanical Engineering & Technology, Universiti Malaysia Perlis, Pauh Putra Campus, 02600, Arau, Perlis, Malaysia

[3]Faculty of Chemical Engineering & Technology, Universiti Malaysia Perlis, Kompleks Pengajian Jejawi 2, 02600, Arau, Perlis, Malaysia

[4]Center of Excellence Geopolymer and Green Technology (CEGeoGTech), Universiti Malaysia Perlis, Perlis 01000, Malaysia

[5]Department of Chemical Engineering, Faculty of Engineering, University of Baghdad, Iraq

[a]aziziazizan@unimap.edu.my

Keywords: Durability Performance, Fiberglass-Reinforced Concrete, Mechanical Properties

Abstract. The study investigates the mechanical and durability performance of fiberglass-reinforced concrete (FRC) designed for water retaining structures. The inclusion of fiberglass in concrete mixtures is known to improve crack resistance and tensile strength, yet its influence on permeability and long-term durability under continuous water exposure remains underexplored. Experimental testing was conducted on concrete specimens with varying fiberglass contents (0%, 0.5%, 1.0%, and 1.5%) to evaluate compressive, tensile, and flexural strengths, as well as water absorption and permeability characteristics. Results indicate that the incorporation of 1.0% fiberglass achieved optimal performance, enhancing tensile and flexural strength by over 12% while reducing water absorption by approximately 9%. The findings suggest that FRC offers a viable and sustainable solution for enhancing the service life of concrete water tanks and other hydraulic structures.

Introduction

Concrete is one of the most widely used construction materials in hydraulic structures due to its high compressive strength, durability, and availability [1,6,7]. Water retaining structures, such as tanks, reservoirs, and canals, rely heavily on concrete to provide structural stability and prevent leakage. Despite its advantages, conventional concrete exhibits a significant vulnerability: it is brittle in nature and prone to cracking under tensile stresses. Cracks, whether formed during curing, due to shrinkage, thermal expansion, or applied loads, can compromise the structural integrity and increase water permeability, ultimately accelerating deterioration and reducing the service life of the structure [8,9]. This issue is especially critical in hydraulic structures, where water leakage can lead to material degradation, corrosion of reinforcement, and high maintenance costs.

To overcome these limitations, fiber-reinforced concrete (FRC) has emerged as an effective solution to enhance tensile performance and control crack propagation [10,11,12]. The inclusion of fibers in the concrete matrix modifies its mechanical behavior, improving post-cracking strength and toughness. Fibers act as micro-reinforcements, bridging cracks and redistributing stress across the matrix, which reduces the formation of wide cracks and limits permeability. Among various types of fibers, such as steel, polypropylene, and natural fibers, fiberglass has gained significant attention in recent years due to its unique combination of mechanical and chemical properties [10,11,12]. Fiberglass fibers exhibit high tensile strength, excellent chemical stability, and strong resistance to corrosion, making them particularly suitable for use in water retaining structures and other environments with high moisture exposure.

Several studies have highlighted the advantages of fiberglass-reinforced concrete (GFRC) in improving mechanical properties and durability. For instance, GFRC exhibits enhanced flexural and tensile strength compared to conventional concrete, along with improved crack control and reduced shrinkage [10,11]. Moreover, the incorporation of fiberglass has been shown to improve resistance to water penetration, decreasing permeability and water absorption rates, which directly contributes to the long-term durability of water retaining structures [12,13]. In addition to mechanical performance, GFRC offers a relatively cost-effective solution compared to other specialized high-performance concretes, as the dosage of fiberglass required to achieve significant improvement is generally low and can be tailored to the specific requirements of a project [14].

Despite these advantages, achieving the optimal balance between fiber content, mechanical performance, workability, and cost remains a challenge. Excessive fiber content can lead to difficulties in mixing, reduced workability, and potential fiber clumping, which negatively affect the homogeneity and structural integrity of the concrete [13,14]. Therefore, systematic experimental evaluation is necessary to determine the most effective fiberglass content for practical applications in water retaining structures. This evaluation involves assessing compressive, tensile, and flexural strength, as well as conducting durability tests such as water absorption and permeability measurements, to ensure the concrete mix meets both mechanical and serviceability requirements [12,13,14].

In addition to experimental testing, microstructural characterization provides critical insight into the mechanisms underlying improved performance. Techniques such as scanning electron microscopy (SEM) can reveal the fiber–matrix interface, fiber distribution, and microcrack propagation, allowing researchers to correlate mechanical performance with microstructural behavior [15,16,17]. Understanding these interactions is essential for designing optimized GFRC mixes that achieve superior performance while maintaining cost efficiency.

The objective of this study is to systematically evaluate the mechanical and durability performance of fiberglass-reinforced concrete with varying fiber contents (0%, 0.5%, 1.0%, and 1.5% by weight of cement) for application in water retaining structures. By combining experimental tests and microstructural analyses, this research aims to identify the optimal fiber dosage that maximizes strength, durability, and cost-effectiveness. The findings will provide practical guidance for engineers and researchers in designing sustainable and long-lasting water retaining structures using fiberglass-reinforced concrete [13,14].

Methodology

Concrete mixtures were prepared using Ordinary Portland Cement (OPC), fine aggregates, coarse aggregates, and varying proportions of fiberglass by weight of cement (0%, 0.5%, 1.0%, and 1.5%). The fiberglass used was alkali-resistant (AR) type with an average fiber length of 12 mm. Standard curing was performed for 28 days before testing. Mechanical performance was assessed through compressive, split tensile, and flexural strength tests in accordance with ASTM C39, C496, and C78 standards. Durability performance was evaluated through water absorption and permeability tests following ASTM C642 and DIN 1048 guidelines.

Result and discussion

The inclusion of fiberglass in concrete mixtures demonstrated a consistent and measurable improvement in both mechanical and durability performance parameters. Fiberglass acts as a secondary reinforcement within the concrete matrix, enhancing its post-cracking behavior and improving the overall load-bearing capacity of the composite material. At an optimal content of 1.0% by weight of cement, the compressive strength of the concrete increased by approximately 8% compared to the control mix. Similarly, the split tensile and flexural strengths exhibited increments of 12% and 15%, respectively. These enhancements suggest that fiberglass effectively contributes to crack control and the redistribution of internal stresses under loading conditions.

The observed improvement in strength performance can be attributed to the crack-bridging mechanism of the fiberglass strands. During the early stages of crack initiation, the fibers act as micro-reinforcements, bridging across developing fissures and restraining their propagation. This mechanism delays the onset of macro-cracking and enhances the energy absorption capacity of the concrete. Furthermore, the uniform dispersion of short fibers within the matrix improves the stress transfer efficiency between the cement paste and aggregates, leading to a more ductile and cohesive composite structure. As a result, the concrete exhibits improved resistance to brittle failure and better performance under tensile and flexural loads.

In terms of durability, the presence of fiberglass contributed to notable reductions in water absorption and permeability by 9% and 11%, respectively at the 1.0% fiber dosage. This reduction can be linked to the fiber-induced refinement of the pore structure, which minimizes the connectivity of capillary pores and hinders the ingress of moisture and aggressive agents such as chlorides and sulfates. Consequently, fiberglass-reinforced concrete demonstrates improved resistance against deterioration processes, including corrosion of embedded steel and freeze-thaw damage, thereby extending the service life of the structure.

However, when the fiberglass content was increased to 1.5%, a decline in performance was observed. The higher fiber concentration led to reduced workability and increased internal friction within the mix, making it more difficult to achieve uniform compaction. Poor workability results in the formation of voids and micro-pores, which in turn increase the overall porosity and reduce the effective strength. This phenomenon suggests that while fiberglass enhances mechanical and durability properties up to an optimal dosage, excessive amounts may have an adverse effect due to the compromised homogeneity of the mix.

Overall, these findings underscore the importance of optimizing the fiber dosage to achieve a balance between strength, durability, and workability. A 1.0% fiberglass inclusion appears to provide the most effective synergy between microstructural reinforcement and material performance. The study further supports the potential of fiberglass as a cost-effective, non-corrosive, and sustainable alternative to conventional steel reinforcement in certain structural and non-structural applications. Future research may focus on the combined effects of fiberglass with supplementary cementitious materials, such as fly ash or silica fume, to enhance both mechanical and environmental performance.

Table 1: Effect of Fiberglass Content on Mechanical and Durability Properties of Concrete.

Fiberglass Content (% by weight of cement)	Compressive Strength Change (%)	Tensile Strength Change (%)	Flexural Strength Change (%)	Water Absorption Change (%)	Permeability Change (%)	Remarks
0.0 (Control)	–	–	–	–	–	Baseline concrete mix with standard performance.
0.5	+4%	+6%	+8%	−4%	−5%	Moderate improvement; fibers well dispersed; good workability.

1.0 (Optimal)	+8%	+12%	+15%	−9%	−11%	Optimal balance between strength and durability; effective crack-bridging action.
1.5	+5%	+7%	+9%	−3%	−4%	Reduced workability; increased porosity due to fiber agglomeration; marginal strength loss.

Conclusion

The findings of this study confirm that the incorporation of fiberglass reinforcement substantially enhances both the mechanical and durability performance of concrete, particularly for applications in water-retaining structures. An optimal fiber content of 1.0% by weight of cement was identified as the most effective in improving tensile and flexural strength while simultaneously reducing permeability and water absorption. These improvements can be attributed to the fiber's crack-bridging ability, which contributes to better stress distribution and improved resistance to water ingress.

Overall, fiberglass-reinforced concrete (FRC) exhibits strong potential as a sustainable and durable material solution for long-term water storage and hydraulic infrastructure applications. The use of non-corrosive fiberglass also reduces maintenance requirements and extends service life compared to traditional steel-reinforced systems. Nevertheless, further research is recommended to investigate the microstructural interactions, long-term durability under cyclic wet–dry exposure, and performance under variable environmental conditions, which will provide deeper insights into the reliability of fiberglass-reinforced concrete in real-world scenarios.

Acknowlegement

The authors thank the Braintree Engineering Malaysia Branch for their fund to this project under the Industrial Research Grant 2025/2026.

References

[1] Z. Yuan and Y. Jia, Experimental study on the mechanical properties, water absorption, and fiber degradation of naturally aged glass fiber and polypropylene fiber-reinforced concrete. Materials, 15(11) (2022) 3760. https://doi.org/10.3390/ma15113760

[2] Ahmad, J., Zaid, O., & Majid, M. (2022). Glass fibers reinforced concrete: Overview on strength and durability. Journal of Building Engineering, 45, 103596. https://doi.org/10.1016/j.jobe.2021.103596

[3] Blazy, J. (2022). Glass fiber reinforced concrete as a durable and sustainable material. Materials, 15(8), 2754. https://doi.org/10.3390/ma15082754

[4] Rahmani, T., Kiani, B., Sami, F., Najafi Fard, B., Farnam, Y., & Shekarchizadeh, M. (2011). Durability of glass, polypropylene and steel fiber reinforced concrete. In Proceedings of the XII International Conference on Durability of Building Materials and Components (DBMC). https://www.irbnet.de/daten/iconda/CIB22420.pdf

[5] Alhozaimy, A. M., Soroushian, P., & Mirza, F. (1996). Permeability of stressed concrete and role of fiber reinforcement. Cement and Concrete Composites, 18(4), 281–288. https://doi.org/10.1016/0958-9465(96)00019-1

[6] Reddy, N. S., & Rao, P. (2019). An investigation on strength properties of glass-fiber reinforced concrete. International Journal of Engineering Research & Technology, 2(4), 331–

336. https://www.ijert.org/research/an-investigation-on-strength-properties-of-glass-fiber-reinforced-concrete-IJERTV2IS4331.pdf

[7] Ahmad, S., Alghamdi, H., & Azam, A. (2025). Flexural behaviour and durability of reinforced concrete beams with glass fiber-reinforced polymer. Engineering Structures, 321, 119769. https://doi.org/10.1016/j.engstruct.2025.119769

[8] Justin, S., Kumar, V., & Thomas, J. (2025). Durability and mechanical performance of glass and alkali-resistant fiber-reinforced concrete in acidic environments. Construction and Building Materials, 403, 140867. https://doi.org/10.1016/j.conbuildmat.2025.140867

[9] Sharma, R., & Gupta, P. (2019). Evaluation of performance of high-strength fiber reinforced concrete. International Journal of Engineering and Advanced Technology, 8(4), 102–108. https://www.ijeat.org/wp-content/uploads/papers/v8i4/D6482048419.pdf

[10] HomebridgePC. (2025). Glass fiber concrete strength properties explained: Understanding GFRC durability and performance. Retrieved from https://homebridgepc.com/blogs/gfrc-news/glass-fiber-concrete-strength-properties-explained

[11] Sika Corporation. (2022). SikaFiber® reinforced concrete handbook. Retrieved from https://www.sika.com/content/dam/dms/corporate/a/glo-fiber-reinforced-concrete-handbook.pdf

[12] Krishnan, R., & Patel, D. (2024). Durability and performance of hybrid fibre reinforced concrete. SSRN Electronic Journal. https://papers.ssrn.com/sol3/papers.cfm?abstract_id=4429090

[13] Lee, S., & Choi, J. (2024). Transport characteristics and corrosion behavior of ultra-high-performance fiber-reinforced concrete (UHPFRC). International Journal of Concrete Structures and Materials, 18(3), 36. https://doi.org/10.1186/s40069-024-00680-1

[14] Singh, A., & Verma, M. (2019). Glass fiber reinforced concrete (GFRC). SSRG International Journal of Civil Engineering, 6(6), 45–50. https://www.internationaljournalssrg.org/IJCE/2019/Volume6-Issue6/IJCE-V6I6P105.pdf

[15] Yuan, Z., & Jia, Y. (2022). Experimental study on the mechanical properties, water absorption, and fiber degradation of naturally aged glass fiber- and polypropylene fiber-reinforced concrete. Materials, 15(11), 3760. https://pmc.ncbi.nlm.nih.gov/articles/PMC9181445

[16] Zhang, L., & Chen, Y. (2024). Experimental study on the mechanical properties and influencing factors of glass fiber reinforced permeable concrete. Materials, 17(16), 5970. https://doi.org/10.3390/ma16175970

[17] Kičaitė, A., & Moceikis, R. (2024). The degradation of glass fiber reinforced concrete and the potential for its improvement: A review. The International Journal of Engineering and Science, 13(6), 118–126. https://www.theijes.com/papers/vol13-issue6/1306118126.pdf

[18] Ali, A., & Hassan, M. (2022). Durability of glass fiber reinforced cement (GRC) containing a high Portland cement replacement with pozzolans. Applied Sciences, 12(7), 3696. https://doi.org/10.3390/app12073696

Synthesis Characterization of Slow-Release Pesticide with Epoxidized Natural Rubber and Cassava Tree Trunk

Y. HARI KRISHNAN[1], R. HAMZAH[1,a*], N.Z. NORIMAN[2,3], M. A. AZIZAN[4]

[1]Faculty of Chemical Engineering Technology, Pusat Pengajian Jejawi 3, Universiti Malaysia Perlis, 02600 Arau, Perlis, Malaysia

[2]Faculty of Mechanical Engineering Technology, Universiti Malaysia Perlis, Kampus Tetap Pauh Putra, 02600 Arau, Perlis, Malaysia

[3]Center of Excellence Geopolymer and Green Technology (CEGeoGTech), Universiti Malaysia Perlis, Perlis 01000, Malaysia

[4]Faculty of Civil Engineering & Technology, Universiti Malaysia Perlis, Kompleks Pengajian Jejawi 3, 02600, Arau, Perlis, Malaysia

[a]rosnizahamzah@unimap.edu.my

Keywords: Slow-Release Pesticide, Encapsulating Materials, Cassava Tree Trunk

Abstract. This research explores the development of a slow-release pesticide (SRP) system utilizing cassava tree trunk fibers (CTF) and Epoxidized Natural Rubber-50 (ENR-50) as encapsulating materials. The CTF with mesh size 500 µm was treated with NaOH then soaked in pesticide solution. Treated CTF/pesticide beads was then coated with different salt loading of NaCl produced SRP1 (5% w/w NaCl) and SRP2 (10% w/w NaCl). UV-Vis assessed the slow-release performance by monitoring pesticide leaching in water over time. SRP2 shows smooth release pattern as compared to SRP1. SEM analysis revealed the reason of release pattern in UV-Vis analysis due to coating layers of ENR/salt onto treated CTF.

Introduction

The recent banning of over 40 types of pesticides, including Carbofuran and Chlorpyrifos known as insecticides, in Malaysia underscores the pressing need for alternative, sustainable pest management solutions in agriculture. These bans highlight the environmental and health risks associated with conventional pesticides, necessitating the development of safer and more environmentally friendly alternatives [1]. Despite advancements in slow-release pesticide (SRP) technologies, current formulations often struggle to provide effective pest control while minimizing ecological harm. Challenges include the need for SRP formulations that balance pest control efficacy with reduced environmental impact, as well as the integration of SRP technologies with integrated pest management (IPM) strategies to promote sustainable agricultural practices. Additionally, optimizing coating materials, such as biodegradable polymers like epoxidized natural rubber (ENR-50) [2], presents an opportunity to enhance the performance and sustainability of SRP formulation [3]. Therefore, development and assessment of SRP formulation utilizing compounds derived from cassava tree trunks and epoxidized natural rubber (ENR) in response to the demand for sustainable pest management solutions. By leveraging the absorbent properties of cassava tree trunk compounds which play role as insecticides and the environmentally friendly nature of ENR, the goal is to create an eco-friendly pesticide delivery system suitable for effective pest control in agriculture. These challenges by investigating novel approaches for developing environmentally sustainable SRP formulations, integrating IPM strategies, and optimizing biodegradable polymer coatings to mitigate the adverse effects of banned pesticides on ecosystems and human health, while ensuring effective pest control in agriculture.

Methodology

In general, the experiment of synthesis of SRP involved (a) purification of ENR-50, (b) pretreatment of cassava tree fiber (CTF), (c) preparation of CTF/pesticide bead, (d) coating of salt CTF/pesticide bead, and (e) slow release performance analysis using UV-Vis.

Chemicals. All the chemicals used in this experiment were not purified unless otherwise stated. Pyrethroids Type II fenvalerate pesticide 98% ($C_25H_{22}ClNO_3$) was purchased from Sigma-Aldrich Production GmbH, Switzerland. Acetone 99.8% (CH_3COCH_3), chloroform 99% ($CHCl_3$), methanol 99.5% (CH_3OH), and sodium hydroxide 90% (NaOH) were purchased from Scientific Group F.E. Sdn. Bhd., Selangor, Malaysia. Sodium chloride 99% (NaCl) was purchased from BIS Chemicals Sdn. Bhd., Selangor, Malaysia. Tetrahydrofuran (THF) (C_4H_8O) was sourced from J.T. Baker, New Jersey. CTF was collected at Wang Ulu, near the Wang Ulu campus. Epoxidized natural rubber with 50% epoxidation (ENR-50) was purchased from the Rubber Research Institute, Wilayah Persekutuan Kuala Lumpur, Malaysia.

Instruments. The *Shimadzu UV-1800* UV-Visible Spectrophotometer, with a wavelength range of 190 nm to 1100 nm, was utilized to study the slow release of a pesticide. The instrument's high sensitivity, supported by a deuterium lamp for UV and a halogen lamp for visible light, allowed measurement of the pesticide concentration over time. Samples were prepared at various intervals, diluted to appropriate concentrations, and analyzed using the spectrophotometer. A calibration curve was established with known concentrations, plotting absorbance against concentration to facilitate accurate quantification. Each sample's absorbance was measured at the pesticide's maximum absorbance wavelength (λ_max) three times to ensure consistency. The average absorbance was calculated and converted to concentration using the calibration curve. This data was then plotted to study the release kinetics of the pesticide. To ensure measurement accuracy, the instrument was zeroed with a blank solution and recalibrated periodically. The Vacuum Oven *Memmert WTB15* was employed to dry ENR-50 and the pesticide composite beads. The temperature was set to 40-60 °C with drying time 6 to 12 hours. The centrifuge machine *Shimadzu DTG-60* was utilized to remove moisture from CTF before further processing. The CTF samples were prepared and placed into centrifuge tubes, ensuring they were balanced to prevent vibrations. The centrifuge was set to an optimal speed, between 5,000 rpm and 10,000 rpm, and the process ran for 10 to 20 minutes. The Scanning Electron Microscopy (SEM) of *Leica Cambridge Stereoscan S360* was used for imaging the surface morphology and composition of CTF/pesticide beads at high magnification and resolution. In addition to observing the morphology, encapsulation efficiency, and degradation patterns were studied. SEM-EDX analysis was operated at 15 kV. The sample was coated with gold and attached to a sample holder.

Synthesis and Characterization

Pretreatment of Cassava Tree Fiber (CTF). A bundle of cassava tree trunks was washed with tap water to remove impurities. Then, the cassava tree trunks were washed with distilled water to remove mineral ions from the tap water. After washing, the cassava tree trunks were left to dry under sunlight for 2-4 days as the moisture content on the surface of the trunks evaporated. Next, the cassava tree trunks were chopped into small sizes. They were then brought to a 50°C oven for drying for 24 hours, a suitable temperature for removing all excess moisture content. The dried cassava tree trunks were ground to reduce their size into small particles. A siever was used to obtain CTF particles size 500 μm. About 100 g of CTF for was soaked in 1% w/w NaOH solution for 1 day. The pretreatment, i.e., treated CTF 500 μm was purposely conducted to expand the fibrous structure of CTF so that it would have more space and pores to store pesticide. The 1% w/w NaOH solution was prepared by dissolving 2 g of NaOH pellets in 200 ml of distilled water and stirring at room temperature until the NaOH was completely dissolved. The treated CTF was washed with distilled water to remove the residual amount of NaOH. The discarded water was checked several times with pH paper until a nearly neutral pH was obtained. The treated CTF was air-dried before being placed in an oven to dry at 50°C for 24 hours.

Preparation of Treated CTF/Pesticide Beads. The pesticide used in this research was pyrethroid, which physically existed in liquid form. All preparations involving the pesticide were conducted in a fume hood. About 100 g of treated CTF (mesh size 500 μm) was soaked in 100 mL of pesticide for 24 hours at room temperature without any stirring. After 24 hours, the soaked CTF in pesticide was filtered using a Buchner funnel to collect only the absorbed pesticide in the CTF. The CTF was abbreviated as treated CTF/pesticide. The treated CTF/pesticide was dried in a fume hood before being dried in a vacuum oven at 50°C for 2 hours. It was abbreviated as treated CTF (500 μm)/pesticide beads due to its physical appearance as beads.

Purification of ENR-50 [4]. About 20 g of pristine (raw) ENR-50 was swelled in 400 ml of chloroform ($CHCl_3$) and agitated for 48 hours. During this process, the pristine ENR-50 polymeric chains swelled in the solvent, inducing entanglement amongst the polymeric chains of pristine ENR-50. Consequently, a mixture of long, medium, and low molecular weights of pristine ENR-50 polymeric chains was separated via gravimetric filtration. The swelled mixture of pristine ENR-50 polymeric chains in $CHCl_3$ was filtered through a gauze cloth to separate the gel (high and medium molecular weight ENR-50) from the solution (low molecular weight ENR-50). This filtration process was a simple method for the purification of ENR-50. The gel remained on the filter cloth, while the solution passed through the filter cloth and was collected in a beaker. The solution was then transferred to a Teflon petri dish and dried in a fume hood for 24 hours to reduce the amount of $CHCl_3$ in the solution and concentrate it before being dried in a vacuum oven at 50°C for 48 hours until a constant weight was achieved. This solution was abbreviated as ENR-50. This ENR-50 was used in the preparation of the ENR/salt composite as a coating polymer matrix for the pesticide/treated CTF and pesticide/non-treated CTF beads, respectively.

Preparation of ENR-50/Salt Composite. In a typical preparation of 10% w/w concentrations of NaCl solutions was prepared based on the weight of ENR-50. About 5.00 g of NaCl was dissolved in mixture of 10 mL methanol (CH_3OH) and 10 mL of tetrahydrofuran (THF) and stirred vigorously until all the salt was completely dissolved. Separately, 5 g of ENR-50 was completely swelled in 20 mL of $CHCl_3$. The solutions (NaCl in CH_3OH/THF and ENR-50 in $CHCl_3$) were mixed and stirred at 2000 rpm for another 20 minutes. The ENR-50/salt composite solution was cast onto a Teflon Petri dish and dried in a fume hood before being dried in a vacuum oven at 50°C for 24 hours. In preparation of 5% w/w of NaCL solutions, about 2.50 g of NaCl used to replaced 5.00 g NaCl.

Preparation of Treated CTF/Pesticide Beads Coated with ENR-50/Salt Composite (SRP). In this section, there are two (2) types of SRP prepared (Table 1). The SRP1 and SRP 2 were prepared with 5 and 10 % w/w salt loading respectively.

Table 1: SRP samples with treated and untreated CTF at different salt loading.

SRP samples (mesh size 500μm)	Treated CTF	5% salt loading	10% salt loading
SRP 1	√	√	-
SRP 2	√	-	√

In a typical preparation of SRP1 and SRP2, about 5.00 g of ENR/salt composite was completely swelled in 100 mL of $CHCl_3$. Then, about 5.00 g of treated CTF was added to the swelled ENR/salt composite. The mixture was stirred at room temperature for 4 hours. The product was dried in a fume hood before being dried in a vacuum oven at 50°C for 4 hours. At a minimum, about 6 replicates were prepared for SRP1, with 5 replicates allocated for UV-Vis analysis and 1 replicate reserved for SEM analysis. The replicates for SRP2 was similar to SRP1. SRP 1 uses 5% salt loading while SRP2 used 10% salt loading.

UV-Vis Analysis. The UV-Vis analysis was carried out to determine the concentration of pesticide leached from the groups of samples SRP1 and SRP2. Separately 5 replicates for both samples, About 2 grams of each sample were immersed in 100 mL of distilled water. Then, about 3 mL of water was taken from the container and replaced with another 3 mL of fresh distilled water. The same procedure was repeated for up to 16 experimental days. The wavelength of the 3 mL sample was determined daily. The wavelength of the sample was estimated to be between 190 and 1100 nm. A blank solution (distilled water) was utilized as a control variable for the kinetic research each time the 3 mL sample was analyzed using UV-Vis. The sample was measured three times, and the average was calculated to assess the consistency of the sample's wavelength. Using the Beer-Lambert law, the absorbance from UV-Vis was utilized to calculate pesticide concentration. Before the experiment, a calibration curve was created using pesticide samples containing 2, 4, 6, 8, and 10 mg/L. The wavelength used for pesticides was 440 nm.

$$A=(\epsilon)(L)(C) \quad (1)$$

Where A is Absorbance, ϵ is molar extinction coefficient, L is Path lentgh, C is concentration of the sample. For concentration units of molarity:

$$k = A\log\left(\frac{-Ea}{RT}\right) \quad (2)$$

where ε_λ is wavelength-dependent molar absorptivity coefficient ($L\cdot mol^{-1}.cm^{-1}$).

Experimental Result

UV-Vis. UV-Vis analysis for both SRP1 and SRP2 were shown in Fig. 1.

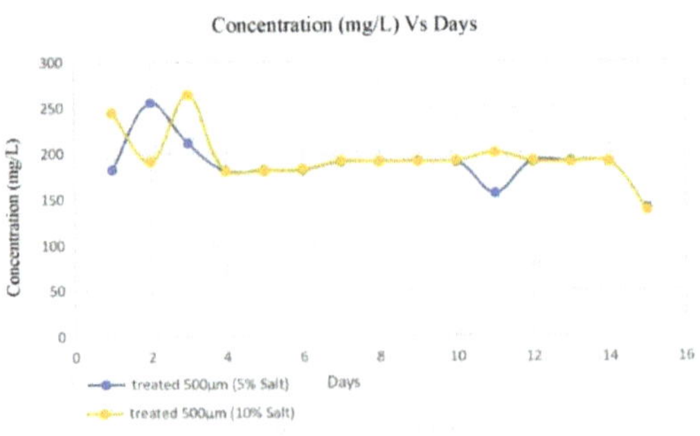

Fig. 1: UV-Vis analysis SRP 1 (treated CTF 500 µm at 5% salt loading) and SRP 2 (treated CTF 500 µm at 10% salt loading).

In general, both SRP show opposite pesticide release behaviour. SRP1 that consist 5% salt loading releases optimum amount of pesticide significantly at 250 mg/L during day 2 and decrease to 180 mg/L on day 4. It shows that during early days of soaking, most of the pesticides easily escapes from SRP either via big pore formed at the ENR/salt coating layer or exposed untreated CTF. Then, during the soaking process, ENR able to swell due to swelling behaviour of the polymeric in the solvents. This creates opportunity of pesticide to leach via swelled layer of the polymeric materials. As known, the swelled polymeric materials has big voids and gaps amongst the polymeric chains thus pesticide tends to each in water. Thus from day 4 until day 10, pesticides leach consistent at range 180-190 mg/L. This indicates leaching of the pesticides via uniform pore size formed at the ENR/salt coating layer. Oppositely, leaching of pesticide in SRP2 decreases significantly at day 2 due to preferences of the pesticides leaches from small pores than follows by big pore sizes at day 3 (270 mg/L) before its stable at day 4 until day 10. From these trends, both SRP1 and SRP2 requires 1-4 days to stabilize leaching rate of the pesticides. During day 11 for SRP1 shows a reverse hump recorded pesticide leaching at 155 mg/L as compared to SRP2 slightly uniform curve at 200 mg/L and follows similar trends from beginning day 4. Thus based on this observation it shows that SRP2 exhibited smoother release patterns than SRP1.

SEM Analysis. The surface morphology of SRP 1 and SRP 2 were analyzed using SEM analysis. In general SEM images for both SRP1 and SRP2 show different morphology in term of distribution coating layer of ENR/salt and agglomeration of the treated CTF in the composites. SRP1 shows thick and uneven coating of ENR/salt layer onto treated CTF. Some of the treated CTF are exposed towards outside surface [5] while some are covered by thick layer of ENR/salt. This supported finding in UV-Vis where a significant release of pesticide during day 2. Its due to the direct released of pesticide from the treated CTF. The thick layer of ENR/salt tends to agglomerates treated CTF when coating process happen at fast coating rate. Oppositely, SRP2 shows even distribution coating of ENR/salt onto treated CTF. It shows most of the CTF are well coated with ENR/salt layer and minimize the agglomeration of CTF of the composites. This supported finding in UV-Vis result of SRP2 on day 11 recorded uniform leaching as compared to SRP1.

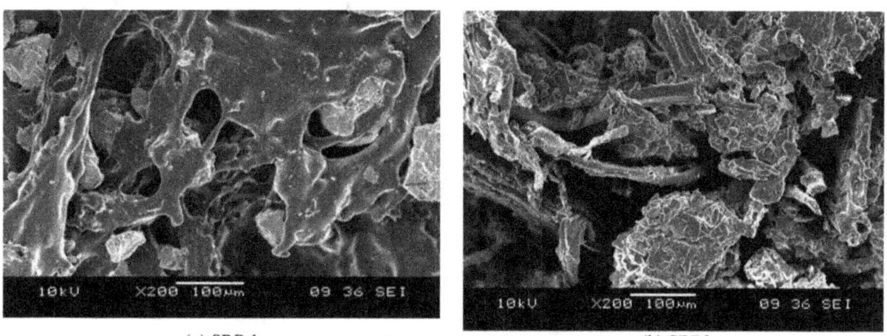

(a) SRP 1 (b) SRP2

Fig. 2: SEM images of SRP 1 (treated CTF 500 μm at 5% salt loading) and SRP 2 (treated CTF 500 μm at 10% salt loading) at 200x magnification.

Conclusion

Overall, treated CTF at 500 μm mesh size is a potential fiber in synthesized slow release pesticide from ENR-50 and CTF. SRP2 shows good performance of SRP based on steady pesticides leaching from UV-Vis results and even distribution of polymeric coating onto treated CTF.

Acknowledgment
The authors acknowledge Faculty of Chemical Engineering Technology, Universiti Malaysia Perlis during the development of this project. Special thanks to those who contributed to this project directly or indirectly.

References

[1] S. Kumar, R. Bhuvaneshwari, S. Jain, S. Nirwan, Z. Fatima, D. Kumar, B.S. Chhikara, B. Rathi, and Poonam, A Systematic Review on Pesticide-loaded Nanocapsules: A Sustainable Route for Pesticide Management to Enhance Crop Productivity. Current Nanoscience, 20(3) (2024) 280–297. https://doi.org/10.2174/1573413719666230417103517

[2] D. Abdollahdokht, Y. Gao, S. Faramarz, A. Poustforoosh, M. Abbasi, G. Asadikaram, and M.H. Nematollahi, Conventional agrochemicals towards nano-biopesticides: An overview on recent advances, Chemical and Biological Technologies in Agriculture, 9(1) (2022). https://doi.org/10.1186/s40538-021-00281-0

[3] A. Singh, N. Dhiman, A.K. Kar, D. Singh, M.P. Purohit, D. Ghosh and S. Patnaik, Advances in controlled release pesticide formulations: Prospects to safer integrated pest management and sustainable agriculture, Journal of Hazardous Materials, 385 (2020) 121525. https://doi.org/10.1016/j.jhazmat.2019.121525

[4] R. Hamzah, M. Abu Bakar, M. Khairuddean, I. Mohammed and R. Adnan, The Structural Studies of Epoxidized Natural Rubber (ENR-50) and its Cyclic Dithiocarbonate Derivative Using NMR Spectroscopy Techniques, Molecules 17 (2012) 10974-10993.

[5] R. Mahadevan, S. Pillai, and G. Rao, SEM studies on the structure of natural bio-fiber composites, Materials Science Review, 47(2) (2022) 188-202.

Green Materials, Processing & Characterization – ICoGMPaC 2025 Materials Research Forum LLC
Materials Research Proceedings 68 (2026) 299-307 https://doi.org/10.21741/9781644904237-40

Design, Finite Element Analysis and Fabrication of a Laboratory Scale Biomass Pelletizer

R. I. Ismail[1,a*], C. Y. Khor[1], A. R. Mohamed[2], M. N. S. Zaimi[1], M. R. Jamalludin[1], L. Y. Lee[1], N. L. Makhtar[1], N. Abdul Razak[1] and W. N. A. Wan Draman[1]

[1]Faculty of Mechanical Engineering & Technology, Universiti Malaysia Perlis, Kampus Tetap Pauh Putra, 02600 Arau, Perlis, Malaysia

[2]Faculty of Chemical Engineering & Technology, Kompleks Pusat Pengajian Jejawi 3, Kawasan Perindustrian Jejawi, Universiti Malaysia Perlis (UniMAP), 02600 Arau, Perlis, Malaysia

[a]rasizzati@unimap.edu.my

Keywords: Biomass Pelletizer, Finite Element Analysis, Screw Mechanism

Abstract. Prior to commencing the fabrication process, it is imperative to address several crucial components, which encompass the identification of purpose, precise design, and detailed analysis. The main focus of this project is to develop a machine that can efficiently convert biomass into compressed pellets. The design phase was conducted with great attention to detail using the Solidworks software. Following this, the stress and displacement analysis specifically concentrated on the important machine components, utilising the capabilities of the Abaqus software. The machine consists of six essential components, including a motor, an iron frame, a pelletizer, a pulley and belting system, and a U clip. The operational procedure encompasses the initial step of introducing biomass into the pelletizer. Within the pelletizer machine, a screw mechanism is employed to propel the biomass towards the plate die. The initiation of motion in this screw mechanism is achieved through the utilisation of an electric motor, which is helped by a network of interconnected pulleys and belting. By employing a pulley ratio of 1:2, the rotational velocity of the screw shaft is effectively reduced by half compared to that of the motor, hence enhancing the efficiency of the pelletization procedure. This pelletizer machine effectively achieves three crucial objectives: the development of a comprehensive design and its subsequent manufacturing, a finite element analysis of stress and displacement characteristics.

Introduction

Tool In recent years, the escalating demand for renewable and sustainable energy sources has intensified the exploration of alternative fuels [1]. Biomass, derived from organic materials such as agricultural residues, forest waste, and dedicated energy crops, stands as a promising candidate in the pursuit of greener energy solutions. One of the pivotal technological advancements in utilizing biomass is the development of biomass pelletizers, which efficiently transform loose biomass materials into standardized pellets. These pellets possess higher energy density, enhanced transportability, and improved combustion characteristics, making them an attractive option for both residential heating and industrial energy generation.

In the single die pelletizer type, the biomass particles exhibit a unidirectional flow along the path of the piston. In contrast, within an industrial-scale pelletizer, biomass is subjected to compression in successive layers, which then proceed to traverse a constricted passageway within the die. This phenomenon enables particles to form bonds with one another in a radial orientation [2].

This research initiative delves into the intricacies of designing and fabricating a laboratory-scale biomass pelletizer, aimed at bridging the gap between theoretical concepts and practical applications. The process involves intricate engineering considerations encompassing mechanical,

Materials Research Proceedings 68 (2026) 299-307 https://doi.org/10.21741/9781644904237-40

thermal, and material science domains. The prime focus lies in devising a compact yet efficient apparatus capable of transforming diverse biomass feedstock into uniform pellets while addressing challenges such as material handling, pellet integrity, and energy efficiency.

The objective of this project was to conceive and construct a laboratory-scale apparatus to produce biomass energy pellets. The pelletizer is specifically engineered to accommodate samples with a weight of less than one kilogram. In the context of industrial applications, biomass pelletizers typically require a minimum input of three kilograms or more in order to effectively generate pellets. The limited sample size necessitates the adoption of a tiny and cost-effective pelletizer. This is achieved by directing attention on the die plate [3,4,5] and roller [6]. Currently, there is a lack of research pertaining to the development of a pelletizer specifically designed for laboratory use. Numerous advancements have been implemented in pelletizer machines since the year 1930 [7]. The enhancements have been implemented with the aim of augmenting the potential for industrial utilisation. The densification of biomass into pellets has been the subject of extensive research, with pellets being one of the resulting products [8]. A pelletizer is the machinery employed in this process. The primary aims of this project encompassed the conceptualization and construction of the laboratory-scale biomass pelletizer machine, assessment of its efficacy, examination of stress and displacement levels on the pivotal component of the laboratory-scale biomass pelletizer using Computational Finite Element analysis [9, 10]. The Finite Element Analysis of the essential component of the laboratory-scale biomass pelletizer was conducted using the Abaqus software. The purpose of simulation analysis is to validate the product's adherence to its operational requirements. This can also offer valuable perspectives on requisite modifications and substantiate the conduction of appropriate real-world experiments, particularly within engineering domains such as mechanical and electrical products [11].

The significance of this project extends beyond the laboratory setting. A well-designed laboratory-scale pelletizer not only serves as a valuable tool for conducting fundamental research on pelletizing various biomass feedstocks but also paves the way for upscaling the technology to industrial levels. Moreover, it contributes to the ongoing efforts to mitigate environmental concerns associated with conventional fossil fuels by promoting the sustainable utilization of biomass resources. In this paper, the design considerations, fabrication processes, and operational principles involved in creating a laboratory-scale biomass pelletizer will be elucidated. Additionally, the underlying scientific principles that govern successful pelletization will be explored, and the potential applications and implications of this technology will be discussed.

Design Concept

The process of concept generation involves the development of a concept that is derived from the functional requirements of a component, as outlined in the conceptual design phase. A total of four concepts, each including distinct concepts, were evaluated. The most promising designs were selected and further refined to create a novel design specifically tailored for a laboratory-scale biomass pelletizer. The aforementioned concepts are visually represented in Fig.1, Fig. 2, Fig. 3 and Fig.4.

Concept 1. This concept is indicated in Fig. 1. The die chosen for this concept was a flat, circular die. The motion was at the roller, where the mechanical energy from the motor was used to drive the roller across the flat die's surface. The die is equipped with a total of 24 holes through which the material is extruded, resulting in the formation of pellets. The pellet exits from the open region beneath the pelletizer after colliding with the bottom surface of the flat die and breaking.

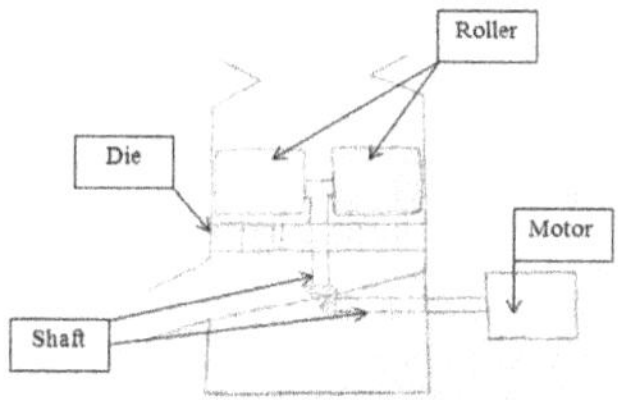

Fig. 1: Concept 1.

Concept 2. The notion in question is depicted in Fig. 2. The die had a rectangular and flat shape. The roller is propelled by a human hand and rotates vertically as it goes. The rotation ratios exhibit a one-to-one correspondence. The roller will exert pressure on the material, causing it to be compacted into twelve cavities on a planar die, resulting in the formation of pellets. The pellets undergo fracture upon impact with the lower region of the pellet mill, afterwards exiting through the hole.

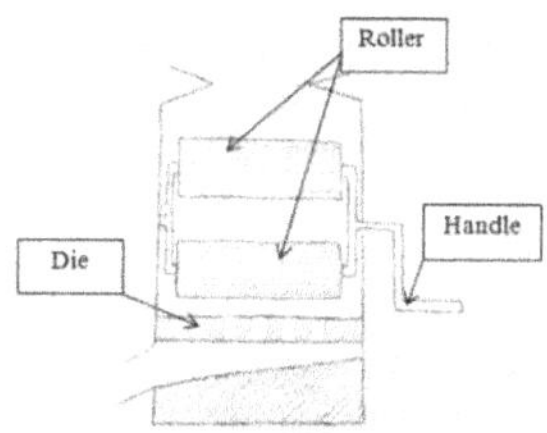

Fig. 2 : Concept 2.

Concept 3. The movement in this concept occurs at the flat die rather than the roller, although being nearly identical to that in Concept 1. This conceptual idea is laid out in Fig. 3. The flat die, including a total of 16 holes, undergoes rotation through the use of motor power. The material will be introduced into the die, where it will undergo pellet formation beneath the flat die, thereafter being cut off by the knife.

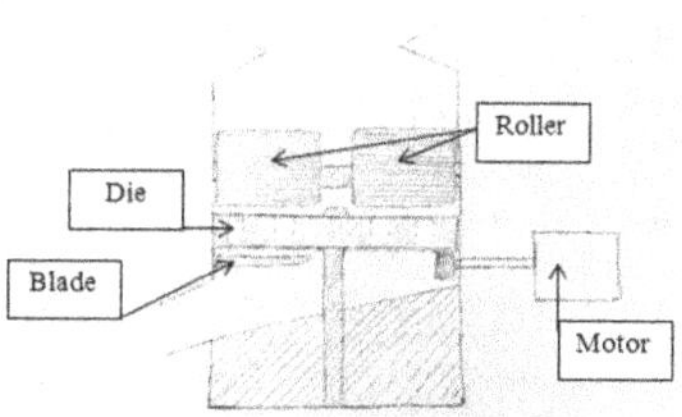

Fig. 3 : Concept 3.

Concept 4. The conceptual design is depicted in Fig. 4. In this design, the biomass is moved into the die plate via a screw shaft. The die plate is equipped with a total of seven holes and is manually operated by a human hand. The pellets were chopped by the blade in front of the die plate as the biomass was fed into the die.

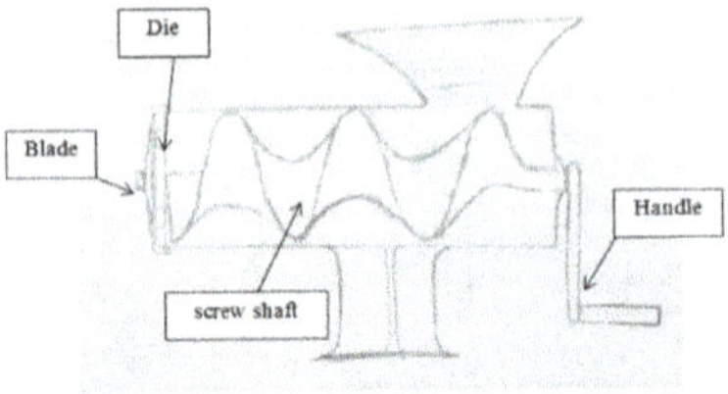

Fig. 4: Concept 4.

Design Analysis and Evaluation

Table 1 lists the analysis and evaluation of the three concepts generated. The analysis is based on the fabrication process and the evaluation is based on the cost and it practical.

Table 1 : Concept Generated Analysis and Evaluation.

Concept	Analysis and Evaluation
1	The design was easy to fabricate, but it required a high cost due to the use of a motor. The design was very practical because it was nearly the same as the flat-die pelletizer.
2	This design was hard to fabricate, but it is very low-cost because it does not use a motor but instead uses hand power. Unfortunately, this design was not practical because the pressure exerted on the material by the die was not uniform.
3	This design was the most practical because it will produce the same length of pellets. It is easy to fabricate but still expensive because of the motor.
4	This design was also practical because it is simple.

Concept Selection. The data were collected based on the concept of generation, and this led to the creation of the morphological chart as presented in Table 2. The laboratory-scale pelletizer was designed and fabricated based on the selected data or qualities from each concept, as indicated in the table. The utilisation of a morphological chart serves as a straightforward method for evaluating the characteristics of various designs or products, facilitating the selection of the most optimal design concept.

Table 2 : Concept Generation Morphology.

No	Characteristic	Concept 1	Concept 2	Concept 3	Concept 4
1.	Shape of die plate	Circle	Rectangular	Circle	Circle
2.	Number of die holes	24	10	16	7
3.	Movement at	Roller	Roller	Die	Screw shaft
4.	Movement angle	Horizontal	Vertical	Horizontal	Horizontal
5.	Power by	Motor	Human hand	Motor	Human hand
6.	Present of Blade	No	No	Yes	Yes

According to the findings presented in Table 2, Concept 4 emerged as the most optimal design concept. The chosen qualities encompass a circular configuration including a 7-hole flat die, material displacement facilitated by a screw shaft operating at a horizontal angle of rotation, powered by a motor, and equipped with a blade. The power source for the operation was modified from manual hand usage to motorised mechanism. The selection of this particular design was based on its inclusion of a blade component, as well as its notable high percentage of pellets possessing uniform lengths.

Engineering Analysis. The examination of the product assumes a crucial function in the process of design development. The drawing was analysed using the finite element analysis technique implemented using the Abaqus Software. The application of Finite Element Analysis (FEA) enables the evaluation of design strength by considering the applied load. The following section outlines the methodology employed in the analysis utilising the Abaqus software in order to determine the magnitudes of stress and displacement for the crucial component of the laboratory-scale biomass pelletizer.

The design portion of the product was completed by employing Solidworks Software, which ensured a high level of precision and accuracy in its conceptualization. In order to ensure smooth interaction with Abaqus Software, the design file was saved in ACIS format. The design used a unifying material category for all components, facilitating a uniform analysis. Accurate simulations were facilitated through precise definition of essential material parameters, such as the yield modulus and Poisson's ratio. The identification of fixed points inside the design was contingent upon the establishment of boundary conditions. The precise positioning of loads was required to match them with their respective load directions. The calculations were conducted to determine the most suitable mesh size, thereby offering valuable insights into the precise region where the load is applied. By including maximum pressure measurements, a full analysis was conducted to determine the magnitudes of stress and displacement. Significantly, the visual depiction of the material underwent a colour shift from blue to red, functioning as a visual cue for the state of the component and the distribution of stress.

Result and Discussion

Design of Laboratory Scale Pelletizer. The SolidWorks software was utilised for the design of the laboratory-scale pelletizer. The design incorporates the material that was deliberated upon during the process of concept selection. The laboratory-scale pelletizer has been constructed using six components. The components of the system include the motor, pulley, V belt, U clip, and pelletizer. The Solidworks design is shown in Fig. 5.

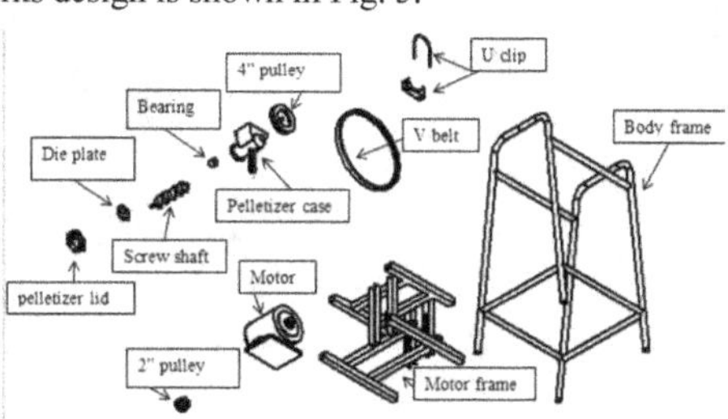

Fig. 5. The exploded view of the lab-scale pelletizer machine.

The schematic depicted in Fig. 5 illustrates the mechanism by which the screw within the pelletizer is propelled. Specifically, the rotational force generated by the motor is transmitted to the screw through the use of a pulley and a V belt. The technical drawing of the final design, created in SolidWorks, is shown in Fig. 6.

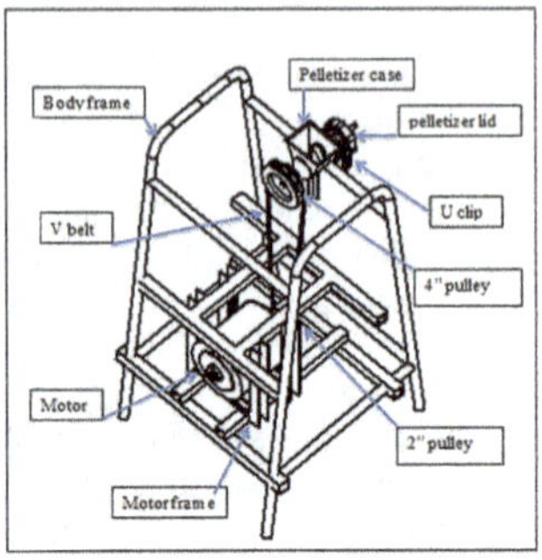

***Fig. 6 :** Design Assembly.*

Structural Analysis. The purpose of structural analysis is to evaluate the integrity of the motor frame and die plate structures. The continuous mode of operation in the pellet press which is similar to the functioning of pelletizers frequently employed in industrial settings. In contrast, a single-pellet press exerts exceptionally high pressures directly onto a specific spot with a limited surface area. The similar mode of operation was reported in a previous study by [1]. The motor frame has a maximum load capacity of 2.8 kilonewtons per square meter (kN/m2), as specified in the given information. Additionally, Fig. 7 indicates that the die plate has a maximum load capacity of 17 meganewtons per square meter (MN/m2). The orange arrow serves to identify the direction of the load, while the red surface signifies the area that has been subjected to pressure.

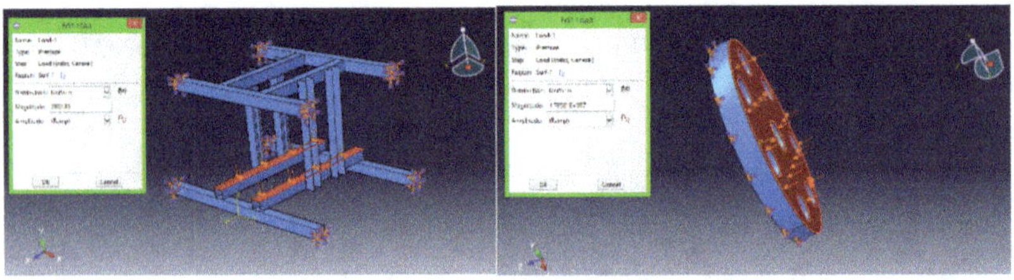

Fig. 7: The position of the load for the motor frame and the die plate.

The load force magnitude acting on the motor frame is 49.05 N. The weight of the motor is estimated to be roughly 5 kg, equivalent to 49.05 N. The motor is regarded as having the greatest weight. The applied force on the die plate was measured to be 29430 N, which is approximately equivalent to a weight of 3000 kg. The quantity of weight employed for the compression of biomass in a singular press machine is 3000 kg.

Fig. 8 illustrates the prescribed boundary conditions for both the die plate and the motor frame. The boundary condition refers to the predetermined and immovable positioning of the components. The red hue serves as an indicator of the stationary surface.

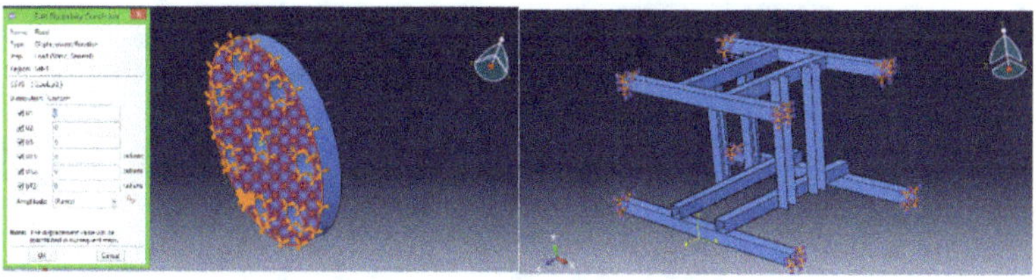

Fig. 8 : The boundary condition of the die plate and the motor frame.

Fig. 9 indicates the magnitude of displacement for the motor frame and the die plate. Fig. 10 and 11 show Von Misses stress taken from the design of the motor frame and the die plate. Both show a good result by showing the blue colour. The blue colour indicates that the frame for the biomass pelletizer can withstand the maximum pressure. While the red colour indicates that the frame is ruptured.

The steel exhibits a yield strength of 350 MPa, indicating its capacity to withstand the highest stress level without permanent deformation. The die plate experiences a pressure of 17 megapascals (MPa). The top left corner of Fig. 10's colour table displays the regions of the figure that are under the lowest and highest stress levels, respectively, with blue representing that area and red representing that one. The analysis reveals that the predominant colour observed is bright blue, indicating that it had a stress of 1.141 e+7 N/mm^2. It is pertinent to note that majority of the force placed on the die plate is located close to its edges.

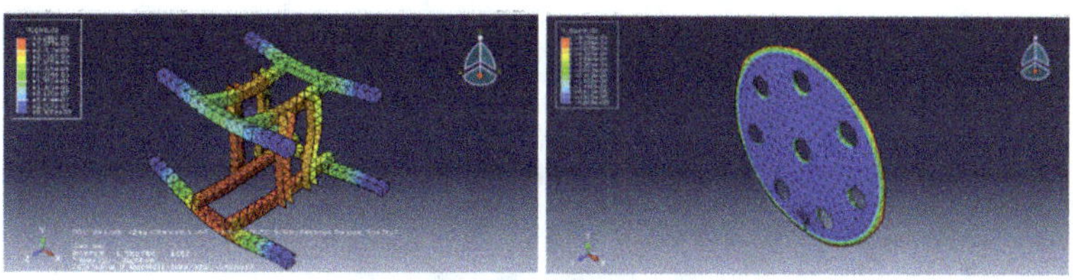

Fig. 9 : Magnitude of motor frame and the die plate.

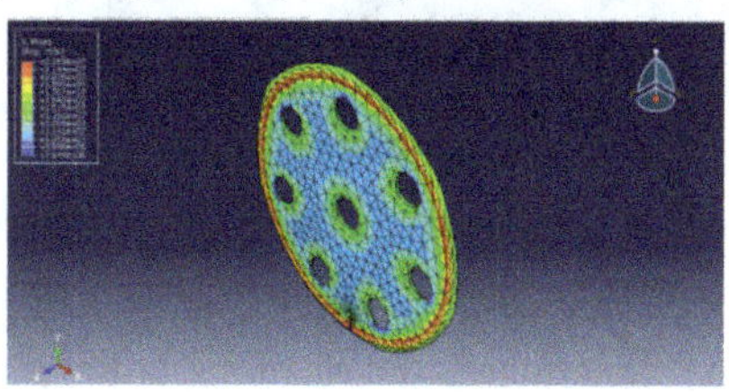

Fig. 10 : The outcome pertaining to the Von Mises stress analysis of the plate.

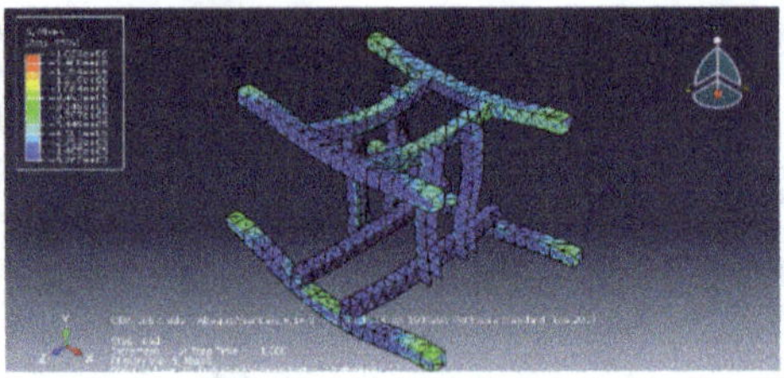

Fig. 11 : Von Mises Stress results for motor frame.

The mild steel material exhibits a maximum stress capacity of 220.03 MPa. The pressure applied to the motor frame measures 2.8 kilopascals (KPa). The colour chart, situated in the upper left corner of Fig. 10, indicates that the blue colour corresponds to the area experiencing the least amount of stress, while the red hue represents the region facing the highest level of stress. The analysis reveals that approximately 75% of the motor frame exhibits a blue coloration, indicating that it experienced a stress of 5.762 e+3 N/mm2. The remaining 25% exhibits a composite hue characterised by a combination of yellow and green. According to Fig. 11, the distribution of frame stress predominantly occurs at the interface of the boundary condition.

Final Product. Following comprehensive analyses of both the conceptual and morphological design stages, the laboratory-scale biomass pelletizer has been successfully fabricated. Its tangible manifestation is visually depicted in Fig. 12.

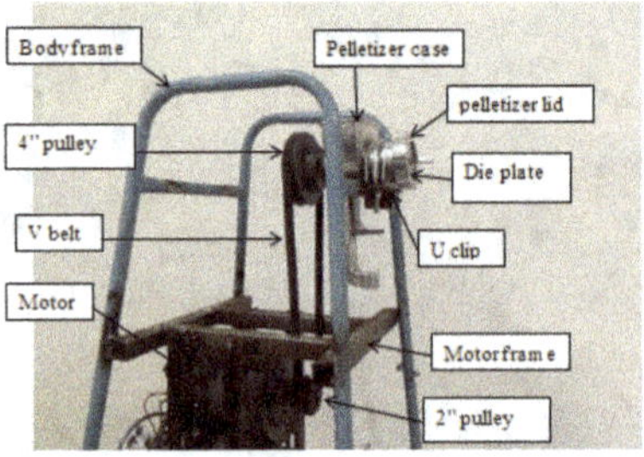

Fig. 12 : Laboratory scale biomass pelletizer without cover.

Conclusion

The machine has successfully accomplished three primary objectives: firstly, to engage in the process of designing; secondly, to conduct an analysis of stress and displacement magnitudes through the utilisation of finite element analysis; and finally, fabricate a biomass pelletizer specifically designed for the laboratory scale fuel energy pellets production. The design was executed via SolidWorks software and subsequently manufactured. The strength and durability of the crucial component are being Importantly, this apparatus maintains a lightweight profile, weighing less than 20 kg. This characteristic stands in contrast to conventional pelletizers, which tend to be heavier. The inherent portability of this lightweight design facilitates its easy transport to diverse locations. Nevertheless, certain challenges have been encountered in measuring the machine's efficiency and capacity due to issues such as biomass accumulation on the die wall. In light of these considerations, it is evident that certain refinements were necessary to tailor the equipment for optimal laboratory utilization.

Acknowledgement
We would like to thank the Higher Education Ministry of Malaysia (FRGS); research grant number FRGS/1/2020/TK0/UNIMAP/03/22 for funding this project, as well as our colleagues, technical experts, and support staff at the Faculty of Mechanical Engineering & Technology, Universiti Malaysia Perlis for their devoted assistance with this research.

References

[1] R. García, M. V. Gil, F. Rubiera and C. Pevida 2019 Fuel 251 pp 739-53 (2019). https://doi.org/10.1016/j.fuel.2019.03.141

[2] A. K. Biswas, M. Rudolfsson, M. Broström and K. Umeki, Appl. Energy 119 pp 79-84. (2014) https://doi.org/10.1016/j.apenergy.2013.12.070

[3] S. Brahmandam, U.S. Patent No. 2013/0078328 A1. (2013)

[4] G. V. Hass, Patent No. US 2012/0244242 A1. (2012)

[5] S. Landers, Patent No. US 2008/0290548 A1 (2008).

[6] G. V. Barbosa-Cánovas, E. Ortega-Rivas, P. Juliano and H. Yan, Food Powders: Physical Properties, Processing, and Functionality ,New York: Kluwer Academic / Plenum Publisher, (2015).

[7] D. Castaldo, History of Pelleting. In C. J. CHARLES STARK €d.), Feed Pelleting Reference Guide, WATT Global Media and K-State University. Accessed August 18, 2023. https://img.feedstrategy.com/files/base/wattglobalmedia/all/document/2019/09/fs.1

[8] Information on Biomass Fuel Densification Processing Methods and Equipment. http://www.gcmec.com/.

[9] M. I. Ishak, A. A. Shafi, M. U. Rosli, C. Y. Khor, M. S. Zakaria, W. M. F. W. A. Rahim and M. R. Jamalludin, Biomechanical evaluation of different abutment-implant connections-a nonlinear finite element analysis, AIP Conference Proceedings 1885 p. 020064 (2017). https://doi.org/10.1063/1.5002258

[10] M. U. Rosli and S. Zulkifli, Finite Element Analysis of Bicycle Crank Arm on the Mechanical Aspect, Advanced and Sustainable Technologies (ASET) 1(2) pp 52-64, (2022). https://doi.org/10.58915/aset.v1i2.19

[11] C. Y. Khor, M. I. Ishak, M. U. Rosli, M. R. Jamalludin, M. S. Zakaria, A. F. M. Yamin and M. Z. Abdullah, Influence of material properties on the fluid-structure interaction aspects during molded underfill process, MATEC Web of Conferences 97 p. 01059 (2017). https://doi.org/10.1051/matecconf/20179701059

Surface Performance Evaluation of Lattice Structure Produce by Fused Deposition Modeling Machine Using Design of Experiment (DOE)

N.A.M. RAZIP[1], M.A. ISMAIL[1,a*], S. SHAHARUDDIN[1], M. JOHAR[1], Z. MOHAMMAD[1], M.S.E. KOSNAN[1]

[1]Advance Facilities Engineering Research Cluster, Malaysian Institute of Industrial Technology, Universiti Kuala Lumpur(UNIKL), 81750 Masai, Johor, Malaysia

[a]manuar@unikl.edu.my

Keywords: Additive Manufacturing, Surface Performance, Fused Deposition Modelling (FDM)

Abstract. Additive manufacturing has evolved as a cutting-edge approach that has the potential to transform traditional production methods. Its layer-by-layer build-up approach allows the creation of complex form and geometric structures, offering remarkable efficiency while minimizing material waste and reducing manufacturing costs compared to traditional subtractive process. Fused Deposition Modelling (FDM) has attracted attention among additive manufacturing technologies for its capabilities to construct complex lattice structures, which enable lightweight de-signs with equivalent or higher mechanical performance to traditional solid structures. However, the presence of geometrical and surface texture imperfections in lattice designs have a substantial impact on their mechanical performance. As a result, achieving dimensional accuracy and surface texture in FDM-based additive manufacturing remains a critical challenge. This research aims to investigate the capabilities of Fused Deposition Modelling machines to create complex lattice structures. FDM will be used to create three different lattice designs with variable diameters. A Design of Experiment (DOE) study and statistical analysis methods will be used to evaluate the quality of surface texture of these structures. The outcomes of this study will enhance the quality of surface finish of lattice structures produced by additive manufacturing machine.

Introduction

Additive manufacturing (AM) presents a unique opportunity to manufacture complex structures, particularly for topology-optimized structures [1]. It has the ability to produce various types of materials, such as metals, ceramics, plastics, alloys, and bioactive glass. In contrast to conventional subtractive manufacturing processes, additive manufacturing is capable of fabricating a wide range of complex shapes and geometries in a short period of time, while significantly reducing material waste and manufacturing costs [1,2]. Given these advantages, additive manufacturing is an ideal candidate for manufacturing complex lattice structures, making it highly beneficial for industries such as biomedical and aerospace [3,6].

Research methodology

A complex lattice structure can be defined as a three-dimensional geometrical arrangement composed of connective links between vertices, which creates a functional structure [7]. The design of lattice structures is typically based on the design of its unit cell. Various designs exist for lattice structures, including diamond-based designs, honeycomb prismatic lattice, and spider web-based lattice [8]. The use of a lattice design is expected to produce a lightweight structure with equiva-lent or better mechanical properties [9]. However, any geometrical variations in the unit cell and lattice structures will affect the mechanical properties of lattice structures.

Fused deposition modelling (FDM) is one of the additive manufacturing tech-niques widely used in industry. This technique involves extruding thermoplastic material through one or more heated nozzles to form a 2D layer. The process is re-peated layer by layer until a 3D structure is produced based on the design [1]. De-spite its ability to manufacture complex structures, FDM has limitations in its po-sitional accuracy, which can result in deviations in the metrological characteristics of the lattice structure [10].The metrological characteristic deviation errors of a lattice structure refer to errors in its dimensional accuracy, geometrical complexi-ty, and surface texture properties In this paper, the capability of Fused Deposition Modelling machine in manufacturing complex lattice structure will be studied. The optimum design parameter for lattice structured that suitable to be manufactured by Fused deposition model-ling (FDM) will also been analyzed.

The research project can be divided into three tasks. The first task involves de-signing the experiment and selecting critical parameters, factors, and levels. This task is crucial because the data analysis will be based on the design of the experiment method, and relevant statistical tools will be employed during data analysis.

The second task entails preparing and manufacturing lattice-structured samples using additive manufacturing techniques. In this task, three different types of lattice designs - Circle unit cell structure, linear unit cell structure, and hexagonal unit cell structure - will be selected. Based on the suggestions from the Design of Experiment (DOE) analysis, various dimensions of lattice-structured samples will be manufactured using FDM machines.

Once the lattice-structured samples have been produced by the FDM machine, their dimensions and surface areal topography will be measured using a non-contact focus variation interferometry 3D measuring machine. The resulting parameters from the first task will be applied during the inspection process.

Design of Experiment

Full factorial design method is employed in designing this experiment. The ad-vantage of using this method is the ability of this statistical Full factorial design to investigate all possible combinations of the level of the factors and hence will help in providing the optimum design parameter for manufacturing lattices structured using FDM machine.

Three factors have been selected in this experiment which is the type of lattice structured, number of unit cell and dimension of unit cells. For each of these factors, three different level are selected. The details of factor and level are presented in Table 1 below.

Table 1: Factors & Levels for design parameter for lattice structure.

Factor	Level		
	1	2	3
Type of Lattice Structure	Linear	Circle	Hexagonal
Number of Unit Cell	3	4	5
Dimension of Unit Cell (cm)	0.3	0.4	0.5

According to the selected variable for three factors and three levels, the total of the 27 lattice structure samples will be produced by FDM machine and. The de-tails arrangement of this inspection is shown in Table 2.

Materials Research Proceedings 68 (2026) 308-314 https://doi.org/10.21741/9781644904237-41

Preparation and Manufacturing of Samples

Material used for these samples is Acrylonitrile Butadiene Styrene (ABS) and the FDM machine employed for this task is from Cube Pro 3D machine. Three different designs for lattice structure have been selected to be manufactured by FDM machine. The samples were fabricated based on the design of the experiment. These designs are Circle unit cell structure, linear unit cell structure, and hexagonal unit cell structure as shown in Fig. 1.

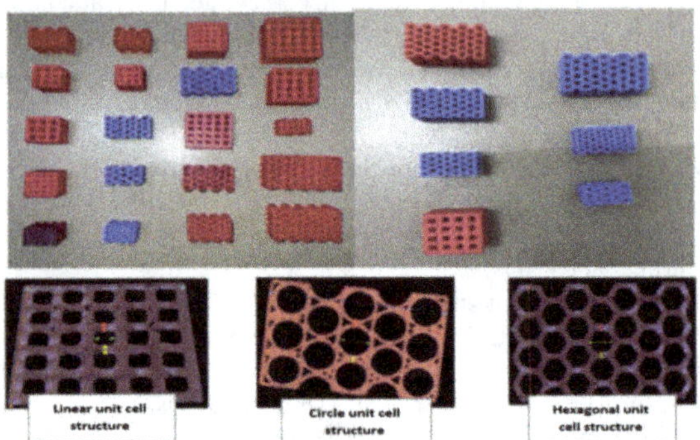

Fig. 1: Three different design of lattice structure sample.

Table 2: Full factorial design for manufacturing of lattice structure sample.

StdOrder	RunOrder	Type of Lattice Structure	Number of Unit Cell	Dimension of Unit Cell, (cm)
12	1	Circle	3	0.5
2	2	Linear	3	0.4
3	3	Linear	3	0.5
4	4	Linear	4	0.3
22	5	Hexagonal	4	0.3
21	6	Hexagonal	3	0.5
1	7	Linear	3	0.3
14	8	Circle	4	0.4
11	9	Circle	3	0.4
10	10	Circle	3	0.3
7	11	Linear	5	0.3
15	12	Circle	4	0.5
8	13	Linear	5	0.4
24	14	Hexagonal	4	0.5
23	15	Hexagonal	4	0.4
9	16	Linear	5	0.5
6	17	Linear	4	0.5
10	18	Hexagonal	3	0.4
18	19	Circle	5	0.5
27	20	Hexagonal	5	0.5
26	21	Hexagonal	5	0.4
16	22	Circle	5	0.3
13	23	Circle	4	0.3
5	24	Linear	4	0.4
17	25	Circle	5	0.4
25	26	Hexagonal	5	0.3
19	27	Hexagonal	3	0.3

Areal Surface Topography Measurement
After preparation and manufacture of the samples using additive manufacturing, the surface condition of the samples are conducted. For these measurements, 3D Infinite focus variation measuring Machine will be employed. Based on the resultant from design of experiments, there are 27 samples need to be inspected and for each samples, surface areal topography, Sa, need to be measured.

Results and Discussions
The surface texture condition of the samples produced by additive manufacturing is a crucial metrological parameter for ensuring high performance. In this re-search, we are measuring the surface areal topography parameter Sa, instead of the surface roughness parameter Ra, to obtain more accurate information about the 2D surface texture. Measuring Sa provides the advantage of obtaining precise surface areal information, which is essential for ensuring accurate surface texture condition. The result of surface areal topography, Sa, is presented in Table 3.

Table 3: Result of Measurement surface areal topography for samples.

Std Order	Run Order	Type of Lattice Structure Design	Areal surface texture ,Sa (μm)
12	1	Circle	347.620
2	2	Linear	241.880
3	3	Linear	52.930
4	4	Linear	340.950
22	5	Hexagonal	170.940
21	6	Hexagonal	476.520
1	7	Linear	303.040
14	8	Circle	285.340
11	9	Circle	241.470
10	10	Circle	295.200
7	11	Linear	1097.800
15	12	Circle	130.000
8	13	Linear	1289.600
24	14	Hexagonal	85.720
23	15	Hexagonal	212.200
9	16	Linear	147.730
6	17	Linear	948.310
20	18	Hexagonal	53.880
18	19	Circle	275.290
27	20	Hexagonal	176.590
26	21	Hexagonal	464.330
16	22	Circle	209.800
13	23	Circle	70.130
5	24	Linear	754.950
17	25	Circle	127.840
25	26	Hexagonal	752.810
19	27	Hexagonal	138.320

The present study will employ three statistical analysis tools to examine the da-ta collected, namely a normality test, analysis of variance (ANOVA), and main effects plot. Initially, a normality test will be conducted to assess the distribution of the results. A P-value greater than 0.05 will indicate a normal distribution, while a P-value of 0.05 or lower will indicate that the data is not normally distributed with a 95% confidence level. ANOVA will then be conducted to explore the significance and relationship among the factors. A P-value below 0.05 will signify a significant factor. Lastly, a main effects plot will be employed to identify the optimal setting values for each factor

Materials Research Proceedings 68 (2026) 308-314 https://doi.org/10.21741/9781644904237-41

Fig. 2 presents the results of a series of statistical analyses performed on the surface areal topography measurements parameter Sa for Linear lattice structure. The normality test reveal that the surface texture data conforms to a normal distribution at a 95% confidence level, as indicated by a P-value higher than 0.05. The ANOVA results show that none of the factors are significant to the measurement. The main effect plot reveals the optimal values of the number and dimension of the unit cell factors are identified as 5 and 0.4 cm, respectively.

Similar trend of resultant analysis can be observed in these surface areal topography measurement for other type of lattice structure, which is on the circle and hexagonal lattice structure. As shown in Fig. 3 and Fig. 4, all surface texture condition data has a normal distributed data at a 95% confidence level, as indicated by a P-value higher than 0.05. However, similar to linear type of lattice structure measurement, based on the ANOVA analysis, none of the factors are significant to the surface texture condition data for both circle and hexagonal data. In contrast, the main effect plot reveals the optimal values of the number and dimension of the unit cell factors are identified as 3 and 0.5 cm, respectively for the circle lattice structure type and 3 and 0.5 cm, respectively, for the hexagonal lattice structure type.

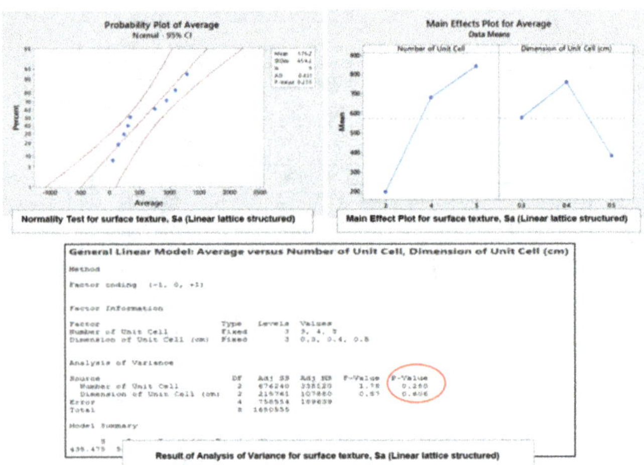

Fig. 2: Result of normality test, analysis of variance (ANOVA), and main effects plot for Surface areal topography, parameter Sa for Linear lattice structure.

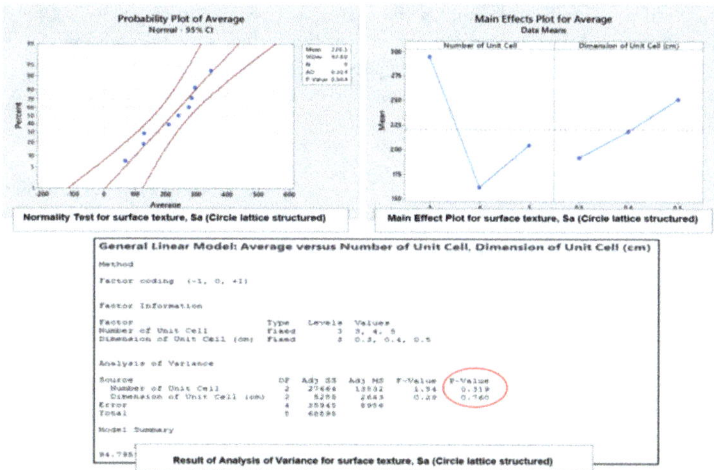

Fig. 3: Result of normality test, analysis of variance (ANOVA), and main effects plot for Surface areal topography, parameter Sa for Circle lattice structure.

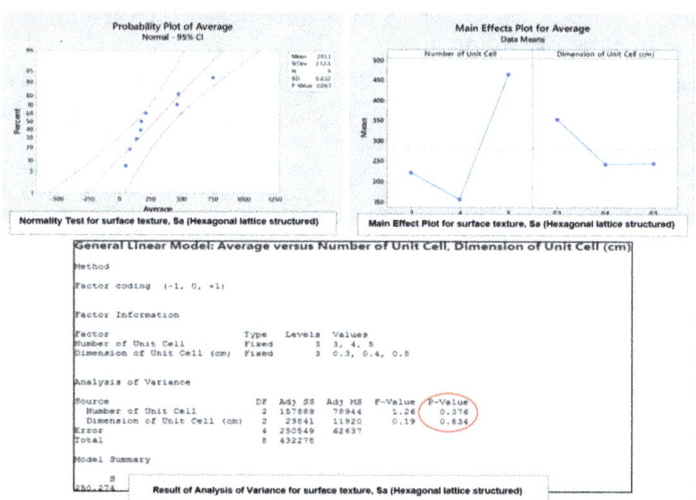

Fig. 4: Result of normality test, analysis of variance (ANOVA), and main effects plot for Surface areal topography, parameter Sa for Hexagonal lattice structure.

Conclusion

This study aimed to investigate the behavior of Fused Deposition Modelling (FDM) machines in producing lattice structures with various designs. Metrological measurements were conducted to determine surface texture conditions of the lattice structures at critical locations. The results indicate that the design of the lattice structure and the number and dimensions of the unit cells in the structure did not significantly impact the surface texture condition of the lattice structures during manufacturing. Nevertheless, the analysis of the surface texture condition can be further improved by enhancing the FDM machine's capability to produce better surface finish. In addition, more surface performance investigation such as the porosity inspection of the surface need to be conducted in the future.

References

[1] S. Daminabo, S. Goel, S. Grammatikos, H. Nezhad, and V.K. Thakur, Mater. Today Chem. 16, 100248 (2020). https://doi.org/10.1016/j.mtchem.2020.100248

[2] C.G. Ferro, S. Varetti, G. De Pasquale, and P. Maggiore, Mater. Today Commun. 15, 185-189 (2018). https://doi.org/10.1016/j.mtcomm.2018.03.007

[3] Z.S. Bagheri, D. Melancon, L. Liu, R.B. Johnston, and D. Pasini, J. Mech. Behav. Biomed. Mater. 70, 17-27 (2017). https://doi.org/10.1016/j.jmbbm.2016.04.041

[4] A. Gisario, M. Kazarian, F. Martina, and M. Mehrpouya, J. Manuf. Syst. 53, 124-149 (2019). https://doi.org/10.1016/j.jmsy.2019.08.005

[5] W. Zhu, Prog. Aerosp. Sci., (2019).

[6] L. Yuan, S. Ding, and C. Wen, Bioact. Mater. 4, 56-70 (2019). https://doi.org/10.1016/j.bioactmat.2018.12.003

[7] F. Zanini, M. Sorgato, E. Savio, and S. Carmignato, (no journal info available).

[8] A. Kumar, L. Collini, A. Daurel, and J.-Y. Jeng, Addit. Manuf., 101168 (2020). https://doi.org/10.1016/j.addma.2020.101168

[9] V. Sufiiarov, V. Sokolova, E. Borisov, A. Orlov, and A. Popovich, Mater. Today Proc., (2020).

[10] L. Bochmann, C. Bayley, M. Helu, R. Transchel, K. Wegener, and D. Dornfeld, Surf. Topogr. Metrol. Prop. 3, 014002 (2015). https://doi.org/10.1088/2051-672X/3/1/014002

Using the Taguchi Approach for Investigating Mechanical Properties of Recycled Carbon Black Reinforced Acrylonitrile Butadiene Styrene (ABS) and Polylactic Acid (PLA) for Injection Moulding Applications

A.E. KASIM[1], N.A. SHUAIB[1,2,a], C.C. FOOI[3], C.R. QUAN[3], M.H.J.A. HADI[1], U.F. M. ALI[4,5] and A.F. OSMAN[4,6]

[1]Faculty of Mechanical Engineering & Technology, Universiti Malaysia Perlis, 02600 Arau, Perlis, Malaysia

[2]Centre of Excellence for Frontier Materials Research, Universiti Malaysia Perlis, 02600 Arau, Perlis, Malaysia

[3]Eco Power Synergy Sdn Bhd, No. 1A, Jalan Kenari 9, Bandar Puchong Jaya, Batu 8, 47100 Puchong, Selangor, Malaysia

[4]Faculty of Chemical Engineering & Technology, Universiti Malaysia Perlis, 02600 Arau, Perlis, Malaysia

[5]Centre of Excellence for Biomass Utilization, Universiti Malaysia Perlis, 02600, Jejawi, Perlis, Malaysia

[6]Center of Excellence Geopolymer & Green Technology, Universiti Malaysia Perlis, 02600, Jejawi, Perlis, Malaysia

[a]norshahaizat@unimap.edu.my

Keywords: Recycled Carbon Black, Taguchi Approach, Mechanical Testing

Abstract. Carbon black production and demand are expanding globally due to the continued production of automotive, aerospace and many other plastic industries. This is concerning because carbon black waste has the potential to harm the environment and public health. Efforts must be taken to limit the consumption. Recovered carbon black (rCB) is currently a good alternative product. However, using re-claimed carbon black can be difficult, particularly in injection moulding processing. Lack of study and information about processing conditions makes assessing the end product's mechanical properties difficult. This study focused on processing parameters such as thermoplastic type, rCB weight loading, and injection moulding pressure. The L4 Taguchi orthogonal design was utilised to reduce experimental runs while still gathering quantitative results. From the results, type of plastic is the most dominant factor followed by weight loading and injection pressure. Optimisation of the parameters should depend on the product's final application. This study highlights the potential use of rCB in reinforcing pure thermoplastics as an alternative material to virgin carbon black (vCB).

Introduction

Carbon black (CB) is a carbonaceous material that has long been used as a reinforcement material. Its unique structure and properties have helped it evolve into a large category of filler materials with outstanding mechanical properties due to its higher electrical conductivity, high surface area, and stability. Thus, CB is generally used as a strengthening filler in vehicle tyres and other rubber automotive products. Inks, paints, plastics, and coatings are other common everyday products that include CB [1]. The effect of reinforcing has to do with how the elastomer molecules interact with the carbon particles and how the carbon particles interact with the elastomer matrix. In addition to the reinforcement, CB is ultraviolet (UV) resistant. It is an ozone scavenger stabilising the tyre rubber against UV light and oxidation, reducing fissuring and cracking.

As commercial CB prices have steadily decreased, attention has switched to utilising alternative low-cost and readily available agricultural by-products. Agricultural by-products are frequently inexpensive, making their optimal use desirable. Many researchers have attempted to extract natural CB from agricultural by-products such as coconut shells, oil palm shells, and bamboo. This offers an opportunity to use recovered CB (rCB) as an alternative to original or vCB.

CB composites have been used in many applications, including electrical conductivity and electromagnetic shielding [2]. Melt-mixing and injection moulding have been the most extensively used processing processes for these composites. Injection moulding is the most efficient method for obtaining the proper net form for thermoplastic polymers.

The growth of CB demand leads to an increase in its waste from manufacturing processes. This also necessitates the urge to recycle CB. The rCB can be used as filler in the thermoplastic composite, increasing the properties of the new product. However, the impact of adding rCB to polymeric material's mechanical properties is rarely considered in the literature. As a result, the filler is less likely to be used instead of virgin CB filler.

This study investigates the mechanical properties of rCB reinforced acrylonitrile butadiene styrene (ABS) and polylactic acid (PLA) for injection moulding applications. The findings can help producers decide on the best parameters for utilising rCB waste for optimal mechanical qualities. The rCB waste commercial worth may be fully comprehended with the Taguchi optimisation method in this study. The method has become a widely accepted technique for optimising processes and improving part quality. This study used the method as the experimental design procedure to examine the influential processing parameters on the rCB polymer composites.

Methodology

The experimental trials in this study were based on the L4 Taguchi orthogonal design. An L4 experiment necessitates four trial runs. The array can handle up to three components, each with two levels. The number of samples required can be reduced using the Taguchi technique in experimental design. This method uses a unique orthogonal array architecture to analyse all parameters with a small number of experiments. The Taguchi L4 orthogonal array was utilised in the experimental design for three factors and two levels, as shown in Table 1.

Table 1: Experimental trials based on L4 Taguchi orthogonal array design.

Sample Product	Level		
	Type of thermoplastic	rCB weight loading (%)	Injection moulding pressure (MPa)
Product A	PLA	3.0	15
Product B	PLA	1.5	13
Product C	ABS	3.0	13
Product D	ABS	1.5	15

The research methodology used in this study is shown in Fig. 1. The procedure can be classified into two stages which are the sample preparation stage and the testing stage. The sample preparation stage can be classified into three sub-stages: distribution of rCB, mixing of raw material and production processes. The second stage deals with testing, characterisation, and analysis of the results.

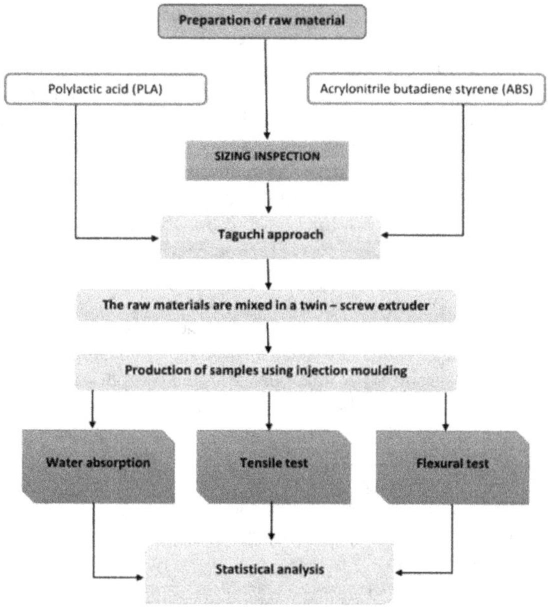

Fig. 1: Research methodology flowchart.

Analysis of Variance (ANOVA) was used to assess the influence of the input parameters: the type of plastics, weight loading, and injection moulding pressure studied at a 95% confidence level. This analysis aims to understand the link between the dependent and independent variables. Minitab software was used to conduct the statistical analysis. Compared to all the other criteria, the factor that made up the most significant contribution to the overall experiment was placed first in relative relevance and significant impact on the results. Mean effect plots were used to explain the relationship between the variables and the impacts of each input parameter. The plot determines the arrangement of factors affecting the product properties.

Materials. The rCB was obtained from a pyrolysis process of tire waste at a facility owned by Eco Power Synergy Sdn Bhd, located in Selangor, Malaysia. The composites created in this study used rCB as a filler. The filler comes in sizes ranging from 63 μm to 75 μm.

The ABS pellets supplied by Sheng Foong Plastic Industries Sdn Bhd in Tronoh. Perak. Thermoplastic material ABS is frequently employed in injection moulding processes. Due to its low manufacturing cost and ease of processing by plastic producers. Polylactic acid (PLA), in contrast to most thermoplastic polymers, is derived from renewable resources such as maise starch or sugar cane. Its production is relatively low-cost. The Pebbles 3D Sdn Bhd, located in Subang Jaya, Selangor supplied the PLA pellets used in this study.

Methods. In order to create a continuous profile, the mixing of ABS and PLA with rCB was done using a Lab Tech twin-screw extruder. The process is to ensure that the thermoplastic and rCB are mixed homogeneously. The products were mixed individually using the twin-screw extruder with a screw speed of 160 rpm, 20 rpm feeding screw speed and temperature of 210 °C for 20 minutes. Following the extrusion procedure, a BOY 35 E injection moulding machine, was used to make the tensile and flexural test samples. The equipment injected the sample at a temperature of 180 °C for both ABS and PLA products. The sample was released from the mould after 15 seconds of the cooling phase. The pressures selected for this study were 13 MPa and 15 MPa. With the selection, the products' mechanical properties may differ, making it easy to compare and analyse.

Mechanical tests were carried out based on tensile and flexural tests. A water absorption test was carried out to learn more about the samples' water absorptivity. Tensile testing is when a sample is put through a destructive engineering and materials science process called tensile testing, in which controlled tension is applied to it until it completely fails. This is one of the most often used mechanical testing methods for determining the strength and maximum stretch material can sustain before failing. The Shimadzu Universal Testing Machine AG-XD, and ASTM D638, was used to conduct the tensile tests. A 50 mm gauge length was used. The crosshead speed was 5 mm/min, and the distance between the two grips was 115 mm. Three dog bone-shaped samples (165 mm x 19 mm x 3.2 mm) were used, and the average values of the tensile properties were calculated. This test produced results for tensile strength, elongation at sample break, tensile modulus, and tensile failure strain for the injection-moulded composites.

Flexural testing determines material flexibility or the ability to bend. In 3-Point Bend testing or 4-Point Bend testing, a sample is placed between two points or supports and a load is applied using a third or two points, as appropriate. The flexural tests in this study were performed based on a 3-Point Bend test at 5 mm/min speed by ASTM D790.The Shimadzu Universal Testing Machine (AG-XD plus) instrument, was used to conduct the test in this study. Utilizing Trapezium X Materials Testing Software, the flexural characteristics can be determined. Three block-shaped samples (120 mm x 13 mm x 3 mm) were used, and the average values of the flexural properties were calculated. The gauge length of the specimen was 15 mm, and a gap of three points was produced. The distance between the two grips was 115 mm, and the testing speed for the flexural tests was 5 mm/min. The results of this test include flexural strength, sample elongation at break, and elastic modulus. The sample material's maximum strength and flexural modulus can be used to evaluate its ability to withstand bending force.

Water absorption test measures polymers' relative water absorption rate upon immersion under given circumstances. Each product's water absorption test sample was 10 mm long and 10 mm wide. The test was carried out in accordance with ASTM D570. Before being immersed in distilled water, the samples were weighed using a weight balance. Afterwards, the samples were placed in a container and kept in distilled water at a constant room temperature of 26 °C for two weeks. The samples were removed, dried using a dry cloth, and weighed. The initial weight (W_0) ratio to final weight (W_1) was used to calculate the water absorptivity. Eq. 1 was used to get the percentage of water absorption:

$$\text{Water Absorption \%,} = \frac{W_1 - W_0}{W_0} X\, 100 \quad (1)$$

Result and discussion

Table 2 presents the F-value and p-value for each parameter and output variable. The F-value indicates whether there is a significant difference between the means of multiple groups, the p-value indicates whether the results are statistically significant, and the rank indicates the relative position of an observation within a dataset. Together, these measures provide a comprehensive understanding of the significance of a statistical result.

Materials Research Proceedings 68 (2026) 315-321 https://doi.org/10.21741/9781644904237-42

Table 2: F-value, p-value, and rank with result interpretation for all parameters and output variables.

Parameter		F-value, p-value, and rank with result interpretation				
		Tensile strength	Tensile modulus	Flexural strength	Flexural modulus	Water absorptivity
Type of thermoplastic	F-value	56.79	97.06	9.30	25.50	4.89
	p-value	0.000 (S)	0.000 (S)	0.012 (S)	0.000 (S)	0.051 (S)
	Rank	1	1	1	1	1
Weight loading	F-value	0.27	0.00	4.43	1.45	0.04
	p-value	0.612 (NS)	0.946 (NS)	0.062 (NS)	0.256 (NS)	0.837 (NS)
	Rank	3	3	2	2	2
Injection pressure	F-value	0.44	0.28	0.91	0.16	0.00
	p-value	0.524 (NS)	0.607 (NS)	0.362 (NS)	0.698 (NS)	0.979 (NS)
	Rank	2	2	3	3	3

According to Table 2, the statistical significance of each parameter varies across all variables. The type of thermoplastic has a significant influence on all variables, including tensile strength, tensile modulus, flexural strength, flexural modulus, and water absorption, which represented by low p-values. In comparison to injection pressure, which affected tensile strength and tensile modulus, weight loading was the second most significant parameter, influencing flexural strength, flexural modulus, and water absorption.

Main effect plot of mean between each factor and output variables are illustrated in Fig. 2. Slope of the line represents degree of influence of each factor for the variable. The type of thermoplastic shows great influence on the flexural and flexural properties, as presented in Fig. 2(a), Fig. 2(b), Fig. 2(c), Fig. 2(d) and Fig. 2(e).

Statistical analysis using Minitab software yielded the main effect plot from mean. The plot is used to establish how factors affecting product attributes are arranged. Large amount of rCB deteriorated the mechanical properties of PLA/CB composites. This indicated that CB agglomerated in the polymer matrix, resulting in increased stress concentration locations and poor mechanical characteristics. This also revealed that there was some compatibility between rCB and PLA. According to Lohar et al [3] the substantial increase in tensile strength for rCB/ABS due to the lowest droplet diameter of ABS dispersed phase and interparticle distance in between dispersed. As a result, when reinforced with rCB, the ABS still improved in tensile strength, although only slightly. As according to Zainudin et al [4] the increased tensile strength of ABS is related to the strength of a composite changing due to microscopic damage caused by interfacial debonding fractured or broken resins.

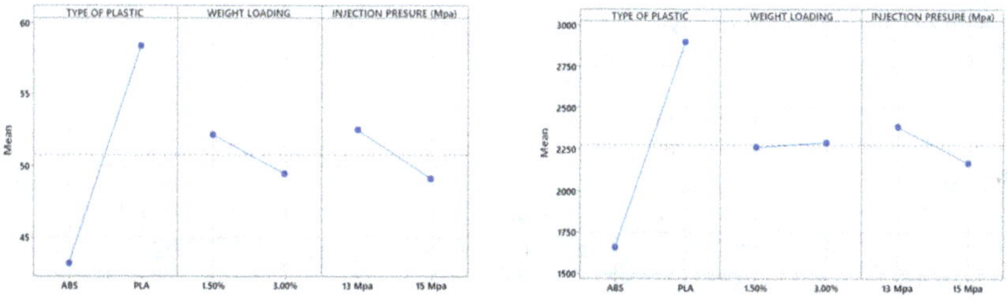

a) b)

Materials Research Proceedings 68 (2026) 315-321 https://doi.org/10.21741/9781644904237-42

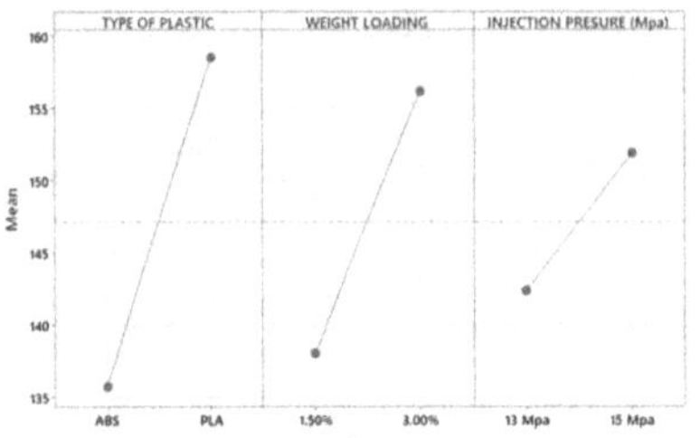

c)

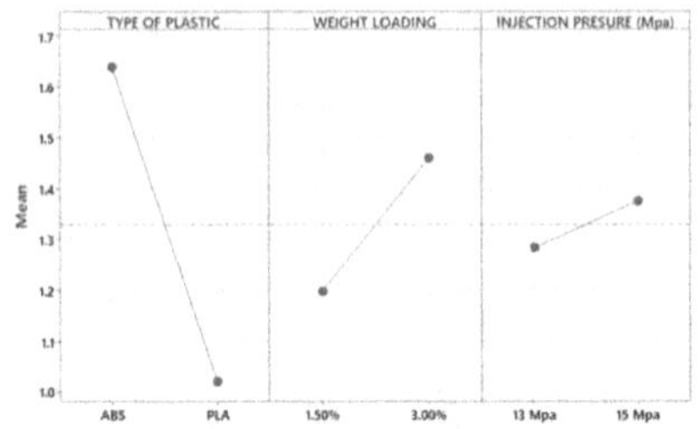

d)

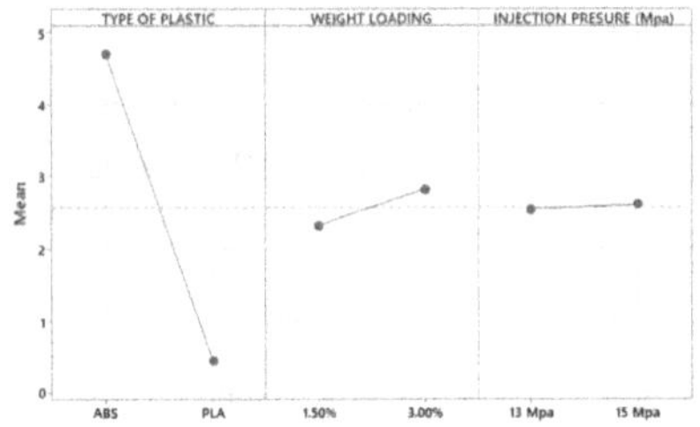

e)

Fig. 2: Main effect plots for a) tensile strength, b) tensile modulus, c) flexural strength, d) flexural modulus and e) water absorption.

The decrease in tensile modulus for rCB/PLA with higher rCB percent content indicated that PLA was brittle after being filled with the CB filler. Weight loading content does not show a significant influence towards the value of tensile modulus. Theoretically, tensile strength should increase with the increment of injection pressure, which force the material to flow throughout the die cavity. The air inside the cavity was forced out because of this forceful motion, which reduced air inclusion within the cavity. According to the study by Yadav et al [5], high injection pressure leads to enhanced intramolecular bonding inside the material, which increases the material's strength. However, the fact that both thermoplastic with lower injection pressure achieve the greatest tensile modulus value indicates that elastic modulus grows with CB loading, as supported by Savetlana et al [6].

When the filler material rCB is reduced, the flexural strength of manufactured composites is reduced. This could be due to less stress transfer between the reinforcement material as the matrix material increases. Inadequate matrix filler assembly and weak interfacial bonding between matrix and filler materials could also be the reason. Flexural modulus is a simple method for calculating composite stiffness. The findings also show that rCB reinforcement improves the modulus properties of the products significantly.

High absorptivity leads to poor performance of the material. The presence of rCB, regardless of the nature of the dispersion polymer, increases the amount of water absorbed at saturation while decreasing the rate of evaporation. The filled polymers's water absorption capacity is primarily determined by the oxygen content of the CB particles. The rate of penetration also increases with the presence of filler loading [7]. The type of thermoplastic used is the most important factor, followed by weight loading and injection pressure. To ensure that thermoplastic products are water resistant, low values for water absorptivity are preferred. According to the findings, rCB/PLA with either higher or lower injection pressure is suitable for use in moist environments. The injection pressure has little impact on how much water is absorbed.

The study's most obvious finding is that the type of thermoplastic has a significant influence on the changes in the properties. PLA outperforms all other mechanical properties tests, apart from the flexural modulus and water absorption tests, where ABS performs better. It is difficult to determine the optimal parameters for filler weight loading and injection pressure. As a result, the parameters should be chosen based on the final application of the rCB reinforced product. For instance, rCB/ABS with a weight loading of 3.0 wt% should be used instead of rCB/PLA in products that are subjected to prolonged bending stress. The filler content has the greatest influence on the modulus of elasticities. When the filler weight loading content is high, the elasticity increases. Injection pressure influences resistance to maximum strain. Meanwhile only the type of plastic affects water absorptivity, while rCB weight loading and injection pressure are essentially irrelevant.

Conclusion

This study set out to investigate effect of parameters in fabricating rCB reinforced thermoplastics on mechanical properties of the products. The most obvious finding in this study is that type of thermoplastic has significant influence on the changes of the properties compared to the weight loading, and injection pressure. Excellent mechanical qualities were obtained from tensile testing with PLA thermoplastic. However, ABS performs better for flexural modulus.

As the filler weight loading increases, the flexural strength of the specimen increases as the addition of rCB can improve the mechanical properties of a material. The specific effect on tensile and flexural strength will depend on the type and amount of weight loading used, as well as the properties of the base material. However, there is a limit to the maximum strength improvement that can be achieved, and at high loading levels, the strength may decrease due to the formation of defects and poor adhesion between the filler and the matrix.

It is challenging to determine the best parameters as each property is influenced by different factors. This study can thus be used to inform the selection of the appropriate material, the adaptation of practical injection moulding parameters, and the consideration of other mechanical properties for the specific application.

Acknowledgement

The author would like to acknowledge the support from the Malaysian Technical University Network (MTUN) Pre-commercialization grant from the Ministry of Higher Education Malaysia.

References

[1] Tofighy, M. A. and Mohammadi, T., Carbon-Based Nanofillers and Their Rubber Nanocomposites: Fundamentals and Applications (Elsevier), 253–285, (2019). https://doi.org/10.1016/B978-0-12-817342-8.00009-3

[2] Nan, X., Zhang, Y., Shen, J., Liang, R., Wang, J., Jia, L., Yang, X., Yu, W. and Zhang, Z., Polymers, 16, 2539, (2024).

[3] Lohar, G. S. and Jogi, B. F., Procedia Manuf., 20, 85–90, (2018).

[4] Zainudin, S., Shuaib, N. A., Omar, N. W. Y. and Azmi, A. I., Int. J. Mod. Manuf. Technol., 13(2), 149–154, (2021).

[5] Yadav, R., Pancharya, A. and Kant, R., Mater. Today: Proc., (2020). https://doi.org/10.1016/j.matpr.2020.09.486

[6] Savetlana, S., Zulhendri, Sukmana, I. and Saputra, F. A., "The effect of carbon black loading and structure on tensile property of natural rubber composite," in Innovation in Polymer Science and Technology 2016 (IPST 2016), IOP Conf. Ser.: Mater. Sci. Eng., 223, 012009, (2016).

[7] Siva, R., Nemali, S. R., Kunchapu, S. K., Gokul, K. and Arun Kumar, T., Mater. Today: Proc., 47, 4382–4386, (2021). https://doi.org/10.1016/j.matpr.2021.05.189

Monitoring and Prediction of Tool Wear in Turning Process using Machine Learning Technique

S.Y. QUEK[1], W.K. LEE[1,a *], N. TALIB[1,b], H. ABDULLAH[1,c], A.M SABRI[1,3,d] and A. SUNG[2]

[1]Faculty of Mechanical and Manufacturing Engineering, Universiti Tun Hussein Onn Malaysia, Batu Pahat, Johor, Malaysia

[2]Vitrox Academy, 746, Persiaran Cassia Selatan 3, Taman Perindustrian Batu Kawan, 14110 Bandar Cassia, Penang, Malaysia

[a]wklee@uthm.edu.my, [b]fazillah@uthm.edu.my, [c]haslina@uthm.edu.my, [d]ainaamardhiahsabri@gmail.com

Keywords: Tool Wear, Machine Learning Technique, Performance Analysis

Abstract. The benefit of an intelligent monitoring system is which can detect the changes that could indicate a developing fault of cutting tool more earlier to avoid the catastrophic failure of turning machine. The worn cutting tool can be replaced in time to ensure the accuracy of the cutting process. The objective of this study is to apply support vector machine prediction model to predict the flank wear during turning process. Experiments have been conducted to measure tool wear based on the factorial design technique in a turning of AISI 1045 carbon steel using carbide insert. Then, the cuttings conditions and statistical features from vibration signal in time domain were used as the input of regression model corresponding to the output, flank wear for the tool wear prediction by using MATLAB. The results showed that polynomial kernel prediction model performed outstanding which had the highest accuracy with 93.57% among the three models. The R^2 value for gaussian and polynomial kernel were stated 0.9956 and 0.9866 respectively. While the linear kernel had the lowest of R^2 value which was only 0.6297.

Introduction

Research in the tool wear monitoring and prediction based on machine learning has been conducted for decades with the sustained development of industry 4.0 and the demand of intelligent manufacturing. Cutting tool condition has been proven significantly affects the surface quality and dimensional precision, while tool fracture can cause machine downtime and reduce the machining efficiency. It is reported that time lost due to tool breakage account 20% of the total downtime. For these reasons, it is imperative to build an accurate and effective tool wear monitoring and prediction method to prevent the catastrophic failure [1].

Tool wear monitoring generally can be categorized into direct method and indirect method. The direct method usually conduct the measurement of tool wear using tool maker's microscope or computer vision. However, these methods are not feasible to perform online due to the continuous cutting and presence of coolant and chips [2,3]. The indirect methods predict the tool wear state by using a machine learning model with the online acquisition of the sensor signals [4]. For example, cutting force signal [5,6], vibration [7], sound signal [8, 9] and other sensor signals as the input. The features from sensor signals are extracted then establish the model to identify the wear condition. The indirect method allows the cutting tool to be monitored without interrupt the machining process. With the development of signal processing technology based on machine learning to monitor the condition of tool wear in real time during machining process has become a current mainstream method. The signal acquired from the machining process are pre-processed and analyzed to obtain the important information about the tool condition. The extraction of signal features associated with tool state is the key issue in tool condition monitoring system. Only those signal features that show a high sensitivity to tool state should be used to achieve the best performance of tool condition monitoring system [10].

Machine learning is gaining more and more attentions in tool wear monitoring with the improvement of data-processing capability of computer. It generally consists of two processes, viz learning of the existing training data and classification or regression of the test data. Extensive research has been done using machine learning in predicting tool wear. Most of the earlier work focused on prediction of tool wear using artificial neural networks (ANN). Kaya et al.[11] developed ANN-based flank wear monitoring system based on force signal. The results showed that ANN model manifested a better statistical performance with a high correlation between the actual and predicted values of flank wear. Paul and Varadarajan [12] introduced a multisensor fusion model to predict tool wear in turning processes using ANNs. Drouillet et al. [13] applied spindle power with ANN to predict remaining useful life (RUL) of end milling tools. ANN has good performance complex behavior but it requires long computational time. The prediction accuracy is restricted to the inherent weakness of neural network, such as over-fitting, local minimum value, and poor generalization when fewer samples is used.

Other machine has also widely applied in tool wear monitoring such as Markov models [14], K-nearest neighbours [15] and etc. However, these methods still has their limitation. Markov models require extensive training data sets. Non-homogenous behaviors will be difficult to predict using K-nearest neighbours [16]. Support vector machine are supervised, classifying and regression algorithm that has been utilized in tool wear monitoring. Previous study showed that SVM models gave better results compared to ANN [17]. Gomes et al. [18] applied vibration and sound signals to monitor the tool wear with the help of SVM. High prediction accuracy is achieved (97.54%) which demonstrates SVM's ability in tool wear prediction. SVM has good performance with complex behavior and high dimensionality datasets [16]. In this study, a tool condition monitoring system is developed based on vibration sensor signal using using support vector regression models with three kernel function: linear, Gaussian and polynomial kernel.

Methodology

HAAS SL-20T CNC lathe machine is used in the experiment. AISI 1045 carbon steel workpiece material is turned with a carbide insert (TNMG160408PS, Kyocera). Table 1 shows the value of cutting parameters used in turning process. The vibration signals were collected by using Movipack vibration analyser during the turning process. Flank wear is measured by using Nikon MM-60 Toolmaker's Microscope according to ISO 3685:1993.

Table 1: Machining conditions.

No	Parameter	Level		
		1 (Low)	2 (Medium)	3 (High)
1	Cutting Speed, V(m/min)	300	400	500
2	Feed Rate, f (mm/rev)	0.3	0.4	0.5
3	Depth of Cut, h (mm)	0.1	0.2	0.3

Feature extraction is usually performed before regression. Statistical features such as mean and standard deviation were the parameters that extracted from the vibration signal in time-frequency domain. The standard deviation and mean of vibration signal as well as the cutting parameters (cutting speed, feed rate and depth of cut) as an input data to predict the tool flank wear. Support vector regression (SVR) developed in MATLAB is used to predict the flank wear. SVR applied the kernel function to map the input data in the original space to the high dimensional space non-linearly. The linear, Gaussian and polynomial kernel were used in this study. A total of 27 sample features based on factorial design technique from vibration signal and three cutting parameters are acquired. The feature set is divided into two groups which were 26 training sample and 1 test sample was used for testing the performance of the trained model. This step was repeated until a total of 27 prediction wear value is obtained. The accuracy of the prediction is analyzed with the absolute percentage error (APE), the mean absolute percentage error (MAPE) and the coefficient of determination (R^2). The higher the R^2 reflects the better the regression model fits the data.

Result and Discussion

The measured and predicted values of tool flank wear along the percentage error are tabulated in Table 2. It can be seen in Table 2 that the prediction values mostly were very close to the experimental values for flank wear. The minimum error of prediction was 0.77%. However, there have few huge absolute percentages error were found. This may be due to the linear kernel occurred some errors during manage the data for prediction of tool wear value. As an overall, the performances of the three SVR kernel functions were high with the minimum average accuracy of 76.43% which obtained by the linear kernel. Nevertheless, the polynomial function kernel outperformed other SVM kernels with the highest average accuracy of 93.57%. Descending order of their accuracy performance, the polynomial kernel function is followed closely by the Gaussian kernel function, then the linear kernel function. It can be observed that the value for mean absolute percentage error of linear kernel is 23.57%, Gaussian kernel is 10.11% and polynomial kernel is 6.43%. As we know, the smaller the errors occurred, the more accuracy of the model performed. Similar finding is found in [19], polynomial kernel SVR method outperforms the linear and Gaussian kernel SVR method in terms of overall accuracy.

Table 2: Prediction efficiency for linear, Gaussian and polynomial kernel.

Experiment run	Actual wear (μm)	Linear Kernel		Gaussian Kernel		Polynomial Kernel	
		Predicted wear(μm)	APE	Predicted wear(μm)	APE	Predicted wear(μm)	APE
1	1.6	2.94	84.06	3.42	113.87	2.11	31.88
2	3.4	3.28	3.65	3.91	14.96	2.89	14.99
3	4.00	4.20	5.08	4.51	12.76	4.51	12.75
4	4.80	10.11	110.63	5.31	10.62	5.31	10.61
5	5.30	11.82	123.08	5.81	9.63	5.81	9.61
6	5.80	13.63	134.96	6.31	8.79	5.96	2.76
7	6.10	6.73	10.29	6.61	8.35	6.05	0.77
8	6.50	7.00	7.70	7.01	7.84	7.01	7.84
9	6.60	8.15	23.43	7.11	7.71	6.98	5.73
10	7.00	5.72	18.35	7.51	7.28	6.49	7.28
11	7.50	8.00	6.65	8.01	6.80	8.01	6.79
12	8.50	8.91	4.86	9.01	5.99	9.01	6.00
13	9.50	9.08	4.40	9.80	3.12	10.01	5.37
14	11.00	10.57	3.90	10.49	4.64	10.49	4.63
15	11.60	10.38	10.51	11.09	4.40	11.09	4.39
16	11.90	11.43	3.99	11.39	4.28	11.39	4.28
17	12.20	11.13	8.79	11.69	4.19	12.71	4.18
18	12.50	11.98	4.13	11.99	4.07	12.18	2.53
19	12.70	13.53	6.52	12.19	4.01	13.21	4.01
20	13.00	12.76	1.82	12.49	3.93	12.49	3.92
21	13.10	13.68	4.41	12.59	3.88	12.90	1.53
22	13.40	10.37	22.62	12.89	3.80	12.89	3.80
23	13.60	11.53	15.24	13.09	3.74	14.11	3.75

24	13.70	13.23	3.45	13.19	3.72	13.19	3.72
25	14.10	13.58	3.70	13.59	3.61	13.59	3.61
26	14.50	13.14	9.38	13.99	3.51	13.99	3.52
27	14.80	14.68	0.79	14.29	3.44	15.31	3.44
		MAPE	23.57		10.11		6.43

Fig. 1 to Fig. 3 illustrated the scatter diagrams of the predicted value and actual values of flank wear obtained from experiment for linear kernel function, Gaussian kernel function and polynomial kernel function respectively. As illustrated in Fig. 1, it can be observed that the plotted graph for linear kernel looks scattered compared to Gaussian kernel and polynomial kernel. The R^2 value for linear kernel was lowest which stated only 0.6297. A low R^2 is usually a bad signal for a predictive model.

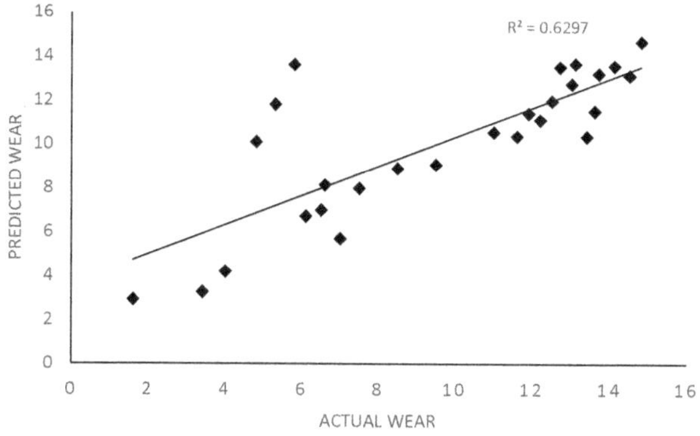

Fig. 1: Predicted versus actual flank wear for linear kernel.

As a comparison, the more perfect graphs for Gaussian and polynomial kernel which is shown in Fig. 2 and Fig. 3 respectively. The correlation between the actual and predicted values of flank wear followed the 45 degree line very closely. In other words, the predicted values were not far from actual measurement values obtained from the experiment. The correlation coefficient R^2 value for both graphs were higher which stated 0.9956 for Gaussian kernel and 0.9866 for polynomial kernel. In short, the higher the R^2 value, the more precisely the predictor can forecast the value of the response variable. Thus, it can be concluded that the SVR model with Gaussian and polynomial had a very good relationship and fit which indicates that the proposed SVR model was capable in prediction of tool flank wear accurately. The result for linear kernel was also less accurate compared to polynomial kernel because it is the most simplest kernel which just takes less time to process the data. It used non-linear equation as feature space transition inputs. This caused it to take longer time and be more complex than the linear kernel. Therefore, polynomial kernel can be produced the results that are more accurate and dependable compared to linear kernel [20].

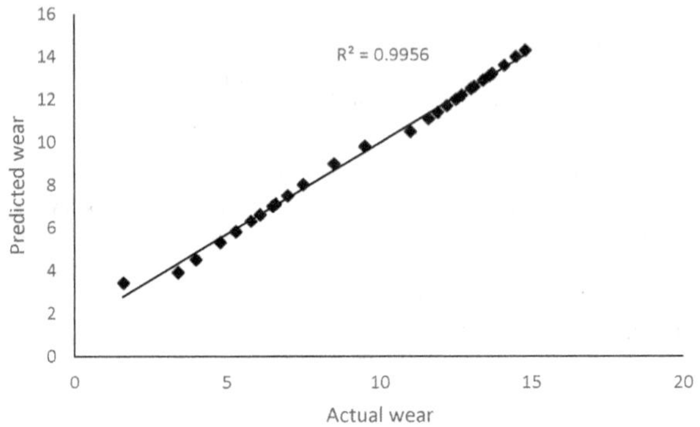

Fig. 2: Predicted versus actual flank wear for Gaussian kernel.

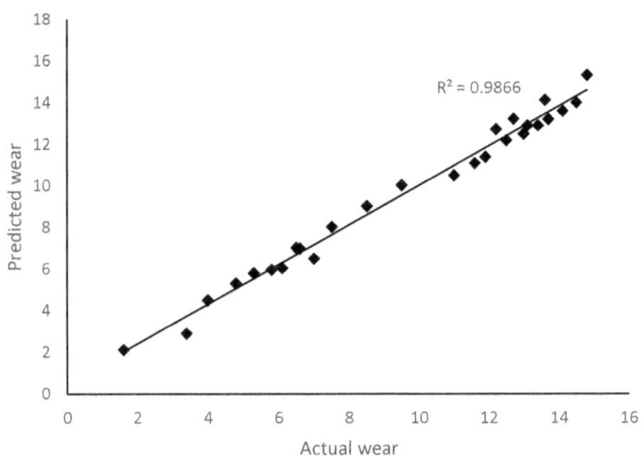

Fig. 3. Predicted versus actual flank wear for polynomial kernel.

Conclusion

In this paper, the performance analysis comparing different kernel methods of SVR in tool wear prediction is investigated. With the results obtained, the polynomial kernel had the highest accuracy (93.57%), followed by Gaussian kernel (89.89%) and linear kernel (76.43%). SVR with Gaussian and polynomial kernel also demonstrates a very good relationship and fit between the experimental and predicted results with R^2 of 0.9956 and 0.9866 respectively. The results of this study indicated that the proposed SVR could be used to predict flank accurately in turning process and kernel function significantly influence the accuracy of flank wear prediction.

Acknowledgment

This research was supported by Ministry of Higher Education (MOHE) through Fundamental Research Grant Scheme (FRGS/1/2018/TK03/UTHM/03/6). We also want to thank to the Government of Malaysia which provide MyBrain15 programme for sponsoring this work under the self-funded research grant and L00022 from Ministry of Science, Technology and Innovation (MOSTI).

References

[1] Y. Zhou, C. Liu, X. Yu, B. Liu and Y. Quan, SN. Appl. Sci. 4, (2022). https://doi.org/10.1007/s42452-022-05114-9

[2] M.S.H. Bhuiyan and I.A. Choudhury, Comprehensive Materials Processing 13, 539-569 (2014). https://doi.org/10.1016/B978-0-08-096532-1.01330-3

[3] S. Dutta, S.K. Pal, S, Mukhopadhyay and R. Sen, J Manuf. Sci. Tech. 6, 212-232 (2013). https://doi.org/10.1016/j.cirpj.2013.02.005

[4] M. Kuntoglu, A. Aslan, H. Saglam, D.Y. Pimenov, K. Giasin and T. Mikolajczyk, Sensors 20, 4377 (2020) https://doi.org/10.3390/s20164377

[5] D. Kong, Y. Chen and N. Li, J. Mech. Eng. Sci., 203-210, 1989-1996 (2019)

[6] M.E. Korkmaz, N. Yasar and M. Gunay, Eng. Sci. Tech. 23, 664-673 (2020). https://doi.org/10.1016/j.jestch.2020.02.001

[7] C. Xu, Y. Chai, H. Li and S. Shi, J. Adv. Mech. Des. Syst. Manuf. 12(2), (2018).

[8] F. Huda, K. Karjuni, and M. Rusli, IOP Conf. Ser.: Mater. Sci. Eng., 830 (2020). https://doi.org/10.1088/1757-899X/830/4/042028

[9] M.K. Liu, Y. Tseng and M.Q. Tran, Int. J. Dav Manuf. Technol. 114, 3793-3802 (2019).

[10] C.H. Lauro, L.C. Brandao, D. Baldo, R.A. Reis and J.P. Davim, Measurement 58, 73-86 (2014). https://doi.org/10.1016/j.measurement.2014.08.035

[11] B. Kaya, C. Oysu and H.M. Ertunc, Adv. Eng.Softw 42, 76-84 (2011). https://doi.org/10.1016/j.advengsoft.2010.12.002

[12] P.S. Paul, and A. Varadarajan A, Proc. Inst. Mech. Eng 226(5), 853-860 (2012). https://doi.org/10.1177/0954405411432381

[13] C. Drouillet, J. Karandikar, C. Nath, A.C. Journeaux, M. E. Mansori and T. Kurfess, J. Manuf. Processes 22, 161-168 (2016). https://doi.org/10.1016/j.jmapro.2016.03.010

[14] N.N. Bhat, S. Dutta, S.K. Pal and S. Pal, Measurement 90,500-509 (2016). https://doi.org/10.1016/j.measurement.2016.05.022

[15] P. Junior, D.M. D'Addona, P. Aguitar and R. Teti, Sensors 18, 4435 (2018). https://doi.org/10.3390/s18124453

[16] D.Y. Pimenov, A. Bustillo, S. Wojciechowski, V.S. Sharma, M.K. Gupta and M. Kuntoglu, J Intell Manufac., (2022)

[17] U. Çaydas, and S. Ekici, J Intell Manuf 23, 639-650 (2012) https://doi.org/10.1007/s10845-010-0415-2

[18] M.C. Gomes, L.C. Brito, M.B. da Silva, and M.A.V Duarte, Precis Eng 67, 137-151 (2021) https://doi.org/10.1016/j.precisioneng.2020.09.025

[19] C. Savas, and F. Dovis, Sensors 19, 1-16 (2019). https://doi.org/10.1109/JSEN.2019.2912694

[20] B. Yekkehkhany, A. Safari, S. Homayouni and M.A. Hasanlou, ISPRS Archives 2014, 281-285 (2014). https://doi.org/10.5194/isprsarchives-XL-2-W3-281-2014

Mechanical Characterization of Solder/Intermetallic Compound (IMC) Interface: The Role of Temperature and Strain Rate

S. F. M. ASASAARI[1], M. N. TAMIN[1], M. A. ISMAIL[2, a*], S. SHAHARUDDIN[2], M. AZMI[2], M. JOHAR[2], Z. MOHAMMAD[2], M. S. E. KOSNAN[2], K. J. WONG[3]

[1]School of Mechanical Engineering, Faculty of Engineering, Universiti Teknologi Malaysia

[2]Advance Facilities Engineering Research Cluster, Malaysian Institute of Industrial Technology, Universiti Kuala Lumpur, 81750 Masai, Johor, Malaysia

[3]Department of Mechanical Engineering, Faculty of Engineering and Science, Curtin University Malaysia

[a]sitifaizah_sfma@yahoo.com, [f]manuar@unikl.edu.my

Keywords: Mechanical Characterization, Solder Shear Push Test, Solder/Intermetallic Compound

Abstract. This research paper comprehensively investigates the cohesive zone model (CZM) interface properties of SAC405 solder joints under solder shear push tests. The accurate extraction of interface properties is achieved using a computational-experimental approach. The study quantifies the temperature dependence of lead-free solder joints through analysis of load-displacement curves at various straining rates. Solder ball shear tests are conducted to assess the strength of SAC405 solder joints under dynamic loading conditions. Utilizing a 3D finite element model incorporating the Unified Inelastic Strain Constitutive Model and cohesive zone model, the research predicts damage initiation, propagation, and solder/intermetallic compound interface fracture during the solder shear push test. The CZM parameter values are analyzed as a function of temperature and strain rate, revealing a degradation trend as temperature increases from 25 °C to 220 °C. Furthermore, the study demonstrates that temperature changes influence the initiation and evolution of damage in SAC405 solder joints. The findings provide valuable insights into the behavior of SAC405 solder joints and contribute to developing more reliable microelectronic packages by accurately predicting solder interconnect failure.

Introduction

In microelectronic packaging, ball grid array (BGA) solder interconnects play a crucial role in ensuring the reliability of electronic devices[1]. These solder interconnects experience thermomechanical load cycles during fabrication and service life, requiring a comprehensive understanding of their behavior under varying conditions [2]. Recently, lead-free solder joints have been adopted to address environmental concerns. However, these joints exhibit temperature- and strain-rate dependency, posing challenges in accurately predicting the package's reliability [3].

The reliability of solder joints is influenced by several factors, including solder joint geometry, interface metallurgy, and aging conditions [4]. To identify the most reliable package design, researchers have studied different solder joint geometries, such as barrel-shaped, column-shaped, and hourglass-shaped, considering varying aspect ratios and shape factors [5]. Critical solder locations are determined based on the maximum equivalent creep strain, with barrel-shaped solder joints often experiencing crack initiation at the substrate-solder joint interface due to stress concentration at the edge. In previous research, the cohesive zone model (CZM) has been employed to simulate the behavior of solder/intermetallic compound (IMC) interfaces during drop impact on BGA packages [6]. While effective in capturing solder joint behavior, the CZM parameters used in earlier studies did not account for strain rate effects [7, 8].

Materials Research Proceedings 68 (2026) 328-334 https://doi.org/10.21741/9781644904237-44

Addressing this limitation, a recent study investigated the rate-dependent behavior of BGA packages with SAC405 solder joints [9]. Strain rate and phase angle variations were considered in the simulations, covering a range of experimental strain rates. A 3D finite element model, incorporating a cohesive element at the board side, was utilized to predict fracture load under different strain rates at room temperature. The research also involved solder ball shear tests on SAC405 solder joints at room temperature and varying speeds. The study discussed the process of extracting solder/IMC interface properties and presented the SAC405 solder/IMC interface responses to high-speed solder ball shear tests. The study successfully determined the interface properties of SAC405 lead-free solder joints under dynamic loading conditions by combining computational modeling and experimental testing.

The findings from this research emphasize the significance of considering temperature effects in predicting the reliability of microelectronic packages. The observed rate-dependent behavior of solder/IMC interfaces highlights the need for robust damage-based models to achieve more accurate reliability predictions for advanced microelectronic packaging technologies. As such, this study contributes valuable insights into the ongoing development of such models and paves the way for enhanced performance and reliability of future microelectronic devices.

Material and experimental procedure

Experimental testing on SAC405 solder joints followed the JEDEC JESD22-B117A test standard [10]. Displacement-controlled solder ball shear push tests on BGA joints were conducted at room temperature using the Dage 4000 bond tester. The setup included a prescribed load cartridge to ensure precise and repeatable results, facilitating a comprehensive assessment of the solder ball joint's mechanical characterization and reliability. Fig. 1 shows the testing machine placed in a sturdy, vibration-free setup. Analysis software recorded force-displacement responses. The mechanical testing is tested at varying temperature conditions from 25 °C to 125 °C using a controllable hot plate and sheared at loading speed ranges between 0.1 mm/s to 0.8 mm/s. The shear tool stand-off was 5 μm to avoid solder joint deformation. Each shear speed test involved shearing 30 solder balls for repeatability.

Some solder balls on the BGA specimen were removed in the test, following JEDEC standards. The remaining solder balls were renamed based on location. BS5KG load cartridge was used for shear push tests with parameters set at 10 μm shear tool overtravel and 90% peak load fallback. Force-displacement responses were recorded in Excel. Low-speed shear tests showed ductile failure, used as a reference for rate-dependent SAC405 solder joints. Shear tool displacement rates of at least 500 mm/min were needed for interface fracture.

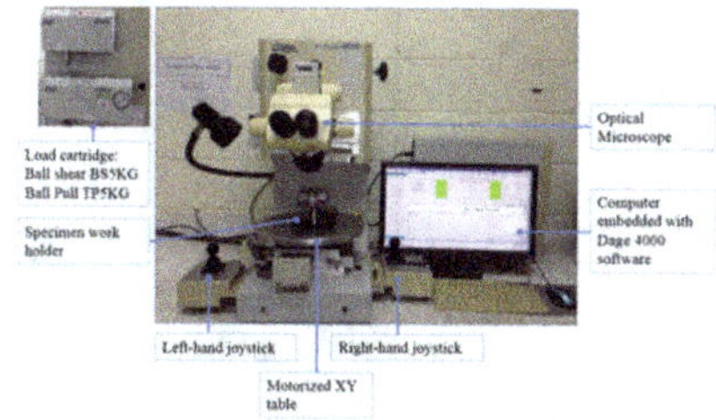

Fig. 1: Dage bond tester in operation during the shear strength test.

Materials Research Proceedings 68 (2026) 328-334 https://doi.org/10.21741/9781644904237-44

Finite element simulation

The mathematical model's specifics included test assembly geometry, material properties, loading, and boundary conditions. This study utilizes the SAC405 solder joint with a refined Anand model parameter in finite element (FE) simulation to accurately represent the creep-viscoplasticity behavior in the SAC405 bulk solder material. Cohesive elements are employed in the simulation to investigate the damage resulting from traction-relative displacement at the solder/IMC interface. These cohesive elements enable a detailed analysis of the solder joint's behavior and provide valuable insights into its mechanical response and failure mechanisms.

The FE model is a single solder, shear push test (SPT) with a 3D representation in Fig 2. It includes a SAC405 reflowed solder ball, a copper pad, and a printed circuit board (PCB). An assumed 2 μm thick IMC layer forms at the solder/copper pad interface. The shear tool is a discrete rigid tool without deformation. Surface-to-surface contact interaction demonstrates contact between the rigid tool and the solder ball. PCB's bottom surface has zero displacements and rotation. All contact surfaces are perfectly bonded throughout the FE simulation. The rigid tool speed is 3,000 mm/s in the Y-axis with a 10 μm shear tool stand-off, simulated at 30 °C.

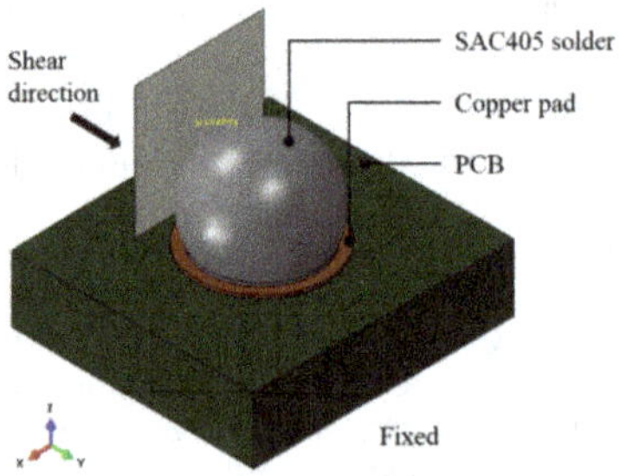

Fig. 2: FE model of single solder SPT.

Details on the material properties of the SAC405 solder joint and other materials used in SPT simulation are given in Table 1.

Table 1: Material properties used in SPT simulation.

Material	*E* (MPa)	*G* (MPa)	*ν*
SAC405 [11, 12]	41,050	-	0.36
Cu layer [13]	128,750	-	0.35
FR-4 substrate [14]	16,898	7,623.4	0.11
(isotropic in xy-plane)	(x and y)	(xy)	(xy)
	7436	3324.6	0.39
	(z)	(xz and yz)	(xz and yz)

Determination of Cohesive Zone Model Parameter Values

A FE model of a single solder SPT simulated stresses, strains, and displacements at the solder/IMC interface with the same shearing speed as the experiments. The non-damaging interface was assumed. The predicted FE force-displacement curve was compared to the experimental curve in Fig. 3. The critical load, Fi, at point A indicates the initial crack in the solder/IMC interface. The absolute maximum shear stress represents the shear strength of the solder/IMC interface at fracture (point A). Fig. 4 illustrates the highest magnitude of the absolute maximum shear stress. The slope of the shear traction-relative displacement curve gives the penalty stiffness. The FE results

Materials Research Proceedings 68 (2026) 328-334 https://doi.org/10.21741/9781644904237-44

provided the relative displacement at fracture (point A). The same procedures were repeated to quantify the interface shear strength at different test temperatures.

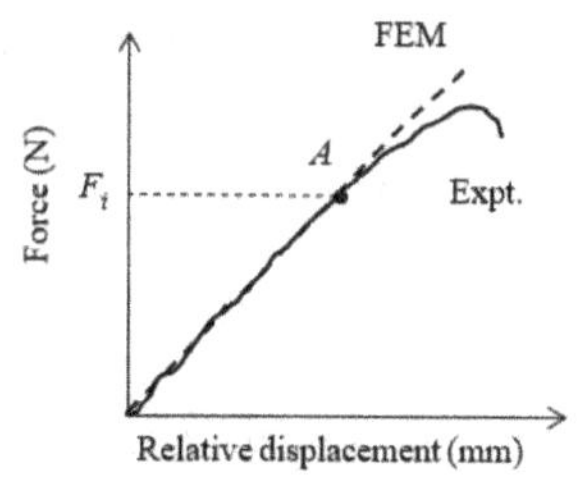

Fig 3: Experimental-FE approach corresponds to the initial crack in the solder/IMC interface.

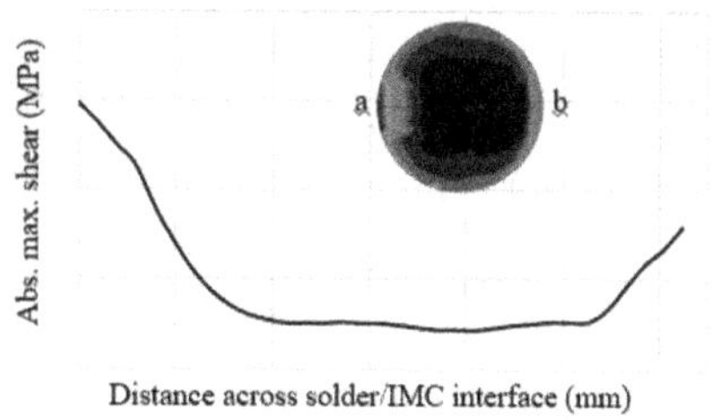

Fig 4: Evaluating absolute maximum shear stress on the solder/IMC interface plane along paths a-b.

The Cohesive Zone Model (CZM) previously described for monotonic loading at room temperature was extended to account for temperature effects. Five CZM parameters describe the damage initiation and evolution of the cohesive interface, as shown in Fig. 5. The effect of temperature on CZM properties was assumed to degrade as temperature increased. Initially, only three CZM parameters (interface strength, penalty stiffness, and critical strain energy release rate) are considered temperature-dependent. Damage initiation and final separation are assumed to be independent of temperature changes. Simplified extended traction-displacement curves, accounting for temperature effects, are illustrated in Fig. 6 for shear loading conditions.

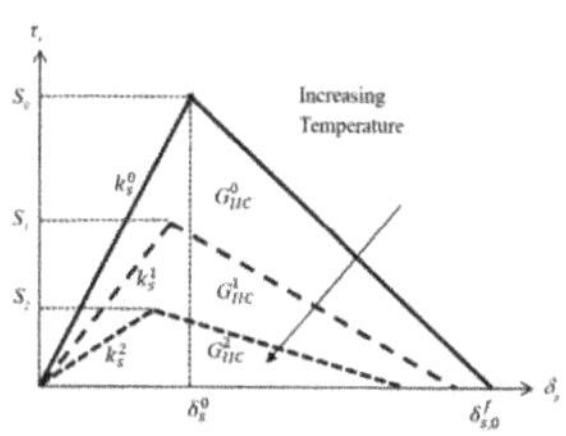

Fig 5: Effect of temperature on shear traction-displacement curves.

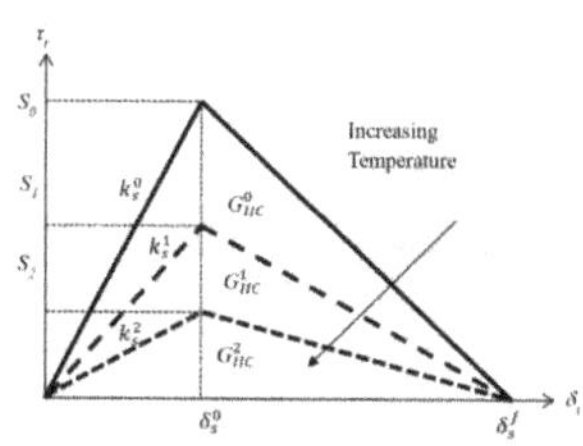

Fig 6: Simplified traction-displacement curves for temperature effects of solder/IMC interface.

Results and Discussions

Fig. 7 shows the comparison between FE-predicted and measured experimental force-displacement curves. The dashed line represents the FE predicted curve, and the solid line represents the measured experimental curve. The FE predictions match well with the experimental data for a shear speed of 3,000 mm/s at 30 °C. The area under the shear force-displacement curve represents the fracture energy or the work done to deform the solder joint at interface fracture. The stress distribution in the SAC405 solder/IMC interface at fracture onset is analyzed to extract the shear strength. The clearance between the shear tool and solder generates a notable bending effect at the interface during shear testing. Both bending and shear stresses influence fracture initiation at the interface edge. In Fig. 8, the von Mises stress distribution (in MPa) in the symmetry plane

Materials Research Proceedings 68 (2026) 328-334 https://doi.org/10.21741/9781644904237-44

of a single solder is depicted, revealing elevated stress values near the leading edge of the solder/IMC interface, particularly at the point of contact with the shear tool.

Due to the complex stress state at the solder/IMC interface during the solder ball shear push test, the absolute maximum shear stress (τmax, abs) at the leading edge upon fracture represents the equivalent shear stress of the solder/IMC interface. Fig. 9 depicts the typical distribution of the absolute maximum shear stress on the SAC405 solder. At the leading edge, the absolute maximum shear stress reaches 62.1 MPa, contributing to crack initiation and subsequent opening crack propagation. The stress diminishes upon initiation of interface fracture. The high magnitude of (τmax, abs) is influenced by the geometric stress concentration at the edge of the interface plane.

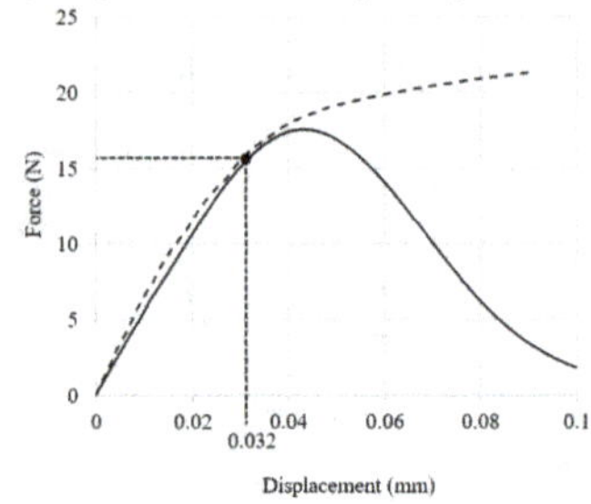

Fig. 7: Comparison of FE-Predicted and Experimental Force-Displacement Curves.

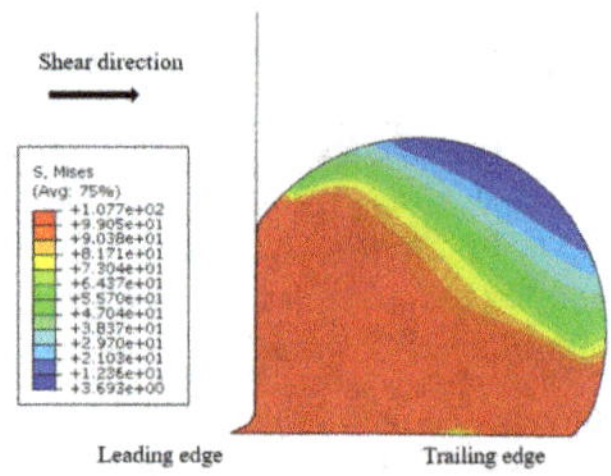

Fig. 8: Von Mises Stress Evaluation in SAC405 Solder After Shear Push Test.

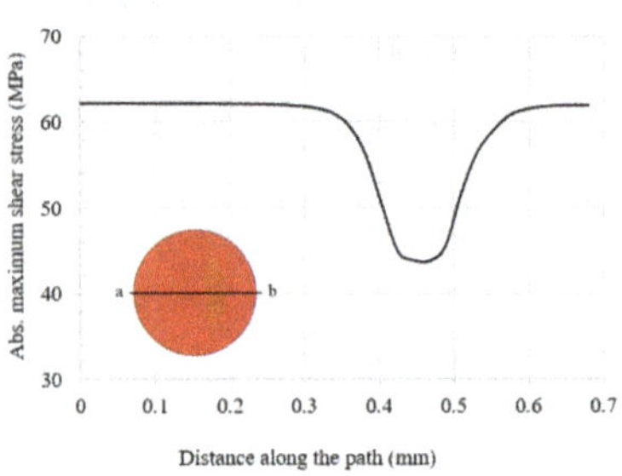

Fig. 9: Mapping Absolute Maximum Shear Stress Distribution along the a-b Path on Solder/IMC Interface Plane.

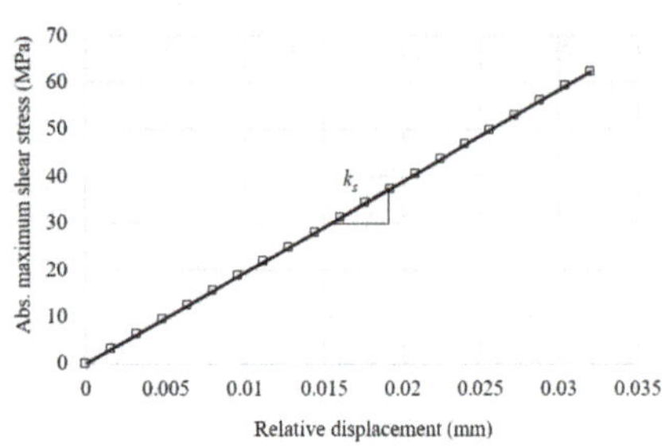

Fig. 10: The shear traction-relative displacement curve.

The shear stiffness, ks, of the solder/IMC interface is determined from the shear traction-relative displacement curve, as shown in Fig. 10. At 25°C, the shear stiffness of SAC405 solder is measured to be 1.94 x 1012 N/m3, consistent with high strain rate CZM parameters [15]. Notably, CZM parameters vary based on loading type and substrate stiffness. The predicted shear strength of the solder/IMC interface aligns well with measured values for SAC387 solder.

Temperature-dependent Solder/IMC Interface Properties. Specific cohesive zone model (CZM) parameters are chosen to represent the solder/IMC interface's shear strength accurately. This selection aims to capture the material degradation process within solder effectively interconnects. These parameters are expressed using logarithmic-type equations (Equations 1).

The shear strength parameters (S) are plotted against temperature (T) and strain rate ($\dot{\varepsilon}$) effects in Fig. 11.

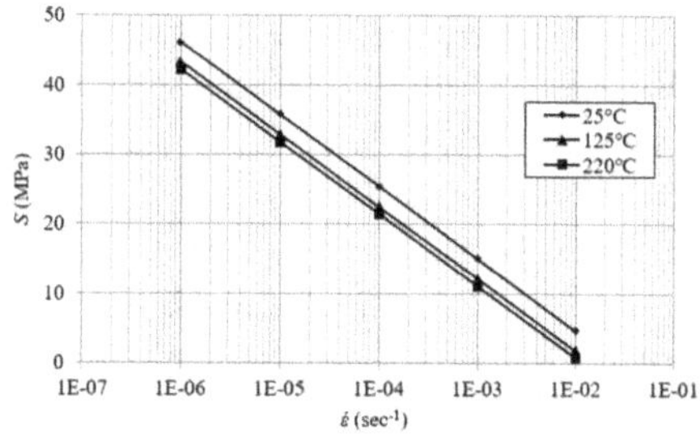

Fig. 11: Solder/IMC interface shear strength against strain rate at different testing temperatures.

The strength of the solder/IMC interface under shear loading is affected by temperature and strain rate variations. As temperature and strain rate increase, the interface experiences reduced shear strength, potentially impacting the reliability and performance of solder interconnects. Understanding this dependence is crucial for optimizing solder joint designs in diverse operating conditions. Other CZM parameter values, such as penalty stiffness and critical energy release rate, are also expected to decrease following a similar trend. Table 2 summarizes solder/IMC interface CZM parameters for shear components at different temperature levels.

Table 2: Temperature-dependent solder/IMC interface properties.

Parameter	25 °C	75 °C	150 °C	220 °C
Shear stiffness, k_s (MPa)	6000	5014.4	4542.8	3792.8
Shear strength, S (MPa)	11.2	9.1	7.9	7.1
Fracture energy under Mode II, G_{IIC} (N/mm)	0.0168	0.0140	0.0128	0.0108
Mode mixity factor, η	1			

Conclusion

The interface behavior between the solder and intermetallic compound (IMC) has been effectively characterized through shear push tests on solder balls, revealing that both temperature and strain rate significantly influence the interface's shear strength. At this temperature, the interface shear strength and penalty stiffness are 62.1 MPa and 1.94×10^{12} N/m^3, respectively, while bending effects at the interface edges are induced by the shear tool stand-off height. Furthermore, cohesive zone model parameters degrade with rising temperature from 25 °C to 220 °C, affecting both the initiation and evolution of damage, highlighting the temperature-dependent nature of solder/IMC interface behavior.

References

[1]M.B. Jeronimo, et al., Microelectron. Rel. 140, 114869 (2023). https://doi.org/10.1016/j.microrel.2022.114869

[2] S. Ghenam, et al., Microelectron. Rel. 146, 115019 (2023). https://doi.org/10.1016/j.microrel.2023.115019

[3] O.E. Gabriel and D.R. Huitink, J. Electron. Packag. 145, 020801 (2023). https://doi.org/10.1115/1.4055774

[4] J.A. Depiver, S. Mallik, and E.H. Amalu, J. Electron. Mater. 50, 263-282 (2021). https://doi.org/10.1007/s11664-020-08525-9

[5] X. Liu, Virginia Polytechnic Inst. State Univ., (2001).

[6] A. Yamin, Master Thesis, Universiti Teknologi Malaysia, Malaysia, (2012).

[7] G. Mirone, et al., Mater. 16, 4808 (2023). https://doi.org/10.3390/ma16134808

[8] N. Shaffiar, Z. Lai, and M.N. Tamin, Key Eng. Mater. 462, 1409-1414 (2011). https://doi.org/10.4028/www.scientific.net/KEM.462-463.1409

[9] S.F. Asasaari, N.A. Fadil, and M.N. Tamin, J. Mech. Eng. Sci. 16, 8718-8729 (2022). https://doi.org/10.15282/jmes.16.1.2022.06.0689

[10] O. Jerhaoui, et al., in IEEE 67th Electron. Compon. Technol. Conf. (ECTC), IEEE, (2017).

[11] A. Schubert, et al., in Proc. Electron. Compon. Technol. Conf., 603-610 (2003).

[12] W. Qiang, et al., in Int. Conf. Thermal, Mech. Multi-Phys. Simul. Exper. Microelectron. Micro Syst., EuroSime, (2007).

[13] X.Q. Shi, et al., ASME Trans. J. Electron. Packag. 124, 85-90 (2002).

[14] B.A. Zahn, in Electron. Compon. Technol. Conf., Proc. 53rd, (2003).

[15] A. Nourani, et al., Eng. Fract. Mech. 186, 119-133 (2017). https://doi.org/10.1016/j.engfracmech.2017.09.027

Keyword Index

3D Printing 105, 113
3D Scaffold 105

A Knee 97
Acid Treatment 134
Activated Carbon 49
Additive Manufacturing 308
Adsorption 1
Advanced Construction Materials 57
Agricultural Waste 204
AI in Materials Science 81
Air Abrasion 232
Akermanite 17
Alkali Treatment 134
Anaerobic Digestion 189
Anti-Termite 33
Autodesk Moldflow Insight 263
Autonomous Healing 246

Bioactive Compounds 156
Biogas Production 189
Biomass Pelletizer 299
Blast Resistance 57
Blend Films 65
Bonding Behaviour 232
Bulk Density 280

Carbon Fiber 121
Cassava Tree Trunk 293
Catalytic Graphitization 182
Cellulose Allomorph 204
Chitosan 65
CO_2 Curing 166
Coal Fly Ash 166
Coconut Husk 121
Coconut Shell Waste 182
Compressive Strength 57
Conventional Sintering 225
Crack Propagation 57
Culantro 156
Curing Time 139

Degradation Plastic 90
Design Expert 113
Design of Experiment 113
Dissimilar Metal Wedding 254
DOE – Mixture 113
Dolomite 73
Double Crucible Method 41
Durability Performance 288
Dye Removal 49
Dynamic Loads 57

Elemental Analysis 218
Encapsulating Materials 293
Enzyme 26
Epoxy Composite 272, 280
Ethylenediamine 204
Explosive Loading 57

Fiberglass-Reinforced Concrete 288
Fiber-Matrix Interphase 272
Finite Element Analysis 299
Finite Element Method 97
Flexural Behaviour 280
Fused Deposition Modelling (FDM) 308

Geopolymer 263
Graphene-Reinforced Concrete 57
Graphitic Carbon 182
Green Mortar 174
Ground Coffee 49

Hybrid Fillers 126

Hydrogel Dressing 148
Hydrophilicity 26
Hydroxyethyl Cellulose 65

Injection Moulding 263

Kenaf Fibre 272, 280
Kernel Shell 126

Lemongrass Essential Oil 65
Lower Limb Socket 97
Lysozyme 26

Machine Learning Technique 322
Macroballoon 1
Magnesium Oxide 166
Mechanical Analysis 254
Mechanical Characterization 328
Mechanical Performance 174
Mechanical Properties 9, 288
Mechanical Testing 315
Merwinite 17
Metalworking Fluid 210
Methylene Blue 49
Microencapsulation 246
Microstructural Analysis 225, 254, 280
Modified Vegetable Oil 210
Moisture Exposure 139
Moisture Retention 148
Monticellite 17

Nanosilica Surface Modification 272, 280
Natural Fiber Composites 81
Natural Fiber 97, 134

Oil Palm Fruit Bunches 126
Optical Microscopy 272

Paddy Straw Waste 189
Palm Oil 33, 105
Performance Analysis 322
Perlis Dolomite 17
Physical and Biological Degradation 90
PLA 121
POFA 174
Polycaprolactone (PCL) 9
Polyethylene Glycol 65
Polymer Blends 9
Polymer Composites 73
Polyol-to-Isocyanate Ratios 139
Polyurethane 139
Polyvinyl chloride (PVC) 9
Porous Carbon 1
Porous 49
Powder Metallurgy 225
Predictive Modeling 81
Pyrolysis 41

Radiation Curable Resin 33
Radiation Curing 33
Rapid Tooling 263
Recycled Carbon Black 315
Recycled Copper 263
Remazol Red Dye 1
Rice Husk Ash 1
Rice Husk 41, 126
RSM 73

Sand-Cement Brick 166
Screw Mechanism 299
Self Healing Concrete 246
Silica 17, 41
SLA 105, 113
Slow-Release Pesticide 293
Smart Composites 81
Sol Gel Alkalinity 218

Solder Shear Push Test 328
Solder/Intermetallic Compound 328
Steam Curing 174
Stirring Temperature 9
Stress Analysis 97
Structural Analysis 204
Surface Performance 308
Sustainability 81
Sustainable Manufacturing 263

Taguchi Approach 315
Tea Bag 156
Tensile Properties 272
Tensile Strength 134
Thermoplastic Starch Film 126
Tool Wear 322
Total Polyphenol Content 156

UHMWPE 73
UV Radiation Crosslinking 105
UV-Vis 26

Viscosity Properties 210

Warpage 263
Waste 90
White Chicken Egg 26
Wood Coatings 33
Wound Hydration 148

Zirconia-Resin 232
ZnO Nanostructure 218

About the Editor

ASNA RASYIDAH BINTI ABDUL HAMID

Dr. Asna Rasyidah Andul Hamid is an academic staff member of the Machining Engineering Technology Programme, Department of Manufacturing Engineering Technology, Faculty of Mechanical Engineering and Technology, Universiti Malaysia Perlis. She serves as a senior lecturer in the field of Materials Science and Engineering. Her academic and research interests encompass polymer matrix composites, natural fiber-reinforced composites, sustainable materials, materials characterization, and mechanical behavior of engineering materials. Her teaching and research activities focus on the development, processing, testing, and characterization of advanced composite materials, with particular emphasis on environmentally friendly and high-performance materials for engineering applications. She is actively involved in research, publication, student supervision, and academic collaboration at both national and international levels. As an editor, reviewer, and academic contributor, she has participated in maintaining the scientific quality, accuracy, and integrity of research publications and academic outputs in the field of materials engineering.

FIRUZ BINTI ZAINUDDIN

Assoc. Prof. Dr. Firuz Zainuddin is an academic staff member of the Polymer Engineering Programme, Materials Department, Faculty of Chemical Engineering and Technology, Universiti Malaysia Perlis. Her research interests encompass cellular solids, polymer matrix composites, materials characterization, and failure analysis. Her academic and research activities are focused on advancing sustainable and high-performance materials through experimental investigation and characterization techniques. She has actively contributed to research, publication, and academic collaboration in the field of materials engineering. As one of editors, she was involved in coordinating the peer-review process and ensuring the scientific quality and integrity of the published manuscripts.

MOHD KAHAR BIN AB WAHAB

Assoc. Prof. Dr. Mohamad Kahar Ab Wahab is an academic staff member of the Polymer Engineering Programme, Materials Department, Faculty of Chemical Engineering and Technology, Universiti Malaysia Perlis. His research interests primarily encompass biodegradable polymers, thermoplastic starch blends, polymer composites, sustainable materials, and materials characterization. His academic and research activities are focused on the development of environmentally sustainable polymeric materials through formulation, processing, and physicochemical characterization techniques. He has actively contributed to research, publication, and academic collaboration in the field of polymer and materials engineering, particularly in advancing green and bio-based polymer technologies. His scholarly contributions are evidenced by numerous publications in indexed journals and conference proceedings, with an h-index of 17. In addition, he has been involved in publication management and editorial coordination activities, contributing to the maintenance of scientific quality, technical rigor, and publication integrity in conference proceedings and scholarly manuscripts.

MUHAMMAD FAHEEM BIN MOHD TAHIR

Muhammad Faheem Mohd. Tahir is a dedicated academician at the Faculty of Chemical Engineering Technology, Universiti Malaysia Perlis (UniMAP). He holds a Bachelor's Degree in Civil Engineering from Universiti Tun Hussein Onn Malaysia, a Master of Science and PhD in Material Engineering from UniMAP. His primary research interests encompass geopolymers, rigid pavement, composite materials, and sustainable construction technologies. As a highly active contributor to the scientific community, he has established a strong scholarly footprint with a Scopus h-Index of 22 and a Google Scholar h-Index of 26, having authored numerous papers in high-impact journals and international conference proceedings. Bringing valuable peer-review and editorial expertise to this publication, he frequently serves as a reviewer for prominent scientific journals. In his capacity as an editor for this proceeding, he has been instrumental in shaping the academic discourse by meticulously curating submissions, guiding the peer-review process, and ensuring the scientific rigor and quality of the published works. Furthermore, he is a registered professional with the Board of Engineers Malaysia and the Malaysia Board of Technologists, reflecting his dedication to maintaining high standards in engineering research and innovation.

www.ingramcontent.com/pod-product-compliance
Lightning Source LLC
LaVergne TN
LVHW052244200926
848481LV00002B/262
9781644904220